Methods in Enzymology

Volume 392
RNA INTERFERENCE

METHODS IN ENZYMOLOGY

EDITORS-IN-CHIEF

John N. Abelson   Melvin I. Simon

DIVISION OF BIOLOGY
CALIFORNIA INSTITUTE OF TECHNOLOGY
PASADENA, CALIFORNIA

FOUNDING EDITORS

Sidney P. Colowick and Nathan O. Kaplan

Methods in Enzymology

Volume 392

RNA Interference

EDITED BY

David R. Engelke

PROGRAM IN BIOMEDICAL SCIENCES
UNIVERSITY OF MICHIGAN
ANN ARBOR, MICHIGAN

John J. Rossi

DIVISION OF MOLECULAR BIOLOGY
GRADUATE SCHOOL OF BIOLOGICAL SCIENCES
BECKMAN RESEARCH INSTITUTE OF THE CITY OF HOPE
DUARTE, CALIFORNIA

ELSEVIER
ACADEMIC
PRESS

AMSTERDAM • BOSTON • HEIDELBERG • LONDON
NEW YORK • OXFORD • PARIS • SAN DIEGO
SAN FRANCISCO • SINGAPORE • SYDNEY • TOKYO

Table of Contents

Contributors to Volume 392

Article numbers are in parentheses and following the names of contributors.
Affiliations listed are current.

HIDEO AKASHI (6), *Department of Chemistry and Biotechnology, School of Engineering, The University of Tokyo, Tokyo 113-8656, Japan*

JOSÉ ALBEROLA-ILA (12), *Division of Biology, Cal Tech, Pasadena, California 91125*

SUSAN ARMKNECHT (4), *Department of Genetics, Harvard Medical School, Drosophila RNAi Screening Center, Boston, Massachusetts 02115*

HYE JUNG BACK (21), *National Cancer Institute, National Institutes of Health, Bethesda, Maryland 20892*

MAUREEN BARR (3), *University of Wisconsin School of Pharmacy, Pharmaceutical Sciences Division, Madison, Wisconsin 53705-2222*

QUETA BOESE (5), *Dharmacon Inc., Lafayette, Colorado 80026*

MICHAEL BOUTROS (4), *German Cancer Research Center, 69120 Heidelberg, Germany*

ROBERT M. BRAZAS (7), *Mirus Bio Corporation, Madison, Wisconsin 53719*

ZUEZHONG CAI (22), *Howard Hughes Medical Institute, Department of Molecular Genetics and Microbiology, Duke University Medical Center, Durham, North Carolina 27710*

DANIELA CASTANOTTO (10), *City of Hope Medical Center, Division of Molecular Biology, Beckman Research Institute, Duarte, California 91010-3000*

VICKI CHANDLER (1), *Department of Plant Sciences, University of Arizona, Tucson, Arizona 85721*

KAREN CONE (1), *Division of Biological Sciences, University of Missouri, Columbia, Missouri 65211*

BRYAN R. CULLEN (22), *Howard Hughes Medical Institute, Department of Molecular Genetics and Microbiology, Duke University Medical Center, Durham, North Carolina 27710*

JONATHAN R. DAVEY (24), *Diabetes and Obesity Research Program, Garvan Institute of Medical Research, Darlinghurst, Sydney NSW 2010, Austalia*

BEVERLY L. DAVIDSON (9), *Department of Internal Medicine, Neurology, Physiology and Biophysics, University of Iowa College of Medicine, Iowa City, Iowa 52242*

ANTONIN DE FOUGEROLLES (16), *Alnylam Pharmaceuticals, Cambridge, Massachusetts 02142*

DAVID DORRIS (15), *Ambion Inc., Austin, Texas 78744-1832*

CHRISTOPHE ECHEVERRI (15), *Cenix BioScience GmbH, 01307 Dresden, Germany*

MARK FEITELSON (14), *Department of Pathology and Cell Biology, Jefferson Medical College, Philadelphia, Pennsylvania 19107-5587*

WITOLD FILIPOWICZ (19), *Friedrich Miescher Institute for Biomedical Research, 4002 Basel, Switzerland*

ANNE GRABNER (15), *Cenix BioScience GmbH, 01307 Dresden, Germany*

MICHAEL W. GRAHAM (24), *Benitec Ltd., University of Queensland, St. Lucia QLD 4072, Australia*

PETER J. GOSS (24), *Benitec Ltd., University of Queensland, St. Lucia QLD 4072, Australia*

DIRK GRIMM (23), *Department of Pediatrics, Stanford University, Stanford, California, 94305-5208*

JAMES E. HAGSTROM (7), *Mirus Bio Corporation, Madison, Wisconsin 53719*

MICHAEL HANNUS (15), *Cenix BioScience GmbH, 01307 Dresden, Germany*

SCOTT Q. HARPER (9), *Department of Internal Medicine, University of Iowa College of Medicine, Iowa City, Iowa 52242*

BRUCE T. HARRISON (24), *Benitec Ltd., University of Queensland, St. Lucia QLD 4072, Australia*

CHRIS A. HELLIWELL (2), *CSIRO Plant Industry, Canberra ACT 2601, Australia*

GABRIELA HERNÁNDEZ-HOYOS (12), *Division of Biology, Cal Tech, Pasadena, California 91125*

TOM C. HOBMAN (19), *Department of Cell Biology, University of Alberta, Edmonton, Alberta, Canada T6G 2E1*

DAVID E. JAMES (24), *Diabetes and Obesity Research Program, Garvan Institute of Medical Research, Darlinghurst, Sydney NSW 2010, Austalia*

KASIA JARONCZYK (19), *Department of Cell Biology, University of Alberta, Edmonton, Alberta, Canada T6G 2E1*

HEIDI KAEPPLER (1), *Department of Agronomy, University of Wisconsin, Madison, Wisconsin 53705*

SHAWN KAEPPLER (1), *Department of Agronomy, University of Wisconsin, Madison, Wisconsin 53705*

ANDREW J. KASSIANOS (24), *Benitec Ltd., University of Queensland, St. Lucia QLD 4072, Australia*

YOSHIO KATO (6), *Gene Function Research Center, National Institute of Advanced Industrial Science and Technology (AIST), Tsukuba Science City 305-8562, Japan*

MARK A. KAY (23), *Department of Pediatrics and Genetics, Stanford University, Stanford, California, 94305-5208*

ARTHUR KERSCHEN (1), *Department of Plant Sciences, University of Arizona, Tucson, Arizona 85721*

ANASTASIA KHVOROVA (5), *Dharmacon Inc., Lafayette, Colorado 80026*

AMY KIGER (4), *Division of Biological Sciences, University of California, San Diego, La Jolla, California 92093*

FABRICE A. KOLB (19), *Friedrich Miescher Institute for Biomedical Research, 4002 Basel, Switzerland*

EBERHARD KRAUSZ (15), [†]*Cenix BioScience GmbH, 01307 Dresden, Germany*

ANDREA KRONKE (15), *Cenix BioScience GmbH, 01307 Dresden, Germany*

DEVIN LEAKE (5), *Dharmacon Inc., Lafayette, Colorado 80026*

[†]*Current address: Max Planck Institute of Molecular Cell Biology and Genetics, 01307 Dresden, Germany*

DAVID L. LEWIS (20), *Mirus Corporation, Madison, Wisconsin 53719*

MING-JIE LI (13), *Division of Molecular Biology, Beckman Research Institute of the City of Hope, Duarte, California 91010-3000*

CONCETTA LIPARDI (21), *National Cancer Institute, National Institutes of Health, Bethesda, Maryland 20892*

MUTHIAH MANOHARAN (16), *Alnylam Pharmaceuticals, Cambridge, Massachusetts 02142*

WILLIAM S. MARSHALL (5), *Dharmacon Inc., Lafayette, Colorado 80026*

BERNARD MATHEY-PREVOT (4), *Department of Genetics, Harvard Medical School, Drosophila RNAi Screening Center, Boston, Massachusetts 02115*

ALEXEY A. MATSKEVICH (14), *Department of Pathology and Cell Biology, Jefferson Medical College, Philadelphia, Pennsylvania 19107-5587*

NARELLE J. MAUGERI (24), *Benitec Ltd., University of Queensland, St. Lucia QLD 4072, Australia*

ROBIN C. MAY (18), *Hubrecht Laboratory, 3584 CT Utrecht, The Netherlands**

KAREN MCGINNIS (1), *Department of Plant Sciences, University of Arizona, Tucson, Arizona 85721*

RACHEL MEYERS (16), *Alnylam Pharmaceuticals, Cambridge, Massachusetts 02142*

MAKOTO MIYAGISHI (6), *Gene Function Research Center, National Institute of Advanced Industrial Science and Technology (AIST), Tsukuba Science City 305-8562, Japan; 21st Century Center of Excellence Program, Graduate School of Medicine, The University of Tokyo, Tokyo 113-8656, Japan*

DANESH MOAZED (17), *Department of Cell Biology, Harvard Medical School, Boston, Massachusetts 02115*

ANDREW N. MUIRHEAD (24), *Benitec Ltd., University of Queensland, St. Lucia QLD 4072, Australia*

KENT NYBAKKEN (4), *Department of Genetics, Harvard Medical School, Boston, Massachusetts 02115*

DMITRIY OVCHARENKO (15), *Ambion Inc., Texas 78744-1832*

KUSUM PANDEY (23), *Stanford University, Department of Pediatrics, Stanford, California, 94305-5208*

JACK M. PARENT (11), *Program in Neuroscience, University of Michigan, Ann Arbor, Michigan 48109-0669*

BRUCE M. PATERSON (21), *National Cancer Institute, National Institutes of Health, Bethesda, Maryland 20892*

CYNTHIA P. PAUL (8), *Department of Biological Chemistry, University of Michigan, Ann Arbor, Michigan 48109*

NORBERT PERRIMON (4), *Department of Genetics, Howard Hughes Medical Institute and Harvard Medical School, Boston, Massachusetts 02115*

CRAIG PIKKARD (1), *Department of Biology, Washington University, St. Louis, Missouri 63130*

RONALD H. A. PLASTERK (18), *Hubrecht Laboratory, 3584 CT Utrecht, The Netherlands*

STEVEN READ (5), *Dharmacon Inc., Lafayette, Colorado 80026*

ANGELA REYNOLDS (5), *Dharmacon Inc., Lafayette, Colorado 80026*

*Current address: School of Biosciences, University of Birmingham, Birmingham B15 2TT, United Kingdom

ROBERT R. RICE (24), *Benitec Ltd., University of Queensland, St. Lucia QLD 4072, Australia*

ERIC RICHARDS (1), *Department of Biology, Washington University, St. Louis, Missouri 63130*

JOHN J. ROSSI (13), *Division of Molecular Biology, Beckman Research Institute of the City of Hope, Duarte, California 91010-3000*

CHRISTOPH SACHSE (15), *Cenix BioScience GmbH, 01307 Dresden, Germany*

MASAYUKI SANO (6), *Gene Function Research Center, National Institute of Advanced Industrial Science and Technology (AIST), Tsukuba Science City 305-8562, Japan*

STEPHEN A. SCARINGE (5), *Dharmacon Inc., Lafayette, Colorado 80026*

LISA SCHERER (10), *City of Hope Medical Center, Division of Molecular Biology, Beckman Research Institute, Duarte, California 91010-3000*

PETRA L. SEDLAK (24), *Benitec Ltd., University of Queensland, St. Lucia QLD 4072, Australia*

LYUDMILA SIDORENKO (1), *Department of Plant Sciences, University of Arizona, Tucson, Arizona 85721*

TODD SMITH (1), *Department of Biology, Washington University, St. Louis, Missouri 63130*

BIRTE SONNICHSEN (15), *Cenix BioScience GmbH, 01307 Dresden, Germany*

NATHAN SPRINGER (1), *Department of Plant Biology, University of Minnesota, St. Paul, Minnesota 55108*

DAVID S. STRAYER (14), *Department of Pathology and Cell Biology, Jefferson Medical College, Philadelphia, Pennsylvania 19107-5587*

BILL SUN (14), *Department of Pathology and Cell Biology, Jefferson Medical College, Philadelphia, Pennsylvania 19107-5587*

NASSER TAHBAZ (19), *Department of Cell Biology, University of Alberta, Edmonton, Alberta, Canada T6G 2E1*

KAZUNARI TAIRA (6), *Department of Chemistry and Biotechnology, University of Tokyo, Tokyo 113-8656, Japan; Gene Function Research Center, National Institute of Advanced Industrial Science and Technology (AIST), Tsukuba Science City 305-8562, Japan*

CLAUDE TRUDEL (15), *Ambion Inc., Austin, Texas 78744-1832*

DAVID L. TURNER (11), *University of Michigan, Mental Health Research Institute, Ann Arbor, Michigan 48109-0669*

ANDRÉ VERDEL (17), *Department of Cell Biology, Harvard Medical School, Boston, Massachusetts 02115*

ANNE B. VOJTEK (11), *Department of Biological Chemistry, University of Michigan, Ann Arbor, Michigan 48109-0669*

HANS-PETER VORNLOCHER (16), *Alnylam Europe AG, 95326 Kulmbach, Germany*

ANDREW WALSH (15), *Cenix BioScience GmbH, 01307 Dresden, Germany*

JUAN WANG (3), *University of Wisconsin School of Pharmacy, Pharmaceutical Sciences Division, Madison, Wisconsin 53705-2222*

TSU-WEI WANG (11), *Program in Neuroscience, University of Michigan, Ann Arbor, Michigan 48109-0669*

PETER M. WATERHOUSE (2), *CSIRO Plant Industry, Canberra ACT 2601, Australia*

QIN WEI (21), *National Cancer Institute, National Institutes of Health, Bethesda, Maryland 20892*

JON A. WOLFF (20), *Department of Pedicatrics, Waisman Center, University of Wisconsin-Madison, Madison, Wisconsin 53705*

TUYA WULAN (1), *Department of Biology, Washington University, St. Louis, Missouri 63130*

JENN-YAH YU (11), *Program in Neuroscience, University of Michigan, Ann Arbor, Michigan 48109-0669*

YAN ZENG (22), *Department of Molecular Genetics and Microbiology, Howard Hughes Medical Institute, Duke University Medical Center, Durham, North Carolina 27710*

HAIDI ZHANG (19), *Friedrich Miescher Institute for Biomedical Research, 4002 Basel, Switzerland*

METHODS IN ENZYMOLOGY

VOLUME 72. Lipids (Part D)
Edited by JOHN M. LOWENSTEIN

VOLUME 73. Immunochemical Techniques (Part B)
Edited by JOHN J. LANGONE AND HELEN VAN VUNAKIS

VOLUME 74. Immunochemical Techniques (Part C)
Edited by JOHN J. LANGONE AND HELEN VAN VUNAKIS

VOLUME 75. Cumulative Subject Index Volumes XXXI, XXXII, XXXIV–LX
Edited by EDWARD A. DENNIS AND MARTHA G. DENNIS

VOLUME 76. Hemoglobins
Edited by ERALDO ANTONINI, LUIGI ROSSI-BERNARDI, AND EMILIA CHIANCONE

VOLUME 77. Detoxication and Drug Metabolism
Edited by WILLIAM B. JAKOBY

VOLUME 78. Interferons (Part A)
Edited by SIDNEY PESTKA

VOLUME 79. Interferons (Part B)
Edited by SIDNEY PESTKA

VOLUME 80. Proteolytic Enzymes (Part C)
Edited by LASZLO LORAND

VOLUME 81. Biomembranes (Part H: Visual Pigments and Purple Membranes, I)
Edited by LESTER PACKER

VOLUME 82. Structural and Contractile Proteins (Part A: Extracellular Matrix)
Edited by LEON W. CUNNINGHAM AND DIXIE W. FREDERIKSEN

VOLUME 83. Complex Carbohydrates (Part D)
Edited by VICTOR GINSBURG

VOLUME 84. Immunochemical Techniques (Part D: Selected Immunoassays)
Edited by JOHN J. LANGONE AND HELEN VAN VUNAKIS

VOLUME 85. Structural and Contractile Proteins (Part B: The Contractile Apparatus and the Cytoskeleton)
Edited by DIXIE W. FREDERIKSEN AND LEON W. CUNNINGHAM

VOLUME 86. Prostaglandins and Arachidonate Metabolites
Edited by WILLIAM E. M. LANDS AND WILLIAM L. SMITH

VOLUME 87. Enzyme Kinetics and Mechanism (Part C: Intermediates, Stereo-chemistry, and Rate Studies)
Edited by DANIEL L. PURICH

VOLUME 88. Biomembranes (Part I: Visual Pigments and Purple Membranes, II)
Edited by LESTER PACKER

VOLUME 89. Carbohydrate Metabolism (Part D)
Edited by WILLIS A. WOOD

VOLUME 244. Proteolytic Enzymes: Serine and Cysteine Peptidases
Edited by ALAN J. BARRETT

VOLUME 245. Extracellular Matrix Components
Edited by E. RUOSLAHTI AND E. ENGVALL

VOLUME 246. Biochemical Spectroscopy
Edited by KENNETH SAUER

VOLUME 247. Neoglycoconjugates (Part B: Biomedical Applications)
Edited by Y. C. LEE AND REIKO T. LEE

VOLUME 248. Proteolytic Enzymes: Aspartic and Metallo Peptidases
Edited by ALAN J. BARRETT

VOLUME 249. Enzyme Kinetics and Mechanism (Part D: Developments in Enzyme Dynamics)
Edited by DANIEL L. PURICH

VOLUME 250. Lipid Modifications of Proteins
Edited by PATRICK J. CASEY AND JANICE E. BUSS

VOLUME 251. Biothiols (Part A: Monothiols and Dithiols, Protein Thiols, and Thiyl Radicals)
Edited by LESTER PACKER

VOLUME 252. Biothiols (Part B: Glutathione and Thioredoxin; Thiols in Signal Transduction and Gene Regulation)
Edited by LESTER PACKER

VOLUME 253. Adhesion of Microbial Pathogens
Edited by RON J. DOYLE AND ITZHAK OFEK

VOLUME 254. Oncogene Techniques
Edited by PETER K. VOGT AND INDER M. VERMA

VOLUME 255. Small GTPases and Their Regulators (Part A: Ras Family)
Edited by W. E. BALCH, CHANNING J. DER, AND ALAN HALL

VOLUME 256. Small GTPases and Their Regulators (Part B: Rho Family)
Edited by W. E. BALCH, CHANNING J. DER, AND ALAN HALL

VOLUME 257. Small GTPases and Their Regulators (Part C: Proteins Involved in Transport)
Edited by W. E. BALCH, CHANNING J. DER, AND ALAN HALL

VOLUME 258. Redox-Active Amino Acids in Biology
Edited by JUDITH P. KLINMAN

VOLUME 259. Energetics of Biological Macromolecules
Edited by MICHAEL L. JOHNSON AND GARY K. ACKERS

VOLUME 260. Mitochondrial Biogenesis and Genetics (Part A)
Edited by GIUSEPPE M. ATTARDI AND ANNE CHOMYN

VOLUME 261. Nuclear Magnetic Resonance and Nucleic Acids
Edited by THOMAS L. JAMES

VOLUME 299. Oxidants and Antioxidants (Part A)
Edited by LESTER PACKER

VOLUME 300. Oxidants and Antioxidants (Part B)
Edited by LESTER PACKER

VOLUME 301. Nitric Oxide: Biological and Antioxidant Activities (Part C)
Edited by LESTER PACKER

VOLUME 302. Green Fluorescent Protein
Edited by P. MICHAEL CONN

VOLUME 303. cDNA Preparation and Display
Edited by SHERMAN M. WEISSMAN

VOLUME 304. Chromatin
Edited by PAUL M. WASSARMAN AND ALAN P. WOLFFE

VOLUME 305. Bioluminescence and Chemiluminescence (Part C)
Edited by THOMAS O. BALDWIN AND MIRIAM M. ZIEGLER

VOLUME 306. Expression of Recombinant Genes in Eukaryotic Systems
Edited by JOSEPH C. GLORIOSO AND MARTIN C. SCHMIDT

VOLUME 307. Confocal Microscopy
Edited by P. MICHAEL CONN

VOLUME 308. Enzyme Kinetics and Mechanism (Part E: Energetics of
Enzyme Catalysis)
Edited by DANIEL L. PURICH AND VERN L. SCHRAMM

VOLUME 309. Amyloid, Prions, and Other Protein Aggregates
Edited by RONALD WETZEL

VOLUME 310. Biofilms
Edited by RON J. DOYLE

VOLUME 311. Sphingolipid Metabolism and Cell Signaling (Part A)
Edited by ALFRED H. MERRILL, JR., AND YUSUF A. HANNUN

VOLUME 312. Sphingolipid Metabolism and Cell Signaling (Part B)
Edited by ALFRED H. MERRILL, JR., AND YUSUF A. HANNUN

VOLUME 313. Antisense Technology (Part A: General Methods, Methods of
Delivery, and RNA Studies)
Edited by M. IAN PHILLIPS

VOLUME 314. Antisense Technology (Part B: Applications)
Edited by M. IAN PHILLIPS

VOLUME 315. Vertebrate Phototransduction and the Visual Cycle (Part A)
Edited by KRZYSZTOF PALCZEWSKI

VOLUME 316. Vertebrate Phototransduction and the Visual Cycle (Part B)
Edited by KRZYSZTOF PALCZEWSKI

[1] Transgene-Induced RNA Interference as a Tool for Plant Functional Genomics

By Karen McGinnis, Vicki Chandler, Karen Cone,
Heidi Kaeppler, Shawn Kaeppler, Arthur Kerschen,
Craig Pikaard, Eric Richards, Lyudmila Sidorenko,
Todd Smith, Nathan Springer, and Tuya Wulan

Abstract

RNA interference (RNAi) is a powerful tool for functional genomics in a number of species. The logistics and procedures for doing high-throughput RNAi to investigate the functions of large numbers of genes in *Arabidopsis thaliana* and in *Zea mays* are described. Publicly available plasmid vectors that facilitate the stable chromosomal integration of inverted repeat transgenes that trigger RNAi have been used to generate more than 50 independent transgenic lines each in *Arabidopsis* and maize. Analysis of mRNA abundance of the targeted genes in independent lines transformed with distinct constructs indicates that the success of RNAi-induced silencing is gene dependent. mRNA levels were not detectably reduced for some genes, but were dramatically reduced for a number of genes targeted. A common pattern was that multiple independent lines transgenic for the same construct showed the same extent of silencing. This chapter describes the procedures used to generate and test transgenic lines mediating RNAi in *Arabidopsis* and maize.

Introduction

In recent years, RNA interference (RNAi) has been exploited as a tool for investigating gene functions in numerous organisms (reviewed by Agrawal *et al.*, 2003; Bailis and Forsburg, 2002; Carpenter and Sabatini, 2004; Dawe, 2003; Fraser *et al.*, 2000; Martienssen, 2003; Matzke and Matzke, 2003; Paddison *et al.*, 2004; Simmer *et al.*, 2003). Gene silencing by transgene-induced RNAi is useful because the loss or reduction of gene function is dominant, circumventing the need to generate homozygous loss-of-function mutations. This not only saves time, which is especially useful in plants with long generation times, but also allows gene knockdown studies to be conducted in F_1 hybrids that inherit the RNAi-inducing transgene from only one parent. Likewise, because RNAi mechanisms can target multiple closely related mRNAs, dominant loss-of-function phenotypes can be generated for entire gene families or for multiple orthologous genes in polyploids (Lawrence and Pikaard, 2003).

Methods for targeted RNAi-mediated gene silencing in plants have been reviewed recently (Waterhouse and Helliwell, 2003), and RNAi has proven to be a useful method in a number of plant studies (Ketelaar *et al.*, 2004; Lawrence *et al.*, 2004; Wesley *et al.*, 2001; Xiao *et al.*, 2003; Zentella *et al.*, 2002). In this chapter, we focus on the logistics of doing high-throughput RNAi to investigate the functions of large numbers of targeted genes in plants, specifically in *Arabidopsis thaliana* and *Zea mays*. The approaches summarized here stem from the authors' communal efforts to generate RNAi lines targeting several hundred genes. This effort has been funded by the United States National Science Foundation as part of its Plant Genome Research Program and represents one of the first large-scale efforts to use RNAi as a tool for plant functional genomics. The goal of our project is to investigate the function of chromatin-modifying genes in maize and *Arabidopsis* and to provide the research community with tools to study these genes, including RNAi lines that knock down their expression (detailed information available at http://ChromDB.org/). Central to the effort has been the design and construction of publicly available plasmid vectors that facilitate the stable chromosomal integration of transgenes that trigger RNAi. Each RNAi-inducing transgene consists of a portion of a targeted gene's cDNA cloned in inverted orientation downstream from a strong, widely expressed promoter, a design based on the pioneering work of the Waterhouse laboratory (Waterhouse *et al.*, 1998). When expressed, the inverted repeats cause the transcript to fold back on itself to generate a double-stranded RNA (dsRNA) that acts as a substrate for the Dicer and RISC complexes, ultimately leading to the reduction or elimination of highly homologous mRNAs in the cell (Hannon, 2002). The salient features of the vectors and the approaches we have used to generate large numbers of transgenic RNAi lines as efficiently as possible are described in the following two sections, the first focused on *A. thaliana* and the second on maize.

High-Throughput Generation of *A. thaliana* RNAi Lines

Vectors for Generating Transgenic Plants and Inducing RNAi of Targeted Genes

The method of choice for genetically engineering *A. thaliana* is *Agrobacterium tumefaciens*-mediated genetic transformation. *A. tumefaciens* is a common soil bacterium that can transmit a segment of DNA, known as the T-DNA (transferred DNA), from a large resident plasmid, called the Ti plasmid, to a recipient plant cell through a process that closely resembles bacterial conjugation (Gelvin, 2000, 2003; Zupan *et al.*, 2000). The T-DNA is delimited by short direct repeats, known as the left border (LB) and right border (RB), which are acted on by the machinery involved in generating the transferred DNA. Most of the machinery required for T-DNA transfer is encoded elsewhere on the Ti plasmid and acts on the border sequences in *trans*. As a result, any segment of DNA flanked by LB and RB sequences can be transferred from *A. tumefaciens* to a plant cell, even if the T-DNA is located on a plasmid distinct from the Ti plasmid. In fact, a two-plasmid, or binary, system is commonly used for *A. tumefaciens*-mediated plant transformation. The plasmid that carries the T-DNA (the binary vector) can replicate in both *E. coli* and *A. tumefaciens*, which allows all cloning steps to be conducted by using *E. coli* before mobilization of the vector plasmid into *A. tumefaciens* for plant transformation.

The binary vector designed by our Functional Genomics of Chromatin (FGC) consortium, which serves as our vector of choice for generating transgenic *Arabidopsis* RNAi lines, is pFGC5941 (Fig. 1). The complete nucleotide sequence of pFGC5941 can be retrieved from GenBank (AY310901), restriction enzyme cleavage maps can be found at the FGC project Web site (http://ChromDB.org/), and the plasmid can be obtained from the Arabidopsis Biological Resource Center (ABRC, stock number CD3-447). The backbone of pFGC5941 is the pCAMBIA1300 binary vector plasmid developed by the Center for Application of Molecular Biology to International Agriculture (CAMBIA; http://www.cambia.org). pCAMBIA1300 has two origins of replication, a wide-host-range origin that facilitates plasmid replication in *A. tumefaciens* and the pBR322 origin for replication in *E. coli* (Hajdukiewicz *et al.*, 1994), as well as an aminoglycoside phosphotransferase gene that confers kanamycin resistance in both *E. coli* and *A. tumefaciens*. Within the T-DNA region of pFGC5941, adjacent to the LB, is a phosphinothricin acetyl transferase gene driven by the mannopine synthase 2' (Mas2') promoter. This gene confers herbicide resistance for bialaphos and phosphinothricin and is commonly known as the *BAR* (Basta resistance) gene. Adjacent to the *BAR* gene is the cauliflower mosaic virus (CaMV) 35S promoter region, which includes enhancer sequences upstream of the gene promoter. The 35S promoter is expressed in most plant cells, and this strong promoter drives the expression of the various inverted repeat (IR) transgenes engineered into the pFGC5941 vector. The tobacco mosaic virus omega leader sequence, which is known to act as a translational enhancer in plants (Gallie, 2002; Sleat *et al.*, 1987), serves as the 5' UTR for the various transgenes. The omega sequence most likely is irrelevant for RNAi, which depends on RNA production rather

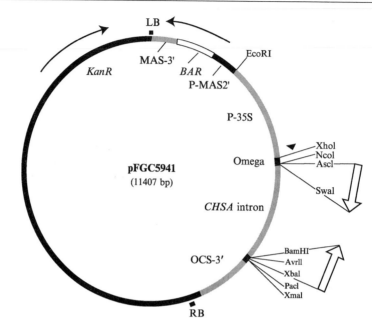

Fig. 1. Organization of the *Agrobacterium* T-DNA transformation vector pFGC5941. The T-DNA region transferred to the plant cell lies between the left border (LB) and the right border (RB) sequences located between 12 and 5 o'clock on the plasmid map. The T-DNA includes a selectable marker gene conferring Basta resistance (the *BAR* gene) under the control of the MAS2′ upstream and downstream regions. In addition, the T-DNA region contains the strong, widely expressed CaMV 35S promoter (P-35S) upstream of the *CHSA* intron and an octopine synthase 3′ end. The arrowhead indicates the orientation and position of transcription initiation from P-35S. Two multiple cloning sites flank the *CHSA* intron as insertion sites for target gene sequences (open arrows); unique restriction sites in these regions are shown. The *KanR* gene confers kanamycin resistance in bacterial hosts; Omega refers to the tobacco mosaic virus omega leader sequence.

than protein translation, but makes the vector adaptable to other purposes such as overexpression of protein.

To construct an RNAi-inducing transgene in pFGC5941, cDNA fragments (typically 500–700 bp) of targeted genes are cloned in two steps into multiple cloning sites that flank a 1352-bp intron obtained from the petunia chalcone synthase A (*CHSA*) gene. The two cDNA fragments are cloned such that they are in opposite orientations relative to one another, with the *CHSA* intron acting as a spacer between them (Fig. 1). Expression of this IR transgene results in a transcript that terminates within 3′ sequences of the octopine synthase (OCS) gene and folds back on itself by virtue of the

Transcript foldback to produce a dsRNA stem-loop

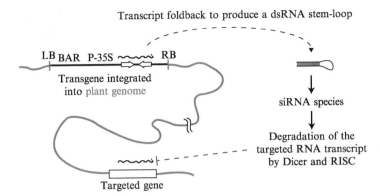

FIG. 2. Strategy for transgene-induced RNAi of targeted genes. After random integration of the transgene into the plant genome, the strong 35S promoter drives expression of transcripts that are self-complementary through the inverted orientation of target gene sequences (open boxes), generating a double-strand RNA stem, which is processed by the Dicer complex into 22- to 26-nt short interfering RNA (siRNA) species. The siRNAs, in turn, direct degradation of highly homologous RNA transcripts by the RISC complex.

inverted repeats, thus generating a dsRNA (Fig. 2). The dsRNA is then a substrate for the Dicer and RISC complexes that process the dsRNA into short interfering RNAs (siRNAs) that target homologous mRNAs for degradation.

Procedure for Generating RNAi Lines

Design an RNAi-Inducing Transgene for Each Gene Targeted for Knockdown. The starting point for generating a given IR vector is a bioinformatics effort. By using known cDNA sequences or predicted gene sequences corresponding to the target gene of interest, primers are designed to amplify a portion of the cDNA by using reverse transcription-PCR (RT-PCR). If the target gene is a member of a multigene family, multiple alignments of family members are needed to help guide the design of PCR primers. The region of the gene to be amplified and its similarity to other genes dictates whether the resulting IR construct is likely to target a single mRNA or transcripts of multiple related genes.

Clone cDNA Fragments Corresponding to Targeted mRNAs. Using RT-PCR, we typically amplify a 500- to 700-bp portion of a given cDNA and then clone this fragment twice, in inverted orientation, within pFGC5491. The construction of the inverted repeat transgene involves a two-step ligation procedure facilitated by pairs of restriction endonuclease sites included in the primer sequences that constitute the ends of the PCR fragment. In the first cloning step, the PCR product is cleaved at the innermost restriction sites (*Asc*I and *Swa*I) at each end of the PCR product and is ligated into *Asc*I/*Swa*I-digested pFGC5941 (see Fig. 1). The resulting plasmid then serves as the template for a second PCR amplification by using the same primers. The resulting PCR product is cleaved with *Bam*HI and *Xba*I, which are distal to the *Asc*I and *Swa*I sites at the termini of the PCR fragment, and this fragment is cloned into the *Bam*HI/*Xba*I-digested plasmid. Because of the orientation of the restriction sites in the pFGC5491 plasmid, the second ligation inserts the PCR product in inverted orientation with respect to the first cloned fragment, yielding an inverted repeat separated by the CHSA intron (see Fig. 1). Suggested primer sequences and other details are available at the FGC Web site (http://ChromDB.org/).

Both *Asc*I and *Swa*I recognize infrequent 8-bp sites, whereas *Bam*HI and *Xba*I have 6-bp recognition sites and are therefore more likely to occur in target genes. It is usually possible to identify a span of DNA lacking these 6-bp recognition sites in the target transcript. However, if these sites are present within the cDNA region to be amplified, other unique restriction sites are available for use within the multiple cloning sites of pFGC5941, e.g., *Pac*I, *Xma*I, *Xho*I, and *Nco*I (see Fig. 1).

Verify Engineered Plasmid Vector. The two-step cloning of the inverted repeats described previously is carried out by using transformation of *E. coli* following each ligation reaction, selection of kanamycin-resistant colonies, preparation of plasmid DNA from small-scale cultures (2–3 ml), and identification of recombinant plasmids by restriction mapping (Sambrook and Russell, 2001). To be sure that the correct fragment of DNA has been cloned, which is a potential problem in a large project in which numerous cDNA fragments are being cloned in parallel and mix-ups could occur, DNA sequencing is used to verify each final vector.

Mobilize Plasmid Vector into A. tumefaciens. Once an IR plasmid vector has been verified, purified plasmid DNA isolated from *E. coli* is used to transform *A. tumefaciens* by electroporation (den Dulk-Ras and Hooykaas, 1995; Mersereau *et al.*, 1990). Kanamycin-resistant colonies are isolated and plasmid DNA is prepared from small-scale cultures (2–5 ml) grown to stationary phase in LB medium (Sambrook and Russell, 2001). Purified plasmid DNA is then subjected to restriction mapping to verify that the plasmid appears to be correct and has not suffered any obvious rearrangements due to growth in *A. tumefaciens*. Large-scale cultures (~500 ml) are then grown for plant transformation.

Transform A. thaliana *Plants (Ecotype WS) by Minor Modification of the Floral Dip Method of Clough and Bent (1998).* In brief, 5–10 plants for each T-DNA to be introduced are grown individually in 3-in. pots (20°, continuous light) in a growth chamber until they begin to flower. The plants are then inverted and submerged for 5 min in a solution of *A. tumefaciens* that had been grown to mid-log phase, pelleted by centrifugation, and resuspended in a solution that includes detergent (0.02% Silwet L-77). The detergent allows the solution to thoroughly coat the plants cells and infiltrate into the intercellular spaces through the stomata. After dipping, plants are too waterlogged to stand upright and are therefore laid horizontally in shallow plastic trays with fitted clear plastic lids to maintain high humidity. The following day, plants are returned to the growth chamber. At some point following dipping, the agrobacteria transfer into plant cell T-DNAs that integrate into *Arabidopsis* chromosomes in one or more copies. If one or more transformed cells happen to be in a meristematic region that gives rise to gametes, the T-DNA is transmitted to the progeny (T_1 generation) of the dipped plant. For unknown reasons, the female gametophyte rather than the pollen (the male gametophyte) is transformed and transmits the T-DNA to the T_1 progeny (Desfeux *et al.*, 2000). As a result of fertilization by untransformed pollen, T_1 progeny are generally hemizygous for the T-DNA.

Select Transformed Plants. Transgenic T_1 plants arise at a frequency that typically ranges from 0.2 to 1.4% (e.g., see Table I). To identify these relatively infrequent transformants, approximately 1000 seeds collected from dipped plants are sown in flats of soil. Germinated seedlings are then sprayed (~10 ml/flat) with Finale herbicide (AgrEvo Environmental Health, Montvale, NJ) diluted 1:200 with water. Finale herbicide, as purchased, is 5.78% (w/v) glufosinate-ammonium (butanoic acid, 2-amino-4-hydroxymethylphosphinyl, monoammonium salt). The *BAR* gene confers resistance to this herbicide. Because the T-DNA includes the *BAR* gene, transgenic plants that express this selectable marker gene survive herbicide spraying whereas nontransgenic plants are killed. Seedlings are typically sprayed twice, first when they are approximately 1 week old and again when they are 2 weeks old. Enough seeds (5000 or more) are sown such that 50–70 herbicide-resistant T_1 plants are selected for each construct.

Identify Transgenic Plants Possessing Single T-DNA Insertions. The goal of our FGC project is to produce high-quality, stable RNAi lines that are publicly available to the research community. Because multicopy transgenes are often associated with transcriptional silencing, we decided to identify and distribute only transgenic lines that have single-copy T-DNA insertions. This decision was based on the premise that expression from such single-copy transgenes is likely to be more stable and reliable than from multicopy lines, especially after several generations. There is a report that single-copy RNAi lines silence target genes more

TABLE I

COMPARISON OF *A. TUMEFACIENS* STRAINS LBA4404 AND GV3101 FOR *A. THALIANA* TRANSFORMATION

Targeted gene	Agro. strain	No. of T_1 seeds screened	No. of Basta-resistant T_1 plants	Transformation efficiency	No. of T_1 plants screened (Southern)	No. of lines with single LB fragments	Percentage of lines with single LB fragments
CHR18	LBA4404	10,000	19	0.19	19	7	36.8
	GV3101	5000	70	1.40	56	12	21.4
CHR19	LBA4404	5000	37	0.74	37	14	37.8
	GV3101	5000	65	1.30	54	11	20.4
CHR21	LBA4404	5000	18	0.36	18	7	38.8
	GV3101	5000	61	1.22	52	9	17.3

efficiently than multicopy lines (Kerschen *et al.*, 2004). Unfortunately, single-copy T-DNA insertions are infrequent when using common *A. tumefaciens* strains (LBA4404 and GV3101) in combination with pCAMBIA-derived vectors such as pFGC5491 (see Table I). As a result, we typically need to screen 50–70 transgenic T_1 plants to find at least 3–5 plants that carry only a single T-DNA.

DNA purified from single leaves (<50 mg of tissue) of T_1 transformants is screened by DNA gel blot analysis to identify plants that possess only a single copy of the T-DNA. To increase throughput, leaves of 192 plants are processed simultaneously in a 96-well format, using the QIAGEN DNeasy™ 96 Plant DNA Extraction Kit and a bead-beater homogenization system (Mixer Mill) marketed by Qiagen. Resulting DNA is sufficient for three Southern blots. The DNA is then digested with a restriction endonuclease that cleaves internal to the T-DNA LB and RB and is subjected to agarose gel electrophoresis. Following DNA (Southern) blotting to a nylon membrane, the blot is hybridized sequentially to LB and RB probes. Genomic DNA fragments that hybridize to these probes are generated by one cleavage event internal to the T-DNA and one cleavage event at a site in the genomic DNA flanking the T-DNA. As a result, the fragments are unique in size for each T-DNA integration event and the number of bands observed reflects the number of T-DNA copies integrated (Fig. 3). A single T-DNA integration will yield a single band on using both LB and RB probes and plants displaying single LB and RB fragments are retained. Plants that have multiple T-DNA insertions are discarded.

A caveat to the screening procedure for single-copy lines is that if the restriction enzyme site is sensitive to cytosine methylation, the efficiency of cleavage can be reduced and multiple hybridizing bands can be observed by using LB or RB probes, or both, even if the T-DNA is single copy. This potential pitfall is circumvented by using restriction endonucleases whose recognition sites lack cytosines and are not known to be sensitive to methylation at flanking cytosines, such as *Ase*I (recognition site 5'-ATTAAT-3').

Generate Homozygous Transgenic Lines. As mentioned previously, primary transformants (T_1 generation) are hemizygous for the T-DNA. To identify homozygous transgenic lines, 10 T_2 progeny of a single self-pollinated T_1 plant are grown; 25% of these should be homozygous for the T-DNA. To identify these homozygous lines, T_3 seeds are collected from each T_2 plant and aliquots of these seeds are tested for herbicide resistance. If 100% of the T_3 progeny of a single T_2 plant are resistant to the herbicide, it indicates that the T_2 parent (as well as all T_3 sibs) is homozygous for the T-DNA.

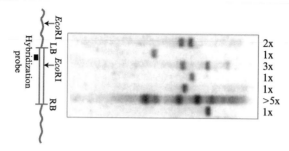

FIG. 3. Genomic Southern blot screen for RNAi-inducing transgene copy number. Genomic DNA purified from T_1 transformants was digested with a restriction enzyme, size-fractionated by gel electrophoresis, transferred to a nylon membrane, and hybridized with a radioactively labeled DNA fragment corresponding to the region adjacent to the T-DNA left border (LB). The schematic below the autoradiogram shows the expected organization of an intact single-copy T-DNA insertion. The position of the hybridization probe (black box) and the restriction endonuclease digestion site used in this version of the screen (EcoRI) is shown. Our preference is to use the cytosine methylation-insensitive restriction endonuclease AseI (5'-ATTAAT-3') rather than EcoRI (5'-GAATTC-3') to avoid partial endonuclease cleavage caused by cytosine methylation. Single-copy insertions will yield a unique hybridizing band corresponding to the junction fragment between the T-DNA; LB and RB and the flanking plant DNA (gray wavy line). In the examples shown, four T_1 transformants carried single (designated 1×) LB junction fragments whereas two T_1 transformants carried multiple LB junctions (3× and >5×). T_1 lines with a single LB band are then screened in a similar manner with a hybridization probe corresponding to the RB (not shown). Lines with single LB and RB junction fragments contain single-copy T-DNA insertions.

Amplify Seed Stocks for RNAi Lines Bearing Single-Copy T-DNAs. To obtain sufficient quantities of seed for public distribution, approximately 20 T_3 plants are grown and their T_4 seeds harvested. One can expect at least 1000 seeds per plant unless fertility is severely affected by the RNAi-inducing transgene.

Perform Quality Control Tests on T_4 Lines and Submit Seeds to the Arabidopsis Biological Resource Center (ABRC). Before submitting seed stocks to the ABRC, DNA gel blot analysis as described for T_1 plants is repeated on homozygous T_4 lines to double-check the T-DNA copy number. In some cases, this quality control test identifies lines that possess more than one T-DNA. If so, the lines are released to the ABRC with the designation of being multicopy lines. Basta resistance is also confirmed for T_4 seed by plating seed on a bialaphos-containing semisolid plant growth medium, a test that simultaneously tests seed viability.

Test for Knockdown of Targeted mRNAs. To determine whether the RNAi-inducing transgenes affect mRNA levels of targeted genes, RT-PCR is performed by using two primer pairs: one pair amplifies a control gene not targeted for RNAi (e.g., *GAPC*; the glyceraldehyde-3-phosphate de-hydrogenase C subunit; *A. thaliana* locus At3g04120) and the other pair is specific for the mRNA of interest. The FGC consortium has analyzed target transcript abundance in 77 independent, single-copy RNAi lines by using RNA pooled from multiple T_4 plants. Of these 77 lines, 42 lines (14 target genes) exhibit strong reduction or elimination of target transcript and 14 lines (8 target genes) exhibit no apparent reduction of target transcript. The remaining 21 lines (8 target genes) show an intermediate degree (two- to threefold) of target transcript reduction. In our experience, target mRNA reduction is fairly consistent among independent lines targeting the same gene (see summary).

Factors Affecting the Generation of RNAi Lines and the
  Success of RNAi in Arabidopsis

In the course of doing high-throughput transformation of *A. thaliana*, we have found that the strain of *A. tumefaciens* affects transformation efficiency. Data resulting from a direct comparison of two commonly used strains, LBA4404 and GV3101, are shown in Table I. For this comparison, IR constructs targeting the putative chromatin-modifying proteins CHR18, CHR19, or CHR21 were transformed into *A. tumefaciens* strains LBA4404 or GV3101 and grown in appropriate selective media. For each IR con-struct, five plants were transformed by using each of the *A. tumefaciens* strains and transgenic T_1 plants were screened for single-copy LB DNA fragments. As seen in Table I, *A. thaliana* transformation efficiency is three- to sevenfold higher when the T-DNA construct is delivered from GV3101 than from LBA4404. However, the frequency of single T-DNA fragments spanning the left border is 50–100% higher by using LBA4404 than by GV3101. These data suggest that high transformation efficiency is likely to reflect an increased efficiency of delivering T-DNAs into plant cells, but the disadvantage is that multiple T-DNAs per cell tend to be delivered. Because DNA blot screening is more laborious than dipping *A. thaliana* plants or selecting transformed plants, we recommend the use of LBA4404 if the goal is to isolate single-copy T-DNA insertions.

Smith *et al.* (2000) reported that using an intron as the spacer between the inverted repeats improved the efficiency of target gene silencing. Our FGC consortium also tested the ability of inverted repeat constructs with different spacer sequences, including two different introns or a 335-bp nonintronic fragment of a GUS reporter gene. In our hands, no increase in silencing efficiency was observed by using vectors containing introns as compared to the GUS fragment. The reasons for these different results are unclear.

Promoter strength appears to be correlated with silencing efficiency. For instance, the Meyerowitz laboratory (Chuang and Meyerowitz, 2000) showed that the strong CaMV 35S promoter is more efficient at inducing RNAi than is the weaker nopaline synthase promoter. Members of our FGC consortium have performed equivalent experiments and obtained results consistent with those of Meyerowitz and colleagues.

A potential problem with an RNAi-based gene knockdown strategy is that a given target gene may be an essential gene whose knockdown is lethal. In this case, it may be impossible to generate and maintain a stably silenced line expressing the inverted repeat from a strong promoter. To circumvent such problems, an inducible gene expression system for inducing RNAi is desirable. Thus, FGC vectors based on a dexamethasone-inducible promoter system described by Aoyama and Chua (1997) have been engineered. These vectors, pFFC4302 and pFGC8431, and protocols for their use, are described at the FGC Web site (http://www.ChromDB.org).

High-Throughput Generation of Maize RNAi Lines

Vectors for Generating Transgenic Plants and Inducing RNAi of Targeted Genes

To transform maize, immature embryos are subjected to particle bombardment, or biolistic transformation, using microscopic gold or tungsten particles onto which DNA has been precipitated (discussed in detail later). The base construct used for biolistic transformation of maize is plasmid pMCG161 (Fig. 4), which contains a vector backbone similar to that used for the *Arabidopsis* constructs. pMCG161 is based on pCAMBIA 1200 (CAMBIA; http://www.cambia.org). The vector is a T-DNA binary vector for use with *Agrobacterium*-mediated transformation of plants, although in our hands this binary vector does not work well for *Agrobacterium*-mediated maize transformation (H. Kaeppler, unpublished data). A full description of pMCG161, including its nucleotide sequence and features, are described at the FGC Web site (http://www.ChromDB.org/). The complete sequence is also available from GenBank (accession number AY572837). The expression cassette in pMCG161 includes the maize *ubiquitin* promoter and first intron (Christensen and Quail, 1996) driving expression of the *BAR* gene, which confers resistance to the herbicide bialaphos (Thompson *et al.*, 1987). The CaMV35S promoter and maize *adh1* intron1 drive expression of the target gene inverted repeat. The rice *waxy-A* intron1, which was amplified from genomic DNA based on sequence published by Li *et al.* (1995) serves as the spacer between the

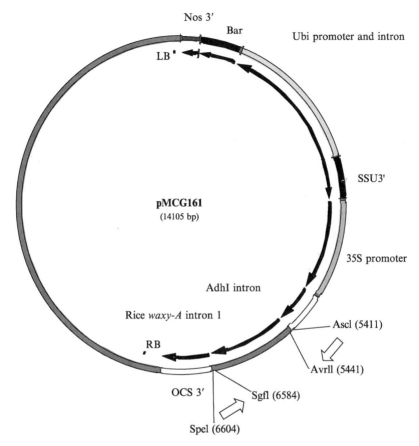

FIG. 4. Organization of the maize transformation vector, pMCG161. The expression cassette includes a selectable marker gene conferring Basta resistance (the *BAR* gene) under the control of the ubiquitin (ubi) promoter and intron. In addition, the T-DNA region contains the CaMV 35S promoter (35S promoter) upstream of the maize *adhI* and rice *waxy-A*1 introns and an octopine synthase 3′ end. The arrowheads indicate the orientation of each feature. Two multiple cloning sites flank the rice *waxy-A* intron, providing sites for cloning target gene sequences in an inverted orientation (open arrows). Only sites used for the two-step cloning procedure for our constructs are indicated in this map. The SSU 3′ sequence is a small subunit ribosomal RNA sequence that was included as a spacer between the two expressed sequences, which are in reverse orientation relative to each other. We initiated the project by using a T-DNA binary vector backbone to enable either biolistics or *Agrobacterium*-mediated transformation; the left border (LB) and right border (RB) sequences are indicated. This vector proved to not be efficient for *Agrobacterium*-mediated transformation of maize (H. Kaeppler, unpublished).

inverted repeats. The stock number for pMCG161, which can be ordered from ABRC, is CD3-459.

Procedure for Generating Maize RNAi Lines

Design an RNAi-Inducing Transgene for Each Gene Targeted for Knockdown. As in *Arabidopsis*, the first step in our process is target gene identification, which involves database searches of maize sequences. Because genomic sequence is not available for most maize genes, we use known target genes from other organisms to search through the available sequenced maize cDNAs, which are publicly available. We verify each clone by DNA sequencing and use 5' RACE as necessary to generate full-length cDNA sequences. Sequence information then guides the design of PCR primers to amplify a 500- to 700-bp segment of a given cDNA clone, which in some cases can target multiple homologous genes. cDNA intervals that contain *Avr*II, *Asc*I, *Spe*I, and *Sgf*I restriction sites are avoided because these enzymes are used in our two-step cloning process.

Clone Amplified Fragments Corresponding to Targeted mRNAs. A two-ligation cloning strategy similar to that used for the *Arabidopsis* constructs is used to generate constructs for maize. Our PCR primers include tails with two restriction endonuclease recognition sequences at each end (*Asc*I and *Avr*II are the inner sites; *Spe*I and *Sgf*I are the outer sites) for directional cloning of the PCR product in inverted orientation (Fig. 4). For the first cloning step, the PCR product is digested with *Avr*II and *Asc*I and ligated into pMCG161. The second step of cloning is to digest the same PCR product with *Spe*I and *Sgf*I and ligate into the plasmid generated by the first ligation. Each construct is then verified by sequencing the entire inverted repeat fragment and aligning the resulting sequence against the target gene sequence.

Transform Maize. Microprojectile biolistic delivery of DNA is used to produce transgenic maize lines expressing the inverted repeat constructs. Bombardments are carried out as described in Frame *et al.* (2000), except that smaller microcarrier particles (0.6 μm) are used and they are coated with a lower concentration of plasmid DNA (0.4 μg plasmid DNA/1 mg microcarriers) so as to achieve lower copy number transgene insertions. Modifications also include the use of 1100-psi rupture discs, a rupture disc-to-macrocarrier distance of 6.5 mm, a macrocarrier-to-stopping screen distance of 8.0 mm, and a stopping screen-to-target tissue distance of 10 cm.

The target tissues for bombardment-mediated maize transformation are F_2 immature embryos derived from self-pollinated, greenhouse-grown HiIIA $\times$ HiIIB (Armstrong *et al.*, 1991) F_1 hybrid maize plants. Twenty precultured, immature embryos (1.5–1.8 mm, isolated 10–14 days postpollination) are targeted per bombardment. At least 20 plates (400 embryos) are bombarded for each IR vector. Immature embryos are isolated continuously and precultured for 7–10 days such that bombardment experiments can be carried out two to three times per week.

Embryos are precultured axis-side down on N6-based initiation medium (NIM) containing N6 basal salts and vitamins (Chu *et al.*, 1975) supplemented with 25 mM proline, 100 mg/l casein hydrosylate, 50 μM silver nitrate (Armstrong and Green, 1985; Songstad *et al.*, 1991), 2 mg/l 2,4-dichlorophenoxyacetic acid (9 μM), 3% sucrose, and 0.35% Phytagel (Sigma Chemical Co., St. Louis, MO).

Selecting Transgenic Calli and Plant Regeneration. Bombarded tissues are transferred to callus induction medium and cultured for 7 days and then individually transferred to selection medium containing 5 mg/ml bialaphos (Dennehey *et al.*, 1994). Selection medium consists of NIM minus glycine vitamin stock. Following five to six biweekly subculture and subdivision steps, independent transgenic callus lines are identified and placed on the selection medium (one line per plate) and grown for 3 weeks. Callus of each transgenic line is then subdivided into three portions, the first to be used for DNA isolation and Southern blot analysis to test for number of insertions and integration of an intact IR region (see the next section for DNA preparations and digestions used), the second for plant regeneration, and the third placed on the selection medium to be saved as a backup.

Production of T_1 Seed. Approximately five transgenic maize plants (T_0 generation) are regenerated for each independent callus line. T_1 seed is produced by crossing regenerated transgenic T_0 plants as the male or female parent with maize inbred line B73, producing on average $\sim$3 ears with seed for each independent event. Plants are outcrossed rather than self-pollinated to keep the transgene hemizygous and thus to potentially increase the stability of transgene expression. Based on initial bombardment data, the mean maize transformation frequency (fertile transgenic maize lines recovered per 100 embryos bombarded) for chromatin target gene constructs has been 4.5%, with a range of 0.3–20%. Substantial effects on transformation frequency are sometimes observed for specific target gene constructs when compared with controls.

T_1 Analysis and Seed Production. Approximately 20 T_1 seed obtained from plants regenerated from each independent transgenic callus line are planted and tested for herbicide resistance and the quantity and integrity of the IR transgene. Comparisons of these two tests provide an indication of the number of transgene loci, the complexity of the transgene insertions at each locus, and whether the herbicide resistance and the IR transgene loci cosegregate. Typically, we only carry forward lines in which the IR

transgene and herbicide resistance cosegregate, as this greatly facilitates efforts to track the transgenic plants by using visual scoring (see later) when propagating lines representing many different transgenic events.

Plants are tested for herbicide resistance as described by Dennehey *et al.* (1994) and Spencer *et al.* (1992). Briefly, a fully emerged leaf is marked about 6 in. from the tip with a black permanent marking pen. Then, herbicide (1:100 dilution of a stock containing 200 mg/ml glufosinate, 0.01% Tween 20) is painted in a 1-in.-long stripe with a sponge-tipped applicator immediately distal to the mark. After 3–5 days, herbicide resistance or susceptibility can be scored. Young susceptible plants and seedlings can be killed by this application, but susceptible plants treated at the eight- to nine-leaf stage or older generally continue to grow even though the treated leaf is damaged. A susceptible plant will exhibit a yellowing or browning of the painted leaf and other signs of necrosis. A resistant leaf will retain its healthy, green appearance. Examples of the range of phenotypes observed can be seen at the FGC Web site (http://ChromDB.org/).

DNA is extracted from young leaves and tested to see whether the IR transgene remains intact by digesting the genomic DNA with *Hind*III/ *Swa*I. (See Fig. 5 for location of these restriction enzymes.) The size of the resulting bands on a DNA gel blot is compared to the predicted size. To produce three to five independent transgenic lines expressing each construct, we analyze as many T_1 lines as possible and carry forward to the T_2 generation up to 10 lines per construct. To meet this goal, lines with multiple copies of insertions or lines that do not have the size fragment predicted for an intact inverted repeat are occasionally carried forward. To increase the seed available, T_1 plants are backcrossed to B73, resulting in T_2 seed that is hemizygous for each transgene insertion.

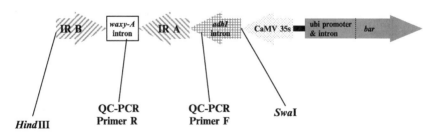

FIG. 5. Location of QC-PCR primers and digestion sites for gel blot analysis in pMCG161. This simplified schematic of the pMCG161 expression cassette indicates the position of the PCR primers (QC PCR Primer F and R) used to amplify and sequence one inverted repeat (IR A) to confirm transgene identity. Positions of the restriction sites used for T_1 Southern blot analysis are also indicated (*Hind*III and *Swa*I).

Before submitting lines to the Maize Genetics Stock Center, quality control tests are conducted to ensure cosegregation of the transgene with herbicide resistance, examine reduction of target gene expression, and verify that the correct transgene is present. These tests are described in the following sections, and detailed protocols are available online (http:// ChromDB.org/v3/transuserguide.html).

Test for Transgene Activity and Cosegregation of the Transgene with Herbicide Resistance in T_2 Plants. T_2 plants for each transgenic line are planted in the greenhouse and tested for herbicide resistance to verify that the transgene is active and cosegregating with herbicide resistance. Leaves are collected from resistant plants for DNA and RNA extractions. Some lines fail to segregate the expected 50% herbicide-resistant individuals. Lines in which all plants are herbicide sensitive could derive from a lack of transmission of the transgene or they could contain a completely silenced or rearranged transgene; these lines are not carried forward. Lines with significantly less than 50% of the plants showing herbicide resistance could result from reduced transmission of the transgene or sporadic silencing of the transgene. In these instances, if the number of independent lines for a given construct is low, only herbicide-resistant individuals are carried forward.

Test for Knockdown of Targeted mRNAs. RT-PCR is performed to determine whether the target gene mRNA is reduced. For logistical reasons, four individuals are analyzed separately for each maize line rather than pools of T_4 plants as in the case of *Arabidopsis* lines. This protocol increases the number of RT-PCR reactions by fourfold; however, eliminating the dilution series from the procedure described by Kerschen *et al.* (2004) and using maize-specific *GAPC* control primers and inhibitors increase throughput. Details of the protocol and primer sequences are available at the FGC Web site. Sixty-five unique transgenic maize lines have been analyzed by our FGC consortium, representing 18 constructs targeting 21 genes. mRNAs are strongly reduced in 25 maize lines (39%), representing 13 constructs targeting 16 genes. An additional four lines (6%), representing six constructs targeting six genes, showed a two- to threefold reduction of the mRNA. The remaining 36 maize lines (55%) showed no detectable reduction of the targeted mRNA. Similar to the observations with *Arabidopsis*, target mRNA reduction is fairly consistent among independent lines targeting the same gene (see summary).

Verify Transgene Identity in Each Transgenic Line. To confirm that the transgenic plant contains the correct construct, one inverted repeat is amplified from genomic DNA of transgenic plants by using primers to the *adh1* and *waxy-A* introns that flank the IR (QC Primer F and R in Fig. 5) and the product is sequenced. Each line is also verified to be a unique transgenic event by Southern blot analysis using genomic DNA digested with *Bam*HI, which cuts once in pMCG161, resulting in a distinct banding pattern for transgenes inserted in different genomic locations. The DNA gel blot is hybridized with a probe containing the *BAR* gene, the Nos3', and part of the T-DNA LB, which corresponds to base pairs 66 to 774 of pMCG161. Only those lines passing all quality control tests are submitted to the stock center for distribution.

Submission of Seed to Stock Center. T_2 seed that meets our quality control standards is submitted to the Maize Genetics Cooperative Stock Center for distribution to the public (http://maizegdb.org).

Factors Affecting the Generation of RNAi Lines and the Success of RNAi in Maize

The biolistic bombardment approach used for maize transformation results primarily in transgenic lines with more than one insertion. Of the 568 T_1 lines examined to date, most possess fewer than five introduced DNA fragments, but at least 75% of the lines contain more than one IR-hybridizing fragment. Many lines also contain fragment sizes that indicate transgene DNA rearrangements. These results are not unexpected for transgenes introduced by particle bombardment. Segregation tests have demonstrated that 84% of the 568 T_1 lines analyzed contain a single transgenic locus, indicating that the multiple transgenes tend to integrate at a single site. In 71% of these lines, the IR-hybridizing fragments and the *bar* gene cosegregate in outcrosses of the transgenic plant to B73, and herbicide resistance segregates at the expected frequency of 50%. In the remaining 29% of these lines, the IR cross-hybridizing fragments appear to segregate as a single locus, but fewer than 50% herbicide-resistant plants are observed (could be caused by reduced transmission, reduced plant survival, or transgene silencing).

Although the bombardment method produces a high percentage of lines with more than one copy of the IR sequences, the frequency of transgene silencing that occurs between the T_1 and T_2 generations appears to be low. Of 140 lines that have been examined at the T_2 stage, 134 continued to segregate herbicide resistance at the frequency expected for a single locus, whereas 6 lines show all herbicide-susceptible plants. An efficient protocol for *Agrobacterium*-mediated maize transformation by using publicly available binary vectors has been published (Frame *et al.*, 2002). The use of the *Agrobacterium*-mediated approach rather than the biolistics approach should result in more efficient generation of RNAi lines, because transgenic lines with simple insertions occur at a higher frequency.

The lower number of maize lines that showed silencing phenotypes relative to *Arabidopsis* may have resulted from the different transformation methodologies used in the two species. The *Arabidopsis* lines were transformed by using *Agrobacterium*-mediated T-DNA transfer and were screened for simple, single-copy insertions with intact left and right borders resulting in essentially all lines having an intact IR construct. In contrast, the maize lines were generated by particle bombardment, which results in an increased frequency of rearranged transgenes. A higher frequency of rearranged transgenes could result in lower yield of lines that efficiently express the inverted repeat.

One of the justifications for generating RNAi lines is that highly similar genes can be simultaneously silenced. Our results with three different constructs targeting pairs of highly similar genes reveal efficient silencing of duplicate genes (Fig. 6). For example, both NFA104 and NFA103 mRNAs are consistently reduced in the transgenic line illustrated in Figure 6A and B. These two genes share 93% identity at the nucleotide level over the available cDNA sequences, including the target sequence region. A similar result was observed with HDT103 and HDT102, which share ~85% nucleotide identity in the IR (Fig. 6C and D) and NFC101 and NFC102, which share ~95% nucleotide identity in the IR (Fig. 6E and F). There is a high level of duplication in the genome of maize, which is an ancient allotetraploid (Gaut and Doebley, 1997); therefore, these results bode well for using RNAi in maize.

Summary

Large-scale efforts to generate transgenic RNAi lines targeting hundreds of genes are clearly feasible in both *Arabidopsis* and maize. The vectors designed enabled high-throughput generation of nearly 200 constructs by a small number of technicians for the FGC. In the majority of cases for both plants, multiple independent lines transgenic for the same construct showed the same extent of silencing. In *Arabidopsis*, a uniform silencing phenotype (either all lines silenced or no lines silenced) was observed for 20 out of 25 IR constructs tested (77 lines total). In maize, 9 out of 18 constructs (65 lines total) showed consistent silencing phenotypes for all lines carrying the same construct. In both *Arabidopsis* and maize, when multiple silencing phenotypes are observed, the variability is low, with usually only one exceptional line observed.

The observation that multiple independent integration events for the same construct most often result in the same level of silencing suggests that silencing is gene dependent. The reason why some mRNAs are highly sensitive to RNAi and others are not is unclear. One contributor could

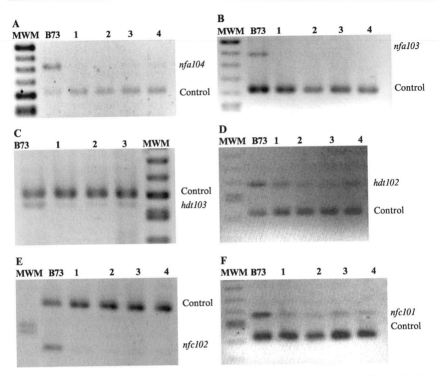

Fig. 6. Duplicate homologous genes can be targeted by single IR constructs. Semiquantitative RT-PCR results analyzing lines transformed with three different constructs highly homologous to three distinct gene pairs. In each panel, molecular weight markers are indicated (MWM); B73 is DNA from a nontransgenic control plant, and the numbered lanes are herbicide-resistant transgenic siblings for the same transgene locus. The band expected for amplification of the maize *GAPC* transcript (Control) and the bands expected for each pair of gene-specific primers are indicated. All the transgenic lines showed reduction of the target genes relative to the control. Panels A and B illustrate the ability of the construct 100% identical to *nfa104* to reduce the mRNA for both *nfa104* and *nfa103*, panels C and D illustrate the ability of the construct 100% identical to *hdt103* to reduce the mRNA for both *hdt103* and *hdt102*, and panels E and F illustrate the ability of the construct 100% identical to *nfc102* to reduce the mRNA for *nfc102* and *nfc101*.

be mRNA abundance, as a trend that we observe is that genes expressed at higher levels tend to be among those more readily silenced (all data at http://ChromDB.org/). Another possibility could be that different sequences within the distinct IR regions generate 21- to 23-nt fragments that function with distinct efficiencies. Different overlapping 21- to 23-nt fragments have been demonstrated to show different efficiencies of silencing in mammals (Shirane *et al.*, 2004). However, we think that the latter explanation is unlikely as our inverted repeats contain 500–700 bp of target gene sequence, which should generate a large number of distinct 21- to 23-nt fragments to target a given mRNA. Thus, we favor the idea that certain genes may be intractable to RNAi regardless of the construct design. Thus, if one 500- to 700-bp construct is not effective at silencing, it may not be productive to design a new inverted repeat sequence to another part of the gene; instead, alternative mutational approaches should be tried.

There are examples in the literature in which the primary silencing step is to block translation rather than to degrade the mRNA (Ambros *et al.*, 2003; Doench *et al.*, 2003). Thus, it is possible that some lines expressing the IR may show silencing and phenotypic effects even though the mRNA levels are not detectably decreased. For almost all the genes we have targeted, antibodies are not available to directly examine protein levels. However, we will test whether phenotypic effects in several functional assays always correlate with reduced target mRNA levels.

Acknowledgments

This project was funded by a grant from the National Science Foundation (DBI-9975930) to Richard Jorgensen (principal investigator), Judith Bender, Vicki Chandler, Karen Cone, Stanton Gelvin, Heidi Kaeppler, Shawn Kaeppler, David Mount, Craig Pikaard, and Eric Richards (co-principal investigators). We thank Carolyn Napoli and Rich Jorgensen for vector design; Rayeann Archibald and Virginia O'Connell for vector construction; Judith Bender for participating in the generation and testing of *Arabidopsis* transgenic lines; Andreas Muller for supervising the *Arabidopsis* RT-PCR experiments, Ross Atkinson for generating and testing the inducible silencing vectors; Nives Kovacevic for assistance with maize T_1 analyses; and the following technicians, specialists, and students who contributed their efforts to this project—Heather Basinger, Dean Bergstrom, Erin Berry, Kari Hesselbach Chambers, Heather Ferguson, Miriam Hankins, William Haun, Kirsten Lovette, Jill Mahoy, Annie McGill, Justin Rincker, Robert Sandoval Jr., Laura Schmitt, Alan Smith, Kevin VerWeire, and Jessica Yoyokie.

References

Agrawal, N., Dasaradhi, P. V., Mohmmed, A., Malhotra, P., Bhatnagar, R. K., and Mukherjee, S. K. (2003). RNA interference: Biology, mechanism, and applications. *Microbiol. Mol. Biol. Rev.* **67,** 657–685.

Ambros, V., Lee, R. C., Lavanway, A., Williams, P. T., and Jewell, D. (2003). MicroRNAs and other tiny endogenous RNAs in *C. elegans*. *Curr. Biol.* **13,** 807–818.

Aoyama, T., and Chua, N. H. (1997). A glucocorticoid-mediated transcriptional induction system in transgenic plants. *Plant J.* **11,** 605–612.

Armstrong, C. L., and Green, C. E. (1985). Establishment and maintenance of friable, embryogenic maize callus and the involvement of L-proline. *Planta* **164,** 207–214.

Armstrong, C., Green, C., and Phillips, R. (1991). Development and availability of germplasm with high Type II culture formation response. *Maize Genet. Coop. Newslett.* **65,** 92–93.

Bailis, J., and Forsburg, S. (2002). RNAi hushes heterochromatin. *Genome Biol.* **3** reviews 1035.1–1035.4.

Carpenter, A. E., and Sabatini, D. M. (2004). Systematic genome-wide screens of gene function. *Nat. Rev. Genet.* **5**, 11–22.

Christensen, A. H., and Quail, P. H. (1996). Ubiquitin promoter-based vectors for high-level expression of selectable and/or screenable marker genes in monocotyledonous plants. *Transgenic Res.* **5**, 213–218.

Chu, C., Wang, C., Sun, C., Hsu, C., Yin, K., Chu, C., and Bi, F. (1975). Establishment of an efficient medium for anther culture of rice through comparative experiments of the nitrogen source. *Sci. Sin.* **18**, 659–668.

Chuang, C. F., and Meyerowitz, E. M. (2000). Specific and heritable genetic interference by double-stranded RNA in *Arabidopsis thaliana*. *Proc. Natl. Acad. Sci. USA* **97**, 4985–4990.

Clough, S. J., and Bent, A. F. (1998). Floral dip: A simplified method for *Agrobacterium*-mediated transformation of *Arabidopsis thaliana*. *Plant J.* **16**, 735–743.

Dawe, R. K. (2003). RNA interference, transposons, and the centromere. *Plant Cell* **15**, 297–301.

den Dulk-Ras, A., and Hooykaas, P. J. (1995). Electroporation of *Agrobacterium tumefaciens*. *Methods Mol. Biol.* **55**, 63–72.

Dennehey, B. K., Petersen, W. L., Fordsantino, C., Pajeau, M., and Armstrong, C. L. (1994). Comparison of selective agents for use with the selectable marker gene *bar* in maize transformation. *Plant Cell Tissue Organ Cult.* **36**, 1–7.

Desfeux, C., Clough, S. J., and Bent, A. F. (2000). Female reproductive tissues are the primary target of *Agrobacterium*-mediated transformation by the *Arabidopsis* floral-dip method. *Plant Physiol.* **123**, 895–904.

Doench, J. G., Petersen, C. P., and Sharp, P. A. (2003). siRNAs can function as miRNAs. *Genes Dev.* **17**, 438–442.

Frame, B. R., Shou, H., Chikwamba, R. K., Zhang, Z., Xiang, C., Fonger, T. M., Pegg, S. E., Li, B., Nettleton, D. S., Pei, D., and Wang, K. (2002). *Agrobacterium tumefaciens*-mediated transformation of maize embryos using a standard binary vector system. *Plant Physiol.* **129**, 13–22.

Frame, B. R., Zhang, H., Cocciolone, S. M., Sidorenko, L. V., Dietrich, C. R., Pegg, S. E., Zhen, S., Schnable, P. S., and Wang, K. (2000). Production of transgenic maize from bombarded type II callus: Effect of gold particle size and callus morphology on transformation efficiency. *In Vitro Dev. Biol. Plant* **36**, 21–29.

Fraser, A. G., Kamath, R. S., Zipperlen, P., Martinez-Campos, M., Sohrmann, M., and Ahringer, J. (2000). Functional genomic analysis of *C. elegans* chromosome I by systematic RNA interference. *Nature* **408**, 325–330.

Gallie, D. R. (2002). The 5′-leader of tobacco mosaic virus promotes translation through enhanced recruitment of eIF4F. *Nucleic Acids Res.* **30**, 3401–3411.

Gaut, B. S., and Doebley, J. F. (1997). DNA sequence evidence for the segmental allotetraploid origin of maize. *Proc. Natl. Acad. Sci. USA* **94**, 6809–6814.

Gelvin, S. B. (2000). *Agrobacterium* and plant genes involved in T-DNA transfer and integration. *Annu. Rev. Plant Physiol. Plant Mol. Biol.* **51**, 223–256.

Gelvin, S. B. (2003). *Agrobacterium*-mediated plant transformation: The biology behind the "gene-jockeying" tool. *Microbiol. Mol. Biol. Rev.* **67**, 16–37.

Hajdukiewicz, P., Svab, Z., and Maliga, P. (1994). The small, versatile pPZP family of *Agrobacterium* binary vectors for plant transformation. *Plant Mol. Biol.* **25**, 989–994.

Hannon, G. J. (2002). RNA interference. *Nature* **418**, 244–251.

Kerschen, A., Napoli, C. A., Jorgensen, R., and Müller, A. E. (2004). Effectiveness of RNA interference in transgenic plants. *FEBS Lett.* **566** (1–3), 223–228.

Ketelaar, T., Allwood, E. G., Anthony, R., Voigt, B., Menzel, D., and Hussey, P. J. (2004). The actin-interacting protein AIP1 is essential for actin organization and plant development. *Curr. Biol.* **14,** 145–149.

Lawrence, R. J., Earley, K., Pontes, O., Silva, M., Chen, Z. J., Neves, N., Viegas, W., and Pikaard, C. S. (2004). A concerted DNA methylation/histone methylation switch regulates rRNA gene dosage control and nucleolar dominance. *Mol. Cell* **13,** 599–609.

Lawrence, R. J., and Pikaard, C. S. (2003). Transgene-induced RNA interference: A strategy for overcoming gene redundancy in polyploids to generate loss-of-function mutations. *Plant J.* **36,** 114–121.

Li, Y. Z., Ma, H. M., Zhang, J. L., Wang, Z. Y., and Hong, M. M. (1995). Effects of the first intron of rice waxy gene on the expression of foreign genes in rice and tobacco protoplasts. *Plant Sci.* **108,** 181–190.

Martienssen, R. A. (2003). Maintenance of heterochromatin by RNA interference of tandem repeats. *Nat. Genet.* **35,** 213–214.

Matzke, M., and Matzke, A. J. (2003). RNAi extends its reach. *Science* **301,** 1060–1061.

Mersereau, M., Pazour, G. J., and Das, A. (1990). Efficient transformation of *Agrobacterium tumefaciens* by electroporation. *Gene.* **90,** 149–151.

Paddison, P. J., Silva, J. M., Conklin, D. S., Schlabach, M., Li, M., Aruleba, S., Balija, V., O'Shaughnessy, A., Gnoj, L., Scobie, K., Chang, K., Westbrook, T., Cleary, M., Sachidanandam, R., McCombie, W. R., Elledge, S. J., and Hannon, G. J. (2004). A resource for large-scale RNA-interference-based screens in mammals. *Nature* **428,** 427–431.

Sambrook, J., and Russell, D. (2001). "Molecular Cloning: A Laboratory Manual," 3rd Ed., Cold Spring Harbor Laboraory Press, Cold Spring Harbor, NY.

Shirane, D., Sugao, K., Namiki, S., Tanabe, M., Iino, M., and Hirose, K. (2004). Enzymatic production of RNAi libraries from cDNAs. *Nat. Genet.* **36,** 190–196.

Simmer, F., Moorman, C., Van Der Linden, A. M., Kuijk, E., Van Den Berghe, P. V., Kamath, R., Fraser, A. G., Ahringer, J., and Plasterk, R. H. (2003). Genome-wide RNAi of *C. elegans* using the hypersensitive rrf-3 strain reveals novel gene functions. *PLoS Biol.* **1,** E12.

Sleat, D. E., Gallie, D. R., Jefferson, R. A., Bevan, M. W., Turner, P. C., and Wilson, T. M. (1987). Characterisation of the 5'-leader sequence of tobacco mosaic virus RNA as a general enhancer of translation *in vitro*. *Gene* **60,** 217–225.

Smith, N. A., Singh, S. P., Wang, M. B., Stoutjesdijk, P. A., Green, A. G., and Waterhouse, P. M. (2000). Total silencing by intron-spliced hairpin RNAs. *Nature* **407,** 319–320.

Songstad, D. D., Armstrong, C. L., and Petersen, W. L. (1991). Agno3 increases Type-II callus production from immature embryos of maize inbred B73 and its derivatives. *Plant Cell Rep.* **9,** 699–702.

Spencer, T. M., O'Brien, J. V., Start, W. G., Adams, T. R., Gordon-Kamm, W. J., and Lemaux, P. G. (1992). Segregation of transgenes in maize. *Plant Mol. Biol.* **18,** 201–210.

Thompson, C. J., Movva, N. R., Tizard, R., Crameri, R., Davies, J. E., Lauwereys, M., and Botterman, J. (1987). Characterization of the herbicide-resistance gene *Bar* from *Streptomyces hygroscopicus*. *EMBO J.* **6,** 2519–2523.

Waterhouse, P. M., Graham, M. W., and Wang, M. B. (1998). Virus resistance and gene silencing in plants can be induced by simultaneous expression of sense and antisense RNA. *Proc. Natl. Acad. Sci. USA* **95,** 13959–13964.

Waterhouse, P. M., and Helliwell, C. A. (2003). Exploring plant genomes by RNA-induced gene silencing. *Nat. Rev. Genet.* **4**, 29–38.

Wesley, S. V., Helliwell, C. A., Smith, N. A., Wang, M., Rouse, D. T., Liu, Q., Gooding, P. S., Singh, S. P., Abbott, D., Stoutjesdijk, P. A., Robinson, S. P., Gleave, A. P., Green, A. G., and Waterhouse, P. M. (2001). Construct design for efficient, effective and high-throughput gene silencing in plants. *Plant J.* **27**, 581–590.

Xiao, H., Wang, Y., Liu, D., Wang, W., Li, X., Zhao, X., Xu, J., Zhai, W., and Zhu, L. (2003). Functional analysis of the rice *AP3* homologue OsMADS16 by RNA interference. *Plant Mol. Biol.* **52**, 957–966.

Zentella, R., Yamauchi, D., and Ho, T. H. (2002). Molecular dissection of the gibberellin/abscisic acid signaling pathways by transiently expressed RNA interference in barley aleurone cells. *Plant Cell* **14**, 2289–2301.

Zupan, J., Muth, T. R., Draper, O., and Zambryski, P. (2000). The transfer of DNA from *Agrobacterium tumefaciens* into plants: A feast of fundamental insights. *Plant J.* **23**, 11–28.

[2] Constructs and Methods for Hairpin RNA-Mediated Gene Silencing in Plants

By CHRIS A. HELLIWELL and PETER M. WATERHOUSE

Abstract

Double-stranded RNA (dsRNA) induces an endogenous sequence-specific RNA degradation mechanism in most eukaryotic cells. The mechanism can be harnessed to silence genes in plants by expressing self-complementary single-stranded (hairpin) RNA in which the duplexed region has the same sequence as part of the target gene's mRNA. We describe a number of plasmid vectors for generating hairpin RNAs, including those designed for high-throughput cloning, and provide protocols for their use.

Introduction

Double-stranded RNA (dsRNA) is an effective trigger of gene silencing in vertebrate, invertebrate, and plant systems (Sharp, 2001; Waterhouse *et al.*, 2001). This silencing operates by sequence-specific RNA degradation. In plants, a particularly effective method of silencing an endogenous gene is to transform the plant with a gene construct encoding a hairpin RNA (hpRNA) consisting of an inverted repeat of a fragment of the gene sequence separated by a spacer to allow easier plasmid construction. Using an intron as this spacer fragment increases the frequency of obtaining silenced plants (Smith *et al.*, 2000; Wesley *et al.*, 2001). The efficiency of intron-spliced hpRNA-mediated gene silencing makes it particularly attractive for functional genomics applications, in which large groups of genes are under study. In these cases, a method to reduce or remove the function of these genes can reveal information about their function in the normal growth and development of plants. The sequence specificity of hpRNA-mediated gene silencing allows the use of unique sequences to target specific genes and to use conserved sequences to simultaneously target more than one member of a multigene family. hpRNA constructs can also target two (and probably more) distinct sequences to give silencing of both genes, providing a method for silencing two genes of redundant function that do not have a high degree of sequence conservation (Helliwell *et al.*, 2002).

A major limitation to the use of hpRNA-mediated silencing for high-throughput applications, such as functional genomics studies, is the number of cloning steps needed to produce hpRNA constructs. However, a vector system (pHELLSGATE) that uses the Gateway recombination alleviates this problem. By using pHELLSGATE vectors, the inversely orientated gene-specific arms of the hairpin construct are inserted in a single recombination step, removing the need to use restriction enzymes, greatly streamlining the construction procedure.

Simple Generic Hairpin RNA (hpRNA) Vectors: pHANNIBAL and pKANNIBAL

A pair of simple constructs, pHANNIBAL (with bacterial ampicillin resistance) and pKANNIBAL (with bacterial kanamycin resistance), have been designed (Fig. 1) so that a PCR fragment can be inserted, using conventional restriction enzyme digestion and DNA ligation techniques, in the sense orientation into the *XhoI.EcoRI.KpnI* polylinker and in the antisense orientation in the *ClaI.HindIII.BamHI.XbaI* polylinker. This can be accomplished either by two separate PCR reactions with the appropriate single sites introduced with each primer or by a single PCR reaction using primers each introducing two restriction sites (e.g., primer 1 *XbaI.XhoI.*xxx; primer 2 *ClaI.KpnI.*xxx). The pKANNIBAL vector is particularly useful because the PCR fragments from the target gene can be directly cloned, without prior restriction enzyme digestion, into a commercially prepared 3′-T-overhang ampicillin-resistant vector such as pGEM®-T Easy (Promega) and then subcloned into pKANNIBAL by using differential antibiotic selection.

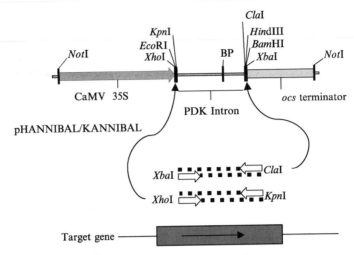

Fig. 1. Outline of the use of the pHANNIBAL or pKANNIBAL vectors for gene silencing. A PCR product is amplified from the target gene with restriction sites incorporated into the PCR primers. These products are digested with appropriate restriction enzymes and ligated into the appropriately cut vector. The two PCR products are usually ligated into the vector sequentially rather than simultaneously.

The *Not*I fragment from pH/KANNIBAL, containing the hpRNA cassettes, can then be subcloned into a convenient binary vector such as pART27 (Gleave, 1992) and used to transform plants. This pH/KANNIBAL system has worked very efficiently and effectively for a number of genes (Wesley *et al.*, 2001) and is appropriate for the silencing of one or several genes, but is somewhat cumbersome when one wishes to individually silence more than 10 target genes. The construction of each hairpin construct usually takes around 2 weeks.

High-Throughput hpRNA Vectors: The pHELLSGATE Series

A high-throughout cloning system to generate hpRNA constructs requires a vector, such as pHANNIBAL, to give directional insertion of a gene fragment into an inverted repeat conformation separated by an intron. Alternatively, the intron-spaced inverted repeat can be assembled by pull-though PCR or by a ligation of four fragments (vector, two target gene fragments, and intron spacer). In practice, the pH/KANNIBAL system is relatively slow and the other approaches are not sufficiently efficient or robust for routine use. Therefore, the pHELLSGATE vector series was created to take advantage of the commercially available Gateway unidirectional *in vitro* cloning system (Invitrogen, Carlsbad,

CA). Details about the recombinase enzymes, their recognition sites, and the underlying mechanism of the system are available at http://www. invitrogen.com/content/sfs/manuals/gatewayman.pdf.

The concept of the pHELLSGATE vectors is based on our finding that a PCR product flanked by *att*B1 and *att*B2 sites would recombine with a plasmid carrying two *att*P1–*att*P2 cassettes separated by an intron sequence (Fig. 2) when incubated with BP Clonase™ (Invitrogen, Carlsbad, CA), giving rise to inverted-repeat constructs with high efficiency (Wesley *et al.*, 2001). This is also true for an LR Clonase™ reaction between a plasmid carrying an insert flanked by *att*L1 and *att*L2 sites and a plasmid containing two *att*R1–*att*R2 cassettes (Fig. 2). These concepts were used in designing pHELLSGATE4 and pHELLSGATE8 (Helliwell *et al.*, 2002), which were used to test the effects of the resulting *att*L and *att*B sites on silencing efficiencies. These constructs are described in detail in Helliwell *et al.*

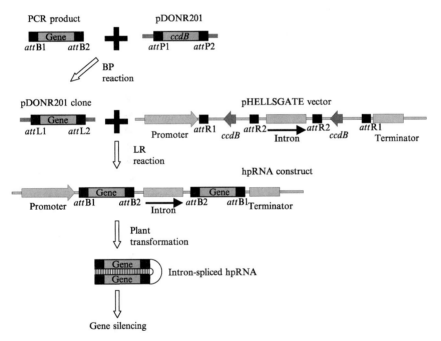

FIG. 2. Outline of the use of pHELLSGATE vectors for gene silencing. A PCR product is amplified from the target gene with *att*B1 and *att*B2 sites incorporated into the PCR primers. This product is then inserted into the pDONR201 vector by recombination between *att*B1/*att*B2 and *att*P1/*att*P2 mediated by BP Clonase to yield an intermediate clone. The gene fragment from the intermediate clone is then inserted into the pHELLSGATE vector by recombination between *att*L1/*att*L2 and *att*R1/*att*R2, mediated by LR Clonase. When the construct is expressed in plants, an intron-spliced hairpin RNA (hpRNA) is produced.

(2002). When constructing these vectors, the *ccdB* genes are orientated as a direct repeat within the inverted repeat of the *att*P/R1–*att*P/R2 cassettes to reduce the length of inverted repeat sequence in the nonrecombined plasmids, with the aim of increasing the stability of the plasmid in *E. coli*. Results with pHELLSGATE4 and pHELLSGATE8 suggested that hpRNA constructs containing *att*B sites gave more efficient silencing than those containing *att*L sites (Helliwell *et al.*, 2002). We suspect that the longer *att*L sites may contain a secondary structure that interferes with the transcription or processing of the hpRNA molecule, but that this is not the case for constructs containing *att*B sites.

Products of recombination into pHELLSGATE vectors include not only plasmids in which the intron has retained its forward orientation with respect to the promoter but also plasmids in which the intron has been reversed with respect to the promoter. This reversal is because of recombination of the *att*B/L1 and *att*B/L2 sites of an incoming fragment with *att*P/R1 and *att*P/R2 sites with separate *att*P/R cassettes in the pHELLS-GATE vector (Fig. 3). The frequency of intron inversion appears to be dependent on the insert sequence, but we have not been able to determine any relationship between insert size and the ratio of the two possible intron orientations. In high-throughput applications, it is desirable to have the highest possible efficiency of useable clones to reduce the number of bacterial colonies that need to be screened following a recombination reaction. We have addressed this by constructing the vector pHELLS-GATE12 (Fig. 4) in which the spacer fragment consists of two introns in opposite orientations such that one of the introns is in the forward, and therefore functional, orientation with respect to the promoter irrespective of the orientation of the spacer fragment resulting from the recombination reaction.

pHELLSGATE12 was created by inserting the catalase-I intron of castor bean as used in the p35SH-iH vector (Wang *et al.*, 1997) into the *Hin*dIII site of pHELLSGATE8. The presence and orientation of the insert were confirmed by restriction enzyme digestion analysis and DNA sequencing. LR Clonase recombination reactions, with various gene fragment inserts flanked by *att*L sites, gave results similar to those of pHELLSGATE8 in terms of accuracy of insertion of the fragments and recovery of recombined clones.

The efficiency and extent of silencing by pHELLSGATE12 constructs have been tested against two *Arabidopsis* genes, *FLC* (flowering locus C) and *PDS* (phytoene desaturase). Silencing of the *FLC* gene gives rise to early flowering in the C24 ecotype of *Arabidopsis*, and the extent of silencing shows an inverse relationship to flowering time (Helliwell *et al.* 2002; Wesley *et al.*, 2001). Silencing of *PDS* gives a range of phenotypes

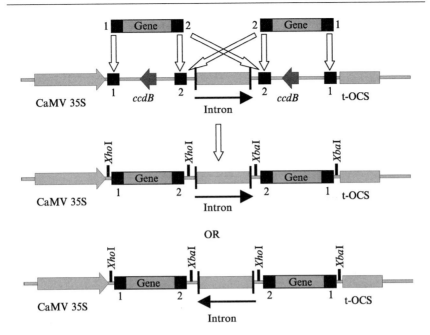

FIG. 3. Recombination products with pHELLSGATE vectors. The *attL2* sites flanking the gene fragment can interact with the *attR2* sites on either side of the intron, giving rise to two possible constructs with the intron in either the sense or antisense orientation with respect to the promoter.

from bleaching of cotyledons to complete bleaching of the plant (Helliwell *et al.*, 2002). Constructs were made with both orientations of the two-intron spacer. Phenotypes of T1 plants obtained with these constructs suggest an effectiveness in silencing similar to that of pHELLSGATE8 constructs, using the same gene fragments, with no significant difference between the two orientations of the intron (Fig. 5; Table I).

Gene Silencing by Using the pHELLSGATE Vectors

In this section, we outline the steps and factors to consider when using the pHELLSGATE system. The considerations with respect to choice of gene fragment are equally applicable to use of the pH/KANNIBAL vectors.

Choice of Gene Fragment

We have used gene fragments ranging from 50 bp to 1 kb to successfully silence genes. Two factors can influence the choice of length of the fragment. The shorter the fragment, the less likely it is that effective silencing

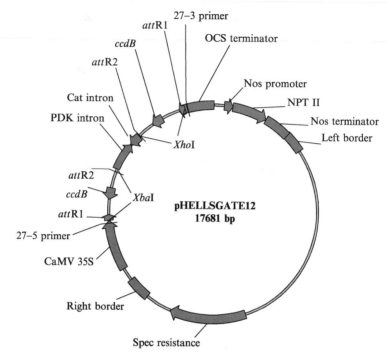

FIG. 4. Map of pHELLSGATE12. pHELLSGATE12 is identical to pHELLSGATE8 with the catalase-1 intron (8) inserted between the PDK intron and the second *att*R cassette in the antisense orientation with respect to the promoter.

will be achieved. However, very long hairpins increase the chance of recombination in bacterial host strains. The effectiveness of silencing also appears to be gene dependent and could reflect accessibility of target mRNA or the relative abundances of the target mRNA and the hpRNA in cells in which the gene is active. We recommend a fragment length of 300–800 bp as a suitable size to maximize the efficiency of silencing obtained.

The other consideration is the part of the gene to be targeted. We have obtained equally good results with the 5′UTR, coding region, and 3′UTR fragments. As the mechanism of silencing depends on sequence homology, there is potential for cross-silencing of related mRNA sequences. Where this is not desirable, a region with low sequence similarity to other sequences, such as a 5′ or 3′UTR should be chosen. The general rule for avoiding cross-homology silencing appears to be to use sequences that do not have blocks of sequence identity of more than 18 bases in length between the construct and nontarget gene sequences.

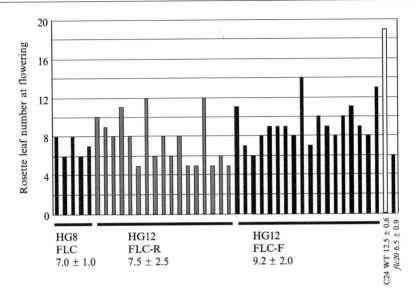

FIG. 5. Silencing of *FLC* by using pHELLSGATE12. Rosette leaf number at flowering is shown for T1 plants of *Arabidopsis* ecotype C24 transformed with HG8-FLC, HG12-FLC-R (PDK intron reversed with respect to promoter), and HG12-FLC-F (PDK in forward orientation with respect to promoter); C24 wild type and *flc20* (an *flc* null mutant) are also shown. Mean and standard deviation of rosette leaf numbers at flowering are also shown.

TABLE I
PHENOTYPES OF T1 PDS-SILENCED PLANTS

Construct	No. of transformants	No phenotype	Bleached cotyledons	Some leaf bleaching	Total bleaching
HG8-PDS	8	1 (13%)	5 (63%)	2 (25%)	0
HG12-PDS-F	15	0	6 (40%)	7 (47%)	2 (13%)
HG12-PDS-R	5	0	3 (60%)	2 (40%)	0

PCR Amplification with attB Primers

Primer sequences can be designed by using normal design rules with the addition of *att*B1 and *att*B2 sites to the 5′ end of the forward and reverse primers, respectively (Fig. 6). Addition of the *att*B sequences does not appear to affect the amplification of PCR products. PCR reactions should be carried out by using standard protocols and the products should be checked by agarose gel electrophoresis for yield and product size. The

attB1 + recognition sequence (XXXX...)
5'GGGGACAAGTTTGTACAAAAAAGCAGGCTXXXXXXXXXXXXXXXXXXXXX3'

attB2 + recognition sequence (XXXX...)
5'GGGGACCACTTTGTACAAGAAAGCTGGGTXXXXXXXXXXXXXXXXXXXXX3'

P27–5
5' GGGATGACGCACAATCC 3'

P27–3
5' GAGCTACACATGCTCAGG 3'

FIG. 6. Primer sequences. Sequences of the 5' extensions used to incorporate attB1 and attB2 sites into gene-specific primers and the sequencing primers from the CaMV35S promoter (P27–5) and ocs terminator (P27–3) used to sequence inserts in pHELLSGATE vectors are shown.

primers should be removed from the PCR product as they can interfere with subsequent recombination reactions. This is done by diluting the PCR reaction with three volumes of TE and precipitation with two volumes of 30% PEG 8000, 30 mM MgCl$_2$. The precipitate is collected by centrifugation at >13,000g for 15 min and the supernatant removed with a pipette. The DNA pellet is then resuspended in one volume of TE.

Recombination Reactions

The PCR fragment is first inserted into a plasmid containing attP1 and attP2 sites. We use pDONR201 (Invitrogen) for this step, but in theory any plasmid containing attP1 and attP2 sites flanking a negatively selectable marker such as the ccdB gene can be used. The reaction conditions used are based on the Invitrogen protocol: 2 μl BP Clonase buffer, 1–2 μl PCR product, 1 μl (150 ng) pDONR201, TE to 8 μl, 2 μl BP Clonase (Invitrogen). Incubate at room temperature (25°) for 1 h, add 1 μl proteinase K mix (supplied with BP Clonase), and incubate 10 min at 37°. This protocol can also be used with half the reaction components. Use 1–10 μl of the reaction mixture to transform *E. coli* DH5α cells. The competent cells should have a transformation efficiency of at least 10 colonies per μg plasmid DNA (Waterhouse *et al.*, 1993). We generally use cells prepared by using an RbCl method (Hanahan, 1985) or electrocompetent cells for this step. Colonies are selected overnight on kanamycin-containing plates when using pDONR201. Colonies selected following a BP clonase reaction normally have the correct insert in at least 90% of cases; therefore, very few colonies need be screened for each BP reaction.

LR reactions into pHELLSGATE vectors take PCR fragments inserted into pDONR201 and recombine them into the two attR cassettes forming the arms of the hairpin. As the insertion into two *att*R cassettes is less efficient than into a single cassette, the incubation time for the recombination reaction is increased and the transformation efficiency of the DH5α cells used to select recombined plasmids is more critical. The standard LR reaction mix is 2 μl LR Clonase buffer, 2 μl pDONR201 PCR clone (100–200 ng), 2 μl pHELLSGATE vector (300 ng), 2 μl TE, and 2 μl LR Clonase. The mix is incubated for 1–16 h at room temperature (25°), with longer incubations being preferable. At the end of the incubation the reaction is treated with 1 μl proteinase K mix for 10 min at 37°. Aliquots of 1–10 μl of the reaction are then used to transform DH5α, and colonies are selected on plates containing 100 mg/l spectinomycin. The plates generally require a 24-h incubation at 37° before colonies are visible.

Identification and Verification of Hairpin Constructs

The pHELLSGATE vectors contain convenient restriction sites that can be used to check the presence of inserted sequences and diagnose the orientation of the intron spacer fragment. Incubation with *Xho*I will excise the first arm and incubation with *Xba*I will excise the second arm of the inverted repeat, unless the intron has been flipped during the recombination process. In this case, each arm will be excised with the intron (Fig. 7).

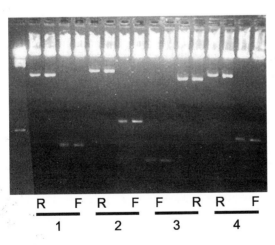

FIG. 7. Restriction digests of various PCR products following recombination into pHELLSGATE12. Each pair of lanes shows digests with *Xba*I and *Xho*I, which release the first and second arm of the hairpin, respectively. Plasmids with the intron cassette in the original orientation (F) release fragments only slightly larger than the original PCR products, whereas those in which the intron cassette orientation is reversed (R) give larger fragments that include the intron cassette as well as the PCR product.

Alternatively, PCR reactions using primers flanking the insertion sites can be used to show the presence of inserts. Sequencing can be carried out by using primers based on the promoter and terminator sequences (Fig. 6). Reduction in the sequence trace signal is often seen in duplexed parts of constructs. This can be alleviated by linearizing the plasmid at the *Bgl*II site in the intron before sequencing. Sequencing through *att*R, P, or L sites is difficult, probably because of secondary structures formed in these sites; however, *att*B sites are not refractory to sequencing.

Maintenance and Manipulation of Plasmid Vectors

Vectors containing the negatively selectable marker *ccdB* (pHELLS-GATE8, 12 and pDONR201) must be maintained in the DB3.1 *E. coli* strain. Competent cells can be purchased from Invitrogen; alternatively, electrocompetent cells prepared by standard methods can be used for transformation following ligation steps with these vectors.

Summary and Conclusions

The efficiency and effectiveness of hpRNA as a tool for silencing plant genes make this the method of choice over antisense or cosuppression methods. The pH/KANNIBAL vectors are useful for silencing a small number of target genes, whereas the pHELLSGATE vectors are designed for making hpRNA constructs in large numbers (e.g., members of a gene family, or a pathway, or genes of unknown function in a genome—even whole genomes). This large-scale silencing has been very successful in nematodes (Fraser *et al.*, 2000; Kamath *et al.*, 2003), using injection or feeding with dsRNA, and is showing great promise in the AGRIKOLA project, which aims to silence every gene in *Arabidopsis* (Hilson *et al.*, 2003). The PCR, recombination, and transformation steps to produce hpRNA constructs in pHELLSGATE vectors can be performed in a 96-well format, making it suitable for the automation required for large-scale projects. At present, *Arabidopsis* is the most suitable candidate species for application of this technology in view of the availability of its complete genome sequence and its ease of transformation. The application of hpRNA silencing is dependent only on the ability to deliver the hpRNA construct efficiently into the plant cell and on the availability of gene sequences, making it a functional genomics tool that can be applied to understanding the biology of virtually any plant species.

Availability of Plasmid Vectors

The plasmids pHANNIBAL, pKANNIBAL, pHELLSGATE2, and pHELLSGATE8 are available to academic institutions under a simple material transfer agreement. See http://www.pi.csiro.au/tech_licensing_biol/GeneSilencingVectors.htm.

References

Fraser, A. G., Kamath, R. S., Zipperlen, P., Martinez-Campos, M., Sohrmann, M., and Ahringer, J. (2000). Functional genomic analysis of *C. elegans* chromosome I by systematic RNA interference. *Nature* **408**, 325–330.

Gleave, A. P. (1992). A versatile binary vector system with a T-DNA organisational structure conducive to efficient integration of cloned DNA into the plant genome. *Plant Mol. Biol.* **20**, 1203–1207.

Hanahan, D. (1985). Techniques for transformation of *E. coli* in DNA cloning. In "DNA Cloning: A Practical Approach" (D. Glover, ed.), Vol. 1, pp. 109–138. IRL Press, London.

Helliwell, C. A., Wesley, S. V., Wielopolska, A. J., and Waterhouse, P. M. (2002). High throughput vectors for efficient gene silencing in plants. *Func. Plant Biol.* **29**, 1217–1225.

Hilson, P., Small, I., and Kuiper, M. T. (2003). European consortia building integrated resources for *Arabidopsis* functional genomics. *Curr. Opin. Plant Biol.* **6**, 426–429.

Kamath, R. S., Fraser, A. G., Dong, Y., Poulin, G., Durbin, R., Gotta, M., Kanapin, A., Le Bot, N., Moreno, S., Sohrmann, M., Welchman, D. P., Zipperlen, P., and Ahringer, J. (2003). Systematic functional analysis of the *Caenorhabditis elegans* genome using RNAi. *Nature* **421**, 231–237.

Sharp, P. A. (2001). RNA interference—2001. *Genes Dev.* **15**, 485–490.

Smith, N. A., Singh, S. P., Wang, M. B., Stoutjesdijk, P. A., Green, A. G., and Waterhouse, P. M. (2000). Total silencing by intron-spliced hairpin. *Nature* **407**, 319–320.

Wang, M. B., Upadhyaya, N. M., Brettell, R. I. S., and Waterhouse, P. M. (1997). Intron mediated improvement of a selectable marker gene for plant transformation using *Agrobacterium tumefaciens. J. Genet. Breed.* **51**, 325–334.

Waterhouse, P., Griffiths, A. D., Johnson, K. S., and Winter, G. (1993). Combinatorial infection and *in vivo* recombination: A strategy for making large phage antibody repertoires. *Nucleic Acids Res.* **21**, 2265–2266.

Waterhouse, P. M., Wang, M. B., and Lough, T. (2001). Gene silencing as an adaptive defence against viruses. *Nature* **411**, 834–842.

Wesley, S. V., Helliwell, C. A., Smith, N. A., Wang, M. B., Rouse, D. T., Liu, Q., Gooding, P. S., Singh, S. P., Abbott, D., Stoutjesdijk, P. A., Robinson, S. P., Gleave, A. P., and Waterhouse, P. M. (2001). Constructs for efficient, effective and high throughput gene silencing in plants. *Plant J.* **27**, 581–590.

[3] RNA Interference in *Caenorhabditis elegans*

By JUAN WANG and MAUREEN M. BARR

Abstract

RNA interference (RNAi) was first discovered in the nematode *Caenorhabditis elegans* (Fire *et al.*, 1998; Guo and Kemphues, 1995). The completion of the *C. elegans* genome in 1998 coupled with the advent of RNAi techniques to knock down gene function ushered in a new age in the field of functional genomics. There are four methods for double-stranded RNA (dsRNA) delivery in *C. elegans*: (1) injection of dsRNA into any region of the animal (Fire *et al.*, 1998), (2) feeding with bacteria producing dsRNA (Timmons *et al.*, 2001), (3) soaking in dsRNA (Tabara *et al.*, 1998), and (4) *in vivo* production of dsRNA from transgenic promoters (Tavernarakis *et al.*, 2000). In this chapter, we discuss the molecular genetic mechanisms, techniques, and applications of RNAi in *C. elegans*.

Introduction

C. elegans is a free-living, self-fertilizing hermaphroditic roundworm (Brenner, 1974). The rapid generation time (3 days), large brood size (300–350 self progeny), small size (adult 1.5 mm), simple anatomy (<1000 cells), and ease of laboratory culturing make "the worm" a favorite genetic model organism (Barr, 2003). The constantly expanding molecular toolkit includes transgenesis, green fluorescent protein (GFP) to study gene expression and protein localization, primary cell culture, microarrays, and RNAi. For information about *C. elegans*, two outstanding books have been published: "The Nematode *Caenorhabditis elegans*" (Wood and the Community of *C. elegans* Researchers, 1988) and "*C. elegans* II" (Riddle *et al.*, 1997). The former provides an excellent description of *C. elegans* developmental biology and the latter tackles *C. elegans* molecular biology. For technical aspects, "*C. elegans*: A Practical Approach" (Hope, 1999) is a useful reference. Table I lists the online resources for *C. elegans* biology and data mining.

The *C. elegans* genome project served as a format for other sequencing projects (1998). Likewise, *C. elegans* systematic genomewide approaches culminating in effective public data archiving at WormBase for *C. elegans* genome and biology (www.wormbase.org) provide an exemplar for high-throughput functional genomics in more complex biological systems.

TABLE I
ONLINE RESOURCES FOR *C. ELEGANS*

Annotation	URL	Description
WormBase	www.wormbase.org	Accessible information on the genetics, genomics, and biology of *C. elegans* and some related nematodes
C. elegans RNAiDB	www.rnai.org	Provides results from RNAi phenotypic analysis of genes in *C. elegans*
C. elegans WWW server	http://elegans.swmed.edu	Essential link to all *C. elegans* databases or Web sites
Caenorhabditis Genetics Center	http://biosci.umn.edu/CGC/CG Chomepage.htm	Responsible for collecting, maintaining, and distributing stocks of *C. elegans*
C. elegans genomics	www.sanger.ac.uk/Projects/C_elegans/ http://genome.wustl.edu/	*C. elegans* and *C. briggsae* sequence data from the U.K. and U.S. genome projects
Fire lab vector kits	ftp://www.ciwemb.edu/pub/FireLabInfo/	Documentation for all Fire lab reagents, including *C. elegans* expression and RNAi vectors
Nematode Expression Pattern Database	http://nematode.lab.nig.ac.jp/db/index.html	DNA database of Japan; Dr. Yuji Kohara provides *C. elegans* yk cDNAs
MRC Geneservice	www.hgmp.mrc.ac.uk/geneservice/ reagents/products/rnai/index.shtml	Commercial distributor of the Ahringer RNAi feeding library
C. elegans Knockout Consortium	http://elegans.bcgsc.bc.ca/knockout.shtml	Provides and takes requests for gene knockouts
C. elegans National Bioresource Project	http://shigen.lab.nig.ac.jp/C. *elegans*/mutants/index.jsp	Provides and takes requests for gene knockouts
DNA microarray (Kim lab)	http://cmgm.stanford.edu/~kimlab	Contains searchable microarray data from the Kim lab
C. elegans literature search	www.textpresso.org	*C. elegans* literature search engine
WormPD	www.incyte.com/proteome/databases.jsp	Commercial *C. elegans* proteome database (requires a licensing fee)
The *C. elegans* ORFeome project	worfdb.dfci.harvard.edu/	Vidal lab project to clone all predicted protein-encoding open reading frames (ORFs) in *C. elegans*
The *C. elegans* Interactome	http://vidal.dfci.harvard.edu/interactomedb/ i-View/interactomeCurrent.pl	Vidal lab project to generate comprehensive protein-protein interaction mapping strategies for *C. elegans*

Mechanisms of RNA Interference (RNAi)

RNAi is an evolutionarily conserved process leading to posttrancriptional gene silencing (PTGS) triggered by double-stranded RNA (dsRNA) or small interfering dsRNA (siRNA) (Hannon, 2002). RNAi has been proposed to function as a naturally occurring cellular antiviral defense mechanism against foreign dsRNA invasion (Tijsterman *et al.*, 2002). A model for RNAi is shown in Fig. 1. First, a dsRNA complementary to an endogenous mRNA is processed by the RNase III enzyme Dicer into 21- to 23-nucleotide (nt) siRNAs (Carmell and Hannon, 2004). Dicer also processes microRNAs (miRNAs), which function as negative regulators of specific target mRNAs (Carrington and Ambros 2003). In *C. elegans*, the RDE-4 dsRNA RNA-binding protein interacts with Dicer (DCR-1), the argonaute-like protein RDE-1, and a DExH-box helicase to regulate RNAi (Tabara *et al.*, 2002). *rde-1* and *rde-4* mutants are completely

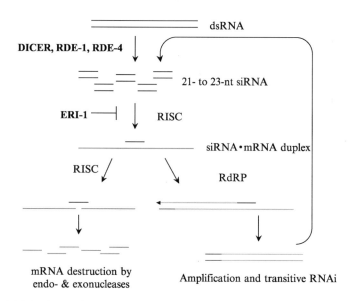

FIG. 1. Molecular mechanism of RNAi in *C. elegans*. dsRNA triggers the RNAi machinery, starting with Dicer enzymatic cleavage of the dsRNA into 21- to 23-nt small interfering RNAs (siRNAs). In certain cell types (such as neurons), ERI-1 may degrade 2-nt 3′ overhangs on dsRNA such that the siRNAs cannot enter the RNA-induced silencing complex (RISC) or are unstable *in vivo*. RISC unwinds siRNA duplexes. The siRNA binds to a target messenger RNA (mRNA) with precise complementarity and then forms an siRNA • mRNA duplex. The siRNA • mRNA duplex may either be targeted for destruction or may be amplified through an RNA-dependent RNA polymerase (RdRP). The latter pathway is thought to account for the systemic and transitive nature of RNAi in *C. elegans* and plants.

defective in maternal and zygotic RNAi and provide a useful genetic background to bypass maternal effects of RNAi (Table II).

The 21- to 23-nt siRNAs are recognized by the RNA-induced silencing complex (RISC) (Caudy *et al.*, 2003; Hammond *et al.*, 2000). The RISC RNAi effector complex functions to unwind the siRNA duplexes to allow precise binding to the complementary mRNA target, followed by siRNA • mRNA duplex cleavage and mRNA degradation. *C. elegans* ERI-1 and its human counterpart degrade siRNAs *in vitro*, thereby inhibiting RISC complex formation and negatively regulating RNAi (Kennedy *et al.*, 2004). *eri-1* (enhanced *RNAi*) mutants show increased sensitivity to the effects of RNAi and particularly neuronal RNAi. As RNAi is inefficient in neurons (Tavernarakis *et al.*, 2000), *eri-1* mutants may prove to be a valuable tool for investigating the function of genes acting in the nervous system (Table II).

On RISC association and siRNA • mRNA duplex formation, one of two events is thought to occur: either the mRNA is cleaved and targeted for disruption or is amplified. The former appears to be an evolutionarily conserved process whereas the latter may be specific to *C. elegans* and plants. In the first scenario, the mRNA is immediately cleaved by Dicer and degraded by endo- and exonucleases.

Alternatively, the siRNA may act as a primer for replication by a dsRNA-dependent RNA polymerase (RdRP), resulting in a new dsRNA molecule that is then subsequently cleaved by Dicer. The *C. elegans* genome encodes four predicted RdRPs: *rrf-(1–3)* and *ego-1* (Sijen *et al.*, 2001; Simmer *et al.*, 2002; Smardon *et al.*, 2000). The germline of *ego-1* mutants is insensitive to RNAi. In contrast, *rrf-3* mutants are hypersensitive to RNAi by dsRNA feeding or injection. The increased RNAi sensitivity of *rrf-3* mutants has been exploited to uncover gene function in genomewide systematic screens as well as in directed assays for function in life span, signal transduction, and neurons (Lee *et al.*, 2003a; Simmer *et al.*, 2003; Yau *et al.*, 2003).

RdRP amplication of a new dsRNA molecule may result in transitive RNAi in plants and *C. elegans* but apparently not in *Drosophila* or mammalian systems (Dillin, 2003). In *C. elegans*, new siRNAs may be generated to a 5′-upstream region of the initial target region (Alder *et al.*, 2003; Sijen *et al.*, 2001). In plants, new siRNAs may be generated to both adjacent upstream and downstream regions of the initial target region (Vaistij *et al.*, 2002). These new siRNAs unrelated to the original target sequence silence nearby transcripts. Although transitive RNAi appears to be restricted to no more than 180-nt traveling distances, this is an important consideration in *C. elegans*, in which approximately 15% of the genes are located in operons (Blumenthal and Gleason, 2003).

TABLE II

USEFUL GENETIC BACKGROUNDS FOR RNAi EXPERIMENTS

Strain	Mutant defect	Protein product	RNAi uses	Ref.
N2	None	N/A	Wild type	Brenner, 1974
eri-1(mg366)	Enhanced RNAi	Nucleic-acid binding and exonuclease	Increasing sensitivity to dsRNA; neuronal RNAi; knockdown of GFP transgenes	Kennedy et al., 2004
rde-1(ne219)	RNAi defective	Argonaute-like protein	Bypassing maternal effects of RNAi: expose RNAi-resistant rde-1 hermaphrodites to dsRNA and cross to wild-type males; resulting cross progeny will be RNAi sensitive	Tabara et al., 1999b
rrf-3(pk1426)	Increased sensitivity to RNAi; silences transgenes	RNA-directed RNA polymerase	Increasing sensitivity to dsRNA; neuronal RNAi; cannot be used for GFP knockdown as rrf-3 mutants also silence transgenes	Sijen et al., 2001; Simmer et al., 2002, 2003
sid-1(qt2)	Systemic RNAi defective but sensitive to autonomous RNAi	Transmembrane protein	Examining tissue-specific RNAi phenotypes by blocking systemic RNAi	Winston et al., 2002
rsd-2 rsd-3 rsd-6	RNAi spreading defective: silencing of ingested dsRNA in germline only	Novel, interacts with RSD-6; ENTH domain protein; novel, interacts with RSD-2	Examining tissue-specific RNAi phenotypes by blocking systemic RNAi	Tijsterman et al., 2004

Note: All strains are available from the *Caenorhabditis* Genetics Center (CGC).

A second feature of RNAi that operates in *C. elegans* but apparently not in *Drosophila* or mammalian cells is systemic RNAi. Systemic RNAi refers the introduction of dsRNA into one tissue, leading to RNA silencing in distant cells. There are four methods for dsRNA delivery in *C. elegans*: (1) injection of dsRNA into any region (Fire *et al.*, 1998), (2) feeding with bacteria producing dsRNA (Timmons *et al.*, 2001), (3) soaking in dsRNA (Tabara *et al.*, 1998), and (4) *in vivo* production of dsRNA from transgenic promoters (Tavernarakis *et al.*, 2000). Injecting, feeding, and soaking dsRNA may result in systemic RNA silencing. With the exception of one report (Winston *et al.*, 2002), systemic spread of RNAi from dsRNA-producing transgenes is not observed (Tijsterman *et al.*, 2004; Timmons *et al.*, 2003). The systemic effects of RNAi feeding are advantageous for large-scale genomewide RNAi screens. An RNAi feeding library consisting of 16,757 bacterial strains representing more than 87% of *C. elegans* genes is available through MRC Geneservice (Table I).

There appear to be specific systemic RNAi mechanisms for each mode of dsRNA spreading (ingesting, soaking, and injection), suggesting that cell-specific factors regulate dsRNA entry and export (Timmons *et al.*, 2003). Several genes required for systemic but not autonomous RNAi have been identified (Tijsterman *et al.*, 2004; Timmons *et al.*, 2003; Winston *et al.*, 2002). Mutants in these genes may potentially provide a useful genetic background for examining tissue-specific RNAi phenotypes by blocking systemic RNAi (Table I). *sid-1/rsd-8*, *rsd-4*, and *fed* (for *s*ystemic RNA*i* *d*efective, *R*NA*i* *s*preading *d*efective, *f*eeding *d*efective for RNAi, respectively) mutants are resistant to the systemic effects of dsRNA feeding for germline or somatic genes but remain sensitive to dsRNA injection (Tijsterman *et al.*, 2004; Timmons *et al.*, 2003; Winston *et al.*, 2002). *rsd-2, rsd-3*, and *rsd-6* are resistant to the systemic effects of dsRNA feeding for germline but not somatic genes and remain sensitive to dsRNA injection (Tijsterman *et al.*, 2004). *sid-1* encodes a transmembrane protein that transports dsRNA (Feinberg and Hunter, 2003). Heterologous expression of SID-1 sensitizes *Drosophila* cells to soaking RNAi, opening up the possibility that systemic RNAi may be similarly induced in mammalian cells. For a more thorough discussion of the molecular mechanisms and uses of systemic RNAi in *C. elegans*, refer to Chapter 18.

Description of Methods

Selection of Double-Stranded RNA (dsRNA) Fragment

In *C. elegans*, long dsRNA fragments (more than 100 bp) trigger RNAi. This is in contrast to mammalian systems, which require siRNAs to be effective in eliciting sequence-specific mRNA cleavage yet do not trigger a dsRNA response (such as interferon responses). There are four recommendations for fragment selection used for *C. elegans* RNA interference (Tabara *et al.*, 1998): (1) The insert should be cDNA or genomic DNA containing exons. Introns will not induce RNAi. (2) The fragment should specifically target one gene. If there is high similarity or match to coding regions of other genes, RNAi phenotypes may be the result of nonspecific interference. However, degenerative RNAi may actually be desirable when examining redundancy and determining the function of multigene families. (3) For most genes, dsRNA stretches from 200 to 1000 nt or longer appear to effectively induce interference. However, some specific gene segments are ineffective at inducing interference and it is suggested that dsRNAs from several segments of a gene be tried. (4) About 15% of the genes (~2500) in the *C. elegans* genome are organized into operons (Blumenthal *et al.*, 2002). Transitive RNAi of one gene can interfere with the function of another upstream gene in the same operon (Bosher *et al.*, 1999). If your target gene is in an operon, it is important to design controls that exclude cross-interference by transitive RNAi.

Selection of C. elegans *Strain*

Table II lists various strains that may be useful for different RNAi purposes. Bristol strain N2 is wild type (Brenner, 1974). For genes that do not show an RNAi phenotype in wild type, *eri-1(mg366)* IV and *rrf-3(pk1426)* II mutants exhibit increased sensitivity to dsRNA and neuronal RNAi (Kennedy *et al.*, 2004; Simmer *et al.*, 2002). RNAi phenotypes in *rrf-3* may be stronger and more closely resemble a null phenotype as compared to a wild-type background (Simmer *et al.*, 2003). However, *rrf-3(pk1426)* II mutants also silence transgenes and are therefore not useful for examining properties of transgenes (such as GFP expression). To bypass maternal effects of RNAi, RNAi-resistant *rde-1(ne219)* V hermaphrodites (Tabara *et al.*, 1999b) may be exposed to dsRNA and then crossed to wild-type males. Resulting cross-progeny [*rde-1(ne219)/+*] will be RNAi sensitive.

RNAi Protocols

RNAi efficacy and efficiency vary spatially and temporally. For example, resistance to RNAi-by-feeding effects are observed in late-stage embryos whose eggshell may be impermeable to dsRNA, neurons, males, and early larval stages (Kamath *et al.*, 2001; Timmons *et al.*, 2001). Stable mRNA is more susceptive to RNAi than nonstable messages (Fire, 1999). RNAi phenotypes vary in gene-dependent and method-dependent manners. dsRNA feeding, soaking, or injecting methods are generally ineffective in neurons. In some neurons, only injected plasmids producing dsRNA knockdown gene function reproducibly (Tavernarakis *et al.*, 2000). Thus, note that RNAi does not necessarily produce a null mutant phenotype. Moreover, Plasterk and colleagues have found that false negatives contribute to interexperimental variability (10–30%) both within and between laboratories (Simmer *et al.*, 2003).

Four RNAi methods have been published: (1) feeding *C. elegans* with bacteria that express dsRNA, (2) soaking *C. elegans* with *in vitro* synthesized dsRNA, (3) microinjecting *in vitro* synthesized dsRNA into the *C. elegans* adult hermaphrodite, and (4) microinjecting and generating transgenic animals with plasmids expressing inverted repeats (IRs) of a target gene under the control of a constitutive or tissue-specific promoter. Each of these methods possesses distinct advantages and disadvantages. In the following section, we describe and provide protocols and references for each RNAi method. Detailed protocols and explanations for standard *C. elegans* methods are found in Wood (1988) and Hope (1999).

Feeding with E. coli *Expressing dsRNA*

Timmons and Fire (1998) first described an effective method for RNAi in which bacteria expressing dsRNA are fed to *C. elegans*. On ingestion, dsRNA is amplified and systemically spread through cells, resulting in RNAi phenotypes in the immediate dsRNA recipients (P0) as well as resulting offspring (F1 progeny). RNAi by feeding can be titrated to study a range of hypomorphic phenotypes, providing the RNAi equivalent of an allelic series (Kamath *et al.*, 2001). Feeding multiple genes can potentially reduce individual RNAi phenotypes, warning that RNAi cannot always be used as the reverse genetics equivalent of genetic epistasis tests (Gonczy *et al.*, 2000; Parrish *et al.*, 2000).

A fragment corresponding to the gene of interest is cloned into a feeding vector (L4440) between two T7 promoters in inverted orientation. The plasmid is transformed into the RNase III-defective *E. coli* strain HT115(DE3). The double T7 promoter-containing plasmid as well as control plasmids for use in feeding experiments are available in the 1999 Fire Lab vector kit. This kit also provides another vector for expressing dsRNA, using T3 and T7 phage promoters flanking the insert multiple cloning site. Vector and sequence information and kit request forms can be accessed at www.ciwemb.edu.

Dr. Julie Ahringer's group optimized the efficiency and techniques of the feeding method, thereby enabling systematic genomewide RNAi screens (Kamath *et al.*, 2001, 2003). The Ahringher RNAi feeding library consists of 16,757 bacterial strains, which cover 87% of *C. elegans* genes. Either individual clones or the entire library in chromosomal sets is available for purchase through MRC Geneservice at www.hgmp.mrc.ac.uk/geneservice/reagents/products/rnai/index.shtml.

Detailed methodology for the library construction and screening procedure has been published (Kamath and Ahringer, 2003). This library has been used in numerous systematic genomewide RNAi screens (Kamath *et al.*, 2003) to identify genes required for longevity (Dillin *et al.*, 2002; Lee *et al.*, 2003b; Murphy *et al.*, 2003), genome stability (Pothof *et al.*, 2003), fat regulation (Ashrafi *et al.*, 2003), development (Fraser *et al.*, 2000; Simmer *et al.*, 2003; Zipperlen *et al.*, 2001), and signal transduction (Keating *et al.*, 2003; Tewari *et al.*, 2004).

Bacterial Strain

The plasmid containing a target gene insert is transformed into *E. coli* HT115(DE3) by standard techniques (Timmons *et al.*, 2001) or using a modification of a one-step bacterial transformation (Kamath and Ahringer, 2003). The genotype of *E. coli* HT115 is F⁻, *mcrA, mcrB, IN(rrnD-rrnE)1, lambda-, rnc14::Tn10* (DE3 lysogen: *lacUV5* promoter-T7 polymerase). T7 polymerase gene expression is driven by the *lacUV5* promoter, which is IPTG inducible. *rnc14* encodes RNase III. On disruption by Tn10, *rnc14* is unable to degrade the dsRNA expressed *in vivo*. Tn10 carries a tetracycline-resistant gene; therefore, there are two antibiotic selections—ampicillin for the plasmid and tetracycline for the transposon.

Feeding Protocol 1 (Timmons et al., *2001)*

• *Day 1*: Inoculate an overnight culture of HT115(DE3) containing feeding plasmid in LB plus antibiotics (75–100 μg/ml ampicillin for amp-resistant plasmids and 12.5 μg/ml tetracycline). Incubate at 37° with shaking overnight.

• *Day 2*:
  1. Dilute overnight culture 1:100 in 2× YT with antibiotics and grow to an OD_{595} of 0.4. Induce by adding sterile IPTG to 0.4 mM. Incubate at 37° with shaking ∼4 h. Spike culture with additional antibiotics (another 100 μg/ml ampicillin and 12.5 μg/ml tetracycline) and IPTG (to a final total concentration of 0.8 mM).
  2. Seed NGM plates with the induced culture. For small plates containing a few hand-picked worms, the culture can be used directly. For large plates containing large numbers of chunked worms, the cells can be concentrated by centrifugation and spotted onto plates.

    i. The ratio of bacteria to worms is important. RNAi will not be effective if the worms eat all the bacteria and starve.

    ii. Bacterial lawns should not be allowed to grow after induction to avoid contaminants or cells that have lost the plasmid or the ability to produce T7 polymerase.

    iii. Add worms by hand picking or by adding chunks to wet, freshly seeded plates or to plates that have been allowed to dry after seeding.

- *Day 3 and on*: Phenotypes can be observed at temperatures of 16–25°. It can take up to 3 days before an RNAi phenotype is observed. Results vary depending on the dsRNA and the worm strains used. Freshly seeded plates versus older seeded plates are also considerations.

Feeding Protocol 2 (Kamath et al.*, 2001)*

    To improve RNAi efficacy, Ahringer and colleagues describe an alternative feeding by using a different method to induce the expression of dsRNA. Tetracycline was omitted based on two findings: the Tn10 transposon is stable and, in some cases, inclusion of tetracycline during feeding significantly decreases the RNAi effect (Kamath *et al.*, 2001).

- *Day 1*: Pick and grow bacteria for 6 h to overnight (but no longer than 18 h) in LB with 50 μg/ml ampicillin liquid culture.
- *Day 2*:
  1. Seed onto NGM agar plates including 50 μg/ml ampicillin and 1 mM IPTG.
         i. The lawn quality is improved if the culture is dried quickly by leaving the lids off for about 20 min after seeding.
         ii. Bacterial cultures grown for shorter times (6 h) sometimes give better results.
         iii. Let it dry and induce bacterial expression of dsRNA at room temperature overnight.
- *Day 3 and on*:
  1. Transfer an L4 larval stage hermaphrodite onto an induction plate. Minimize the amount of normal *C. elegans* culturing OP50 bacteria transferred by washing worms in M9 buffer and then aliquoting directly onto induction plates.
  2. Culture for 72 h at 15° (or 36–40 h at 22°) for RNAi to take effect.
         i. Replica plate the adult onto fresh plate seeded with the same bacteria.
         ii. After 24 h, remove the adult from the replica and score the F1 progeny for phenotypes. Alternatively, aliquot embryos or larvae onto the feeding plates and score them later.

Empirical note: Some genes give different phenotypes at 15° and 22°. Both feeding protocols are provided because, for certain genes, different methods and experimental condition should be taken into account. One consideration is the feeding stage of the worm. P0 worms may be fed at the L4 stage for analysis of F1 progeny or at earlier embryonic or larval stages for P0 phenotyping. Other factors include IPTG concentration, culturing temperature, and age of the bacterial lawn, all of which may affect RNAi phenotypes.

Injecting with dsRNA

The first description of the efficacy of dsRNA knocking down gene function in *C. elegans* came from the Fire and Mello labs (Fire *et al.*, 1998). RNAi is achieved by injecting *in vitro* synthesized dsRNA into the gonad or intestine of the adult hermaphrodite. The observation that RNAi effects spread from the intestine to other tissues (including F1 and sometimes even subsequent generations) led to the hypothesis of dsRNA amplification and systemic dsRNA transport between cells. The dsRNA injection method has been widely used for both individual gene and genomewide approaches (141 references found by using the keyword "dsRNA injection RNAi" at the information retrieval and extraction system for *C. elegans* literature at http://www.textpresso.org/).

dsRNA is produced by separately transcribing sense and antisense RNAs *in vitro* and then annealing to form dsRNA. Single-strand RNA can be synthesized from phagemid clones by using T3 and T7 polymerase. Ambion (Austin, TX) manufactures the widely used kit MEGAscript™. DNA templates can be removed by DNase treatments. Sense strand and antisense annealing is carried out in injection buffer (2% polyethylene glycol, molecular weight 6000–8000; 20 mM potassium phosphate, pH 7.5; 3 mM potassium citrate, pH 7.5) at 37° for 10–30 min. Formation of predominantly dsRNA can be checked by examining migration on an agarose gel. Injection mixes are constructed such that animals receive an average of 0.5×10^6 to 1×10^6 RNA molecules. Although more technically challenging and labor consuming than the feeding method, injection of dsRNA may be more effective for certain genes. For specifics about *C. elegans* microinjection technique, refer to Dr. Yishi Jin's chapter on transformation (Jin, 1999).

After injection, recovery, and transfer to standard solid media, injected animals are transferred to fresh culture plates for 8- to 16-h intervals. This yields a series of semisynchronous cohorts in which it is straightforward to identify phenotypic differences. Fire and colleagues (1998) noted a characteristic temporal pattern of phenotypic severity among progeny: a short clearance interval with unaffected progeny from impermeable fertilized eggs, an RNA interference phenotype interval, and finally a reversion interval with incompletely affected or phenotypically normal progeny.

Soaking with In Vitro *Prepared dsRNA*

Simply soaking *C. elegans* in dsRNA can also induce specific interference (Tabara *et al.*, 1998). dsRNA is synthesized as described previously. RNAi by soaking is improved by adding spermidine to the soaking buffer. [Refer to Maeda *et al.* (2001) for detailed protocol.] To characterize postembryonic phenotypes, early developmental staged worms are soaked and P0 phenotypes observed in these animals (Kuroyanagi *et al.*, 2000). Alternatively, P0 hermaphrodites are soaked and F1 progeny examined (Tabara *et al.*, 1999a). Soaking is useful for large-scale RNAi applications (Maeda *et al.*, 2001), although the Ahringer RNAi feeding library makes RNAi by feeding the method of choice for high-throughput screening.

Transgenic Expression of Plasmid(s) Producing Hairpin dsRNA

RNAi is not effective in neurons. However, transgenes that express IR hairpin-forming RNA either with an inducible or a tissue-specific promoter are effective in neurons (Tavernarakis *et al.*, 2000). dsRNA transcribed from transgenic promoters does not elicit robust systemic RNAi phenotypes, thereby allowing analysis of tissue-specific function (Tijsterman *et al.*, 2004; Timmons *et al.*, 2003).

Tavernarakis *et al.* (2000) have described an RNAi method relying on dsRNA produced from a plasmid containing a heat-shock-inducible or tissue-specific promoter driving expression of an IR of a target gene (Fig. 2). Exon-rich genomic DNA (or cDNA) is amplified by using two primers that introduce unique restriction sites at the fragment ends. One restriction site is used to generate the IR and is situated at the inversion point (B in Fig. 2). The other restriction site (end A in Fig. 2) is used to join the IR to the vector. Amplified fragments are digested with the enzyme present at the IP restriction site and ligated. Digestion at the end restriction site enables the fragment to be cloned into a similarly digested, CIAP-treated *C. elegans* expression vector pPD49.78 (available in the 1995 Fire laboratory vector kit), which includes the strong heat-shock-inducible promoter *hsp16-2* and the 3' UTR of muscle myosin *unc-54* for transcript stability. A tissue-specific promoter may also be used to express the IR RNA. To examine the effects of RNAi in male-specific neurons, we have

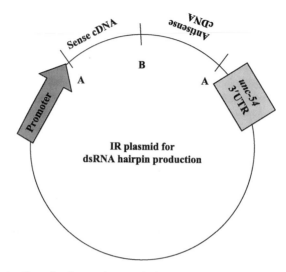

FIG. 2. Construction of an inverted repeat (IR) plasmid for dsRNA hairpin production. A constitutive (such as heat-shock) or tissue-specific promoter is positioned upstream of a multiple cloning site. Sense (5′→3′) and antisense (3′→5′) fragments of the target gene are inserted into the multiple cloning site. The *unc-54* 3′UTR is included for stability.

used the *pkd-2* promoter (Barr *et al.*, 2001) to drive IR transgenes and observe RNAi phenotypes of male mating behavior defects (Hu and Barr, 2004). We have also found that the dsRNA feeding method in an *rrf-3* mutant background is effective for knocking down the function of some neuronal genes (Hu and Barr, 2004).

A significant drawback of this method is that the IR is difficult to manipulate in *E. coli*. Stability of IR plasmids differs on a case-by-case basis. N. Tavernarakis recommends two hosts for these IR plasmids, SURE (Stratagene) and STBL2 (BRL). A plasmid may be unstable in SURE but stable in STBL2 and vice versa. There are no general rules and stability is determined empirically. Repeats longer than 1 kb are less stable and sometimes a spacer DNA of 50–100 bp between repeats greatly improves stability, although making plasmid construction more difficult (N. Tavernarakis, personal communication). For GFP hairpin RNA, the Fire laboratory uses a 900-bp unrelated spacer sequence between two copies of GFP DNA in IR orientation (Timmons *et al.*, 2001). With the promoter of *myo-3, vit-2,* or *unc-119* to drive IR expression, this configuration effectively silences GFP in muscle, gut, but not neurons, respectively (Timmons *et al.*, 2003).

Another approach for hairpin RNAi involves the use of sense and antisense plasmids (Gupta *et al.*, 2003), thereby bypassing the problem of instability of IR constructs in bacteria. A *lin-11* cDNA fragment was separately cloned in sense and antisense orientations into pPD49.83. The Jorgensen and Rothman laboratories have been successful in using similar sense and antisense constructs to induce RNAi (unpublished observations).

To establish heritable transgenic lines, IR plasmids are injected with a suitable cotransformation marker. IR transgenes are maintained as extrachromosomal arrays. For specifics about the *C. elegans* microinjection technique, the reader is referred to the chapter on transformation (Jin, 1999). In addition to IR plasmid instability, we also observe an unusually high loss of the IR-containing extrachromosomal with multiple passaging (J. Hu, J. Wang, and M. Barr, unpublished observations). Therefore, we routinely (1) freeze transgenic stable lines and (2) perform IR RNAi experiments before multiple passages.

Phenotypic Scoring

Reverse genetics starts with a gene identified by its sequence and aims to elucidate function by eliminating or reducing gene activity, otherwise known as gene knockout or knockdown, respectively. Accurate and thorough phenotypic scoring is essential to study functional genomics. Phenotypes are empirical descriptions and rely on experimenter-established criteria and visual expertise. For example, the *emb*ryonic lethal (Emb) phenotype is a general term that encompasses 84 individual phenotypic descriptions that may be used in a phenotype search at the *C. elegans* RNAi Database (RNAiDB; Gunsalus *et al.*, 2004). For a full list of RNAi phenotypes, refer to PhenoBlast at RNAiDB, which includes 78 early embryonic, 6 embryonic, 6 larval, and 25 adult phenotypes (http://nematoda.bio.nyu. edu/cgi-bin/rnai3/ace/pheno_search.cgi), or WormBase, which sorts by maternal phenotype (sterile), embryonic lethality, and 4 general or 27 specific postembryonic phenotypes (www.wormbase.org/db/searches/rnai_search). Table III lists published postembryonic RNAi phenotypes. Adding to the complexity is the range of RNAi phenotypes from full, partial, tissue-specific, or no mRNA silencing.

Screen Design

The success of functional genomics, and particularly high-throughput approaches, depends greatly on screen design. Screen design entails (1) identification of genes to be targeted, (2) assembly of necessary reagents

TABLE III
POSTEMBRYONIC PHENOTYPES

Bli, blistered	Him, high incidence of males	Pvl, protruding vulva
Bmd, body morphology defect	Hya, hyperactive movement	Rol, roller
Clr, clear	Let, larval lethal	Rup, ruptured
Daf, dauer formation defect	Lon, long body	Sck, sick
Dpy, dumpy body	Lva, larval arrest	Sma, small body
Egl, egg-laying defect	Lvl, larval lethal	Ste, sterile
Fem, feminization	Mlt, molt defective	Stp, sterile progeny
Fog, feminization of germline	Muv, multivulva	Unc, uncoordinated
Gro, slow growth	Prl, paralyzed	Vul, vulvaless

(*C. elegans* strain, specific dsRNA, and dsRNA delivery method), (3) phenotypic screening criteria, (4) screen execution, (5) confirmation of hits by repeating the RNAi experiment and phenotypic assays, and (6) detailed analysis of confirmed hits (Fig. 3). For the genomewide feeding RNAi screening procedure followed by the Ahringer group, refer to Kamath and Ahringer, 2003.

The use of reverse genetics or RNAi to examine *in vivo* gene function may range from analysis of a single gene to the entire genome. A large-scale systematic screen requires an easy and effective dsRNA delivery method (such as RNAi feeding or soaking), whereas a single gene or small groups of genes may be targeted by any or a combination of the four dsRNA delivery methods. The hypersensitive *rrf-3* or *eri-1* strain or transgenic IR dsRNA delivery method may reveal gene function in the nervous system.

Forward or classical genetics screens have been traditionally designed with a small range of phenotypic defects in mind. Genomewide RNAi screens have been performed by using general as well as extremely specialized phenotypic criteria. Once the phenotypic assays are determined, screening begins. One important factor to consider is the schedule for dsRNA delivery, animal generation time, and phenotypic scoring. Small-scale screens are done manually, whereas large-scale screens may rely on a combination of manual and automated approaches.

Reduction-of-function phenotypes should be confirmed by repeating the RNAi procedure. Significant variability is observed, with experiments done in the same laboratory varying from 10 to 30%. Variability appears to arise from the high frequency of false negatives but not false positives (only 0.4%; Simmer *et al.*, 2003). Candidates may be subjected to more detailed and rigorous analysis. For example, two independent systematic RNAi

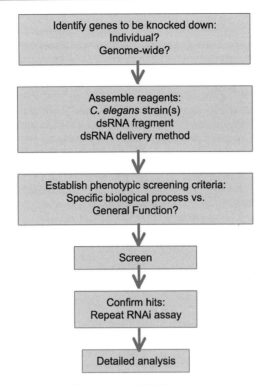

Fig. 3. Flow chart for RNAi screen design.

feeding screens for genes required for aging identified components of the mitochondrial respiratory chain (Dillin *et al.*, 2002; Lee *et al.*, 2003b). From here, the Ruvkun and Kenyon groups went on to show that impairing mitochondrial function extends *C. elegans* lifespan.

In Silico RNAi: Data Mining

Online *C. elegans* resources are listed in Table I. Of particular note are WormBase and RNAiDB, which archive published RNAi phenotypes that can be searched according to phenotype or specific RNAi result.

Acknowledgments

This work was supported by grants from the NIH and PKD Foundation.

References

The *C. elegans* Sequencing Consortium (1998). Genome sequence of the nematode *C. elegans*: A platform for investigating biology. *Science* **282**, 2012–2018.

Alder, M. N., Dames, S., Gaudet, J., and Mango, S. E. (2003). Gene silencing in *Caenorhabditis elegans* by transitive RNA interference. *RNA* **9**, 25–32.

Ashrafi, K., Chang, F. Y., Watts, J. L., Fraser, A. G., Kamath, R. S., Ahringer, J., and Ruvkun, G. (2003). Genome-wide RNAi analysis of *Caenorhabditis elegans* fat regulatory genes. *Nature* **421**, 268–272.

Barr, M. M. (2003). Super models. *Physiol. Genom.* **13**, 15–24.

Barr, M. M., DeModena, J., Braun, D., Nguyen, C. Q., Hall, D. H., and Sternberg, P. W. (2001). The *Caenorhabditis elegans* autosomal dominant polycystic kidney disease gene homologs *lov-1* and *pkd-2* act in the same pathway. *Curr. Biol.* **11**, 1341–1346.

Blumenthal, T., Evans, D., Link, C. D., Guffanti, A., Lawson, D., Thierry-Mieg, J., Thierry-Mieg, D., Chiu, W. L., Duke, K., Kiraly, M., and Kim, S. K. (2002). A global analysis of *Caenorhabditis elegans* operons. *Nature* **417**, 851–854.

Blumenthal, T., and Gleason, K. S. (2003). *Caenorhabditis elegans* operons: Form and function. *Nat. Rev. Genet.* **4**, 110–118.

Bosher, J. M., Dufourcq, P., Sookhareea, S., and Labouesse, M. (1999). RNA interference can target pre-mRNA: Consequences for gene expression in a *Caenorhabditis elegans* operon. *Genetics* **153**, 1245–1256.

Brenner, S. (1974). The genetics of *Caenorhabditis elegans*. *Genetics* **77**, 71–94.

Carmell, M. A., and Hannon, G. J. (2004). RNase III enzymes and the initiation of gene silencing. *Nat. Struct. Mol. Biol.* **11**, 214–218.

Carrington, J. C., and Ambros, V. (2003). Role of microRNAs in plant and animal development. *Science* **301**, 336–338.

Caudy, A. A., Ketting, R. F., Hammond, S. M., Denli, A. M., Bathoorn, A. M., Tops, B. B., Silva, J. M., Myers, M. M., Hannon, G. J., and Plasterk, R. H. (2003). A micrococcal nuclease homologue in RNAi effector complexes. *Nature* **425**, 411–414.

Dillin, A. (2003). The specifics of small interfering RNA specificity. *Proc. Natl. Acad. Sci. USA* **100**, 6289–6291.

Dillin, A., Hsu, A. L., Arantes-Oliveira, N., Lehrer-Graiwer, J., Hsin, H., Fraser, A. G., Kamath, R. S., Ahringer, J., and Kenyon, C. (2002). Rates of behavior and aging specified by mitochondrial function during development. *Science* **298**, 2398–2401.

Feinberg, E. H., and Hunter, C. P. (2003). Transport of dsRNA into cells by the transmembrane protein SID-1. *Science* **301**, 1545–1547.

Fire, A. (1999). RNA-triggered gene silencing. *Trends Genet.* **15**, 358–363.

Fire, A., Xu, S., Montgomery, M. K., Kostas, S. A., Driver, S. E., and Mello, C. C. (1998). Potent and specific genetic interference by double-stranded RNA in *Caenorhabditis elegans*. *Nature* **391**, 806–811.

Fraser, A. G., Kamath, R. S., Zipperlen, P., Martinez-Campos, M., Sohrmann, M., and Ahringer, J. (2000). Functional genomic analysis of *C. elegans* chromosome I by systematic RNA interference. *Nature* **408**, 325–330.

Gonczy, P., Echeverri, G., Oegema, K., Coulson, A., Jones, S. J., Copley, R. R., Duperon, J., Oegema, J., Brehm, M., Cassin, E., Hannak, E., Kirkham, M., Pichler, S., Flohrs, K., Goessen, A., Leidel, S., Alleaume, A. M., Martin, C., Ozlu, N., Bork, P., and Hyman, A. A. (2000). Functional genomic analysis of cell division in *C. elegans* using RNAi of genes on chromosome III. *Nature* **408**, 331–336.

Gunsalus, K. C., Yueh, W. C., MacMenamin, P., and Piano, F. (2004). RNAiDB and PhenoBlast: Web tools for genome-wide phenotypic mapping projects. *Nucleic Acids Res.* **32**, (database issue), D406–D410.

Guo, S., and Kemphues, K. J. (1995). *par-1*, a gene required for establishing polarity in *C. elegans* embryos, encodes a putative Ser/Thr kinase that is asymmetrically distributed. *Cell* **81**, 611–620.

Gupta, B. P., Wang, M., and Sternberg, P. W. (2003). The *C. elegans* LIM homeobox gene *lin-11* specifies multiple cell fates during vulval development. *Development* **130**, 2589–2601.

Hammond, S. M., Bernstein, E., Beach, D., and Hannon, G. J. (2000). An RNA-directed nuclease mediates post-transcriptional gene silencing in *Drosophila* cells. *Nature* **404**, 293–296.

Hannon, G. J. (2002). RNA interference. *Nature* **418**, 244–251.

Hu, J., and Barr, M. M. (2004). The PLAT domain of LOV-1 interacts with ATP-2 to regulate polycystin signaling in *C. elegans*. *Curr Biol.*

Jin, Y. (1999). *In* "Transformation in *C. elegans*: A Practical Approach" (I. A. Hope, ed.), pp. 69–96. Oxford University Press, New York.

Kamath, R. S., and Ahringer, J. (2003). Genome-wide RNAi screening in *Caenorhabditis elegans*. *Methods* **30**, 313–321.

Kamath, R. S., Fraser, A. G., Dong, Y., Poulin, G., Durbin, R., Gotta, M., Kanapin, A., Le Bot, N., Moreno, S., Sohrmann, M., Welchman, S. P., Zipperlen, P., and Ahringer, J. (2003). Systematic functional analysis of the *Caenorhabditis elegans* genome using RNAi. *Nature* **421**, 231–237.

Kamath, R. S., Martinez-Campos, M., Zipperlen, P., Fraser, A. G., and Ahringer, J. (2001). Effectiveness of specific RNA-mediated interference through ingested double-stranded RNA in *Caenorhabditis elegans*. *Genome Biol.* **2**, 1–10.

Keating, C. D., Kriek, N., Daniels, M., Ashcroft, N. R., Hopper, N. A., Siney, E. J., Holden-Dye, L., and Burke, J. F. (2003). Whole-genome analysis of 60 G protein-coupled receptors in *Caenorhabditis elegans* by gene knockout with RNAi. *Curr. Biol.* **13**, 1715–1720.

Kennedy, S., Wang, D., and Ruvkun, G. (2004). A conserved siRNA-degrading RNase negatively regulates RNA interference in *C. elegans*. *Nature* **427**, 645–649.

Kuroyanagi, H., Kimura, T., Wada, K., Hisamoto, N., Matsumoto, K., and Hagiwara, M. (2000). SPK-1, a *C. elegans* SR protein kinase homologue, is essential for embryogenesis and required for germline development. *Mech. Dev.* **99**, 51–64.

Lee, S. S., Kennedy, S., Tolonen, A. C., and Ruvkun, G. (2003a). DAF-16 target genes that control *C. elegans* life-span and metabolism. *Science* **300**, 644–647.

Lee, S. S., Lee, R. Y., Fraser, A. G., Kamath, R. S., Ahringer, J., and Ruvkun, G. (2003b). A systematic RNAi screen identifies a critical role for mitochondria in *C. elegans* longevity. *Nat. Genet.* **33**, 40–48.

Maeda, I., Kohara, Y., Yamamoto, M., and Sugimoto, A. (2001). Large-scale analysis of gene function in *Caenorhabditis elegans* by high-throughput RNAi. *Curr. Biol.* **11**, 171–176.

Murphy, C. T., McCarroll, S. A., Bargmann, C. I., Fraser, A., Kamath, R. S., Ahringer, J., Li, H., and Kenyon, C. (2003). Genes that act downstream of DAF-16 to influence the lifespan of *Caenorhabditis elegans*. *Nature* **424**, 277–283.

Parrish, S., Fleenor, J., Xu, S., Mello, C., and Fire, A. (2000). Functional anatomy of a dsRNA trigger: Differential requirement for the two trigger strands in RNA interference. *Mol. Cell* **6**, 1077–1087.

Pothof, J., Van Haaften, G., Thijssen, K., Kamath, R. S., Fraser, A. G., Ahringer, J., Plasterk, R. H., and Tijsterman, M. (2003). Identification of genes that protect the *C. elegans* genome against mutations by genome-wide RNAi. *Genes Dev.* **17**, 443–448.

Riddle, D. L., Blumenthal, T., Meyer, B. J., and Priess, J. (eds.) (1997). "*C. elegans* II." Cold Spring Harbor Laboratory Press, Cold Spring Harbor, NY.

Sijen, T., Fleenor, J., Simmer, F., Thijssen, K. L., Parrish, S., Timmons, L., Plasterk, R. H., and Fire, A. (2001). On the role of RNA amplification in dsRNA-triggered gene silencing. *Cell* **107**, 465–476.

Simmer, F., Moorman, C., Van Der Linden, A. M., Kuijk, E., Van Den Berghe, P. V., Kamath, R., Fraser, A. G., Ahringer, J., and Plasterk, R. H. (2003). Genome-wide RNAi of *C. elegans* using the hypersensitive *rrf-3* strain reveals novel gene functions. *PLoS Biol.* **1**, E12.

Simmer, F., Tijsterman, M., Parrish, S., Koushika, S. P., Nonet, M. L., Fire, A., Ahringer, J., and Plasterk, R. H. (2002). Loss of the putative RNA-directed RNA polymerase RRF-3 makes *C. elegans* hypersensitive to RNAi. *Curr. Biol.* **12**, 1317–1319.

Smardon, A., Spoerke, J. M., Stacey, S. C., Klein, M. E., Mackin, N., and Maine, E. M. (2000). EGO-1 is related to RNA-directed RNA polymerase and functions in germ-line development and RNA interference in *C. elegans*. *Curr. Biol.* **10**, 169–178.

Tabara, H., Grishok, A., and Mello, C. C. (1998). RNAi in *C. elegans*: Soaking in the genome sequence. *Science* **282**, 430–431.

Tabara, H., Hill, R. J., Mello, C. C., Priess, J. R., and Kohara, Y. (1999a). *pos-1* encodes a cytoplasmic zinc-finger protein essential for germline specification in *C. elegans*. *Development* **126**, 1–11.

Tabara, H., Sarkissian, M., Kelly, W. G., Fleenor, J., Grishok, A., Timmons, L., Fire, A., and Mello, C. C. (1999b). The *rde-1* gene, RNA interference, and transposon silencing in *C. elegans*. *Cell* **99**, 123–132.

Tabara, H., Yigit, E., Siomi, H., and Mello, C. C. (2002). The dsRNA binding protein RDE-4 interacts with RDE-1, DCR-1, and a DExH-box helicase to direct RNAi in *C. elegans*. *Cell* **109**, 861–871.

Tavernarakis, N., Wang, S. L., Dorovkov, M., Ryazanov, A., and Driscoll, M. (2000). Heritable and inducible genetic interference by double-stranded RNA encoded by transgenes. *Nat. Genet.* **24**, 180–183.

Tewari, M., Hu, P. J., Ahn, J. S., Ayivi-Guedehoussou, N., Vidalain, P. O., Li, S., Milstein, S., Armstrong, C. M., Boxem, M., Butler, M. D., Busiguina, S., Rual, J. F., Ibarrola, N., Chaklos, S. T., Bertin, N., Vaglio, P., Edgley, M. L., King, K. V., Albert, P. S., Vandenhaute, J., Pandey, A., Riddle, D. L., Ruvkun, G., and Vidal, M. (2004). Systematic interactome mapping and genetic perturbation analysis of a *C. elegans* TGF-beta signaling network. *Mol. Cell* **13**, 469–482.

Tijsterman, M., Ketting, R. F., and Plasterk, R. H. (2002). The genetics of RNA silencing. *Annu. Rev. Genet.* **36**, 489–519.

Tijsterman, M., May, R. C., Simmer, F., Okihara, K. L., and Plasterk, R. H. (2004). Genes required for systemic RNA interference in *Caenorhabditis elegans*. *Curr. Biol.* **14**, 111–116.

Timmons, L., Court, D. L., and Fire, A. (2001). Ingestion of bacterially expressed dsRNAs can produce specific and potent genetic interference in *Caenorhabditis elegans*. *Gene* **263**, 103–112.

Timmons, L., and Fire, A. (1998). Specific interference by ingested dsRNA. *Nature* **395**, 854.

Timmons, L., Tabara, H., Mello, C. C., and Fire, A. Z. (2003). Inducible systemic RNA silencing in *Caenorhabditis elegans*. *Mol. Biol. Cell* **14**, 2972–2983.

Vaistij, F. E., Jones, L., and Baulcombe, D. C. (2002). Spreading of RNA targeting and DNA methylation in RNA silencing requires transcription of the target gene and a putative RNA-dependent RNA polymerase. *Plant Cell* **14**, 857–867.

Winston, W. M., Molodowitch, C., and Hunter, C. P. (2002). Systemic RNAi in *C. elegans* requires the putative transmembrane protein SID-1. *Science* **295**, 2456–2459.

Wood, W. B., and the Community of *C. elegans* Researchers, (eds.) (1988). "The Nematode *Caenorhabditis elegans*." Cold Spring Harbor Laboratory Press, Cold Spring Harbor, NY.

Yau, D. M., Yokoyama, N., Goshima, Y., Siddiqui, Z. K., Siddiqui, S. S., and Kozasa, T. (2003). Identification and molecular characterization of the G alpha12-Rho guanine nucleotide exchange factor pathway in *Caenorhabditis elegans*. *Proc. Natl. Acad. Sci. USA* **100,** 14748–14753.

Zipperlen, P., Fraser, A. G., Kamath, R. S., Martinez-Campos, M., and Ahringer, J. (2001). Roles for 147 embryonic lethal genes on *C. elegans* chromosome I identified by RNA interference and video microscopy. *EMBO J.* **20,** 3984–3992.

[4] High-Throughput RNA Interference Screens in *Drosophila* Tissue Culture Cells*

By Susan Armknecht, Michael Boutros, Amy Kiger, Kent Nybakken, Bernard Mathey-Prevot, and Norbert Perrimon

Abstract

This chapter describes the method used to conduct high-throughput screening (HTs) by RNA interference in *Drosophila* tissue culture cells. It covers four main topics: (1) a brief description of the existing platforms to conduct RNAi-screens in cell-based assays; (2) a table of the *Drosophila* cell lines available for these screens and a brief mention of the need to establish other cell lines as well as cultures of primary cells; (3) a discussion of the considerations and protocols involved in establishing assays suitable for HTS in a 384-well format; and (A) a summary of the various ways of handling raw data from an ongoing screen, with special emphasis on how to apply normalization for experimental variation and statistical filters to sort out noise from signals.

Introduction

A full grasp of the biological underpinnings of an organism such as *Drosophila* requires not only knowledge of the inherent function, expression profile, and subcellular localization of every gene product in the organism but also an awareness of the dynamic range of each gene product's interactions with other gene products over time and development. Although much information has accrued from the application of classical genetics, biochemistry, and developmental biology, the time and labor involved in these approaches makes their use in whole-genome assays a distant possibility. The completion of the *Drosophila* genome sequence in

---

*S. A., M. B., and A. K. contributed equally to this work.

2000 (Adams *et al.*, 2000) ushered the possibility of developing genome-wide approaches to systematically explore gene function in this organism. Newly emerged technologies were quickly exploited to extract maximal information encrypted in the raw sequence of the *Drosophila* genome. Among them, RNA interference (RNAi) is proving to be immensely valuable. RNAi refers to the ability of double-stranded RNA (dsRNA) or short-interfering RNA (siRNA) to silence a target gene through the specific destruction of the gene's mRNA. Unlike other genomic-based approaches such as microarrays and proteomics, RNAi provides a direct link from gene to function and, to date, offers the best tool for realizing the full potential of the genome project. The mechanisms by which RNAi silences gene expression are not discussed here as they can be found in a series of excellent reviews (Dykxhoorn *et al.*, 2003; Fire, 1999; Hannon, 2002; McManus and Sharp, 2002; Paddison and Hannon, 2003). Here we describe the method used to conduct high-throughput screening (HTS) RNAi in *Drosophila* tissue culture cells. The concept is based on the seminal observation by Clemens *et al.* (2000) that the single addition of long dsRNAs to *Drosophila* cells leads to an efficient RNAi effect.

Platforms Used for High-Throughout Screening (HTS) RNA Interference (RNAi) Screens

We have established a facility to support the investigation of a wide-range of questions in cell biology through large-scale RNAi screens in *Drosophila* cells [*Drosophila* RNAi Screening Center (DRSC); http://flyrnai.org]. Our long-term goal is to build a comprehensive database of genetic network interactions, protein–protein interactions, cellular phenotypes, and a catalog of gross cellular morphologies that can be linked to particular genes or pathways. In collaboration with the laboratory of Renato Paro and his colleagues (Hild *et al.*, 2003), we have generated a genomewide library of >21,000 dsRNAs that target nearly all (>91%) of the *Drosophila* predicted open reading frames for use in HTS (Boutros *et al.*, 2004). To take advantage of this library, we developed the methodology described in Fig. 1 to rapidly identify all the genes that affect a biological outcome as measured in a specific assay.

1. First, gene-specific dsRNAs are aliquoted into unique wells of multiple 384-well assay plates by using robotics. The concentration of dsRNA in the wells ranges from 25 to 75 nM (~0.2 μg of 500-bp length), a level sufficient with the RNAi bathing method to diminish or deplete endogenous mRNA levels (lesser dsRNA needed in the RNAi transfection method).

2. Cells are uniformly dispensed into the dsRNA-containing 384-well assay plates, using a Multidrop® 384 (Thermo Electron Corporation, Ontario, Canada) liquid dispenser at 1×10^4 to 3×10^4 cells per well, depending on the cell line and assay time course. For RNAi treatment by the bathing method, cells are first exposed to the dsRNAs in serum-free medium. After ≥30 min in serum-free medium, the Multidrop 384 is used again to add additional culture medium containing serum.

3. After the appropriate incubation time, cells can be either exposed first to a specific treatment or induction or directly processed for the assay readout.

The plate reader and automated microscopy are two common detection methods for data acquisition from the 384-well plates. The plate reader quantifies relative luminescence or fluorescence levels to a single numerical readout. The measurement represents the sum signal of all cells per well and is often determined in a biochemical assay performed on lysed cells. Plate reader screens provide a quantitative set of data amenable to normalization and various statistical analyses. At the DRSC, we use the

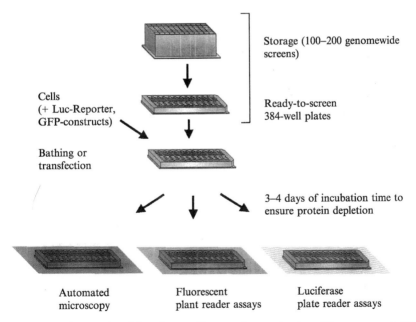

Storage (100–200 genomewide screens)

Cells
(+ Luc-Reporter, GFP-constructs)

Ready-to-screen
384-well plates

Bathing or
transfection

3–4 days of incubation time to ensure protein depletion

Automated   Fluorescent   Luciferase
microscopy   plant reader assays   plate reader assays

Fig. 1. RNAi screens in cell-based assays. There are three major steps: (1) array gene-specific dsRNAs into 384-well assay plates, (2) add cells to assay plates and incubate, and (3) perform assay and measure readout with plate reader or monitor by microscopy. (See color insert.)

Analyst™ GT (Molecular Devices, Sunnyvale, CA) plate reader. It has luminescence, fluorescence intensity, fluorescence polarization, time-resolved fluorescence, and absorbance detection modes in 96-, 384-, and 1536-well formats. A 384-well plate can be read in 1 min, providing rapid screening. It is also equipped with a bidirectional stacker for unattended operation. An application of the plate reader to quantify signals can be found in the assay conducted by Boutros *et al.* (2004), in which levels of luciferase activity were measured in each well and used as a linear measure of total cell number per well.

The inverted fluorescent microscope with automated acquisition software (MetaMorph®, Universal Imaging, Downingtown, PA), is used to track, focus, and capture fluorescent images of the cells within each well across an entire plate, providing cell-by-cell data in image fields of either live or fixed cells. The microscopy-based screens are more qualitative at present and are valuable for detecting inherent spatial information relevant to particular cell biological processes (i.e., protein localization, vesicular trafficking, axonal outgrowth, or host–pathogen interactions). Based on this approach, Kiger *et al.* (2003) designed an assay that followed fluorescently labeled actin filaments and microtubules to monitor cytoskeletal organization, cell attachment, cell spreading, and cell shape. In our screens, we have used the Discovery-1™ (Molecular Devices) high-content screening system. It has automated filter and dichroic wheels and a six-objective turret with high-speed laser auto-focus; it can measure up to eight fluorophores per assay in multiwell plates. In addition, the Discovery-1 is equipped with CataLyst Express™ (Thermo Electron Corporation) robotic arm for continuous, unattended operation. Depending on the assay, commercially available software can extract quantitative information on fluorescence intensity and distribution on a cell-by-cell basis or of the entire field. Automated image analysis is not yet generally applicable to all screens, and this is clearly an area that needs further development.

With both the plate reader- and microscopy-based platforms, there are two current methods for RNAi that involve either bathing or transfecting cells with dsRNAs. In the first method, cells are directly dispensed into 384-well plates containing dsRNA, which is then taken up by the cells (presumably by macropinocytosis), and incubated together for 3–5 days before the assay is read (bathing condition). The second method (transfection condition) is used when the design of the assay dictates that cells be transfected with a reporter gene and normalization constructs, and/or with expression vectors for a particular component of a signaling pathway or cellular process. Detailed protocols for both conditions are found in the section "Designing RNAi screens in HTS Format."

Drosophila Cells Available for RNAi Screens

We have collected a number of *Drosophila* cell lines, but have worked mostly with S2, S2R+, Kc, and clone 8 (cl8) cells in our screens. Table I lists the cell lines available in our laboratory and summarizes their salient characteristics. A number of additional cell lines are also available from DGRC (http://dgrc.cgb.indiana.edu).

Although the number of *Drosophila* cell lines currently available is relatively small, we expect that the next few years will witness the development of additional cell lines. In this regard, there are ongoing efforts to establish cultures of primary myoblastic, neuronal, and glial cells as well as hemocytes (insect hematopoietic cells) to be used in RNAi high-throughput screens. These efforts are not detailed in this chapter, but we mention two future applications that stress their importance and highlight the unique advantage of using *Drosophila* as an organism. The vast collection of mutant embryos that exist in *Drosophila* can serve as a valuable resource for establishing cell lines with sensitized background with which to perform HTS RNAi enhancer or suppressor screens. In addition, the availability of an extensive collection of embryos in which endogenous proteins have been tagged with GFP after exon trapping (Kelso *et al.*, 2004) can be exploited to establish cell lines to conduct visual screens.

Designing RNAi screens in HTS Format

Considerations

To develop an HTS, a cell-based assay must be adapted to high-density cell culture plate formats. (We routinely use 384-well plates rather than 96-well plates as they require a lesser amount of dsRNA.) Different types of plates are available, depending on the assay detection method. For example, white, solid-bottom plates are used for luminescence assays and black, clear-bottom plates for fluorescence assays. Other formats are possible in future screens once the technology is sufficiently advanced. For example, the use of high-density spotted arrays of dsRNAs on glass slides would substantially reduce materials necessary for genomewide screens (Carpenter and Sabatini, 2004). The following is a checklist of important considerations to weigh before embarking on a high-throughput RNAi screen:

Choices for Cell Lines

1. Will existing cell lines or primary cells be used?
2. Does RNAi by bathing or transfection work in the specific cells?
3. Do cells express the target and how refractory is it to RNAi?

TABLE I

LIST OF *DROSOPHILA* CELL LINES AVAILABLE FOR HIGH-THROUGHPUT SCREENING (HTS) RNA INTERFERENCE (RNAi)

Line	Origin	Characterisitics	Reference
Schneider's Line S2 (S2)	Dissociated embryos, near hatching (*Oregon R*)	Hemocyte-like gene expression, phagocytic, semiadherent in colonies, round, granular cytoplasm	Schneider, 1972
Schneider's S2 (S2*)	Dissociated embryos, near hatching (*Oregon R*)	Subclone of S2 cells that is more responsive to LPS, particularly after priming with 1 μM 20-hydroxy ecdysone	Schneider, 1972
Schneider's S2 (S2C)	Dissociated embryos, near hatching (*Oregon R*)	Hemocyte-like gene expression, semiadherent in colonies, round, granular cytoplasm	Schneider, 1972
Schneider's S2 (S2-R+)	Dissociated embryos, near hatching (*Oregon R*)	Hemocyte-like gene expression, phagocytic, adherent, flat cells, Fz+ and Wg-responsive	Schneider 1972; Yanagawa, *et al.,* 1998
Schneider's S2 (DL2)	Dissociated embryos, near hatching (*Oregon R*)	Hemocyte-like gene expression, phagocytic, adherent monolayer of uniformly round, smooth cells	Schneider, 1972
Schneider's Line S3	Dissociated embryos, near hatching (*Oregon R*)	Adherent, spindle-shaped cells, ecdysone responsive, grow in clumps	Schneider, 1972
Kc (Kc 167)	Dissociated embryos, 8–12 h (F2 *ebony × sepia*)	Hemocyte-like gene expression, phagocytic, uniformly round, clump in sheets, ecdysone responsive into adherent, bipolar spindle-shaped cells	Echalier and Ohanessian, 1969

l(2)mbn	Third-instar larvae tumorous hemocytes, [l(2)malignant blood neoplasm]	Larger cells, larger granular, complex cytoplasm, phagocytic, aneuploid, heterogenous size and shape	Haars et al., 1980
ML-DmBG2	Dissociated third-instar larvae brain and ventral ganglia (y v f mal)	Acetylcholine, HRP expression, neuronal-like processes	Ui et al., 1994
ML-DmBG6	Dissociated third-instar larvae brain and ventral ganglia (y v f mal)	Acetylcholine, HRP expression, neuronal-like processes	Ui et al., 1994
Clone 8	Third-instar larvae wing imaginal discs	Derived from wing columnar epithelial cells, adherent, will form multiple layers, conserved signaling pathways	Peel et al., 1990

4. Do cells need to be transfected to express a specific gene product that can serve as a target, modifier, or readout of the pathway or process of interest?
5. Is the phenotype or outcome easily detected in the chosen cell line?
6. Is this phenotype or outcome uniform across the cell population?

Assay Optimization with Genes Known to Belong to a Pathway or a Cellular Process

1. Select a few positive and negative control dsRNAs and confirm the desired phenotype or appropriate biological readout in response to RNAi.
2. When possible, confirm depletion of the gene product after RNAi of the positive and negative controls by immunoblotting or immunofluorescence.
3. Estimate the robustness of the assay in identifying the desired phenotype or outcome.
4. Familiarize yourself with the range of experimental variability in the assay.
5. Determine the optimal range of dsRNA concentration resulting in most penetrant and uniform phenotypes.
6. Determine optimal RNAi incubation time and plating density.

Configure the Assay to the 384-Well Plate Format

Plates. Appropriate plates depend on the type of platform used to read the assay. Confirm that the assay is compatible with these plates. Test and optimize plating, aging, staining, and detection in the 384-well plates to be used in the assay.

- Black, optically clear bottom plates (VWR No. 29444-078, Costar) are used for autoscope- and fluorescence plate reader-based assays.
- White, solid-bottom plates (VWR No. 29444-090, Costar, VWR No. 62409-072, Nunc) are used for luciferase plate reader-based assays.
- Solid black plates are used for fluorescence plate reader screens.

Imaging. Cells are screened with a 20× objective. In general, this resolution offers the best compromise: it facilitates autofocusing and yields a sufficient number of cells per field of study while being high enough to capture meaningful details in cell images. At this resolution, up to 16 sites within each well (covering the entire well surface) can be imaged; typically, images from two sites are collected. *Drosophila* cells are relatively small (10–50 μm), and subtle differences may not be easily resolved with a 20× objective. Depending on the phenotype scored, different resolutions may need to be optimized. When using an automated imaging system, check whether there is a reasonable field of cells within the region imaged with a 20× objective. Cell density also impacts the auto-focusing process, which can in turn interfere with proper detection of cellular phenotypes. In addition, cell density can alter cellular phenotypes. A crisp, planar, bright marker is required for fast and optimal autoscope focusing. Multiple channels can be sequentially imaged (most often a maximum of three). Test positive control dsRNA in 384-well plates to ensure that the resulting RNAi phenotype can be blindly identified among negative controls.

Check Signal/Noise and Well-to-Well Variability. Two important issues to consider during assay development are the signal/background ratio (S/B), and the well-to-well or experimental variability (EV). As assay variability increases, the S/B ratio must increase for the screen to be successful. We recommend preparing a positive control to determine the S/B ratio. To determine EV, add reagents to a plate with the same equipment used in the screen. Spike several wells with a positive control, and determine whether it can be reproducibly detected above the well-to-well variation. This information will give an indication of the false-positive and false-negative rate of the specific assay.

Protocols for Bathing and Transfection Conditions

The protocols have been established for cells dispensed in 384-well plates (used in the HTS RNAi screen). RNAi experiments may be done in plates of a different size, the only difference being that the amounts of reagents listed must be scaled up or down as required.

Bathing Conditions for 384-Well Plates

1. Remove 384-well plates prealiquoted with dsRNA from freezer to thaw. The 384-well plates contain 5 μl of ~0.05 μg/μl dsRNA in water for ~0.25 μg dsRNA/well.
2. Spin plates at ~1200 rpm for 1 min before removing seals.
3. Count cells and then spin to pellet (~1200 rpm, 5 min).
4. Resuspend cells at 5×10^6 cells/ml in serum-free medium.
5. Plate ~10^4 cells 10 μl/well in 384-well plates.
6. Incubate dsRNA with cells at RT for 30 min or more.
7. Add 30 μl of complete medium to each well.
8. Seal or incubate plates in a humid chamber to prevent evaporation and minimize edge effects.
9. Incubate for 3–5 days and analyze. Length of incubation may vary depending on assay.

TABLE II
TRANSFECTION SET-UP FOR 384- AND 96-WELL PLATES

Plate	Diameter (mm)	Area (mm²)	Fold diff. in area from 96 well	Fold diff. in area from 6 well	DNA per well (μg)	Cells per well (× 10⁴)
384	3.5 square	12.25	1/2.6	1/80	0.050–0.175	1–3 × 10⁴
96	6.4	32	1	1/30	0.20–0.30	3–8 × 10⁴

Plate	Diameter (mm)	Area (mm²)	EC buffer (μl)	Amount enhancer (μl)	Amount effectene (μl)	Medium (with cells) added to well (μl)
384	3.5	12.25	8.75	1.0	0.30	30–50
96	6.4	32	21.5	2.5	0.75	100–150

Note: Transfection conditions in 96- and 384-well plates for HTS RNAi screening. Differences in DNA quantity, cell number, and transfection reagent amounts are scaled to well area. See text for details.

Transfection Conditions for 384-Well Plates. This protocol describes the method we have developed to screen in cell lines that require transfection of the dsRNA or for screens involving transient transfection of a transcriptional reporter (or any another construct). This approach was designed to screen for new components of signaling pathways (based on the activation of a luciferase transcriptional reporter), but it can be adapted to survey other phenomena, such as protein–protein interactions, using green fluorescent protein (GFP) fluorescence resonance energy transfer (FRET) constructs. For screening 384- or 96-well plates, it is best to automate as many steps as possible. Good dispensing machines are critical to a smooth running screen, and different dispensing machines are used for different purposes. We use the MultiDrop to aliquot transfection mixes and cells to wells. The MultiDrop relies on a peristaltic pump system and interchangeable heads to dispense fluids, which we find to be easier to keep sterile than the piston pump systems. We use the μFill™ (Bio-Tek, WinoOSRI, VT), piston pump machine for the luciferase assays. It is more precise for assays that are highly sensitive to slight variations in volumes across wells. (Although this is important for luciferase assays, immunostaining procedures are much less sensitive to small volume variations.) When using these machines in a highly automated procedure, it is helpful to familiarize oneself with their use before setting up assay plates. When using 384-well plates, we recommend testing the MultiDrop by aliquoting some of the wash solution into a blank plate as the MultiDrop can occasionally have problems when shifting from the even to the odd rows. Finally, we use Effectene™ (Qiagen, Valencia, CA) as the transfection reagent because it is a good compromise between high efficiency of transfection and relative ease of use (Table II).

1. Count cells and dilute them in fresh medium so that the desired number of cells is in 30–40 μl of medium, which can be dispensed in each well. Fill a sterile container with all the cells/medium needed plus 10 ml to account for priming of the dispensing head. Mix well before dispensing.

2. Centrifuge the plates at 1200 rpm. Remove the sealing membranes and then add 75 ng of control dsRNA in 5 μl to the dedicated empty well in the 384-well plate.

3. Prepare EC + DNA master mix. The master mix for a 384-well plate consists of all DNA constructs that need to be introduced into cells (which may include the experimental reporter, a ligand expression vector, and the control reporter) and EC buffer, a component of the Effectene kit. Dilute the DNA in 8.75 μl EC buffer. It is best if the volumes of all DNA solutions represent less than 3% of the final volume of the master mix. If this is not possible, we recommend precipitating the DNA material and resuspending it directly in EC buffer. It is important to keep the nucleic concentration equal in all wells and use no more than 175 ng of total nucleic acid (DNA + dsRNA) per 384-well plate. Mix EC + DNA well by inversion or pipetting.

4. Prepare the MultiDrop to dispense both transfection solution and cells. This will require two different dispensing cassettes for the Multi-Drop: one exclusively used for cells and one for reagents. For the cell cassette only, wash with 70% ethanol to sterilize. Then wash both cassettes thoroughly by running sterile water and PBS through them. Always wash the cassettes with sterile water before aliquoting reagents or cells. Put the reagent cassette in the machine and set the cell cassette aside in the hood until needed. Check that the number of columns to be aliquoted is correct and then set the volume of media to be dispensed per well. This protocol uses 10 μl of transfection solution per well.

5. Add enough enhancer to the EC + DNA master mix such that each well receives the equivalent of 1.0 μl of enhancer from the transfection solution. Mix well by inversion. Incubate at RT for 2–3 min.

6. Complete the transfection solution by adding enough Effectene to the enhancer + EC + DNA master mix such that finally 0.3 μl of Effectene is present in each well. Mix by pipetting up and down 5–10

times or vortexing on high for 10 sec. Immediately begin dispensing the transfection solution into the assay plates, using as little of the transfection solution as possible to prime the dispensing cassette. Vortex the plate on a low setting for 5–10 sec. using a plate adapter to mix the dsRNA and the transfection solution.

7. Let the dsRNA and transfection solution incubate in the well for 4–8 min. Meanwhile, substitute the cell cassette for the reagent cassette and briefly wash the cell cassette with 10–20 ml cell culture medium and empty the cassette after the wash. After 4–8 min, dispense the desired number of cells in the predetermined volume of medium into the 384-well plate. It is critical to get the transfection solution and the cells into the wells within 20 min of making the master transfection solution or the transfection efficiency well begin to fall rapidly.

8. Incubate the plates in a humidified chamber in a TC incubator for 3–5 days, and then conduct assay.

Standardization of Reporter Gene Assays

Normalization is the process by which raw data returned from an assay is standardized relative to an internal control. This ensures that the quantitative readout is not influenced by well-to-well variability. In this section, we discuss the need for normalization in HTS RNAi transfection screens and describe some of the control luciferase reporter genes that we have developed to carry out normalization for transfection assays in a 384-well plate format.

Transfection assays for HTS RNAi screens require more careful design than for bathing assays because of the higher number of variables involved. Although it is possible to transfect materials other than nucleic acids into cells, the principal application of transfection in HTS screens is to introduce reporter genes designed to reflect the amplitude of a response to a stimulus or a cellular perturbation. Reporter constructs generally consist of an expression construct in which an indicator gene (encoding, for instance, luciferase, GFP, or CAT) is placed under the transcriptional control of promoter and enhancer elements that modulate the indicator gene expression in response to the stimulus or perturbation under study. The strength of the signal measured in such an assay will reflect the biological outcome (i.e., the response to a stimulus), but will also depend on the number of cells successfully transfected with the reporter gene and the strength of expression of the transfected indicator gene. As transfection of cell populations is neither uniform in terms of the percent of cells transfected nor in terms of the amount of material that is introduced into each cell, some method of standardization is required to control for these variables.

Ideally, a standardization control will provide information about both the number of cells that have taken up the transfected material as well as the particular cell type's ability to express the transfected gene. In practice, however, it is difficult to separate these two variables and they are treated as one variable, often referred to as the transfection efficiency. If the variation in transfection efficiency is sufficiently low, a standardization control can also provide information on the number of transfected cells present in a well. Just as it is important in HTS RNAi screens to design an experimental reporter that leads to a high signal-to-noise ratio in response to a stimulus, it is equally critical to have a robust control reporter that accurately reflects the transfection efficiency within each well. We have found that cotransfection of both experimental and control reporters, followed by sequential assay of the two reporters, is the most reliable way to evaluate transfection efficiency used for normalization purposes.

When choosing or designing a control reporter gene, it is important to keep in mind the following criteria. First, as both reporter genes will be expressed in the same cell, one needs to ensure that the control reporter can be assayed independently of the experimental reporter and that its signal will not be influenced by the conditions used to assay the process of interest (e.g., signaling pathway or cell division). Second, the signal from the control reporter gene needs to be robust and should be significantly different from background signal, especially when measured in a 384-well plate format. It has been our experience that many reporter genes (control or experimental) do not perform as well in 384-well format as would be expected based on the signals observed in cells plated in 96- or 6-well plates.

Control Luciferase Reporters. Luciferase assays (Boeckx, 1984) have become the assay of choice in recent years because of their ease of use, large linear signal range, adaptability to plate reader formats, and technical improvements that have increased their reliability and flexibility. The development of dual-luciferase assays (Grentzmann *et al.*, 1998) has been instrumental in allowing the examination of both experimental and control reporters from the same sample, thereby eliminating the need for separate assays. Dual-luciferase assay systems employ the cotransfection of luciferase constructs from phylogenetically diverse organisms that use different substrates and reaction mechanisms. These differences allow the sequential assay of the two luciferases by taking advantage of the fact that one type of luciferase activity can be individually measured and then specifically quenched while the second type of activity is being assayed. The most common commercially available dual-luciferase system uses a beetle (firefly) luciferase and a *Renilla* (a cnidarian) luciferase (Promega, Madison, WI). Most experimental reporter constructs drive firefly luciferase, whereas *Renilla* is used in the control construct. Several *Renilla*-based control reporter genes are commercially available. However, these constructs were designed for use in mammalian systems and the promoters they use do not work well in insect cells. Among them, the thymidine kinase (TK) promoter driven-*Renilla*, SV40-driven *Renilla*, and CMV-driven *Renilla* control vectors give very low signals in 384-well plate luciferase assays when tested in either *Drosophila* S2 or clone 8 cell lines at 24°. Consequently, we sought to design our own *Renilla* control constructs for use in *Drosophila* cell-based HTS assays.

The ideal promoter to drive a *Renilla* control reporter should be capable of directing high levels of expression without being affected by or affecting the cellular process or signaling pathways that are being targeted in a particular HTS assay. With this criterion in mind, we selected five candidate sequences to drive *Renilla* expression in a control reporter: a portion of the *Drosophila* actin5C promoter, the ubiquitin promoter, the mitochondrial large ribosomal protein 49 promoter, the RNA polymerase II 215 subunit promoter (pol II), and the RNA polymerase III 128 subunit promoter (pol III). The first two promoters are known to drive a high level of expression in other heterologous systems, whereas the promoter for mRpL49 (a gene often used as a control in Northern blot analysis of *Drosophila* cells or tissues) was predicted to lead to a robust expression in most cell types. The two RNA polymerase promoters were selected based on our assumption that, as RNA polymerases are likely to be expressed at high levels in continually dividing cells, their promoters were likely to be highly active in cultured cells.

We tested each of these *Renilla* constructs for luciferase activity in a 384-well assay, using both *Drosophila* clone 8 and S2 cells—the cells most often used for transfection assays in our laboratory. As seen in Fig. 2A, the ubiquitin-, mRpL49-, RNA polymerase II-, and SV40-promoter-driven *Renilla* constructs resulted in signals very similar to the background signal observed with the empty vector alone. Their low levels of expression preclude their use in screens carried out in 384-well plate format. Only the Actin–*Renilla* and pol III–Renilla vectors gave sufficient signals to be usable in HTS screens. These two *Renilla* constructs were further tested to ensure that they were not affected by conditions used to study the signaling pathways investigated in our screens. We found that the actin5C–*Renilla* ectopically transactivated at least one experimental pathway reporter in three signaling pathways tested and was therefore ill suited as a control construct in our experiments. The pol III–Renilla had no effect on any of the signaling pathways tested. Testing of the pol III–Renilla also demonstrated that it has a roughly linear dose–response curve in the range used in our 384-well assays (Fig. 2B).

It is important to realize that addition of dsRNA to any transfection assay system reduces the expression of transfected DNA expression constructs between 10- and 20-fold, probably by competing with the intake of plasmid DNA by the cells (data not shown). As a result of the general reduction in signals caused by dsRNA and in light of the transactivation effects of actin5C–Renilla that we observed, we settled on the Pol III–*Renilla* construct as the optimal control reporter for transfection based assays in 384-well formats. The pol III promoter leads to signals that are consistently 10-fold above background and are not affected nor interfered with by any of the signaling pathways examined. (Three different pathways have been formally tested.) Although it is imperative to evaluate the behavior of this promoter in the context of every new assay and cell type, we would recommend use of the pol III–*Renilla* for normalization purposes in transfection assays. This is not to say, however, that other constructs cannot be used for normalization, especially in larger plate formats (Lum *et al.*, 2003). Provided their signals have been adequately studied and remain unchanged under the biological conditions tested in a given assay, other vectors, such as the actin5C–Renilla vector, are also suitable.

Edge Effects. One potential artifact in the process of normalization that we have encountered in the course of luciferase assays is the phenomenon of what has often been called edge effects. The term refers to the observation that artificially high, normalized values are returned for data collected from wells specifically located alongside the edges of a plate. Based on the calculated mean from the other wells in the plate, it appears that the increased value after normalization in these wells is not due to higher firefly luciferase activity. Instead, the edge effect correlates with a reduction in *Renilla* signal in these wells, and this reduction yields normalized values that are artificially high. We have not been able to determine exactly why edge effects occur or how to reliably prevent them. Edge effects are also observed in visual screens of microscopy images, sometimes manifesting as a decrease in cell density (and possibly morphology) in the outermost wells. The basis for these effects is again unclear, but medium evaporation or uneven treatment of the entire plate surface might be contributing factors. However, edge effects are clearly identifiable in the normalized values and they can be accounted for during the analysis.

Data Analysis. To identify bona fide dsRNA-specific phenotypes, the numerical readouts of the reporter gene assay need to be standardized against the appropriate internal control. To ensure meaningful comparisons

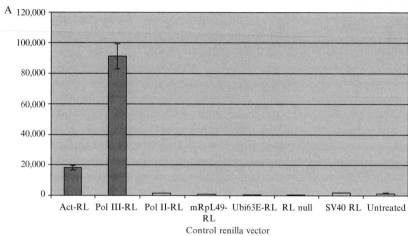

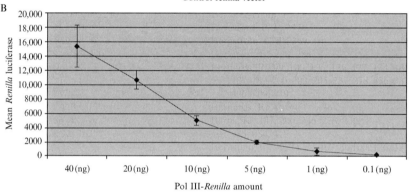

FIG. 2. Analysis of control *Renilla* reporter constructs. (A) Comparison of luciferase expression. A total of 20 ng of the indicated *Renilla* luciferase (RL) control reporter construct was transfected along with 90 ng of experimental reporter into 25,000 clone 8 cells in a 384-well plate, using Effectene. RL activity was determined by using the Dual-Glo kit (Promega) and read by a Analyst plate reader (Molecular Devices) after 5 days of transfection. Each treatment represents the average of 8 wells from a 384-well plate. RL-Null is the parental vector (Promega) into which the promotor regions were subcloned, and SV40 RL is a commercially available SV40 promoter-driven *Renilla* (Promega). All promoter fragments were subcloned into the *Bgl* II–*Spe*1 sites of the pRL-null vector (Promega). Promoter regions corresponding to the pol II and mRpL49 promoters were obtained after PCR amplification of the region between positions 200965 and 200785 for pol II and 282684 and 283154 for mRpl49, according to the genome annotation numbering. The Act5C promoter region is identical to that cloned into the pACW vector available from the DGRC. Details for the pol III construct are available on request. (B) Titration of pol III-*Renilla* expression. Transfection of decreasing amounts of pol III-*Renilla* into clone 8 cells demonstrates that *Renilla* expression is linear from roughly 10 to 40 ng of transfected pol III–*Renilla*. Transfection was done as in (A) except for the indicated amounts of pol III–*Renilla* and the presence of 75 ng of green fluorescent protein (GFP) dsRNA in each well. Each treatment represents the average of 6 wells from a 384-well plate.

between plates, data sets are then mean- or median centered for each individual plate. For each dsRNA condition, the normalized value is assigned a significance or z-score value by calculating the number of standard deviations away from the plate median. In this approach, differences between plates are taken into account to calculate the significance of each value, depending on the variability of each assay plate.

Results from duplicate or multiple screens are averaged before threshold selection. The selection of the threshold used to generate a list of candidate targets depends on overall assay parameters, such as signal-to-noise ratio and dynamic range. A good indicator for threshold selection is usually the performance of spiked-in positive controls. In previous genomewide screens, we set a threshold of three standard deviations and above to select with 99.9% confidence the most statistically significant phenotypes. Data analysis and data representation are usually performed by MATLAB (MathWorks, Nation).

The sequence from each dsRNA that passes the selection threshold is then analyzed for potential off-target effects by BLASTN against all predicted transcripts. Sequences with potential off-site effects are flagged in downstream analysis. The list of predicted gene targets is confirmed by BLASTN searches against the latest published *Drosophila* genome sequence and mapped to specific chromosomes. The identified genes are searched for associated gene ontology, mutant allele, and RNAi phenotype annotations in FlyBase (http://flybase.bio.indiana.edu/) and other public databases. The predicted protein sequences of genes identified by RNAi phenotypes can then be searched for conserved protein domains by using InterPro.

Perspectives and Conclusions

In this chapter, we focused on the most basic issues relevant to high-throughput RNAi screens in *Drosophila* cells in a 384-well plate format. The technology overall is robust and allows the rapid identification of dsRNA candidates that affect a specific readout. In the upcoming years, we expect to see significant improvements and advances in methodology. The range of assays available for this approach will be broadened to include existing readout formats that have not been optimized for the 384-well plate format. Possible applications include cytoblot assays for antibody-based screens, FRET analysis, and HTS multiplexing of mRNA transcripts as signature of a pathway or process. It will also be necessary to develop new cell lines and primary cultures to extend the repertoire of cells in which existing screens can be performed, and, importantly, to enable new screens that address highly specialized functions such as muscle fusion, neuronal outgrowth, or epithelial cell polarity. These advances will require the development of increasingly sophisticated software programs that will automatically scan images to identify or quantify user-defined features (e.g., specific cellular structures, protein localization, dynamic processes, or cell size). To better handle the large amount of data returned from these screens and process the growing load of information as more screens get completed, more robust statistical analyses will need to be developed. These analyses will suggest a variety of criteria on which to base our assignment for potential hits, as opposed to the current empirical and static threshold value. Lastly, there will be added flexibility on how to conduct RNAi screens, as alternative platforms (e.g., RNAi cell microarrays) are likely to emerge as powerful methodologies, rivaling existing ones both in speed and economy of reagents.

Acknowledgments

We thank past and present members of the Perrimon laboratory for discussions on technical issues relevant to RNAi screens in tissue culture cells. K. N. was supported by an NRSA. M. B. was supported by an Emmy-Noether grant from the Deutsche Forschungsgemeinschaft. A. K. was supported by The Jane Coffin Childs Memorial Fund for Medical Research. This work was supported by a grant from the NIGMS and HHMI. N. P. is an investigator of the Howard Hughes Medical Institute.

References

Adams, M. D., Celniker, S. E. *et al.* (2000). The genome sequence of *Drosophila melanogaster*. *Science* **287**(5461), 2185–2195.
Boeckx, R. L. (1984). Chemiluminescence: Applications for the clinical laboratory. *Hum. Pathol.* **15**(2), 104–111.
Boutros, M., Kiger, A. A., Armknecht, S., Kerr, K., Hild, M., Koch, B., Haas, S. A., Consortium, H. F., Paro, R., and Perrimon, N. (2004). Genome-wide RNAi analysis of growth and viability in *Drosophila* cells. *Science* **303**(5659), 832–835.
Carpenter, A. E., and Sabatini, D. M. (2004). Systematic genome-wide screens of gene function. *Nat. Rev. Genet* **5**(1), 11–22.
Clemens, J. C., Worby, C. A., Simonson-Leff, N., Muda, M., Maehama, T., Hemmings, B. A., and Dixon, J. E. (2000). Use of double-stranded RNA interference in *Drosophila* cell lines to dissect signal transduction pathways. *Proc. Natl. Acad. Sci. USA* **97**(12), 6499–6503.
Dykxhoorn, D. M., Novina, C. D., and Sharp, P. A. (2003). Killing the messenger: Short RNAs that silence gene expression. *Nat. Rev. Mol. Cell Biol.* **4**(6), 457–467.
Echalier, G., and Ohanessian, A. (1969). Isolation, in tissue culture, of *Drosophila melangaster* cell lines. *C. R. Acad. Sci. Hebd. Seances Acad. Sci. D.* **268**(13), 1771–1773.
Fire, A. (1999). RNA-triggered gene silencing. *Trends Genet.* **15**(9), 358–363.
Grentzmann, G., Ingram, J. A., Kelly, P. J., Gesteland, R. F., and Atkins, J. F. (1998). A dual-luciferase reporter system for studying recording signals. *RNA* **4**(4), 479–486.
Haars, R., Zentgraf, H., Gateff, E., and Bautz, F. A. (1980). Evidence for endogenous reovirus-like particles in a tissue culture cell line from *Drosophila melanogaster*. *Virology* **101**(1), 124–130.

Hannon, G. J. (2002). RNA interference. *Nature* **418**(6894), 244–251.

Hild, M., Beckmann, B., Haas, S. A., Koch, B., Solovyev, V., Busold, C., Fellenberg, K., Boutros, M., Vingron, M., Sauer, F., Hoheisel, J. D., and Paro, R. (2003). An integrated gene annotation and transcriptional profiling approach towards the full gene content of the *Drosophila* genome. *Genome Biol.* **5**(1), R3.

Kelso, R. J., Buszczak, M., Quinones, A. T., Castiblanco, C., Mazzalupo, S., and Cooley, L. (2004). Flytrap, a database documenting a GFP protein-trap insertion screen in *Drosophila melanogaster. Nucleic Acids Res.* **32**, D418–D420. (Database issue).

Kiger, A., Baum, B., Jones, S., Jones, M., Coulson, A., Echeverri, C., and Perrimon, N. (2003). A functional genomic analysis of cell morphology using RNA interference. *J. Biol.* **2**(4), 27.

Lum, L., Yao, S., Mozer, B., Rovescalli, A., Von Kessler, D., Nirenberg, M., and Beachy, P. A. (2003). Identification of hedgehog pathway components by RNAi in *Drosophila* cultured cells. *Science* **299**(5615), 2039–2045.

McManus, M. T., and Sharp, P. A. (2002). Gene silencing in mammals by small interfering RNAs. *Nat. Rev. Genet.* **3**(10), 737–747.

Paddison, P. J., and Hannon, G. J. (2003). siRNAs and shRNAs: Skeleton keys to the human genome. *Curr. Opin. Mol. Ther.* **5**(3), 217–224.

Peel, D. J., Johnson, S. A., and Milner, M. J. (1990). The ultrastructure of imaginal disc cells in primary cultures and during cell aggregation in continuous cell lines. *Tissue Cell* **22**(5), 749–758.

Schneider, I. (1972). Cell lines derived from late embryonic stages of *Drosophila melanogaster. J. Embryol. Exp. Morphol.* **27**(2), 353–365.

Ui, K., Nishihara, S., Sakuma, M., Togashi, S., Ueda, R., Miyata, Y., and Miyake, T. (1994). Newly established cell lines from *Drosophila* larval CNS express neural specific characteristics. *In Vitro Cell Dev. Biol. Anim.* **30A**(4), 209–216.

Yanagawa, S., Lee, J. S., and Ishimoto, A. (1998). Identification and characterization of a novel line of *Drosophila* Schneider S2 cells that respond to wingless signaling. *J. Biol. Chem.* **273**(48), 32353–32359.

[5] Mechanistic Insights Aid Computational Short Interfering RNA Design

By QUETA BOESE, DEVIN LEAKE, ANGELA REYNOLDS,
STEVEN READ, STEPHEN A. SCARINGE,
WILLIAM S. MARSHALL, and ANASTASIA KHVOROVA

Abstract

RNA interference is widely recognized for its utility as a functional genomics tool. In the absence of reliable target site selection tools, however, the impact of RNA interference (RNAi) may be diminished. The primary determinants of silencing are influenced by highly coordinated RNA–protein interactions that occur throughout the RNAi process, including short interfering RNA (siRNA) binding and unwinding followed by target recognition, cleavage, and subsequent product release. Recently developed strategies for identification of functional siRNAs reveal that thermodynamic and siRNA sequence-specific properties are crucial to predict functional duplexes (Khvorova *et al.*, 2003; Reynolds *et al.*, 2004; Schwarz *et al.*, 2003). Additional assessments of siRNA specificity reveal that more sophisticated sequence comparison tools are also required to minimize potential off-target effects (Jackson *et al.*, 2003; Semizarov *et al.*, 2003). This chapter reviews the biological basis for current computational design tools and how best to utilize and assess their predictive capabilities for selecting functional and specific siRNAs.

RNA Interference (RNAi)

RNA interference (RNAi) is a concerted process of specific interactions between double-stranded RNA (dsRNA) and a dynamic and highly regulated complex of multiple proteins. In invertebrates and plants, a long dsRNA leads to sequence-specific degradation of endogenous mRNA homologues through this enzyme-mediated event (Sharp, 2001). Dicer, a ribonuclease III-like enzyme, processes dsRNA typically derived from viral infection (Hamilton and Baulcombe, 1999), amplification of transposons (Plasterk and Ketting, 2000; Tabara *et al.*, 1999), endogenous transcription of regulatory RNA precursors (Lagos-Quintana *et al.*, 2001; Lau *et al.*, 2001; Lee *et al.*, 1993), or aberrant transcription (Dalmay *et al.*, 2000), into 19- to 23-bp short interfering RNAs (siRNAs) with characteristic two-base 3' overhangs (Bernstein *et al.*, 2001). The siRNAs are then incorporated into an RNA-induced silencing complex (RISC). A helicase(s) within the RISC unwinds the siRNA duplex, enabling the complementary antisense strand to guide target recognition (Nykanen *et al.*, 2001). Finally, an endonuclease(s) within the RISC cleaves the mRNA to induce silencing (Elbashir *et al.*, 2001b). Thus, whether produced naturally or introduced by transfection, siRNAs are involved in multiple protein interactions and participate in a minimum of four major steps: (1) initial duplex recognition by the pre-RISC complex, (2) ATP-dependent RISC activation, (3) target recognition, and (4) target cleavage (Fig. 1). Because of the nature of these RNA–protein interactions, sequence-specific characteristics of functional siRNA duplexes are likely to bias strand selection during siRNA–RISC assembly and activation, contributing to the overall silencing efficiency of RNAi.

Target Site Selection

In nature, target site selection is rendered unnecessary as Dicer processes long dsRNA into a pool of siRNAs. The resulting pool of duplexes comprises a wide range of silencing potencies, but invariably one or more

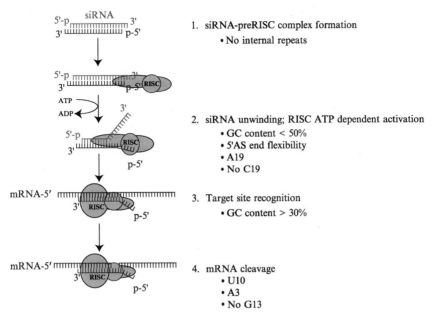

1. siRNA-preRISC complex formation
   • No internal repeats

2. siRNA unwinding; RISC ATP dependent activation
   • GC content < 50%
   • 5'AS end flexibility
   • A19
   • No C19

3. Target site recognition
   • GC content > 30%

4. mRNA cleavage
   • U10
   • A3
   • No G13

FIG. 1. Proposed model for the RNAi mechanism, illustrating key steps in siRNA–protein interactions and their relationship to eight thermodynamic and sequence-specific criteria described by Reynolds *et al.* (2004). Rational design takes into account these criteria in an effort to maximize efficient silencing. (See color insert.)

of these siRNAs exhibit silencing activity. In contrast, the approach taken in many laboratories to silence a gene of interest involves the direct introduction of siRNAs and requires preselection of the optimal 19- to 21-base target site to achieve potent silencing. As not all siRNAs exhibit efficient silencing activity, early efforts focused on identifying basic parameters necessary for predicting functional duplexes (Bernstein *et al.*, 2001; Elbashir *et al.*, 2001a,b,c, 2002). The most widely used conventional method for selecting 21-base mRNA target sequences was derived from these initial investigations and relied on identification of unique targets with AA dinucleotides preceding a 19-base sequence, at least 100 bases downstream of the AUG, with a 30–70% G/C content that resides within the coding region [not within the 5' or 3' untranslated regions (UTRs)]. On average, 60–70% of duplexes selected in this manner silence their target, but with significant variability in silencing efficiency (Aza-Blanc *et al.*, 2003; Holen *et al.*, 2002; Hsieh *et al.*, 2004; Krichevsky and Kosik, 2002; Laposa *et al.*, 2003; Spankuch-Schmitt *et al.*, 2002). However, these guidelines tend to ignore secondary structure of the target RNA, target site accessibility, RNA–RNA or RNA–protein interactions, or other thermodynamic properties and their potential impact on efficient siRNA-mediated gene silencing.

Lessons from Antisense and Ribozyme Technologies

What strategies can be employed to find correlations between sequence and siRNA functionality? Historically, nucleic acid-based gene-silencing strategies that used antisense oligonucleotides or ribozymes were found to depend on the inherent local mRNA structure, target accessibility, and binding affinities. Indeed, binding affinity and target site accessibility are the rate-limiting steps for efficient mRNA degradation by the antisense mechanism. Several reports identified putative correlations between the target mRNA secondary structure and accessibility with the potency of siRNAs (Bohula et al., 2003; Kretschmer-Kazemi Far and Sczakiel, 2003; Lee et al., 2002; Vickers et al., 2003). The design of functional siRNAs that overlay previously identified functional antisense sites lent additional support to a model for target site accessibility and functionality (Vickers et al., 2003). A more recent study characterized the inhibitory effects of the TAR (HIV-1 trans-activation response region), a well-characterized palindromic sequence known for its stable secondary structure, on siRNA functionality when positioned within the local context of the target sequence (Yoshinari et al., 2004). One explanation for these putative correlations among secondary structure, target accessibility, and siRNA potency is that a steady-state equilibrium exists between the complementary RNA for its target mRNA site that is influenced by the stability of competing internal structures for each RNA species (Matveeva et al., 2003; Pancoska et al., 2004).

A more discerning computational approach, Sfold, was developed by using mathematical models for predicting RNA secondary structure and target accessibility and addresses the potential for functional correlations to siRNA potency (Ding and Lawrence, 2003). Sfold incorporates a statistical sampling algorithm to generate a probability profiling of single-stranded regions in RNA secondary structure and can be used to predict the importance of these parameters on silencing. By overlaying the probability profiles, a mutual accessibility plot can be displayed for predicting RNA–RNA interactions and identifying regions of accessibility for potential ribozyme, antisense, or siRNA interactions. To test the most recent version of this predictive algorithm, the theoretical binding energies between the antisense strands from a panel of siRNAs (targeting firefly luciferase and human cyclophilin B) and their cognate mRNA sites were calculated (Ding and Lawrence, 2003). The selected siRNAs, for which silencing efficiencies were known, were sorted according to the predicted values. Data revealed that all functional classes (functionality is defined as

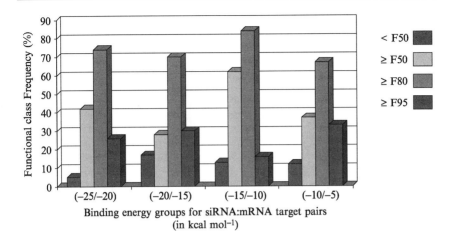

F
I
G
. 2. siRNA functionality classes were sorted according to a predicted mRNA target site accessibility value (reported in kcal mol^{-1}) as determined by Sfold (Ding and Lawrence, 2003). The relatively equal distribution of all functional classes across the four binding energy subgroups (antisense siRNA strand:mRNA target) suggests that there is no strong correlation between siRNA functionality and target site accessibility. (Functionality is defined as an F value, where F80 represents 80% suppression of mRNA levels.) (See color insert.)

an F value for the average level of silencing observed) were distributed relatively equally between four binding energy subgroups (Fig. 2). This suggests that a poor correlation exists between siRNA functionality and mRNA site accessibility or siRNA–mRNA binding affinities as assessed by this methodology.

Finally, functional siRNAs are found at least an order of magnitude more often than antisense oligonucleotides and ribozymes. Furthermore, studies that compared siRNAs and antisense oligonucleotides directed against the same target found that optimal siRNA sites do not necessarily overlap with optimal antisense sites for firefly luciferase, supporting the idea that the rules governing siRNA and antisense target site selection are distinct or unrelated (Xu *et al.*, 2003). When assessing the functionality of a panel of siRNAs composed of duplexes targeting sites juxtaposed in two-base increments, even a two-base shift in the targeting region can affect siRNA activity drastically (Fig. 3). Taken together, these data suggest that if mRNA secondary structure plays a role in siRNA potency, it is not the primary determinant of function.

Lessons from Regulatory MicroRNAs (miRNAs)

One potentially rich source of information regarding attributes essential for siRNA functionality is a class of regulatory RNAs known as

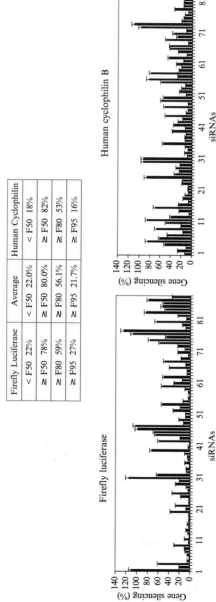

Firefly Luciferase	Average	Human Cyclophilin
< F50 22%	< F50 22.0%	< F50 18%
≥ F50 78%	≥ F50 80.0%	≥ F50 82%
≥ F80 59%	≥ F80 56.1%	≥ F80 53%
≥ F95 27%	≥ F95 21.7%	≥ F95 16%

FIG. 3. Functional characteristics of siRNAs made to every other position of a 197-base region in each of two separate genes: firefly luciferase and human cyclophilin B. Functionality was assessed in HEK293 cells stably expressing luciferase after transfection with 100 n*M* siRNA. Reduction of luciferase was measured by the Steady-Glo assay (Promega). Reduction of cyclophilin B mRNA was measured by using the Quantigene branched-DNA technology (Bayer). Values are reported as % silencing relative to control, nontransfected cells.

microRNA (miRNA). miRNAs are considered to be endogenously ex-pressed counterparts of siRNA. They were originally identified in *C. elegans* (Lee and Ambros, 2001; Lee *et al.*, 1993; Wightman *et al.*, 1993) and have since been found to be highly conserved across phyla (reviewed in Bartel, 2004). These gene-regulatory RNAs are transcribed in the nucleus as unim-olecular precursors that function to suppress heterologous gene targets. In contrast, the related siRNAs are derived from a variety of oligonucleotide sources (viruses, transposons, aberrant transcription products) and ulti-mately function to silence sequences related to their origin. miRNAs begin as primary nuclear miRNA transcripts (pri-miRNA) that are processed by an RNase III endonuclease, Drosha, to a smaller 70-nt precursor hairpin structure, termed pre-miRNA, that is then transported to the cytoplasm by using the Ran-GTP and the exportin-5 receptor (Lee *et al.*, 2003; Lund *et al.*, 2004; Yi *et al.*, 2003). Like siRNA, the cytoplasmic pre-miRNA is cleaved by Dicer to the mature miRNA molecule characterized by a 5′ phosphate and two-nucleotide 3′ overhang (Basyuk *et al.*, 2003; Lee *et al.*, 2003). In animal models, miRNAs enter the RNAi pathway and target specific sequences primarily through imperfect complementarity, mediat-ing gene silencing by translation attenuation (reviewed in Bartel, 2004). Thus, although miRNAs and siRNAs have different origins, in their mature form their paths converge at Dicer and RISC processing steps, indicating that attributes desirable for miRNA processing are likely to be common to functional siRNAs.

To assess the degree to which miRNAs and siRNAs might share key functional characteristics, we systematically compared the thermodynamic profiles (Khvorova *et al.*, 2003) and base preferences (unpublished results) of recently identified miRNAs to a test panel of functional siRNAs. In-terestingly, the one unique feature shared by both groups was the preference for a flexible A/U bp as the closing nucleotide pair at the 5′ antisense (AS) terminus of the duplex, suggesting a bias in strand stability. By using a biochemical approach, Zamore and colleagues made similar observations and demonstrated that the stability of the 5′ end is one of the primary deter-minants for selective strand association and entry into the pre-RISC complex in *Drosophila* cell lysates (Schwarz *et al.*, 2003). Thus, for both miRNA- and siRNA-mediated silencing, a critical determinant for efficient silencing is an asymmetric internal stability profile (Khvorova *et al.*, 2003; Schwarz *et al.*, 2003). Because internal stability profiles are strong predictors of silencing functionality, several computational algorithms now incorporate these ther-modynamic properties in their siRNA design (e.g., si*DESIGN*™ Center, Dharmacon, Inc., Lafayette, CO; Reynolds *et al.*, 2004).

When base preferences were compared for functional siRNAs and miRNAs (aligned by the 5′end of mature miRNAs; Reynolds *et al.*,

2004), no statistically significant similarities were identified in the regions beyond those identified at the 5'AS ends (preference for an A/U closing pair). This suggests that internal siRNA and miRNA characteristics probably represent factors that determine the specific mechanism of target recognition and gene silencing; that is, complete complementarity leads to mRNA cleavage, whereas only partial complementarity results in translation attenuation.

Thermodynamic and Sequence-Based Determinants for Short Interfering RNA (siRNA)

The comparative thermodynamic profile studies described previously demonstrate that low internal stability of the 5'AS end of the siRNA duplex promotes functionality; however, this attribute alone is clearly not sufficient. Further analyses of multiple siRNAs with low internal stability at the 5'AS end revealed that not all were functional for silencing their mRNA target (Khvorova et al., 2003; Reynolds et al., 2004). To identify additional siRNA features that promote efficient silencing and to quantify the importance of certain currently accepted conventional factors such as G/C content and target site accessibility, we analyzed the silencing potential of 180 siRNAs that targeted every other position of a 197-base region in firefly luciferase (fLuc) and human cyclophilin (hCyclo) (90 duplexes each; Reynolds et al., 2004). The 180 sequential duplexes exhibited widely varying silencing abilities, demonstrating that incremental two-base shifts in target position were sufficient to significantly alter siRNA functionality (Fig. 3). These results support the notion that the siRNA sequence plays a key role in determining functionality and that the local mRNA context is a less critical factor. Moreover, these observations further reinforce the distinction between RNAi and antisense or ribozymes technologies in which function is associated with target site accessibility (Ding and Lawrence, 2001).

Based on analysis of the 180 sequences described previously, eight thermodynamic and sequence-related criteria associated with siRNA functionality were defined. Included among these are three thermodynamic criteria: moderate to low G/C content, low internal stability of the 3' end (5'AS), and a lack of internal repeats. Sequence-related determinants were also defined and include three positive and two negative discriminants for specific base positions relative to the sense strand: an A at positions 3 and 19, a U at position 10, the absence of G or C at position 19, and any base other than G at position 13 (Table I; Reynolds et al., 2004).

In addition to their association with functionality, the determinants defined previously may also be considered in the context of their

TABLE I
EIGHT CRITERIA DERIVED FROM CORRELATING BIOPHYSICAL AND
SEQUENCE-RELATED PROPERTIES OF siRNA WITH SILENCING

Criterion	% Functional		Improvement over random
I. 30–52% G/C content	< F50	16.4%	−3.6
	≥ F50	83.6%	3.6
	≥ F80	60.4%	4.3
	≥ F95	23.9%	2.2
II. At least 3 A/U bases at positions 5–19 of the sense strand	< F50	18.2%	−1.8
	≥ F50	81.8%	1.8
	≥ F80	59.7%	3.6
	≥ F95	24.0%	2.3
III. Absence of internal repeats, as measured by T_m of secondary structure $\leq 20°$	< F50	16.7%	−3.3
	≥ F50	83.3%	3.3
	≥ F80	61.1%	5.0
	≥ F95	24.6%	2.9
IV. An A base at position 19 of the sense strand	< F50	11.8%	−8.2
	≥ F50	88.2%	8.2
	≥ F80	75.0%	18.9
	≥ F95	29.4%	7.7
V. An A base at position 3 of the sense strand	< F50	17.2%	−2.8
	≥ F50	82.8%	2.8
	≥ F80	62.5%	6.4
	≥ F95	34.4%	12.7
VI. A U base at position 10 of the sense strand	< F50	13.9%	−6.1
	≥ F50	86.1%	6.1
	≥ F80	69.4%	13.3
	≥ F95	41.7%	20.0
VII. A base other than G or C at position 19 of the sense strand	< F50	18.8%	−1.2
	≥ F50	81.2%	1.2
	≥ F80	59.7%	3.6
	≥ F95	24.2%	2.5
VIII. A base other than G at position 13 of the sense strand	< F50	15.2%	−4.8
	≥ F50	84.8%	4.8
	≥ F80	61.4%	5.3
	≥ F95	26.5%	4.8

Note: These criteria assume that siRNAs consist of 19 bp with two-base 3′ overhangs.
Source: Reprinted with permission from Reynolds *et al.* (2004).

mechanistic roles in the RNAi pathway. For example, for initial duplex recognition, four of the criteria (low GC content, flexible 5′AS end, A at position 19, and any base but G or C at position 19), may be envisioned as having critical roles at the RISC activation and duplex unwinding step, whereas the remaining three (U at position 10, A at position 3, and any base but G at position 13) may play a role in target cleavage catalysis

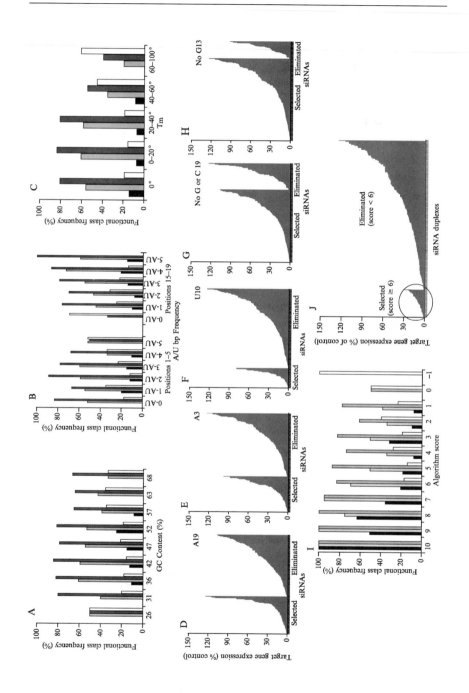

(Fig. 1). When considered in this light, integration of all eight criteria into a single algorithm for evaluation of the panel of 180 duplexes clearly improves reliable selection of functional duplexes (Fig. 4A–J). These studies highlight the power of rational selection as an important molecular tool for facilitating functional genomics studies of virtually any gene and as a potential means to elucidate mechanistic features that underlie the RNAi process at each step.

Rational Design of siRNA

Application of any rational siRNA design algorithm is best implemented by first evaluating an mRNA or cDNA sequence of interest for all possible target sites. For instance, using the strategy described in the previous section as an example, the test panel of 180 duplexes targeting *fLuc* and *hCyclo* were scored and sorted first by individual criteria (Fig. 4A–H) and then with all criteria integrated into a single algorithm (4I–J), using the following weighted criteria. Satisfaction of positive determinants defined earlier earn one point each: low G/C content (Criterion I), a lack of internal repeats (Criterion III), an A at position 19 (Criterion IV), an A at position 3 (Criterion V), and a U at position 10 (Criterion VI). Failure to satisfy the subsequent negative determinants results in a one-point decrease: the absence of C at positions 13 (Criterion VIII) and the absence of G or C at position 19 (Criterion VII). For Criterion II, 1 point may be added for each A or U base present in positions 15–19 (thus, up to 5 points can be earned). By applying this scoring system to the test panel of 180 siRNAs, data revealed that all duplexes scoring at least 6 points were functional (Fig. 4I–J). In contrast, the percentage of nonfunctional siRNAs increased as the selection score decreased from 5 to –1. Among duplexes scoring 5, 20% were found to be nonfunctional (<F50) whereas 100% of those scoring –1 were <F50. As a result, a threshold score for selecting functional duplexes was set at 6. All duplexes scoring 6

FIG. 4. Rational siRNA design: eight thermodynamic and sequence-specific criteria identified by Reynolds *et al.* (2004) and their impact on siRNA functionality on the test panel of 180 siRNAs. siRNAs were first sorted according to each individual criterion: (A) GC content, (B) frequency of A/U bp in positions 1–5 and 15–19, (C) stability of internal repeats measured as T_m, (D) A in position 19, (E) A in position 13, (F) U in position 10, (G) the absence of G or C in position 19, and (H) the absence of G in position 13. (*Note:* All positions are reported for the sense strand.) When integrated into a single weighted algorithm and sorted into subgroups on a scale of 10 to −1 (I), the functional classes (>6) comprised predominately functional siRNAs (≤F50■, ≥F50□, ≥F80□, ≥F95■). (J) The 180 siRNA test panel was subsequently sorted according to their individual functionality into selected siRNAs (left) having a score >6 and eliminated siRNAs (right) having a score of <6.

or more points (15.5% of the test panel) fell into the selected group (Fig. 4J). All remaining duplexes (84.5% of the test panel) were eliminated. Notably, among the low-scoring class, some siRNAs were, in fact, functional (>F70), demonstrating that identification and incorporation of additional key factors would likely improve the reliability of selecting functional duplexes.

Validation of the Selection Algorithm

Validation of the previously described algorithm with the full complement of eight criteria was achieved by applying it to potential target sites derived from the complete mRNA sequences for six genes: human diazepam binding inhibitor (*DBI*), firefly luciferase (*fLuc*), *Renilla* luciferase (*rLuc*), human polo-like kinase (*PLK*), human secreted alkaline phosphatase (*SEAP*), and glyceraldehyde-3-phosphate dehydrogenase (*GAPDH*). For each gene, a minimum of four siRNAs duplexes were selected with scores of 6 or higher and a BLAST analysis was conducted for each. To minimize the potential for off-target silencing effects, only those target sequences with more than three mismatches against unrelated sequences were selected (Semizarov *et al.*, 2003). As control duplexes, several siRNAs targeting each gene were selected at random and assayed in parallel for functionality. The results are indicative of the predictive value of the algorithm, as 29 of 30 rationally designed siRNAs were functional (Reynolds *et al.*, 2004). On average, the algorithm significantly improved the probability of selecting an F50 siRNA from 47% based on random selection to 94% and improved the likelihood of achieving 80% knockdown by more than fivefold (from 12% for random selection to 64% for the algorithm). In a more comprehensive analysis, 360 additional siRNAs for which function was known were assessed according to their algorithm-assigned score. When shown as a scatter plot (Fig. 5), a clear trend correlating score value with functionality was observed. The majority of duplexes ranking higher than 7 were functional. Notably, a number of low-scoring functional duplexes were also eliminated, suggesting that additional critical factors remain to be defined and incorporated into the algorithm.

siRNA Specificity

Reliable selection of potent siRNAs represents one aspect of successful computational design strategies. Another critical facet is the ability to ensure specificity of the siRNA for its cognate target mRNA. Legitimate concerns were raised by several microarray studies that assessed the effects

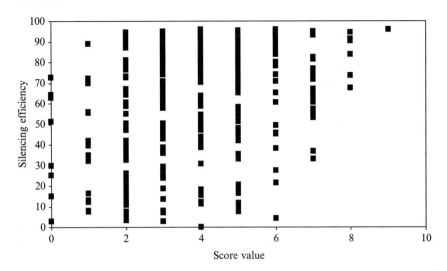

F<small>IG</small>. 5. Validation of the eight-component design algorithm. The 360 independent siRNAs used to validate the algorithm were plotted according to assigned score and silencing functionality. Each dot represents the activity of a single siRNA. All siRNAs are ranked based on the scoring system described here and in Reynolds *et al.* (2004). High scoring duplexes correlate with functionality. Low scores correlate well with poor to moderate silencing, although some functional duplexes also had low scores, indicating room for improving the predictive value of the algorithm.

of siRNA on a genomewide scale (Jackson *et al.*, 2003; Semizarov *et al.*, 2003). In one important study, Jackson and coworkers tested convention-ally designed siRNAs and found that the resulting off-target effects could be attributed to partial complementarity of the duplex to nontargeted mRNAs (Jackson *et al.*, 2003). As few as 11 bases of contiguous identity appeared to be sufficient for inducing off-target recognition and cleavage and in most cases led to only a two- to threefold change in off-target transcript levels. Interestingly, target sites sharing 15, 14, 13, 12, and 11 bases of identity with unrelated sequences might be expected to occur for 3, 11, 45, 179, and 715 sites, respectively, within a genome of approximately 3 billion bases. The actual number of off-target effects observed by Dr. Jackson's group was much less than that expected based on partial complementarity alone, demonstrating that 11 bases of identity are sufficient for gene silencing in the context of some but not all genes.

In concurrent studies, Semizarov *et al.* (2003) combined conventional design methods with a basic bioinformatic approach to select for highly specific candidate siRNAs. The basis of the selection was to identify those siRNAs with a significant thermodynamic differential between the intended targeted site and next best potential off-target sequence. This bioinformatic filter substantially diminished the level of undesired off-target effects. Together with Jackson's results, these studies highlight the limitations of conventional design methods and the necessity for careful bioinformatics to ensure siRNA specificity and potency.

Sequence Comparisons: Key Considerations

The most commonly used analytical tool to ensure siRNA target specificity is the basic local alignment search tool BLASTn, which is used specifically to compare a nucleotide query sequence to a nucleotide database. The BLAST family of programs (BLASTn, BLASTp, BLASTx, tBLASTn, tBLASTx) was initially developed for quick identification of significant sequence similarities between relatively long stretches of nucleic acids and proteins (http://www.ncbi.nlm.nih.gov/BLAST/; Altschul *et al.*, 1990; Madden *et al.*, 1996). Because of the rapid growth of existing databases and the widespread use of BLASTn to screen for potential undesirable matches, a basic understanding of the sequence sources and databases suitable for searching is helpful. In addition, familiarity with key BLASTn parameters helps understand how to optimize similarity searches and to determine the significance of matches for a unique siRNA relative to its specific target and potential off-target sequences.

Sequence Database Selection. The specificity of any given siRNA will only be as good as the database of mRNA sequences against which an analysis by BLASTn is performed. Incomplete or poorly annotated databases may result in the selection of siRNAs that inadvertently have high similarity to other genes. Recommended sources of dependable information are curated, organism-specific databases of expressed sequences, many of which are hosted by the National Center for Biotechnology Information (NCBI), the Sanger Center, the DNA Data Bank of Japan (DDBJ), or the European Molecular Biology Laboratory (EMBL). The Reference Sequence (RefSeq) (http://ncbi.nlm.nih.gov/RefSeq/), Locus-Link (www.ncbi.nlm.nih.gov/LocusLink/), or Unigene (www.ncbi.nlm.nih. gov/Unigene/) databases are representative of such repositories of non-redundant expressed sequence information. Preferred sequence records for mRNAs are the annotated entries defined by a unique alphanumeric code [e.g., NM_012345 (RefSeq)] rather than single, unconfirmed expressed sequence tag (EST) entries. It is also important to have reliable (and complete, if possible) mRNA (or cDNA) sequence information from the genome of choice. Because of codon degeneracy, predicted mRNA sequences derived from protein sequences may not reflect the true expressed transcript and should be avoided. Similarly, mRNA sequences predicted from genomic sequences may lead to duplexes targeting introns or noncoding regions and should also be avoided (Fire *et al.*, 1998). Finally, it is important to understand that sequence databases remain dynamic and are continually updated and refined with the latest information. As a result, comprehensive design tools must be flexible and equally dynamic so as to adjust to the continual flux in information.

Sequence Alignments: BLASTn Parameters and Analysis. The output of a BLASTn comparative analysis is governed by parameters, including those used to set the desired search window for identity (word size) and to establish the significance threshold for matches in a given database (expect value). The word size represents the minimal sequence seed or string used by BLASTn as the basis for identifying and evaluating all possible significant similarities in a particular sequence database. The BLASTn program permits selection of one of three standard word sizes of 7, 11 (default setting), and 15. For siRNA queries, we recommend setting the word size at the lowest setting possible, which for BLASTn is 7, to maximize the search for, and output of, short sequence similarities. Short sequence stretches that are less than 20 nt, like siRNAs, represent a challenge because the number of potential matches by chance alone can be very high for short stretches. Under these circumstances, to optimize the significance of the matches identified by BLASTn, the E or expect value should be set relatively high (e.g., >1000). The E value establishes a threshold of significance for the number of matches in a BLASTn search that may be expected to occur by chance in a database of a given size. Setting a low E value (e.g., 10) is too stringent for short sequences, potentially ignoring sequences with similarities significant for siRNA specificity. Setting a high E value is less stringent but places more significance on suboptimal alignments that may represent potential off-target sequences.

The ability to identify suboptimal alignments, or stretches of imperfect similarity, is a critical consideration as there is evidence for off-target silencing by siRNAs despite the presence of multiple mismatches (mm) or bulges (Jackson *et al.*, 2003). For example, a duplex with a minimal uninterrupted region of identity of less than seven or eight bases (e.g., a 19-mer with mm at positions 7, 13, and 14) might be functional for suppressing expression but would not be identified by conventional BLASTn analysis. It is clear from early studies that mm in position 9 and 10 relative to the 5′AS strand results in almost complete elimination of targeted gene silencing (Elbashir *et al.*, 2001c); thus, potential off-targets sequences with a mm in these positions may be ignored (unpublished observations). Similarly, identities in the 5′ region (antisense strand) were shown to have a greater impact on off-target effects (Jackson *et al.*, 2003). Therefore, identities to the 5′AS region of an siRNA should be assigned higher significance than identities to the 3′AS end of siRNAs, an attribute that current BLASTn analyses underrepresent.

Once the appropriate database and key parameters have been established, one may submit a list of siRNAs to BLASTn as queries. A quick way to evaluate a long list of siRNA candidates is to submit them as a concatenated string, in which each sequence is separated by a string of Ns (20 or more). BLASTn will search and return matches for both sense and antisense orientations (e.g., Plus/Plus and Plus/Minus). The major challenge for directly querying EST databases with siRNAs through the NCBI Web site is the resulting matches to multiple anonymous entries for the same gene. This list will require visual inspection and further manual comparisons to confirm sequence identities. As curation efforts progress, many of these anonymous entries will be assigned to a specific gene and will be annotated appropriately. If BLASTn is performed locally by using downloaded programs and databases, scripts may be written to reconcile related gene clusters (i.e., Unigene entries) so that matches against different sequence entries for the same gene may be easily recognized and ignored. For a locally installed BLASTn, the turnaround time may be a matter of seconds for each siRNA sequence evaluated, depending on the computing resources of the laboratory or research facility. This process can be relatively time consuming when evaluating hundreds to thousands of potential siRNAs candidates for any given gene target or list of targets.

In spite of the popularity of BLASTn as a screen for specificity in siRNA design, its value is somewhat restricted. Because of its limited ability to search for suboptimal alignments or assign significance to mm or bulges in specified positions as described previously, more sophisticated tools may be required. However, standard BLASTn analysis remains useful as a quick method to evaluate and eliminate all significant matches with uninterrupted identities.

Sequence Alignments: Smith–Waterman Dynamic Programming Algorithm. The Smith–Waterman (S–W) dynamic programming sequence alignment algorithm represents a more comprehensive comparison method than BLASTn (Gotoh, 1982; Smith and Waterman, 1981). This algorithm computes a mathematically optimal alignment for a query sequence compared with each sequence record in a given database and assigns a matching score. It ensures that statistically significant matches, no matter how remote, are included and reported. The S–W algorithm is also capable of identifying thermodynamically favorable interactions between siRNA candidates and mRNA sequences throughout the evaluation process. The major limitation of the S–W algorithm is the computational time required to perform the necessary calculations (hours to days). This constraint may be alleviated in part by the development of specialized supercomputers such as GeneMatcher2 (Paracel), designed to accelerate complex computational tasks such as S–W analyses.

Integration of rational design strategies for potency with comparative analyses for specificity promises to improve reliable selection of target-specific siRNAs. One intriguing consideration is the development of first approximation methods to facilitate similarity searches. For example, as illustrated in Fig. 6, a given mRNA target could be evaluated by rational design methods for a list of siRNA candidates. Subsequently, the list of candidates would be subjected to multiple sequence alignment tools that permit sequential (or parallel) analyses of candidates. The end result would be elimination of duplexes found to have similarity to potential off-target sequences that might have been overlooked by a single analytical tool. Thus, the process of combining algorithms into a hybrid tool represents a potentially viable and powerful compromise among rational design, rapid alignment, and thermodynamic comparisons. The final output quality thus offers the best balance in time and cost-efficient solutions for siRNA similarity searches.

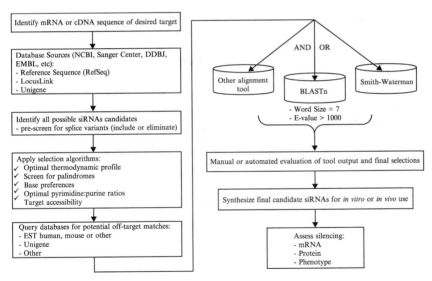

FIG. 6. Flow chart illustrating the process for identifying an siRNA. The desired mRNA or cDNA sequence is found by using a variety of database sources, including entries in RefSeq, LocusLink, or Unigene databases. A list of all potential siRNAs is evaluated by rational selection methods that assess thermodynamic properties, base preferences and composition, and target accessibility. The subsequent list of optimized candidates is used to query the appropriate organism-specific database to further cull the list to unique sequences. This is achieved through bioinformatic analysis, using comparative sequence tools such as BLASTn, S–W, or hybrid algorithms. The final list of siRNA candidates can then be synthesized and tested for silencing.

Available siRNA Design Tools

In addition to NCBI, there are several open domain or publicly available resources at which the BLASTn or BLASTn-style algorithms are available and customized for siRNA searches. In some instances, gene-specific siRNAs have been preselected and analyzed by BLASTn for most, if not all, gene entries in the RefSeq database. Many of the siRNA suppliers, as well as some academic institutions, provide design tools that typically conform to conventional guidelines, providing researchers with quick, simple computational tools for inputting gene sequences (or accession numbers) to quickly identify a list of potential siRNAs (Table II). These design tools rely primarily on conventional guidelines for selecting duplexes (Elbashir *et al.*, 2002). Although many siRNAs selected using these guidelines exhibit silencing, the probability of reliably identifying potent siRNAs is only marginally improved over random selection methods (Hsieh *et al.*, 2004). Of the original guidelines, a refined GC content range remains important (36–52%) (Reynolds *et al.*, 2004). However, the remaining requirements including the dinucleotide leader sequence and the location of the siRNA target sequence within the mRNA, ORF, 5'UTR, or 3'UTR have not been found to contribute significantly to the selection of functional siRNAs (Hsieh *et al.*, 2004; Yoshinari *et al.*, 2004). Most publicly available design tools also eliminate siRNAs with sequence motifs or stretches of repeating nucleotides (more than three) as well as those siRNAs with significant internal structure.

Subsequent screens of sequences that meet the conventional criteria described previously are relatively straightforward. The primary cost is the computational time for automated BLASTn comparisons to identify

TABLE II
URLs FOR siRNA DESIGN TOOLS

Source	Web site
Ambion	www.ambion.com
Dharmacon	www.dharmacon.com
Integrated DNA Technologies	Biotools.idtdna.com
Iris Genetics	www.irisgenetics.com
Qiagen	www.qiagen.com
Oligoengine	www.oligoengine.com
Open Biosystems	www.openbiosystems.com
Sfold	Sfold.wadsworth.org
The Whitehead Institute	jura.wi.mit.edu/pubint/
	www.iona.wi.mit.edu/siRNAext

potential off targets for the candidate siRNAs. Usually, siRNAs are evaluated and assigned a BLASTn score such that all candidates with more than 16 or 17 bases of identity to other unrelated genes are eliminated. Alternatively, a limited number of proprietary custom design services incorporate additional selection criteria, including optimal target accessibility, specific pyrimidine or purine ratios, optimal thermodynamic properties, and base preferences within the duplex followed by extensive bioinformatics analyses. The end result has a far higher success rate than most publicly available tools (Reynolds *et al.*, 2004; Table II).

Concluding Remarks

We now have a preliminary understanding of attributes important for siRNA functionality; however, what fully governs the potency and specificity remains to be fully elucidated. By identifying factors important for siRNA–RISC, siRNA–target, and siRNA–off-target interactions, we can begin to develop more sophisticated and more accurate algorithms that ensure selection of the most specific sequences possible. With these tools in hand, researchers will be able to make educated predictions for the siRNAs most useful for their particular experiments of interest.

The discussion presented here should provide a basis for understanding RNAi and provide insight into attributes that contribute to siRNA functionality and specificity. Incorporating these attributes into predictive algorithms permits reliable selection of potent gene-specific siRNAs. Practical considerations for manipulating existing tools are presented to help educate the researcher with regard to the power and limitations of existing tools. The ultimate goal is to develop a better understanding of the RNAi mechanism and to integrate this information into more advanced sequence comparison tools so as to ensure target specificity and minimize off-target events. With the increase in available datasets and more sophisticated algorithms, more reliable predictive algorithms will be developed.

Experimental Procedures

siRNA Nomenclature

All siRNA duplexes and relative positions are referred to by the sense strand. Position 1 of the 5' sense strand corresponds to position 19 of the antisense strand. Silencing was determined at the mRNA level or by enzymatic activity 24 h following transfection with 100 n*M* siRNA. Transfection efficiencies were at least 95%, and no detectable cellular toxicity was observed. siRNA-silencing functionality is reported as F followed by the minimal degree of knockdown; for example, F50 signifies at least 50% knockdown and F80 means at least 80%. Typically, all sub-F50 siRNAs are considered nonfunctional.

Design and Synthesis of siRNA Duplexes

The test panel of siRNAs was composed of duplexes targeting every other position of the following 197 base regions of human cyclophilin B and firefly luciferase:

Human cyclophilin: 193–390, M60857

GTTCCAAAAACAGTGGATAATTTTGTGGCCTTAGCTACAGGAGAGAAAGGATTTGGCTA
CAAAAACAGCAAATTCCATCGTGTAATCAAGGACTTCATGATCCAGGGCGGAGACTTCAC
CAGGGGAGATGGCACAGGAGGAAAGAGCATCTACGGTGAGCGCTTCCCCGATGAGAACTT
CAAACTGAAGCACTACGGGCCTGGCTGGG

Firefly luciferase: 1434–1631, U47298 (pGL3, Promega, Madison, WI)

TGAACTTCCCGCCGCCGTTGTTGTTTTGGAGCACGGAAAGACGATGACGGAAAAAGA
GATCGTGGATTACGTCGCCAGTCAAGTAACAACCGCGAAAAAGTTGCGCGGAG
GAGTTGTGTTTGTGGACGAAGTACCGAAAGGTCTTACCGGAAAACTCGACGCAAGAAAAAT
CAGAGAGATCCTCATAAAGGCCAAGAAGG

The algorithm was further validated with siRNAs targeting six genes and were selected at random or by rational design. Random selection involves identification of target sites at arbitrary sites in the coding sequence. Ten additional sites were selected for GAPDH. All positions are reported relative to the start codon. All duplexes were synthesized in-house as 21-mers with 3′dTdT overhangs, using a modified method of 2′-ACE chemistry (Scaringe, 2000): *Renilla* luciferase (*rLuc*, AK025845) bp 174, 300, 432, 568, 592, 633, 729, 867, 495, 808, 599, 25; firefly luciferase (*fLuc*, U47296) bp 448, 750, 1196, 1203, 1212, 1314, 206, 893, 1313, 1604; secreted alkaline phosphatase (*SEAP*, NM_031313) 138, 698, 744, 855, 1049, 1212, 1232, 1419, 1230, 1544, 1211, 148; polo-like kinase (*PLK*, X75932) bp 291, 684, 794, 1310, 1441, 245, 554, 751, 1424; diazepam-binding inhibitor (*DBI*, NM_020548.2) bp 255, 355, 472, 478, 287, 261, 254, 263; glyceraldehyde-3-phosphate dehydrogenase (*GAPDH*, NM_002046) bp 337, 355, 375, 395, 415, 435, 455, 475, 495, 515, 401, 479, 343, 347, 389, 407, 409, 417, 419, 421.

Cell Culture and Transfection

Cells were plated in 96-well plates pretreated with 50 μl of 50 mg/ml poly-L-lysine (Sigma, St. Louis, MO) for 1 h, washed thrice with distilled water, and air dried for 20 min. HEK293 cells or HEK293Lucs were trypsinized and plated at 3.5×10^4 cells/well in 100 μl of medium plus serum and were incubated overnight at $37°$, 5% CO_2. Transfection mixtures of 2 ml Opti-MEM I (Gibco-BRL, Carlsbad, CA), 80 μl Lipofectamine™ 2000 (Invitrogen, Carlsbad, CA), 15 μl SUPERNasin at 20 U/μl (Ambion, Austin, TX), and 1.5 μl reporter gene plasmid at 1 μg/μl were prepared in 5-ml polystyrene round-bottom tubes. A 100-μl aliquot of the transfection reagent was mixed with 100 μl of siRNAs in polystyrene deep-well titer plates (Beckman, Fullerton, CA) and incubated for 20–30 min at room temperature. The resulting complexes were diluted with 550 μl of Opti-MEM per well to a final siRNA concentration of 100 nM. Plates were then sealed with parafilm and mixed gently. Media for the HEK293 cells was replaced with 95 μl of transfection mixture and the cells incubated overnight at $37°$, 5% CO_2.

Cytotoxicity Analysis

To each well, 25 μl of AlamarBlue reagent (Trek Diagnostic Systems, Cleveland, OH) was added and the HEK293 cells were incubated for 2 h at $37°$, 5% CO_2. Each well was then read at 570 nm, using a 600-nm subtraction. The optical density (OD) is proportional to the number of viable cells in culture when the reading is in the linear range (0.6–0.9). Thus, wells with an OD reading $\geq 80\%$ of the control were considered nontoxic.

Quantitation of Gene Knockdown

mRNA levels were measured 24 h posttransfection, using the Quantigene® branched-DNA (bDNA) kits (Genospectra, Fremont, CA; Wang *et al.*, 1997) according to manufacturer's instructions. Luciferase activity was measured 24 h posttransfection, using the Steady-GLO® reagent (Promega, Madison, WI) according to manufacturer's instructions.

Calculation of T_m

T_m, the predicted melting temperature of the siRNA hairpin loop, was calculated by the following equation: $T_m = (T_{G0} + 2.1)(\log[\text{salt}])(n - 1)$, where T_{G0} is the temperature at which ΔG of the structure is equal to 0 and n is the length of the hairpin loop stem (Groebe and Uhlenbeck, 1988). The Oligo 6.0 software package was used.

Calculation of mRNA Target Site Accessibility

Predicted mRNA target site accessibility was calculated by using *S*fold (Ding and Lawrence, 2001; http://www.wadsworth.org/resnres/bioinfo/sfold/manual.html). The RNA folding algorithm generates a statistical sample of secondary structures from the Boltzmann ensemble of RNA secondary structures.

Acknowledgments

We thank Amanda Birmingham and Jon Karpilow for their critical reviews and helpful discussions and Julia Kendall for help with manuscript preparation.

References

Altschul, S. F., Gish, W., Miller, W., Myers, E. W., and Lipman, D. J. (1990). Basic local alignment search tool. *J. Mol. Biol.* **215,** 403–410.

Aza-Blanc, P., Cooper, C., Wagner, K., Batalov, S., Deveraux, Q., and Cooke, M. (2003). Identification of modulators of TRAIL-induced apoptosis via RNAi-based phenotypic screening. *Mol. Cell* **12,** 627–637.

Bartel, D. (2004). MicroRNAs: Genomics, biogenesis, mechanism, and function. *Cell* **116,** 281–297.

Basyuk, E., Suavet, F., Doglio, A., Bordonne, R., and Bertrand, E. (2003). Human let-7 stem-loop precursors harbor features of RNase III cleavage products. *Nucleic Acids Res.* **31,** 6593–6597.

Bernstein, E., Caudy, A. A., Hammond, S. M., and Hannon, G. J. (2001). Role for a bidentate ribonuclease in the initiation step of RNA interference. *Nature* **409,** 363–366.

Bohula, E. A., Salisbury, A. J., Sohail, M., Playford, M. P., Riedemann, J., Southern, E. M., and Macaulay, V. M. (2003). The efficacy of small interfering rnas targeted to the type 1 insulin-like growth factor receptor (IGF1R) is influenced by secondary structure in the IGF1R transcript. *J. Biol. Chem.* **278,** 15991–15997.

Dalmay, T., Hamilton, A., Rudd, S., Angell, S., and Baulcombe, D. C. (2000). An RNA-dependent RNA polymerase gene in *Arabidopsis* is required for posttranscriptional gene silencing mediated by a transgene but not by a virus. *Cell* **101,** 543–553.

Ding, Y., and Lawrence, C. E. (2001). Statistical prediction of single-stranded regions in RNA secondary structure and application to predicting effective target sites and beyond. *Nucleic Acids Res.* **29,** 1034–1046.

Ding, Y., and Lawrence, C. E. (2003). A statistical sampling algorithm for RNA secondary structure prediction. *Nucleic Acids Res.* **31,** 7280–7301.

Elbashir, S. M., Harborth, J., Lendeckel, W., Yalcin, A., Weber, K., and Tuschl, T. (2001a). Duplexes of 21-nucleotide RNAs mediate RNA interference in cultured mammalian cells. *Nature* **411,** 494–498.

Elbashir, S. M., Harborth, J., Weber, K., and Tuschl, T. (2002). Analysis of gene function in somatic mammalian cells using small interfering RNAs. *Methods* **26,** 199–213.

Elbashir, S. M., Lendeckel, W., and Tuschl, T. (2001b). RNA interference is mediated by 21- and 22-nucleotide RNAs. *Genes Dev.* **15,** 188–200.

Elbashir, S. M., Martinez, J., Patkaniowska, A., Lendeckel, W., and Tuschl, T. (2001c). Functional anatomy of siRNAs for mediating efficient RNAi in *Drosophila melanogaster* embryo lysate. *EMBO J.* **20,** 6877–6888.

Fire, A., Xu, S., Montgomery, M. K., Kostas, S. A., Driver, S. E., and Mello, C. C. (1998). Potent and specific genetic interference by double-stranded RNA in *Caenorhabditis elegans*. *Nature* **391,** 806–811.

Gotoh, O. (1982). An improved algorithm for matching biological sequences. *J. Mol. Biol.* **162**, 705–708.

Groebe, D. R., and Uhlenbeck, O. C. (1988). Characterization of RNA hairpin loop stability. *Nucleic Acids Res.* **16**, 11725–11735.

Hamilton, A. J., and Baulcombe, D. C. (1999). A species of small RNA in posttranscriptional gene silencing in plants. *Science* **286**, 950–952.

Holen, T., Amarzguioui, M., Wiiger, M. T., Babaie, E., and Prydz, H. (2002). Positional effects of short interfering RNAs targeting the human coagulation trigger tissue factor. *Nucleic Acids Res.* **30**, 1757–1766.

Hsieh, A. C., Bo, R., Manola, J., Vazquez, F., Bare, O., Khvorova, A., Scaringe, S., and Sellers, W. R. (2004). A library of siRNA duplexes targeting the phosphoinositide 3-kinase pathway: Determinants of gene silencing for use in cell-based screens. *Nucleic Acids Res.* **32**, 893–901.

Jackson, A. L., Bartz, S. R., Schelter, J., Kobayashi, S. V., Burchard, J., Mao, M., Li, B., Cavet, G., and Linsley, P. S. (2003). Expression profiling reveals off-target gene regulation by RNAi. *Nat. Biotechnol.* **21**, 635–637.

Khvorova, A., Reynolds, A., and Jayasena, S. (2003). Functional siRNAs and miRNAs exhibit strand bias. *Cell* **115**, 209–216.

Kretschmer-Kazemi Far, R., and Sczakiel, G. (2003). The activity of siRNA in mammalian cells is related to structural target accessibility: A comparison with oligonucleotides. *Nucleic Acids Res.* **31**, 4417–4424.

Krichevsky, A. M., and Kosik, K. S. (2002). RNAi functions in cultured mammalian neurons. *Proc. Natl. Acad. Sci. USA* **99**, 11926–11929.

Lagos-Quintana, M., Rauhut, R., Lendeckel, W., and Tuschl, T. (2001). Identification of novel genes coding for small expressed RNAs. *Science* **294**, 853–858.

Laposa, R. R., Feeney, L., and Cleaver, J. E. (2003). Recapitulation of the cellular xeroderma pigmentosum-variant phenotypes using short interfering RNA for DNA polymerase H. *Cancer Res.* **63**, 3909–3912.

Lau, N. C., Lim, L. P., Weinstein, E. G., and Bartel, D. P. (2001). An abundant class of tiny RNAs with probable regulatory roles in *Caenorhabditis elegans*. *Science* **294**, 858–862.

Lee, N. S., Dohjima, T., Bauer, G., Li, H., Li, M. J., Ehsani, A., Salvaterra, P., and Rossi, J. (2002). Expression of small interfering RNAs targeted against HIV-1 rev transcripts in human cells. *Nat. Biotechnol.* **20**, 500–505.

Lee, R., Feinbaum, R., and Ambros, V. (1993). The *C. elegans* heterochronic gene lin-4 encodes small RNAs with complementarity to lin-14. *Cell* **75**, 843–854.

Lee, R. C., and Ambros, V. (2001). An extensive class of small RNAs in *Caenorhabditis elegans*. *Science* **294**, 862–864.

Lee, Y., Ahn, C., Han, J., Choi, H., Kim, J., Yim, J., Lee, J., Provost, P., Radmark, O., Kim, S., and Kim, V. (2003). The nuclear RNase III Drosha initiates microRNA processing. *Nature* **425**, 415–419.

Lund, E., Guttinger, S., Calado, A., Dahlberg, J. E., and Kutay, U. (2004). Nuclear export of MicroRNA precursors. *Science* **303**, 95–98.

Madden, T. L., Tatusov, R. L., and Zhang, J. (1996). Applications of network BLAST server. *Methods Enzymol.* **266**, 131–141.

Matveeva, O. V., Mathews, D. H., Tsodikov, A. D., Shabalina, S. A., Gesteland, R. F., Atkins, J. F., and Freier, S. M. (2003). Thermodynamic criteria for high hit rate oligonucleotide design. *Nucleic Acids Res.* **31**, 4989–4994.

Nykanen, A., Haley, B., and Zamore, P. D. (2001). ATP requirements and small interfering RNA structure in the RNA interference pathway. *Cell* **107**, 309–321.

Pancoska, P., Moravek, Z., and Moll, U. M. (2004). Efficient RNA interference depends on global context of the target sequence: Quantitative analysis of silencing efficiency using Eulerian graph representation of siRNA. *Nucleic Acids Res.* **32,** 1469–1479.

Plasterk, R. H., and Ketting, R. F. (2000). The silence of the genes. *Curr. Opin. Genet. Dev.* **10,** 562–567.

Reynolds, A., Leake, D., Scaringe, S., Marshall, W. S., Boese, Q., and Khvorova, A. (2004). Rational siRNA design for RNA interference. *Nat. Biotechnol.* **22,** 326–330.

Scaringe, S. A. (2000). Advanced 5'-silyl-2'-orthoester approach to RNA oligonucleotide synthesis. *Methods Enzymol.* **317,** 3–18.

Schwarz, D. S., Hutvagner, G., Du, T., Xu, Z., Aronin, N., and Zamore, P. D. (2003). Unexpected asymmetry in the assembly of the RNAi enzyme complex. *Cell* **115,** 199–208.

Semizarov, D., Frost, L., Sarthy, A., Kroeger, P., Halbert, D. N., and Fesik, S. W. (2003). Specificity of short interfering RNA determined through gene expression signatures. *Proc. Natl. Acad. Sci. USA* **100,** 6347–6352.

Sharp, P. A. (2001). RNA interference—2001. *Genes Dev.* **15,** 485–490.

Smith, T. F., and Waterman, M. S. (1981). Identification of common molecular subsequences. *J. Mol. Biol.* **147,** 195–197.

Spankuch-Schmitt, B., Bereiter-Hahn, J., Kaufmann, M., and Strebhardt, K. (2002). Effect of RNA silencing of polo-like kinase-1 (PLK1) on apoptosis and spindle formation in human cancer cells. *J. Natl. Cancer Inst.* **94,** 1863–1877.

Tabara, H., Sarkissian, M., Kelly, W. G., Fleenor, J., Grishok, A., Timmons, L., Fire, A., and Mello, C. C. (1999). The rde-1 gene, RNA interference, and transposon silencing in *C. elegans. Cell* **99,** 123–132.

Vickers, T. A., Koo, S., Bennett, C. F., Crooke, S. T., Dean, N. M., and Baker, B. F. (2003). Efficient reduction of target RNAs by small interfering RNA and RNase H dependent agents: A comparative analysis. *J. Biol. Chem.* **278,** 7108–7118.

Wang, J., Shen, L., Najafi, H., Kolberg, J., Matschinsky, F. M., Urdea, M., and German, M. (1997). Regulation of insulin preRNA splicing by glucose. *Proc. Natl. Acad. Sci. USA* **94,** 4360–4365.

Wightman, B., Ha, I., and Ruvkun, G. (1993). Posttranscriptional regulation of the heterochronic gene lin-14 by lin-4 mediates temporal pattern formation in *C. elegans. Cell* **75,** 855–862.

Xu, Y., Zhang, H. Y., Thormeyer, D., Larsson, O., Du, Q., Elmen, J., Wahlestedt, C., and Liang, Z. (2003). Effective small interfering RNAs and phosphorothioate DNAs have different preferences for target sites in the luciferase mRNAs. *Biochem. Biophys. Res. Commun.* **306,** 712–717.

Yi, R., Qin, Y., Macara, I. G., and Cullen, B. R. (2003). Exportin-5 mediates the nuclear export of pre-microRNAs and short hairpin RNAs. *Genes Dev.* **17,** 3011–3016.

Yoshinari, K., Miyagishi, M., and Taira, K. (2004). Effects on RNAi of the tight structure, sequence and position of the targeted region. *Nucleic Acids Res.* **32,** 691–699.

[6]   Novel Methods for Expressing RNA Interference in Human Cells

By MASAYUKI SANO, YOSHIO KATO, HIDEO AKASHI, MAKOTO MIYAGISHI, and KAZUNARI TAIRA

Abstract

RNA interference (RNAi) is a conserved process in which a double-stranded RNA (dsRNA) induces sequence-specific gene silencing. Recent developments in the use of the 21-nt small interfering RNA (siRNA) have allowed the specific degradation of mRNA without induction of nonspecific effects in mammalian cells. RNAi provides a method for knocking down genes of interest and a powerful tool for studies on gene functions in various organisms. Although many vector-based siRNA expression systems have been developed for production of siRNAs in mammalian cells, many technical issues for an effective production of siRNAs still need to be resolved. In this chapter, we describe methods for construction of genetically stable and highly active siRNA expression systems and also mention some strategies to overcome serious technical problems.

Introduction

RNA interference (RNAi) was first discovered in *Caenorhabditis elegans* as a gene-silencing phenomenon that involves the double-stranded RNA (dsRNA)-mediated cleavage of a cognate mRNA (Fire *et al.*, 1998; Montgomery *et al.*, 1998). Then, it was observed in various organisms such as plants, fungi, flies, and protozoans. In RNAi, dsRNAs introduced into cells by exogenous or endogenous delivery are processed to small RNAs of 21–25 nt in length. An RNase III-like enzyme, identified as Dicer, catalyzes these endonucleolytic cleavages (Bernstein *et al.*, 2001). Subsequently, cleaved forms of small RNAs are incorporated into a multicomponent nuclease complex known as an RNA-induced silencing complex (RISC; Hammond *et al.*, 2000) and the small RNAs act as guide sequence, directing the complex to the target mRNA (Martinez *et al.*, 2002; Nykanen *et al.*, 2001). The mRNA recognized by the RISC is cleaved by the action of an endonuclease within the RISC (Liu *et al.*, 2004; Martinez and Tuschl, 2004; Song *et al.*, 2004).

RNAi can induce sequence-specific degradation by dsRNAs in a variety of organisms, but it was difficult to prove the effect of RNAi in mammalian cells because of the dsRNA-induced interferon response and the activation of dsRNA-dependent protein kinase (PKR), which leads to nonspecific inhibition of protein synthesis. However, it was demonstrated that 21- to 23-nt RNAs with 2-nt overhangs at 3′ ends, referred to as small interfering RNAs (siRNAs), could induce gene silencing without the nonspecific inhibition of gene expression in mammalian systems (Elbashir *et al.*, 2001).

RNAi has been exploited as a powerful tool to silence gene expression and to analyze the biological functions of genes in a wide range of organisms. Many groups have succeeded in achieving gene silencing by synthetic siRNAs (Hammond *et al.*, 2001; McManus and Sharp, 2002), and vector-based siRNA expression systems have shown a powerful potential in regulating the expression of genes of interest (Dykxhoorn *et al.*, 2003). Vector-based systems appear to hold several advantages for applying RNAi *in vivo* compared with synthetic siRNAs. For example, vector-based expression systems would produce siRNAs for long periods of time inside target cells. Moreover, even if the efficiency of transfection of cells with siRNAs is relatively low, only cells harboring vectors that encode siRNA genes can be easily selected by antibiotics when the vectors encode antibiotic-resistance genes. In addition, viral vectors lead the efficient delivery of siRNA expression cassettes into specific cells and tissues. By using retroviral or lentiviral vectors, it is easy to construct stable knockdown cell lines and knockdown animals by integrating the viral vector into the host genome.

Although the mechanism of RNAi is not fully understood, RNAi technologies are rapidly developing through many efforts to exploit the phenomenon of RNAi in mammalian and other systems. Our group has also developed an efficient vector-based system for expressing siRNAs through a series of studies for optimizing the system. We established a system that produces short hairpin-type RNAs under the control of the U6 promoter, and we determined that the hairpin-type RNA genes with mismatches in their stem could prevent mutations during maintenance and amplification in *Escherichia coli*. In addition, we also confirmed that transcribed hairpin RNAs, which contain a loop derived from the sequence of a human microRNA precursor, exhibited strong activity as a suppressor of the activity of target genes. Furthermore, our vector-based short dsRNAs could reduce the interferon response compared with the widely used synthetic short dsRNAs. This chapter describes the methods to construct U6 promoter-based systems and strategies to overcome some technical problems in constructing the efficient siRNA expression systems.

Construction of Effective Small Interfering RNA (siRNA)
  Expression Systems

Pol III-Based siRNA Expression Systems

In early 2002, seven groups, including our group, reported systems for
the expression of siRNAs by exploiting RNA polymerase III (pol III)-
based promoters (Brummelkamp *et al.*, 2002; Lee *et al.*, 2002; Miyagishi and
Taira, 2002; Paddison *et al.*, 2002; Paul *et al.*, 2002; Sui *et al.*, 2002; Yu *et al.*,
2002). In nature, pol III promoters are mainly responsible for producing
short RNAs such as 5S rRNA, U6 snRNA, and tRNA. Because the level of
the expression of transcripts from the pol III promoter is relatively higher
than that of transcripts from the pol II promoter (Cotten and Birnstiel,
1989; Koseki *et al.*, 1999) and pol III-based transcripts are usually small, it is
likely that pol III-based systems are suitable for producing siRNAs. At
present, two pol III-based promoters—U6 and H1 promoters—have been
preferentially exploited for producing siRNAs. These two promoters contain regulatory motifs, a distal sequence element (DSE), a proximal sequence element (PSE), and a TATA box upstream of the transcribing
region. In general, U6 and H1 promoters have limited nucleotides at the
start site of the transcription. Although G and A residues are located at
the nucleotide position of +1 in native U6 and H1 promoters, respectively,
purine residues seem to be useful for a sufficient level of transcription in
both promoters. In addition to these two promoters, other pol III promoters
such as 5S, 7SL, 7SK, and tRNA promoters have also been exploited for
producing siRNAs (Czauderna *et al.*, 2003; Kawasaki and Taira, 2003; Paul
et al., 2003). We reported that the human tRNA^{Val} promoter-based siRNA
expression system can produce short hairpin-type RNAs effectively,
and this system allowed the constitutive transport of transcripts from
the nucleus to the cytoplasm (Kawasaki and Taira, 2003). As it has been
demonstrated that RNAi might operate in the cytoplasm, a tRNA^{Val}-based
system could lead the appropriate location of the transcripts inside cells.
Indeed, hairpin RNAs transcribed from the tRNA^{Val} promoter exhibited
strong activity in cultured cells. Although each pol III-based promoter
appears to have several advantages and disadvantages, we have created an
effective U6 promoter-based system that has been improved to overcome
the practical problems for producing significant active siRNAs.

Construction of U6-Based siRNA Expression Systems

U6-based siRNA expression systems can be divided into two types:
tandem and hairpin. In the tandem-type system, sense and antisense strands
are separately transcribed from two respective promoters (Fig. 1A; Lee *et al.*,

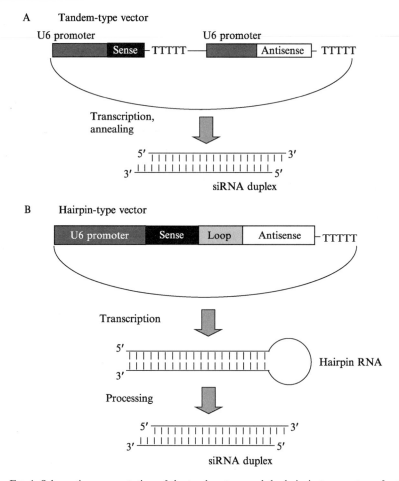

FIG. 1. Schematic representation of the tandem-type and the hairpin-type systems for the expression of siRNAs.

2002; Miyagishi and Taira, 2002). The transcripts can anneal and form siRNA duplexes with overhangs of approximately four U residues at each 3′ end. In contrast, in the hairpin-type system, sense and antisense strands are connected by a loop sequence and are transcribed as a single unit (Brummelkamp *et al.*, 2002; Paddison *et al.*, 2002; Paul *et al.*, 2002; Sui *et al.*, 2002; Yu *et al.*, 2002). The transcripts form a hairpin structure with a stem and a loop, and they can be processed into siRNAs by Dicer within cells (Fig. 1B).

In both tandem-type and hairpin-type systems, we usually use the pi-GENEhU6 plasmid (iGENE Therapeutics, Inc., Japan), which contains a

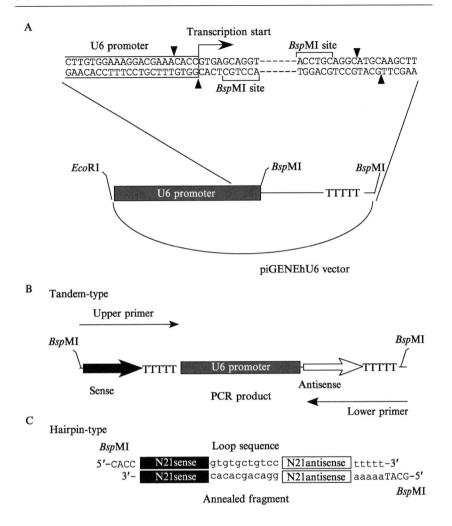

FIG. 2. Construction of U6-based siRNA expression vectors. (A) Cloning sites on the piGENEhU6 vector. The plasmid is digested by *Bsp*MI at specific sites (arrowheads). (B) A DNA insert for construction of the tandem-type siRNA expression vector. The insert is amplified by PCR, using primers and the piGENEhU6 vector as a template. (C) A DNA insert for construction of the hairpin-type siRNA expression vector. Synthesized oligonucleotides are annealed to make the insert.

human U6 promoter and *Bsp*MI recognition sites, to construct siRNA expression vectors (Fig. 2A). For constructing the tandem-type siRNA expression system, specific oligonucleotides are synthesized as primers. For example, when 5'-GTG CGC TGC TGG TGC AAC-3' is used as

a target sequence, the paired primers have the following sequences: 5′-ggc
tct aga ACC TGC cgg cca ccG TGC GCT GCT GGT GCC AAC ttt ttc aat
tca agg tcg ggc ag-3′ and 5′-ggc tct aga ACC TGC tag cgc ata aaa aGT GCG
CTG CTG GTG CCA ACg gtg ttt cgt cct ttc cac aag-3′ (sense or antisense
target sequences underlined). DNA fragments that include the sequences
of the relevant sense and antisense regions and the U6 promoter are
amplified by polymerase chain reaction (PCR), using the piGENEhU6
plasmid as a template and primers, which include the sense or antisense
sequence and the terminator sequence (Fig. 2B). The reaction mixture for
PCR contains 10 μl 10× Pyrobest™ polymerase buffer (Takara Shuzo Co.,
Kyoto, Japan), 0.5 μl Pyrobest polymerase (Takara), 1 μl (100 μM) each set
of primers, 8 μl (2 mM each) dNTP mix, and 79.5 μl distilled water. The
sample is incubated at 95° for 1 min, and then amplification is performed
for 30 cycles (98°, 10 sec; 55°, 1 min; 72°, 1 min). After the products of PCR
are purified, they are digested with the restriction enzyme BspMI (New
England Biolabs Inc., Beverly, MA), and then the fragments are ligated
into the BspMI sites of the piGENEhU6 plasmid to yield a series of siRNA
expression vectors.

   To construct the hairpin-type siRNA expression system, oligonucleotides with the hairpin sequence, terminator sequence, and overhanging
sequences as restriction sites are prepared (Fig. 2C). When the target site
sequence is the same as that given in the example for constructing the
tandem-type system as described previously, oligonucleotides have the
following sequences: 5′-cac cGT GGC TGC TGG TGC CCA ACC Cgt
gtg ctg tcc GGG TTG GGC ACC AGC AGC GCA Ctt ttt-3′ and 5′-gca taa
aaa GTG CGC TGC TGG TGC CCA ACC Cgg aca gca cac GGG TTG
GGC ACC AGC AGC GCA C-3′. Then, 5 μl of each oligonucleotide
(100 μM) with 1 μl of 1 M NaCl is annealed by the following profile: heat
at 95° for 2 min, rapid cool to 72°, and ramp cool to 4° over a period of 2 h.
Subsequently, fragments are diluted 200-fold with TE buffer, and 1 μl of
the diluent is used for ligation with the BspMI-digested piGENEhU6
plasmid. In both constructions, the nucleotide sequences of positive clones
should be determined by standard DNA sequencing techniques to confirm
the nature of constructs.

   When two types of vector-based systems produce siRNAs targeted
against the same site within the mRNA of the firefly gene for luciferase,
the activity of each construct can be estimated by measuring luciferase
activity. Approximately 3 × 10⁴ HeLa S3 cells are seeded in a 48-well
plate. One day after seeding, 3, 30, or 300 ng of each siRNA expression vector, 30 ng of a firefly luciferase expression plasmid (pGL3; Promega, Madison, WI), and 10 ng of a Renilla luciferase expression plasmid
(pRL-RSV; Promega; Miyagishi et al., 2000) is transfected with the

Lipofectamine™ 2000 reagent (Life Technologies, Rockville, MD), according to manufacturer's protocol. Luciferase activities are measured with a Dual Luciferase System (Promega) 24 h after transfection. In our analysis, both types of vectors showed strong activity in cells that had been transfected with a higher concentration of vectors. In contrast, at a lower concentration of vectors, hairpin-type vectors exhibited significantly higher activity than tandem-type vectors (Fig. 3). These different activities at the low concentration might be due to the nature of the two types of construct. It seems that the transcripts produced from the hairpin-type system can effectively form the stem structure through an intramolecular folding. Such rapid folding of the hairpin-type RNAs appears to make them more stable than those of the tandem-type in cells. Besides, the processing of hairpin-type transcripts by Dicer might facilitate their incorporation into the RISC and then RISC with the processed siRNA can suppress the expression of targets efficiently. Moreover, hairpin-type transcripts might be transported from the nucleus to the cytoplasm. Recently, it was demonstrated that pre-microRNAs that form a hairpin structure can be exported to the cytoplasm

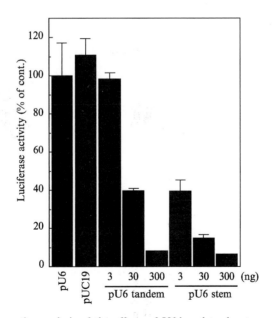

FIG. 3. Comparative analysis of the effects of U6-based tandem-type and hairpin-type siRNA expression vectors. HeLa S3 cells were cotransfected with 10 ng pRL-RSV, 30 ng pGL3, and each amount of siRNA expression vectors that targeted a firefly gene for luciferase. The piGENEhU6 and pUC19 vectors were used as controls. Each bar indicates an average value, and vertical bars indicate standard errors of triple assay.

A   Introduction of C to U or A to G
                     mutations in a sense strand

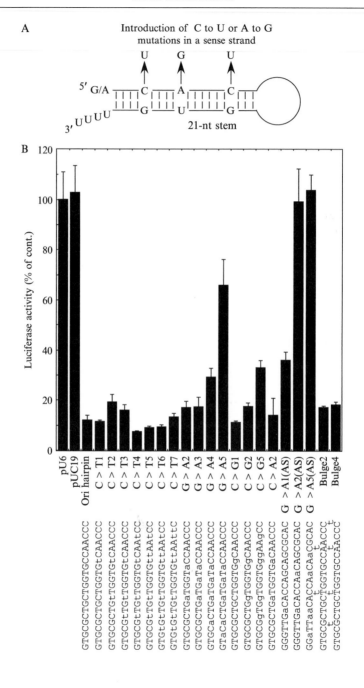

by exportin-5 (Lund *et al.*, 2004; Yi *et al.*, 2003). Thus, it is possible that the exportin-5 recognizes hairpin-type RNAs and transports them into the cytoplasm as pre-microRNAs. Although the reason for why hairpin-type RNAs shows relatively higher activity is unclear, a few factors might contribute to the efficacy of the hairpin-type RNAs in cells.

Improvement of the Hairpin-Type siRNA Expression System

Introducing Mismatch-Sequences on Hairpin-Type siRNA Expression Vector

To construct the hairpin-type siRNA expression vector, in many cases we faced two serious technical problems. The first problem seems to be caused by the structural hairpin region on a construct. Because such a hairpin region makes a tight palindromic structure, it becomes difficult for sequencing the constructs. The other problem is that mutations appear in approximately 20–40% of the constructs during amplification and maintenance in *E. coli*. When we sequenced the mutated constructs, two or more mutations or insertions or deletions within the sense or antisense sequence of the constructs could be found.

To avoid these serious problems, we introduced C to T (or A to G) point mutations in the sense strand when we designed hairpin-type constructs (Fig. 4A). These mutations efficiently restore the sequencing problems and enable us to yield effective constructs. Then, we confirmed that G:U base pairings within the transcripts had no negative effect on the silencing activity of the construct (Fig. 4B). In addition, the introduction of bulge insertions in the sense strand exhibited no loss of activity of transcripts. In contrast, the introduction of five C to G mutations or more than four G to A mutations in the sense strand significantly reduced silencing activity. The reduction of the suppressive activity was also observed in the introduction of a single G to A mutation or two or five G to A mutations in the antisense strand. It seems impossible to introduce

---

FIG. 4. (A) Schematic representation of the hairpin-type RNA with C to U and A to G mutations at its sense strand. (B) Comparative analysis of the activities of the U6-based hairpin-type RNA with a 21-nt stem (indicated as Ori hairpin) and U6-based hairpin-type RNAs with various mutations or bulges in their stems. A single C to T mutation in the sense strand is indicated as C → T1, for example. AS and Bulge indicate mutations in the antisense strand and bulge insertions in the sense strand, respectively. HeLa S3 cells were cotransfected with plasmids that encoded a firefly gene for luciferase, a *Renilla* gene for luciferase, and each short hairpin RNA expression vector targeted against the firefly luciferase. Each column and bar indicates the mean and SEM of results from a triplicate assay.

mutations or bulges in the antisense strand of a hairpin-type RNA without significant or total loss of suppressive activity.

Thus, the introduction of multiple C to T (or A to G) mutations in the sense strand markedly prevents mutations of the construct in *E. coli* and can then enhance the silencing activity of hairpin RNA transcripts. This strategy will take the advantages of constructing an effective hairpin-type system and of stable maintenance of plasmids in the bacterial host cells (Taira and Miyagishi, 2001).

Optimizing a Loop Sequence Within Hairpin RNAs

While designing hairpin-type RNAs, a variety of loop sequences have been exploited. For example, Brummelkamp *et al.* (2002) exploited a 9-nt loop sequence, 5'-UUCAAGAGA-3', for designing the hairpin-type RNA and it showed stronger activity than that with a shorter loop sequence. Moreover, a 4-nt loop sequence, 5'-UUCG-3', was also used to design hairpin-type RNAs, as demonstrated by Paul *et al.* (2002).

To confirm the effect of a loop sequence on the activity of hairpin RNAs, we compared the activity of the original hairpin RNA with the 9-nt loop sequence with that of hairpin RNAs with loop sequences derived from seven human miRNAs by measuring the luciferase activity (Miyagishi and Taira, 2003; Miyagishi *et al.*, 2004b), as described in the previous section. All constructs with more than five C to T mutations were very active as expected; also, the original hairpin RNA with mutations exhibited higher activity than that without mutations (Fig. 5). When we evaluated the activity of constructs with mutations, hairpin RNAs with each loop of miRNAs showed the same suppressive activity as or higher suppressive activity than that of the original hairpin RNA. Loops 2 and 7 seemed to be more effective for the activity of hairpin RNAs. In practice, a several-fold higher concentration of the original hairpin RNA than that of the hairpin RNA with loop 7 was required for similar levels of suppressive activity. Thus, exploiting loops of natural miRNAs is useful to produce active hairpin RNAs.

Searching Specific Target Sequences

The selection of appropriate target sites is one of the essential determinants of effective gene silencing by siRNAs. Many evidences have suggested that the effectiveness of siRNAs depends on target sites within the target mRNA (Holen *et al.*, 2002; Lee *et al.*, 2002; Miyagishi and Taira, 2002; Taira *et al.*, 2003), and we also evaluated it from our systematic analysis of siRNAs, which targeted the gene for luciferase. According to recent reviews, several factors such as GC contents, positional effect, and sequence-dependent effect within the target should be important for

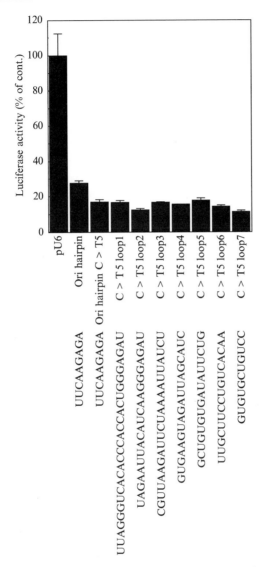

FIG. 5. The effects of loop sequences on the suppressive activity of short hairpin RNAs. Various loop sequences of human microRNAs that were used in the assay are indicated. Details of assay are the same as those in the legend of Fig. 4.

choosing target sites of siRNAs. Interestingly, recent reports have demonstrated that an internal stability at especially the 5′-antisense terminal base pairing determines the activity of siRNAs (Khvorova *et al.*, 2003;

Schwarz *et al.*, 2003). Statistical analysis based on data we have accumulated also indicates that some nucleotides at specific positions are positively or negatively correlated with the efficacy of siRNAs. siRNAs with an A or a U residue at the 19th nucleotide position from the 5' end of the sense strand tended to have relatively strong suppressive activity, whereas siRNAs with a G or C residue at the same position tended to be less effective. In addition, in our analysis, a U residue at the 10th nucleotide position in the sense strand tended to be effective. Moreover, we also confirmed a significant negative correlation between the GC content of 3' half of siRNAs (especially from the 12th to the 19th nucleotide). Thus, these criteria are important and available for selection of desirable target sites. Recently, we demonstrated further that the target sequence itself, rather than its location, is the major determinant of siRNA activity (Yoshinari *et al.*, 2004). Therefore, it is likely that siRNA can efficiently target an untranslated region (UTR) as well as a coding region of a target mRNA.

We also developed an algorithm to predict suitable target sites in mRNAs (Miyagishi and Taira, 2003). The analysis of data sets obtained from many attempts in silencing of specific target sequences led us to identify many factors that strongly correlated with suppressive activity. We used a partial least squares (PLS) method to generate models to predict the effect of siRNAs (Lindberg *et al.*, 1983). More than 700 sets of data that consisted of sequences of siRNAs and their suppressive effects, as well as factors that might be expected to be significantly correlated with the effects of siRNAs such as GC contents and the preference for nucleotide positions (Yoshinari *et al.*, 2004), were analyzed, and a PLS calibration model was generated from the accumulated data. We then determined the optimal conditions for identifying target sites from the lowest standard errors in cross-validation analysis.

Results predicted by applying our algorithm and results from an actual analysis of the effects of siRNAs that directed the EGFP gene showed a significant correlation, the correlation coefficient for the results being 0.7 (Miyagishi and Taira, 2003). Recent fine-tuning of our algorithm has allowed us to achieve a correlation coefficient as high as 0.8.

Reducing Interferon Response

Although it is believed that siRNA can bypass an interferon response, recent reports demonstrated the induction of interferon response by siRNAs in mammalian cells. As demonstrated by Bridge *et al.* (2003), a short hairpin RNA produced from the H1 promoter in a lentiviral vector could induce the expression of 2',5'-oligoadenylated synthetase (OAS1), one factor of the interferon-stimulated gene. This phenomenon was likely to be seen in only vector-based systems and not in synthetic siRNAs.

However, Williams and coworkers demonstrated that synthetic siRNAs could also trigger interferon-mediated activation of the Jak–Stat pathway and an upregulation of interferon-stimulated genes (Sledz et al., 2003). Because the side effect induced by siRNAs through the activation of interferon-related genes makes it difficult to exploit siRNAs as a basic research tool and a therapeutic agent, we must pay attention to such interferon responses carefully and, if possible, avoid it completely while treating siRNAs in mammalian cells.

In our analysis, we could reduce the interferon response by introducing appropriate C to U (or A to G) mutations within the strands of hairpin-type RNAs, as described in the previous section. Although it is believed that dsRNAs more than 30 nt in length can induce the activation of interferon-related genes such as PKR and OAS1, we observed that U6 promoter-driven dsRNAs of 50 bp with appropriate mutations could bypass the interferon response in mammalian cells (Akashi et al., submitted for publication). When U6 promoter-based vectors for the expression of a 50-bp hairpin RNA whose antisense strand was complementary to a part of a firefly gene for luciferase were introduced into cells, they exhibited a specific inhibitory effect on the expression of the luciferase gene without inducing the nonspecific side effect. In contrast, under the same conditions, dsRNAs transcribed *in vitro* that had a sequence identical to that of a vector-based hairpin RNA triggered undesirable nonspecific effects even if they could suppress the expression of target luciferase in a dose-dependent manner. Although the extent of the induction of the interferon response might depend on the sequence of siRNAs, our strategy is clearly advantageous to avoid unfavorable side effects.

Concluding Remarks

Our extensive efforts led us to develop methods to construct an effective siRNA expression system. The introduction of mutations within a stem of hairpin-type RNAs can stabilize the vectors during their maintenance and amplification in bacterial host cells, and adoption of loop sequences of natural miRNAs can enhance the suppressive activity of hairpin RNA transcripts. Furthermore, the appropriate design of hairpin-type expression vectors can significantly reduce nonspecific effects, whereas some short synthetic siRNAs and *in vitro* transcribed long dsRNAs induce undesirable side effects in some cases. These methods are clearly advantageous to apply siRNAs in basic understanding of the gene function and abrogation of aberrant gene expression. Recently, by using our U6-based hairpin-type expression vector, we established a high-quality siRNA expression library for identifying genes of unknown function involved in apoptosis in human cells (Miyagishi and Taira, 2003; Miyagishi et al., 2004a; Futami et al.,

2004). We are at present evaluating the utility of the library in analyzing the human genome. The availability of the library would be helpful to identify the functions of all genes in the human genome.

References

Bernstein, E., Caudy, A. A., Hammond, S. M., and Hannon, G. J. (2001). Role for a bidentate ribonuclease in the initiation step of RNA interference. *Nature* **409**, 363–366.

Bridge, A. J., Pebernard, S., Ducraux, A., Nicoulaz, A. L., and Iggo, R. (2003). Induction of an interferon response by RNAi vectors in mammalian cells. *Nat. Genet.* **34**, 263–264.

Brummelkamp, T. R., Bernards, R., and Agami, R. (2002). A system for stable expression of short interfering RNAs in mammalian cells. *Science* **296**, 550–553.

Cotten, M., and Birnstiel, M. L. (1989). Ribozyme mediated destruction of RNA *in vivo*. *EMBO J.* **8**, 3861–3866.

Czauderna, F., Santel, A., Hinz, M., Fechtner, M., Durieux, B., Fisch, G., Leenders, F., Arnold, W., Giese, K., Klippel, A., and Kaufmann, J. (2003). Inducible shRNA expression for application in a prostate cancer mouse model. *Nucleic Acids Res.* **31**, e127.

Dykxhoorn, D. M., Novina, C. D., and Sharp, P. A. (2003). Killing the messenger: Short RNAs that silence gene expression. *Nat. Rev. Mol. Cell. Biol.* **4**, 457–467.

Elbashir, S. M., Harborth, J., Lendeckel, W., Yalcin, A., Weber, K., and Tuschl, T. (2001). Duplexes of 21-nucleotide RNAs mediate RNA interference in cultured mammalian cells. *Nature* **411**, 494–498.

Futami, T., Miyagishi, M., and Taira, K. (2004). Identification of a network involved in thapsigargin-induced apoptosis using a library of siRNA-expression vectors. *J. Biol. Chem.* Online publication.

Fire, A., Xu, S., Montgomery, M. K., Kostas, S. A., Driver, S. E., and Mello, C. C. (1998). Potent and specific genetic interference by double-stranded RNA in *Caenorhabditis elegans*. *Nature* **391**, 806–811.

Hammond, S. M., Bernstein, E., Beach, D., and Hannon, G. J. (2000). An RNA-directed nuclease mediates post-transcriptional gene silencing in *Drosophila* cells. *Nature* **404**, 293–296.

Hammond, S. M., Caudy, A. A., and Hannon, G. J. (2001). Post-transcriptional gene silencing by double-stranded RNA. *Nat. Rev. Genet.* **2**, 110–119.

Holen, T., Amarzguioui, M., Wiiger, M. T., Babaie, E., and Prydz, H. (2002). Positional effects of short interfering RNAs targeting the human coagulation trigger tissue factor. *Nucleic Acids Res.* **30**, 1757–1766.

Kawasaki, H., and Taira, K. (2003). Short hairpin type of dsRNAs that are controlled by tRNA(Val) promoter significantly induce RNAi-mediated gene silencing in the cytoplasm of human cells. *Nucleic Acids Res.* **31**, 700–707.

Khvorova, A., Reynolds, A., and Jayasena, S. D. (2003). Functional siRNAs and miRNAs exhibit strand bias. *Cell* **115**, 209–216.

Koseki, S., Tanabe, T., Tani, K., Asano, S., Shioda, T., Nagai, Y., Shimada, T., Ohkawa, J., and Taira, K. (1999). Factors governing the activity *in vivo* of ribozymes transcribed by RNA polymerase III. *J. Virol.* **73**, 1868–1877.

Lee, N. S., Dohjima, T., Bauer, G., Li, H., Li, M. J., Ehsani, A., Salvaterra, P., and Rossi, J. (2002). Expression of small interfering RNAs targeted against HIV-1 rev transcripts in human cells. *Nat. Biotechnol.* **20**, 500–505.

Lindberg, W., Persson, J. A., and Wold, S. (1983). Partial least-squares method for spectro-fluorimetric analysis of mixtures of humic acid and lignin sulfonate. *Anal. Chem.* **55**, 643–648.

Liu, J., Carmell, M. A., Rivas, F. V., Marsden, C. G., Thomson, J. M., Song, J. J., Hammond, S. M., Joshua-Tor, L., and Hannon, G. J. (2004). Argonaute 2 is the catalytic engine of mammalian RNA. *Science* **305**, 1437–1441.

Lund, E., Guttinger, S., Calado, A., Dahlberg, J. E., and Kutay, U. (2004). Nuclear export of microRNA precursors. *Science* **303**, 95–98.

Martinez, J., Patkaniowska, A., Urlaub, H., Luhrmann, R., and Tuschl, T. (2002). Single-stranded antisense siRNAs guide target RNA cleavage in RNAi. *Cell* **110**, 563–574.

Martinez, J., and Tuschl, T. (2004). RISC is a $5'$ phosphomonoester-producing RNA endonuclease. *Genes Dev.* **18**, 975–980.

McManus, M. T., and Sharp, P. A. (2002). Gene silencing in mammals by small interfering RNAs. *Nat. Rev. Genet.* **3**, 737–747.

Miyagishi, M., Fujii, R., Hatta, M., Yoshida, E., Araya, N., Nagafuchi, A., Ishihara, S., Nakajima, T., and Fukamizu, A. (2000). Regulation of Lef-mediated transcription and p53-dependent pathway by associating beta-catenin with CBP/p300. *J. Biol. Chem.* **275**, 35170–35175.

Miyagishi, M., Matsumoto, S., and Taira, K. (2004a). Generation of an shRNAi expression library against the whole human transcripts. *Virus Res.* **102**, 117–124.

Miyagishi, M., Sunimoto, H., Miyoshi, H., Kawakami, Y., and Taira, K. (2004b). Optimization of an siRNA-expression system with a mutated hairpin and its significant suppressive effects upon HIV vector-mediated transfer into mammalian cell. *J. Gene Med.* **6**, 715–723.

Miyagishi, M., and Taira, K. (2002). U6 promoter-driven siRNAs with four uridine $3'$ overhangs efficiently suppress targeted gene expression in mammalian cells. *Nat. Biotechnol.* **20**, 497–500.

Miyagishi, M., and Taira, K. (2003). Strategies for generation of an siRNA expression library directed against the human genome. *Oligonucleotides* **13**, 325–333.

Montgomery, M. K., Xu, S., and Fire, A. (1998). RNA as a target of double-stranded RNA-mediated genetic interference in *Caenorhabditis elegans*. *Proc. Natl. Acad. Sci. USA* **95**, 15502–15507.

Nykanen, A., Haley, B., and Zamore, P. D. (2001). ATP requirements and small interfering RNA structure in the RNA interference pathway. *Cell* **107**, 309–321.

Paddison, P. J., Caudy, A. A., Bernstein, E., Hannon, G. J., and Conklin, D. S. (2002). Short hairpin RNAs (shRNAs) induce sequence-specific silencing in mammalian cells. *Genes Dev.* **16**, 948–958.

Paul, C. P., Good, P. D., Li, S. X., Kleihauer, A., Rossi, J. J., and Engelke, D. R. (2003). Localized expression of small RNA inhibitors in human cells. *Mol. Ther.* **7**, 237–247.

Paul, C. P., Good, P. D., Winer, I., and Engelke, D. R. (2002). Effective expression of small interfering RNA in human cells. *Nat. Biotechnol.* **20**, 505–508.

Schwarz, D. S., Hutvagner, G., Du, T., Xu, Z., Aronin, N., and Zamore, P. D. (2003). Asymmetry in the assembly of the RNAi enzyme complex. *Cell* **115**, 199–208.

Sledz, C. A., Holko, M., de Veer, M. J., Silverman, R. H., and Williams, B. R. (2003). Activation of the interferon system by short-interfering RNAs. *Nat. Cell. Biol.* **5**, 834–839.

Song, J. J., Smith, S. K., Hannon, G. J., and Joshua-Tor, L. (2004). Crystal structure of Argonaute and its implications for RISC slicer activity. *Science* **305**, 1434–1437.

Sui, G., Soohoo, C., Affar el, B., Gay, F., Shi, Y., and Forrester, W. C. (2002). A DNA vector-based RNAi technology to suppress gene expression in mammalian cells. *Proc. Natl. Acad. Sci. USA* **99**, 5515–5520.

Taira, K., and Miyagishi, M. (2001). siRNA expression system and method for producing functional gene knock-down cell using the system. *Japanese Patent Application* Heisei-13-363385.

Taira, K., Miyagishi, M., and Matsumoto, S. (2003). Enhancement of RNAi activity through mutations. U.S. Patent Application.

Yi, R., Qin, Y., Macara, I. G., and Cullen, B. R. (2003). Exportin-5 mediates the nuclear export of pre-microRNAs and short hairpin RNAs. *Genes Dev.* **17,** 3011–3016.

Yoshinari, K., Miyagishi, M., and Taira, K. (2004). Effects on RNAi of the tight structure, sequence and position of the targeted region. *Nucleic Acids Res.* **32,** 691–699.

Yu, J. Y., DeRuiter, S. L., and Turner, D. L. (2002). RNA interference by expression of short-interfering RNAs and hairpin RNAs in mammalian cells. *Proc. Natl. Acad. Sci. USA* **99,** 6047–6052.

[7] Delivery of Small Interfering RNA to Mammalian Cells in Culture by Using Cationic Lipid/Polymer-Based Transfection Reagents

By Robert M. Brazas and James E. Hagstrom

Abstract

RNA interference (RNAi) has become a powerful tool for the knockdown of target gene expression and subsequent phenotypic analysis of gene function in mammalian cells in culture. Critical to the success of any small inhibitory RNA (siRNA)-mediated RNAi knockdown in mammalian cells is the efficient delivery of the siRNA to those cells. This chapter describes the use of popular cationic lipid/polymer-based transfection reagents for *in vitro* siRNA delivery and includes a general protocol with special emphasis on key transfection parameters important to the success of siRNA delivery.

Introduction

RNA interference (RNAi) in cultured mammalian cells is facilitated by the direct introduction of small inhibitory RNA (siRNA) into the cytoplasmic compartment (Elbashir *et al.*, 2001). The utility of siRNA-mediated RNAi to inhibit target gene expression depends on many factors, such as siRNA design and the efficient delivery of siRNA to target cells. This chapter focuses on the delivery of siRNA to mammalian cells and the various parameters influencing delivery. Additional information on the design of siRNA can be found elsewhere (Khvorova *et al.*, 2003; Naito *et al.*, 2004; Reynolds *et al.*, 2004) and is not addressed here. To achieve efficient target gene knockdown, siRNA must be delivered to a high percentage of cells in the population. To facilitate the incorporation of siRNA into functional RNA-induced silencing complexes (RISCs) and in turn enhance the inhibition of target gene expression, the delivery method should also introduce the maximal number of siRNA molecules into the cytoplasmic compartment of the cell. In addition, it is important that the delivery method or reagent does not adversely affect cellular health or gene expression. The fewer the number of side effects due to the delivery method, the simpler it is to decipher the result of specific target gene knockdown.

Many different methods have been used successfully to deliver siRNA into mammalian cells. These include physical methods such as microinjection (Usui *et al.*, 2003) and electroporation (Heidenreich *et al.*, 2003; North *et al.*, 2003; Scherr *et al.*, 2003). Both standard and modified electroporation systems, including the Amaxa Nucleofector (Bidere *et al.*, 2003; Ganesh *et al.*, 2003) and the Maxcyte GT, have been used successfully. Various chemical-based transfection methods such as calcium phosphate precipitation (Porter *et al.*, 2002), liposomal reagents (Robert *et al.*, 2003), cationic lipids (Kurisaki *et al.*, 2003) and nonliposomal lipid/polymer formulations (Li *et al.*, 2004) have been used extensively as siRNA delivery reagents. Table I lists many of the commercially available chemical-based siRNA transfection reagents. Because chemical-based transfection reagents and methods represent the most widely used siRNA delivery methodology, this chapter focuses on key issues relating to the use of the very popular cationic lipid/polymer-based siRNA delivery reagents.

Methodology

Because of the vast number of siRNA transfection reagents available (Table I), it will be impossible to describe the specific protocols recommended for each reagent. However, the numerous siRNA transfection reagent protocols share many common themes. For demonstration purposes, this chapter describes in detail the use of one siRNA-specific lipid/polymer transfection reagent, the *Trans*IT-TKO® Transfection Reagent (Mirus Bio, Madison, WI) while highlighting key steps and important variables to keep in mind during any transfection process. Important variations that differ when using other transfection reagents are outlined; however, users should consult the appropriate product manuals to identify variables that are specific and important to a given transfection reagent.

Transfection of Adherent Cells in Six-Well Plates Using the Trans*IT-TKO® Reagent*

The following procedure describes the transfection of adherent cells in one well (9.6 cm^2 surface area) of a six-well tissue culture plate. To transfect cells in different-sized wells or dishes, scale all the components

TABLE I
COMMERCIALLY AVAILABLE CATIONIC LIPID/POLYMER-BASED
siRNA TRANSFECTION REAGENTS

siRNA transfection reagent	Supplier
*Trans*IT-TKO®	Mirus Bio Corporation
*Trans*IT®-siQUEST™	Mirus Bio Corporation
Lipofectamine™ 2000	Invitrogen Corporation
Oligofectamine™	Invitrogen Corporation
siPORT Amine™	Ambion, Inc.
siPORT Lipid™	Ambion, Inc.
TransMessenger™	Qiagen, Inc.
RNAiFect™	Qiagen, Inc.
GeneSilencer™	Gene Therapy Systems, Inc.
jetSI™	PolyPlus-Transfection SAS
jetSI™-ENDO	PolyPlus-Transfection SAS
CodeBreaker™	Promega Corporation
siFECTOR™	B-Bridge International, Inc.
RiboJuice™	Novagen®
GeneEraser™	Stratagene
i-Fect™	MoleculA
SAINT-MIX	Synvolux Therapeutics
X-tremeGENE	Roche Applied Science
siFECTamine™	IC-VEC
siIMPORTER™	Upstate Cell Signaling Solutions

(serum-free medium, transfection reagent, siRNA(s), complete growth medium, and cells) up or down in proportion to the increase or decrease in the surface area of the tissue culture vessel being used.

Cell Preparation. On the day before transfection, plate healthy, actively growing cells at an appropriate cell density so that cells will be 60–80% confluent the following day at the time of transfection. Plate 2.5 ml of this cell suspension per well of the six-well plate and incubate overnight under the appropriate cell culture conditions, usually 37° and 5% CO_2. The actual number of cells plated per well will vary with cell line, depending on the growth characteristics and size of the adhered cell, and must be determined empirically. For example, plating approximately 350,000 CHO-K1 cells per well of a six-well plate will generally yield a well the following day that is 60–80% confluent.

Variations: In general, most protocols recommend cell confluencies between 50 and 80% at the time of transfection. Some protocols such as the Lipofectamine™ 2000 (Invitrogen, Carlsbad, CA) and Oligofectamine™ (Invitrogen) recommend cell confluencies between 30 and 50%. To achieve the highest transfection efficiency with a given cell line and reagent, the optimal cell density should be determined empirically by the investigator. Standard cell culture conditions are 37° and 5% CO_2, but, if necessary, these incubation conditions can be changed to meet the requirements of a particular cell line or experiment.

Additional Comments: The state of the cells at the time of transfection can have a dramatic impact on siRNA transfection and knockdown efficiency. Be certain to monitor cell passage number and cellular morphology as cells are maintained and used. If the cells start to transfect poorly or produce poor knockdown efficiencies, discard the cells and begin with freshly cultured, lower passage number cells (i.e., from a frozen stock). From that point forward, use cells at the optimal cell density and similar passage number to maintain consistent transfection and knockdown results.

Reagent/siRNA Complex Formation

1. Examine the cells to be transfected to confirm that they are at the expected cell density (60–80% confluent).

2. Add 250 µl of serum-free medium to a sterile plastic tube (12 × 75 mm) and add 5–20 µl of the *Trans*IT-TKO Transfection Reagent and vortex gently to mix.

3. Incubate at room temperature for 5–20 min.

Variations: For all transfection reagents, the optimal level of transfection reagent used must be determined empirically for each cell line being transfected. Cell lines differ in their sensitivity to each reagent and the amount of reagent necessary to produce high-efficiency transfection. In addition, slight differences in culturing methods and the subjective nature of measuring cell density can combine to produce different optimal transfection conditions from laboratory to laboratory. When starting with a new cell line, it is preferable to test three or four different reagent levels in the initial transfections, such as 5, 10, and 20 µl of *Trans*IT-TKO® Reagent per well of a six-well plate. Most manufacturers recommend that similar amounts of reagent be initially tested. Once satisfactory results are obtained, the level of reagent can be titrated further to identify the optimal amount of reagent that produces best knockdown with the lowest toxicity.

Additional Comments: Historically, Opti-MEM® I Reduced-Serum Medium (Invitrogen) has been widely used for complex formation in many transfection protocols. Other serum-free media can be used unless they contain polyanions such as heparin or dextran sulfate. Polyanions will compete with siRNA for binding to the transfection reagents and inhibit transfection.

4. Add siRNA (37.5 μl of a 1 μM siRNA stock when following this protocol) to the diluted *Trans*IT-TKO® Reagent such that the final concentration of siRNA in the well after the complexes are added to the cells is 25 nM. Mix gently by pipetting.

5. Incubate at room temperature for 5–20 min to allow complex formation.

Variations: To achieve high-efficiency knockdown of endogenous target gene expression, 25 nM siRNA is usually sufficient. Some manufacturers recommend levels as high as 200 nM. The final siRNA concentration in the well is another variable that can be tested when working with different transfection reagents. The amount of siRNA required to achieve high-level knockdown may also be dependent on the specific siRNA being used. When using higher levels of siRNAs, it is prudent to retitrate the level of transfection reagent to determine the optimal combination of siRNA and transfection reagent required for high-efficiency knockdown. Many other protocols also recommend first diluting the siRNA and transfection reagent separately in serum-free medium and combining the two diluted components to initiate complex formation, but this extra step is not necessary when working with the *Trans*IT-TKO® Transfection Reagent.

Additional Comments: The delivery of high levels of siRNA (>100 nM) has been shown to alter cellular gene expression profiles nonspecifically (Persengiev *et al.*, 2004; Semizarov *et al.*, 2003). Therefore, the lowest level of siRNA necessary to produce the desired knockdown should be used to avoid potential nonspecific effects.

Addition of Complexes to the Cells for Transfection

1. Adjust the amount of serum-containing medium per well to 1.25 ml by removing half (1.25 ml) of the original plating media.

Variations: By removing half of the cell culture media, the amount of siRNA necessary to produce the desired final concentration in the well is reduced by half, thus saving precious siRNA. Alternatively, the entire 2.5 ml of serum-containing medium can be removed from the well and replaced with 1.25 ml of fresh prewarmed serum-containing medium. Depending on the cell line being transfected, one of these approaches to reducing the amount of serum-containing medium per well may enhance transfection and knockdown efficiencies. Other reagent protocols do not reduce the amount of medium in the well, and therefore care must be taken to use the correct amount of siRNA with a particular transfection reagent as per the manufacturer's recommendations. In addition, some reagents require that the serum-containing medium be removed from the wells and replaced with serum-free medium before the complexes are added to the cells. Subsequently, the serum-free medium containing the complexes is removed and replaced with complete growth medium containing serum as per the manufacturer's recommendations.

2. Add the siRNA/*Trans*IT-TKO® Reagent complexes dropwise to the cells, taking care to disperse them over the entire surface area of the well. Rock the plate back and forth and from side to side to disperse the complexes evenly within the well. Do not swirl the plate because the complexes may then accumulate in the center of the well.

3. Incubate the cells under the appropriate cell culture conditions, usually 37° and 5% CO_2, for 24–72 h to activate the RNAi pathway, resulting in cleavage of the target mRNA. The length of time post-transfection necessary to produce sufficient knockdown of gene expression is discussed in detail in the following section.

Variations: If it is necessary to add back additional medium to maintain the health of the cells, an additional 1.25 ml of prewarmed serum-containing medium can be added to the well 4–24 h posttransfection or the medium in the well can be replaced with fresh medium 24 h posttransfection. As mentioned previously, some reagents require transfecting the cells in the absence of serum. This modification requires that the medium be replaced with serum-containing medium 4–24 h post-transfection. If cells are going to be assayed multiple days post-transfection and grown to near confluence before being analyzed, it may be necessary to split the cells and replate them in a larger vessel to maintain the health of the transfected cells.

Detection of Target Gene Knockdown. Once the siRNA has been transfected into the mammalian cells, the level of target gene knockdown must be assessed. This assay generally represents a key step in a knockdown experiment and much thought and care must go into selecting the appropriate assay to perform. Reverse transcription-PCR, Northern blotting, RNase protection, and branched DNA assays can be used to determine the level of knockdown of the target mRNA. Alternatively, the cellular level of the target protein can also be assessed in total cell lysates of the transfected population, using techniques such as Western blotting, immunoprecipitation, or enzymatic assays if the target protein has a specific enzymatic activity associated with it. Immunofluorescence-based assays (fluorescence microscopy and flow cytometry) can also be used to assay for knockdown in single cells within the siRNA-transfected cell population. This type of assay is not only useful for determining the overall level of knockdown, but is also an efficient method for assessing the percentage of cells in the transfected population that has decreased target gene expression. Once efficient knockdown of target gene expression has been verified, the investigator is ready to determine what role that target gene may play within the cell.

Additional Considerations for Effective siRNA-Mediated RNAi Experiments

Quality of siRNA Duplexes. Although the double-stranded nature of siRNAs helps protect them from degradation by common single-strand specific RNases, it is still good practice to ensure the integrity of the siRNA by using RNase-free reagents and procedures. Care must also be taken to prevent siRNA duplex denaturation, which is accomplished by diluting the siRNA in an RNase-free buffer such as 100 mM NaCl, 50 mM Tris, pH 7.5, or one recommended by the manufacturer of the siRNA. Diluting the siRNA in only RNase-free water could lead to duplex denaturation because of the inability of water alone to neutralize the electrostatic repulsion between the strands.

Incubation Time Post-transfection. Many factors affect the length of the posttransfection incubation time necessary to produce a satisfactory knockdown in target gene expression. First, the size of the mRNA pool and the rate of transcription of the target gene can have dramatic effects on knockdown efficiency. If the mRNA pool is large or the rate of transcription is high, a longer period of time post-transfection will be required to produce a significant decrease in the levels of the target mRNA pool. Second, while the initial target of the RNAi pathway is the target mRNA, most investigators are using RNAi to ultimately decrease the steady-state level of the target protein in the cell. Therefore, knowing the half-life of the target protein is critical when performing a knockdown experiment. If the target protein has a long half-life, it will be necessary to increase the time post-transfection before assaying for knockdown and function of the target protein. Third, it is also possible that different cell lines will require different lengths of time post-transfection to produce effective knockdowns. For instance, the RNAi pathways could have different activity levels because of mutations in individual pathway components or lower steady-state levels of those components (i.e., RISC protein components). Decreased activity of the RNAi pathway will therefore require longer incubation times, whereas increased activity in the pathway will allow for shorter incubation times post-transfection. Other important characteristics that could vary between different cells lines are the speed with which the transfection complexes are internalized and the efficiency with which the siRNAs are released into the cytoplasm. These potential cell line differences could change the length of time necessary to see effective knockdown when using the same siRNA to target the same target mRNA in different cells.

Proper Use of Controls. When performing an siRNA-mediated knockdown experiment, it is critical that proper controls be performed in parallel. A very important control is transfection of a nonspecific siRNA that does not target any endogenous mRNA in the cell. This negative control will reveal any changes that are due to the addition of the transfection complexes to the cells and the initiation of the RNAi pathway on incorporation of the siRNA into RISC. However, the choice of a nonspecific siRNA can be daunting because it is laborious and costly to demonstrate that the nonspecific siRNA chosen does not alter gene expression within the cell nonspecifically. Many researchers use scrambled forms of their specific siRNA as their negative siRNA control; however, empirically tested nonspecific siRNA controls are now available commercially (siCONTROL™ Non-Targeting siRNA #1, Dharmacon, Inc., Lafayette, CO). This control siRNA has been prescreened by transfection, followed by microarray analysis, to verify that this siRNA duplex produces minimal changes in the gene expression profile within the transfected cells. It is also desirable to use at least two siRNAs targeted to different regions of the same target mRNA to knockdown target gene expression and then demonstrate that both knockdowns produce the same cellular phenotype. The observation of the same phenotype with two distinct target-specific siRNAs will greatly increase the confidence that the observed phenotype is indeed due to the knockdown in expression of the target gene and not due to the knockdown of an unanticipated gene.

Troubleshooting Poor Knockdown Results

The previous two sections described the general procedure for transfecting siRNAs into mammalian cells and outlined the important points to consider for achieving high-efficiency knockdown. If the preceding recommendations are followed and sufficient knockdown is not obtained, additional experiments can be performed to verify that the delivery method is working well.

Transfection of Fluorescently Labeled siRNAs. Poor knockdown could be the result of inefficient siRNA delivery to the cells of interest because the reagent chosen to deliver the siRNAs is not optimal for the particular cell line, the cells are inherently difficult to transfect, or there is a technical problem with the transfection procedure. A simple and rapid method for testing transfection efficiency is to transfect a fluorescently labeled siRNA into the cells, using the same procedure used to transfect the original unlabeled siRNA. Fluorescent siRNA can be chemically synthesized (i.e., Dharmacon, Inc.; Integrated DNA Technologies, Coralville, IA; Qiagen, Inc., Valencia, CA; RNA-TEC, Belgium), or can be fluorescently labeled postsynthesis (*Label*IT® siRNA Tracker Kit, Mirus Bio, Madison, WI). After transfecting the fluorescently labeled siRNA, the cells can be observed by fluorescence microscopy to verify high-efficiency uptake of the labeled siRNA. For example, in Fig. 1, a fluorescein-labeled siRNA (50 n*M*

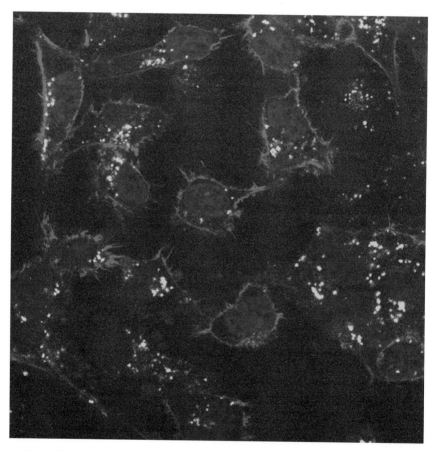

FIG. 1. Transfection of fluorescently labeled siRNA into HeLa cells, using the *Trans*IT-TKO® Transfection Reagent. A siRNA was covalently labeled with the *Label*IT® siRNA Tracker Fluorescein Kit (Mirus Bio), and then HeLa cells were transfected with this fluorescein-labeled siRNA (green) by using the *Trans*IT-TKO® Transfection Reagent (Mirus Bio). The cells were fixed and counterstained 24 h posttransfection with TO-PRO®-3 (Molecular Probes–Invitrogen, Eugene, OR) to stain the nuclei (blue) and Alexa Fluor® 546 Phalloidin (Molecular Probes–Invitrogen) to stain the actin (red). The cells were then visualized by confocal microscopy. (See color insert.)

final concentration per well) was transfected into HeLa cells by using the *Trans*IT-TKO® Reagent, and nearly 100% of the cells show internalized labeled siRNA. If a high percentage (approximately 80%) of the control transfected cells does not show internalization of the labeled siRNA, then the transfection method will need to be modified to increase the transfection efficiency. Changes can range from simple alterations in the current protocol to changing the siRNA transfection reagent. Internalization of the fluorescently labeled siRNA does not necessarily indicate that those cells are actively incorporating the siRNA into the RISC and thus inhibiting target gene expression.

Transfection of a Well-Characterized siRNA Followed by Assessment of Knockdown Efficiency. If transfection of the fluorescently labeled siRNA demonstrates that a high percentage (approximately 80%) of the cells have internalized the siRNA, then it is very likely that the siRNA being used to knockdown target gene expression is not functional, the transfection reagent is not promoting efficient delivery of the siRNA to the cytoplasmic compartment (although it did promote internalization of the siRNA based on the fluorescent-siRNA delivery), or that the RNAi pathway itself is not functioning in the cells being transfected. To test these possibilities, a well-characterized, highly efficient siRNA known to produce high-level knockdown of a target gene should be transfected into the cell line, followed by an assay to measure knockdown efficiency. Many published siRNAs target ubiquitously expressed endogenous genes that can be used for this purpose. The lamin A/C gene is commonly used as such a control (Elbashir *et al.*, 2001), and there are commercially available kits that include a lamin A/C-specific siRNA that functions in human cell lines as well as an anti-lamin A/C primary antibody that can be used to detect the lamin A/C proteins by either Western blotting or single-cell-based immunofluorescence assays (Dharmacon, Inc.). Other potential targets include vimentin and NuMA for which there are both well-characterized commercially available siRNAs (Harborth *et al.*, 2001; i.e., Qiagen; Ambion, Inc., Austin, TX) and commercially available primary antibodies (i.e., Cell Signaling Technology, Beverly, MA; BD Biosciences Pharmingen, San Diego, CA) that can be combined and used to test the functionality of the siRNA transfection procedure being used.

Alternative Methods to Improve or Prolong Knockdown of Target Gene Expression. As mentioned earlier, the steady-state level of the target mRNA and the stability of the target protein have a great influence on the success of RNAi knockdown experiments. Because of these important mRNA and protein characteristics, it is very likely that the efficient delivery of a well-designed, highly active siRNA may not lead to adequate knockdown of target gene expression. To increase knockdown capability, several groups have used a multistep transfection strategy to address and solve this problem (Kullmann *et al.*, 2002; Wagner *et al.*, 2002). Basically, the method that has been used is the sequential transfection of the siRNA into target cells over the course of multiple days before assaying for target gene knockdown. The following brief protocol is adapted from Kullmann *et al.* (2002) and Wagner *et al.* (2002), but variations on this theme should also be tested (additional transfections, incubation time between transfections, and actual transfection conditions). The cells are plated in 24-well plates at the appropriate density. On the following day, Day 1, the cells are transfected with the siRNA, using optimal levels of reagent and siRNA. On Day 2, each well of the 24-well plates is trypsinized and replated in a well of a six-well tissue culture plate (a 1:5 split based on the well surface area of each plate). On Day 3, the cells are transfected with the same siRNA that was transfected on Day 1. The cells are then incubated an additional 24–72 h before assaying for target gene knockdown. In several cases, this has dramatically increased the level of knockdown seen compared with the standard single-hit transfection method.

Depending on the goal of the gene knockdown experiment, it may be desirable to have a sustained knockdown of target gene expression. The length of time for which target gene expression is inhibited will depend on many factors, including the doubling time of the cells, amount of siRNA delivered per cell, activity of the siRNA, and stability and function of the target protein. In our experience, we usually begin to see target mRNA and protein levels begin to return to steady-state levels 3–6 days post-transfection. A second transfection of the siRNA before the target protein begins to rise can be used to maintain the knockdown level from the initial transfection (Kisielow *et al.*, 2002). This method will most likely require that the cells be split because of cell growth before the second transfection, and the timing of the second transfection will have to be determined empirically to obtain the most efficient and sustained gene knockdown. Another option to extend siRNA-mediated gene knockdowns is to use chemically modified siRNA. Several investigators have shown that siRNA containing 2'-*O*-methyl modified ribonucleotides or 2'-fluoro-uridine/2'-fluoro-cytidine substitutions can stabilize the siRNA duplex while maintaining function, and can in fact lead to extended high-level target gene knockdowns (Amarzguioui *et al.*, 2003; Chiu and Rana, 2003; Czauderna *et al.*, 2003). However, note that there are a limited number of examples in which these types of modifications have prolonged target gene knockdowns and there are similar examples in which these types of modifications have inhibited the function of the siRNA and prevented target gene knockdown (Amarzguioui *et al.*, 2003; Chiu and Rana, 2003). Therefore, although these types of modifications show great promise in extending target knockdowns, each new modified siRNA will have to be synthesized and tested empirically in order to identify those that result in efficient and long-lived knockdowns of specific target genes. In addition to custom designing chemically modified siRNA, multiple commercial manufacturers of siRNA offer stabilized versions that are reported to produce longer-lasting target gene knockdowns. These include siSTABLE™ siRNAs (Dharmacon, Inc.) and si² Silencing Duplexes (OligoEngine, Seattle, WA).

Summary

Like most plasmid or DNA transfections, there are many factors that can affect the efficiency of siRNA delivery into mammalian cells in culture. Each researcher must optimize the transfection conditions to obtain the best knockdown of target gene expression. Yet even with the caveats presented here, amazingly in most cell lines 80% or better knockdown of target gene expression can be attained routinely. For most researchers, this level will be sufficient to investigate the role of the target protein (or RNA) within the mammalian cell.

Acknowledgments

We thank Colleen Drinkwater for providing the fluorescein-labeled siRNA transfection data and Hans Herweijer for critically reading the manuscript.

References

Amarzguioui, M., Holen, T., Babaie, E., and Prydz, H. (2003). Tolerance for mutations and chemical modifications in a siRNA. *Nucleic Acids Res.* **31**, 589–595.

Bidere, N., Lorenzo, H. K., Carmona, S., Laforge, M., Harper, F., Dumont, C., and Senik, A. (2003). Cathepsin D triggers Bax activation, resulting in selective apoptosis-inducing factor (AIF) relocation in T lymphocytes entering the early commitment phase to apoptosis. *J. Biol. Chem.* **278**, 31401–31411.

Chiu, Y. L., and Rana, T. M. (2003). siRNA function in RNAi: A chemical modification analysis. *RNA* **9**, 1034–1048.

Czauderna, F., Fechtner, M., Dames, S., Aygun, H., Klippel, A., Pronk, G. J., Giese, K., and Kaufmann, J. (2003). Structural variations and stabilising modifications of synthetic siRNAs in mammalian cells. *Nucleic Acids Res.* **31**, 2705–2716.

Elbashir, S. M., Harborth, J., Lendeckel, W., Yalcin, A., Weber, K., and Tuschl, T. (2001). Duplexes of 21-nucleotide RNAs mediate RNA interference in cultured mammalian cells. *Nature* **411**, 494–498.

Ganesh, L., Burstein, E., Guha-Niyogi, A., Louder, M. K., Mascola, J. R., Klomp, L. W., Wijmenga, C., Duckett, C. S., and Nabel, G. J. (2003). The gene product Murr1 restricts HIV-1 replication in resting CD4+ lymphocytes. *Nature* **426**, 853–857.

Harborth, J., Elbashir, S. M., Bechert, K., Tuschl, T., and Weber, K. (2001). Identification of essential genes in cultured mammalian cells using small interfering RNAs. *J. Cell. Sci.* **114**, 4557–4565.

Heidenreich, O., Krauter, J., Riehle, H., Hadwiger, P., John, M., Heil, G., Vornlocher, H. P., and Nordheim, A. (2003). AML1/MTG8 oncogene suppression by small interfering RNAs supports myeloid differentiation of t(8;21)-positive leukemic cells. *Blood* **101**, 3157–3163.

Khvorova, A., Reynolds, A., and Jayasena, S. D. (2003). Functional siRNAs and miRNAs exhibit strand bias. *Cell* **115**, 209–216.

Kisielow, M., Kleiner, S., Nagasawa, M., Faisal, A., and Nagamine, Y. (2002). Isoform-specific knockdown and expression of adaptor protein ShcA using small interfering RNA. *Biochem. J.* **363**, 1–5.

Kullmann, M., Gopfert, U., Siewe, B., and Hengst, L. (2002). ELAV/Hu proteins inhibit p27 translation via an IRES element in the p27 5'UTR. *Genes Dev.* **16**, 3087–3099.

Kurisaki, K., Kurisaki, A., Valcourt, U., Terentiev, A. A., Pardali, K., Ten Dijke, P., Heldin, C. H., Ericsson, J., and Moustakas, A. (2003). Nuclear factor YY1 inhibits transforming growth factor beta- and bone morphogenetic protein-induced cell differentiation. *Mol. Cell. Biol.* **23**, 4494–4510.

Li, T., Chang, C. Y., Jin, D. Y., Lin, P. J., Khvorova, A., and Stafford, D. W. (2004). Identification of the gene for vitamin K epoxide reductase. *Nature* **427**, 541–544.

Naito, Y., Yamada, T., Ui-Tei, K., Morishita, S., and Saigo, K. (2004). siDirect: Highly effective target-specific siRNA design software for mammalian RNA interference. *Nucleic Acids Res.* **32**, W124–W129.

North, B. J., Marshall, B. L., Borra, M. T., Denu, J. M., and Verdin, E. (2003). The human Sir2 ortholog, SIRT2, is an NAD+-dependent tubulin deacetylase. *Mol. Cell* **11**, 437–444.

Persengiev, S. P., Zhu, X., and Green, M. R. (2004). Nonspecific, concentration-dependent stimulation and repression of mammalian gene expression by small interfering RNAs (siRNAs). *RNA* **10**, 12–18.

Porter, L. A., Dellinger, R. W., Tynan, J. A., Barnes, E. A., Kong, M., Lenormand, J. L., and Donoghue, D. J. (2002). Human Speedy: A novel cell cycle regulator that enhances proliferation through activation of Cdk2. *J. Cell Biol.* **157**, 357–366.

Reynolds, A., Leake, D., Boese, Q., Scaringe, S., Marshall, W. S., and Khvorova, A. (2004). Rational siRNA design for RNA interference. *Nat. Biotechnol.* **22**, 326–330.

Robert, M. F., Morin, S., Beaulieu, N., Gauthier, F., Chute, I. C., Barsalou, A., and MacLeod, A. R. (2003). DNMT1 is required to maintain CpG methylation and aberrant gene silencing in human cancer cells. *Nat. Genet.* **33**, 61–65.

Scherr, M., Battmer, K., Winkler, T., Heidenreich, O., Ganser, A., and Eder, M. (2003). Specific inhibition of bcr-abl gene expression by small interfering RNA. *Blood* **101**, 1566–1569.

Semizarov, D., Frost, L., Sarthy, A., Kroeger, P., Halbert, D. N., and Fesik, S. W. (2003). Specificity of short interfering RNA determined through gene expression signatures. *Proc. Natl. Acad. Sci. USA* **100**, 6347–6352.

Usui, I., Imamura, T., Huang, J., Satoh, H., and Olefsky, J. M. (2003). Cdc42 is a Rho GTPase family member that can mediate insulin signaling to glucose transport in 3T3-L1 adipocytes. *J. Biol. Chem.* **278**, 13765–13774.

Wagner, E. J., and Garcia-Blanco, M. A. (2002). RNAi-mediated PTB depletion leads to enhanced exon definition. *Mol. Cell* **10**, 943–949.

[8] Subcellular Distribution of Small Interfering RNA: Directed Delivery Through RNA Polymerase III Expression Cassettes and Localization by *In Situ* Hybridization

By Cynthia P. Paul

Abstract

Reduction in the expression of specific genes through small interfering RNAs (siRNAs) is dependent on the colocalization of siRNAs with other components of the RNA interference (RNAi) pathways within the cell. The expression of siRNAs within cells from cassettes that are derived from genes transcribed by RNA polymerase III (pol III) and provide for selective subcellular distribution of their products can be used to direct siRNAs to the cellular pathways. Expression from the human U6 promoter, resulting in siRNA accumulation in the nucleus, is effective in reducing gene expression, whereas cytoplasmic and nucleolar localization of the siRNA when expressed from the 5S or 7 SL promoters is not effective. The distribution of siRNA within the cell is determined by fluorescence *in situ* hybridization. Although the long uninterrupted duplex of siRNA makes it difficult to detect with DNA oligonucleotide probes, labeled oligonucleotide probes with 2'-*O*-methyl RNA backbones provide the stability needed for a strong signal. These methods contribute to studies of the interconnected cellular RNAi pathways and are useful in adapting RNAi as a tool to determine gene function and develop RNA-based therapeutics.

Introduction

RNA interference (RNAi) is the silencing of a specific gene based on sequence complementarity between the initiating RNA and the gene. A naturally occurring phenomenon in multicellular eukaryotes, it also provides a useful tool for turning off specific cellular or viral genes for experimental and therapeutic purposes. The presence of double-stranded RNA (dsRNA) in the cell triggers a pathway that leads to a reduction in the amount of the product of the targeted gene. Long dsRNAs are cleaved into fragments of about 21 nt by an enzyme in the RNase III family, called Dicer, leaving 3' single-stranded overhangs at each end (Denli and Hannon, 2003; Zamore, 2002). These small RNA duplexes interact with several proteins, forming the RNA-induced silencing complex (RISC;

METHODS IN ENZYMOLOGY, VOL. 392

Denli and Hannon, 2003; Zamore, 2002). The RNA component provides the specificity for the multisubunit complex, resulting in a decrease in the expression of genes to which it can base pair. Several mechanisms, including mRNA degradation and transcriptional silencing through heterochromatin formation, are implicated in the action of small interfering RNAs (siRNAs; Elgin and Grewel, 2003; Hutvagner and Zamore, 2002; Nelson et al., 2003). The microRNAs (miRNAs), small RNAs generated from hairpins in endogenous RNAs, share elements of the RNAi pathway and also result in the decrease in the expression of specific genes; however, they can exert their effect by turning off translation as well as sometimes increasing mRNA degradation (Cerruti, 2003; Hutvagner and Zamore, 2002; Nelson et al., 2003). The mechanisms by which RNAi occurs are currently being investigated; however, our incomplete understanding of how this phenomenon works does not present a barrier to the current rapid progress in using RNAi in studies of gene function or to developing potential therapeutics based on RNAi.

Immediately after the role of dsRNA was defined, RNAi was adapted for the targeted knockdown of genes in plants, C. elegans, and Drosophila. However, its application in mammalian cells was delayed until a means of circumventing the apoptotic response to the presence of long dsRNAs in these cells was discovered. Elbashir et al. (2001) found that the introduction of siRNAs, duplexes of about 21 nt, resulted in a decrease in expression of the targeted gene without causing cell death. This breakthrough provided the impetus for the current proliferation of papers that use siRNA technology to assign functions for mammalian genes and to developing methods for the use of siRNA in genome screening (Dorsett and Tuschl, 2004). The simplicity, speed, and specificity of this method, in which synthetic siRNAs are introduced into cells, continue to allow the collection of a large amount of information in a very short time. It is an excellent means of determining gene function in transient assays; however, it does not provide for the stable downregulation of the gene in cell lines or transgenic organisms, systems that are needed to explore many phenotypes, including those involved in development. In addition, expression of small RNAs from endogenous promoters in the nucleus provides access to the cellular transport pathways allowing the direction of siRNA to different subcellular locations, unlike RNA delivered to the cell by transfection from the outside.

The ability to stably downregulate specific genes in cell lines and transgenic animals expands the experimental scope of siRNA as a research tool, and keen interest in using this technology for making knockout mutants prompted the development of a number of systems for the stable expression of siRNAs in mammalian cells (Brummelkamp et al., 2002; Lee et al., 2002; McManus et al., 2002; Miyagishi et al., 2002; Paddison et al.,

2002; Paul *et al.*, 2002; Sui *et al.*, 2002; Xia *et al.*, 2002; Yu *et al.*, 2002). Most of these systems rely on constructs that use RNA polymerase III (pol III) promoters to express siRNA in the cell because siRNAs are small, highly structured RNAs resembling natural RNA pol III products. RNA pol III promoters are highly active and produce large amounts of the small RNAs with ends that usually lack modifications. Endogenous pol III products are found in many parts of the cell, and specific subcellular locations can be targeted through the choice of the promoter used to express the small RNA.

The original set of expression cassettes designed for mammalian cells used the U6 and H1 promoters, which do not require sequences in the coding regions for gene expression (Brummelkamp *et al.*, 2002; Lee *et al.*, 2002; McManus *et al.*, 2002; Miyagishi *et al.*, 2002; Paddison *et al.*, 2002; Paul *et al.*, 2002; Sui *et al.*, 2002; Yu *et al.*, 2002). Other constructs were developed by using pol III promoters from the tRNA^{Val} gene (Kawasaki and Taira, 2003) and the genes for 5S ribosomal RNA (rRNA) and the RNA component of the signal recognition particle (SRP; Paul *et al.*, 2003). Successful knockdowns were also made by expressing siRNA hairpins from an RNA pol II promoter (Xia *et al.*, 2002). The genes for the two complementary RNAs that make up the duplex can be introduced on separate plasmids, annealing after expression in the cells, or they can be synthesized in the cell as RNA hairpins in which the two strands are connected by a loop. More recently, expression cassettes that allow synthesis of the siRNA to be regulated in the cell have provided successful downregulation of the targeted gene in response to a signal. These include a tetracycline-inducible U6 promoter (Matsukara *et al.*, 2003), the ecdysone promoter (Gupta *et al.*, 2004), and the Cre-lox system (Kasim *et al.*, 2004). The use of regulated and tissue-specific promoters will be particularly useful for investigations into the roles of particular genes during development. In addition, promoter cassettes have been placed in viral vectors, resulting in higher transfection efficiencies, thus preventing the masking of gene ablation by gene expression in untransfected cells that can occur when plasmids are introduced into cells at low efficiencies. Regardless of how the RNA is delivered to the cell, the small dsRNAs must enter the RNAi pathway, providing the specificity for downregulation of the target gene.

Although the details of the mechanisms through which RNAi results in a decrease in gene expression are not yet understood, it is clear that the various pathways overlap and that some components are shared, but there must also be variability that accounts for regulation at different levels of gene expression (Cerruti, 2003; Nelson *et al.*, 2003). Many of the components of the RNAi pathway have been found in the cytoplasm. Dicer is present predominantly in the cytoplasm, as are several of the RISC components; however, several participants in pathways involved in gene silencing are found at least in part in the nucleus (Cerruti, 2003). There is also evidence that the mRNA may be degraded as it exits the nucleus (Zeng and Cullen, 2002) and at least one type of RNAi, gene silencing through heterochromatin formation, is a nuclear process.

The location of products of pol III expression cassettes can be monitored in the cell by *in situ* hybridization with fluorescent probes that hybridize to the insert. Transcripts expressed from the U6 promoter accumulate in the nucleoplasm, whereas those expressed from the 5S and 7 SL promoters are found in the nucleolus and cytoplasm (Good *et al.*, 1997; Paul *et al.*, 2002, 2003). The siRNA hairpins expressed from the U6 cassette resulted in reduction in gene expression, whereas those from the 5S and 7 SL promoters did not (Paul *et al.*, 2002, 2003). These results suggest that the subcellular location of the transcripts is important to ensure that they can enter the RNAi pathway and that nucleoplasmic expression provides access to the pathway. A set of cassettes derived from the U6, 5S, and 7 SL endogenous RNAs that provide the ability to localize small RNA inserts to various subcellular domains is described. A method for experimentally determining the location of siRNAs within the cell through *in situ* hybridization by using fluorescently tagged oligonucleotide probes with modified backbones is also provided.

Localizing Small Interfering RNA (siRNA) Through Expression
     from RNA pol III Promoters

Expression Cassettes for Small RNAs

The expression of small RNAs from RNA pol III promoters in cells can be achieved by constructing plasmid or viral vectors containing expression cassettes with all the required promoter elements. In addition to upstream sequence elements, sequences within the coding region are often required for the transcription of pol III genes. These internal sequences and those required for transport of the transcript to its natural subcellular location are included in the cassettes. Four different constructs, based on three different pol III genes, were made, allowing the efficacy of the small RNAs to be tested in a variety of subcellular locations (Good *et al.*, 1997; Paul *et al.*, 2003). These constructs were based on the genes for the U6 small nuclear RNA (snRNA), found in nucleoplasmic speckles, and the 7 SL RNA from the SRP and 5S rRNA, both of which travel through the nucleolus and into the cytoplasm.

Each construct contains sequences upstream of the transcription start site and relevant sequences derived from the coding region (Fig. 1). Restriction sites are included in the constructs for the insertion of siRNA sequences or other small RNAs. A stem loop and a polyuridine tract at the 3' end of the cassette signal transcription termination. Transcripts with

A U6 cassette

B 5S cassette

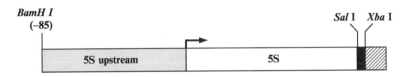

C 7SL cassette

FIG. 1. Expression cassettes for small RNAs that are based on three different RNA polymerase III (pol III) genes. Each cassette includes the upstream and internal promoter elements required for expression and regions required for subcellular localization. Sequences upstream of the transcription start site are represented by shaded boxes and sequences from within the coding region are represented by white boxes. The sites for cleavage by the restriction enzymes *Sal* I and *Xba* I allow insertion of the siRNA sequence into the cassette. The hatched box at the 3′ end of each cassette indicates a stem loop followed by a poly U$_5$ sequence that provides for transcription termination and increases stability of the transcript through increased resistance to 3′ exonucleases. (A) Cassette based on the human U6 snRNA gene. The U6+27 cassette is shown. The U6+1 cassette differs from the U6+27 cassette only by the omission of the first 27 bases of the U6 snRNA sequences from the transcript. (B) Cassette based on the human 5S rRNA gene. (C) Cassette based on the gene that codes for the RNA component of the human signal recognition particle, 7 SL RNA. In this cassette, endogenous RNA sequences flank the siRNA insertion site.

a 3′ stem loop may exhibit enhanced stability in cells because the strong duplex will make them more resistant to exonucleases. The 21-bp duplexes of siRNA hairpins are much more intrinsically stable than ribozymes, aptamers, or antisense RNAs, and therefore a polyuridine tract can be inserted at the 3′ end of the siRNA sequence, providing for transcription termination and resulting in transcripts that do not contain the additional 3′ stem loop.

A   U6 transcripts

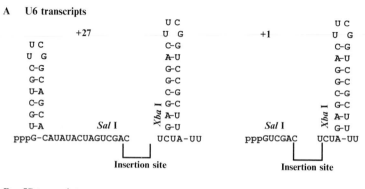

B   5S transcript

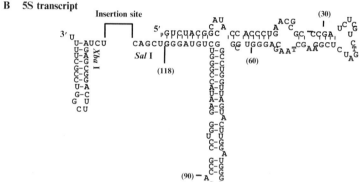

C   7 SL transcript

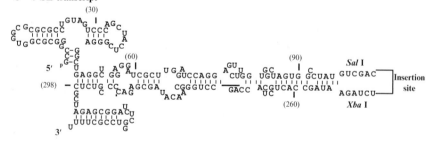

Fig. 2. Transcripts from the pol III cassettes. The sequence and secondary structure of the transcripts are shown. The positions of the restriction sites for insertion of siRNA sequences are indicated. Note the identical stem loop and poly U sequence at the 3' end of each cassette. (A) Transcripts of cassettes based on U6 snRNA. The 5' stem loop formed by the endogenous U6 sequence is present in the U6+27 transcripts, but is absent from the U6+1 transcripts. (B) Transcripts of cassette based on 5S rRNA. The entire 5S rRNA sequence is followed by the sites for siRNA insertion. (C) Transcripts of cassette based on 7 SL RNA. The 5' and 3' portions of the 7 SL RNA that make up the Alu domain interact through base pairing and fold into their native conformation and the siRNA insert replaces the S domain of the 7 SL RNA.

The endogenous U6 RNA is a spliceosomal RNA and accumulates in speckles in the nucleoplasm (Carmo-Fonseca *et al.*, 1992). The vertebrate U6 promoter is entirely upstream of the transcription start site (Danzeiser *et al.*, 1993; Kunkel and Pederson, 1989; Kunkel *et al.*, 1986; Noonberg *et al.*, 1994), unlike most pol III promoters that require elements within the coding region for initiation of transcription, making it suitable for expressing small RNA transcripts that lack endogenous U6 sequences. The inclusion of endogenous U6 sequences in the transcript is not required for nuclear localization. No signal is essential for nuclear retention and all U6-driven transcripts, whether or not they contain U6 sequences, remain in the nucleus just as the endogenous U6 snRNA does (Good *et al.*, 1997).

Two constructs were made by using the U6 promoter, one that expresses the small RNA without any endogenous U6 sequences (U6+1) and one (U6+27) that includes the first 27 bases of the U6 snRNA sequence in the transcript (Good *et al.*, 1997; Figs. 1A and 2A). In the U6+27 transcript, the bases derived from endogenous U6 snRNA fold to form a stem-loop structure, increasing stability and signaling for the addition of a 5′ gamma-monomethyl phosphate cap (Good *et al.*, 1997). Both cassettes contain two restriction sites, *Sal* I and *Xba* I, for insertion of the siRNA sequence (Fig. 1A). The *Xba* I site is followed by the 3′ stem loop for transcription termination (Fig. 1A). Alternatively, transcription can be terminated immediately after the siRNA hairpin by the inclusion of a poly U stretch at the 3′ end of the siRNA insert, just upstream of the *Xba* I site. Four uridine residues can be sufficient for transcription termination (White, 1998), but allowed substantial readthrough in a U6 cassette (Paul *et al.*, 2003). A stretch of five or six uridine residues will provide for more efficient termination. Small RNAs, including siRNAs, expressed from the U6 promoter using the U6+1 and U6+27 cassettes are found in nuclear speckles, just as is endogenous U6 snRNA (Good *et al.*, 1997; Paul *et al.*, 2002, 2003; Fig. 3).

The RNA component of the SRP, 7 SL RNA, is transcribed by RNA pol III and passes through the nucleolus before accumulating in the cytoplasm (Jacobson and Pederson, 1998). The SRP interacts with ribosomes that are translating proteins destined for the endoplasmic reticulum (ER), bringing them to the ER membrane. Promoter elements are located upstream of the transcription start site and within the coding region (Bredow *et al.*, 1990a,b; Kleinert *et al.*, 1988; Ullu and Weiner, 1985). Cellular 7 SL RNA folds into a base-paired structure in which the 3′ and 5′ ends interact. The structure can be divided into two domains that are defined by structure and function. The Alu domain, named because of its similarity to Alu RNAs, contains the 5′ and 3′ ends of the 7 SL RNA and binds the p9/14 protein heterodimer (Strub *et al.*, 1991; Weichenrieder *et al.*, 1997). This domain contains the internal elements required for transcription initiation and is responsible for transport of the SRP from the nucleus to the cytoplasm (Strub *et al.*, 1991; Weichenrieder *et al.*, 1997). The central region of the 7 SL RNA forms the S domain, and, along with associated proteins, interacts with the ribosome and is responsible for stalling translation until the ribosome is bound to the ER. The 7 SL cassette is made up of the sequences that form the Alu domain (Figs. 1B and 2C). The two noncontiguous regions of the Alu domain can interact in the transcript, forming their native conformation when in the context of the full-length molecule (Fig. 2C). The restriction sites are located so that the siRNA sequences are inserted in place of the deleted S domain. The same 3' stem loop used to aid in transcription termination in the U6 cassettes is included at the 3' end of the 7 SL cassette (Figs. 1B and 2C).

The subcellular distribution of the transcripts from the 7 SL cassette is similar to that for endogenous 7 SL RNA (Paul *et al.*, 2003). Both are found in the nucleolus and cytoplasm, although the 7 SL–siRNA chimeras have greater accumulation in the nucleolus, possibly due to inefficient packaging or export of the structurally altered transcripts. The inclusion of a polyuridine sequence at the 3' end of the siRNA insert will result in a transcript containing the 5', but not the 3', portion of the Alu domain RNA. These transcripts are predicted to localize like full-length 7 SL RNA because the 5' portion contains the internal promoter elements and sequences required for export from the nucleus (He *et al.*, 1994; Weichenrieder *et al.*, 2000).

A third RNA pol III gene, for 5S rRNA, provides the promoter and localization signals for another expression cassette. Promoter elements are found within the gene, but upstream sequences are also important (Hallenberg and Frederiksen, 2001; Nielsen *et al.*, 1993). Endogenous 5S rRNA is found in the nucleolus and cytoplasm. The 5S rRNA expression cassette includes the upstream region and the complete 5S rRNA coding sequence, because the intact 5S rRNA is required for proper RNA–protein complex formation and transport to the cytoplasm (Figs. 1C and 2B). The restriction sites for siRNA insertion are at the 3' end of the 5S rRNA sequence. The same 3' stem loop included at the end of the U6 and 7 SL cassettes is included after the restriction sites. Transcripts from the 5S rRNA cassette are found in the same subcellular locations as endogenous 5S rRNA, the nucleolus and cytoplasm, although there is again greater nucleolar accumulation of the chimeric transcript (Paul *et al.*, 2003; Fig. 3).

Designing siRNA Inserts for Expression from pol III Promoters

When expressing siRNAs in cells, the same concerns must be taken into consideration as in the design of synthetic siRNAs for introduction directly into the cell. The target sequence within the gene is an important variable in the design of the siRNA. The rules governing susceptibility of the target to siRNA are not understood completely, although availability to hybridize

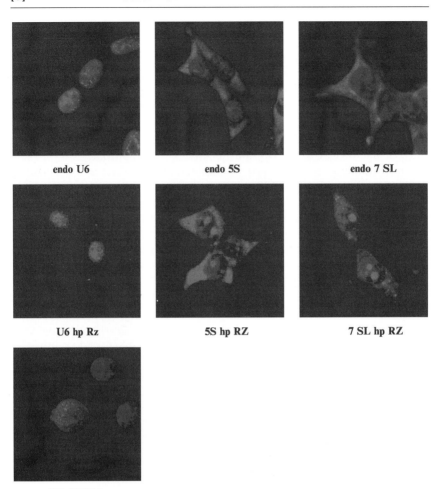

endo U6   endo 5S   endo 7 SL

U6 hp Rz   5S hp RZ   7 SL hp RZ

U6 siRNA hp

FIG. 3. Subcellular distribution of small RNA inserts expressed from the pol III cassettes and the endogenous RNAs from which the cassettes were derived. The localization of each insert corresponds to that of the endogenous RNA from which the cassette used for its expression was derived. The nucleoplasm of the cells is indicated by staining with DAPI (blue). The distributions of hairpin ribozyme inserts expressed from the U6+27, 5S, and 7 SL cassettes are shown beneath those of the corresponding endogenous RNA (red). The pattern of siRNA accumulation when expressed from the U6+27 cassette is also shown (red). Hairpin ribozyme and endogenous RNA probes were DNA oligonucleotides labeled with Cy3 at the 5' terminus. The siRNA was detected with a 2'-O-methyl ribooligonucleotide labeled with Cy3 at the 5' terminus. The sequences of the probes are as follows: endogenous U6 5' CACGAATTTGCGTGTCATCCTTGCGCAGGGGCC 3', endogenous 5S 5' CTTAGCTTCCGAGATCAGACGAGATCGGGCGCG 3', endogenous 7 SL 5' CCGGGAGGTCACCATATTGATGCCGAACTTAGTGCG 3', hairpin ribozyme 5' GCCAGGTAATATACCACAACGTGTGTTTCTCTGGTTGCCTTCTTG 3', and antilamin siRNA 5' AAACUGGACUUCCAGAAGAACACGAA 3'. (See color insert.)

with the siRNA complement appears to play a role (Lee *et al.*, 2002). Until the rules become apparent, several sequences within a target gene should be tested simultaneously to increase the likelihood of finding an effective siRNA. All target sequences should be tested against a database to ensure that they are not present elsewhere in the genome and that the siRNA will be specific for the gene for which the siRNA was designed.

In addition to the same concerns about choosing a target as when applying synthetic siRNAs to cells, consideration must be given to the form of the siRNA for expression in cells. Several approaches have been taken for making small dsRNAs in mammalian cells. The two strands can be expressed separately, forming a duplex with free 5′ and 3′ ends, or as one long strand that can fold back on itself, forming a hairpin with a loop at one end and free ends at the other. Both forms result in successful downregulation of the targeted gene (Brummelkamp *et al.*, 2002; Dykxhoorn *et al.*, 2003; Lee *et al.*, 2002; McManus *et al.*, 2002; Miyagishi *et al.*, 2002; Paddison *et al.*, 2002; Paul *et al.*, 2002; Sui *et al.*, 2002; Xia *et al.*, 2002; Yu *et al.*, 2002). When expressing the siRNA as a hairpin, the two complementary strands can be connected by the loop at either end (Dykxhoorn *et al.*, 2003; Paul *et al.*, 2002). A variety of formats, differing in the size of the loop and length of the stem, have proven successful (Dykxhoorn *et al.*, 2003). Differences in the targets or the expression systems may be responsible for some apparent discrepancies between descriptions of the optimal design. Different designs may result in silencing through distinct mechanisms, turning off the genes at transcription or posttranscriptionally. In our hands, an siRNA hairpin in which the sense strand was connected at the 3′ end to the 5′ end of the antisense strand by a UUCG tetraloop was most efficient (Fig. 4A), reducing lamin expression by more than 90% (Paul *et al.*, 2003). A construct in which the loop was placed at the other end of the duplex resulted in approximately 75% reduction in lamin expression.

Once the target sequence and design of the siRNA insert are selected, complementary DNA oligonucleotides can be purchased, which are then annealed and inserted into an expression cassette. To express the siRNA as a hairpin, the oligonucleotides contain the sense and antisense sequences connected by a loop. To provide for termination of transcription immediately following the siRNA hairpin, five thymidine residues are included in the coding strand at the 3′ end of the hairpin, providing a polyuridine RNA pol III transcription termination signal. The 3′ terminal overhang formed by the poly U sequence in the transcript resembles the single-stranded overhangs that result from cleavage by Dicer and that have been linked to efficient siRNA function (Elbashir *et al.*, 2001). Finally, restriction sites for cleavage with *Sal* I and *Xba* I are included near the ends of the

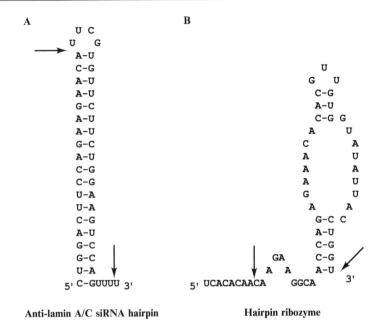

FIG. 4. Small RNA inserts expressed from pol III cassettes. The sequence and secondary structure are shown. Sequences encoding the inserts are placed between the *Sal* I and *Xba* I sites in each of the cassettes. (A) Antilamin siRNA insert (Paul *et al.*, 2002). (B) Hairpin ribozyme insert (Good *et al.*, 1997). The sequences recognized by the probes lie between the arrows.

oligonucleotides when they are synthesized, allowing easy insertion of the annealed oligonucleotides into the expression cassettes.

Determining the Subcellular Distribution of siRNA by *In Situ* Hybridization

Introducing siRNA Constructs into Cells

Expression cassettes can be introduced into cells via a plasmid or viral vector, the choice often depending on the transfection efficiency of the cells. Many lipid-based reagents for the introduction of plasmids are available commercially and cell type is the most important factor when choosing among them. Lipofectamine with Plus™ Reagent and Lipofectamine™ 2000 (Invitrogen, Carlsbad, CA) were effective for transfection of HeLa cells (Paul *et al.*, 2003). Calcium phosphate-mediated transfection is also an efficient means of introducing DNA into many types of cells, including 293 cells (Paul *et al.*, 2003). Splitting the cells the day after transfection allows several assays to be performed on cells from the same sample. Cells can be plated on coverslips for *in situ* hybridization to localize the siRNA or mRNA, or for immunolocalization of protein targets or the proteins involved in the RNAi pathways. Cells from each sample can be expanded to provide sufficient material to perform additional assays, such as Northern or Western blots to determine RNA and protein levels to monitor the efficacy of the siRNA construct. Splitting the cells also provides a way to assess the transfection efficiency of each sample when the cells are cotransfected with a control plasmid expressing β-galactosidase or green fluorescent protein (GFP); (3:1 molar ratio of experimental to control plasmid). The presence of β-galactosidase can be assayed cytochemically (Paul *et al.*, 2003) and GFP can be detected by direct observation with a fluorescence microscope. Immunolocalization and Western blot are alternative assays to detect transfection markers.

For assays using microscopy, coverslips are baked at 150–200° for at least several hours and then cooled before placing them in six-well tissue culture plates by using sterile forceps. Transfected cells are plated on the coverslips after one day and grown for another 1–2 days (2–3 days post-transfection). The plating density varies with the growth rate of the cells and is adjusted so that they are not confluent at the time they are harvested. HeLa and 293 cells are plated at a low density (5–10% confluency) because of their rapid growth rates.

Fixing Cells for siRNA Localization

The use of fluorescent probes requires that the fixation procedure result in low background fluorescence as well as maintain the cell structure, criteria met by fixation in paraformaldehyde. Fixing the cells directly in the media in which the cells are grown prevents changes in cell structure that result from stress responses. Even common procedures for handling cells in the laboratory can cause stress, which results in changes in cell structure. For example, changing the media by spinning down and resuspending yeast cells causes a rearrangement of the actin cytoskeleton (Pringle *et al.*, 1989). Any changes in cellular architecture that occur due to handling can result in inaccurate localization caused by a change in the subcellular distribution of the protein or RNA with which the probe interacts or to a change in the localization of a control. Handling of the cells is kept to a minimum before fixation to avoid this type of artifact.

Optimal conditions for cell fixation must be determined for each cell type and for each set of growth conditions by varying the concentration of the paraformaldehyde and fixation time. The high background fluorescence that has sometimes been reported as a result of fixation with paraformaldehyde can be avoided by using good-quality paraformaldehyde, such as that from EMS Scientific (Ft. Washington, PA). The inclusion of glacial acetic acid during fixation can be beneficial in localizing nuclear contents (http://singerlab.aecom.yu.edu/protocols). To localize small RNAs expressed from the U6, 5S, and 7 SL cassettes, HeLa and 293 cells were fixed by adding acetic acid and paraformaldehyde to the media in which they were grown such that final concentrations of 4% acetic acid and 10% paraformaldehyde were obtained. Cells were incubated for 10 min at room temperature before washing in PBS (58 mM Na$_2$HPO$_4$, 17 mM NaH$_2$PO$_4$, 68 mM NaCl, pH 7.4) twice for 5 min each time. The cells were permeabilized by incubation in 70% ethanol at 4° for at least overnight. Several criteria suggest that the subcellular distribution of components was maintained. Cells treated in this way retained their characteristic cell shape after fixation. Staining with 4′,6-diamidino 2-phenylindole (DAPI) revealed nuclei with obvious nucleoli. *In situ* hybridization of fluorescent probes to cells fixed in this way indicated that the endogenous small RNAs U6 snRNA, 5S rRNA, and 7 SL RNA were found in their expected locations (Paul *et al.*, 2003).

Growth of the cells on 22-mm^2 coverslips in six-well tissue culture plates minimizes the handling of cells before fixation, avoiding possible stress responses. Transfected cells are plated at the appropriate density in 2 ml of media on baked coverslips in six-well plates as described in the previous section. After they have grown for 2–3 days posttransfection, the cells are fixed by adding 440 μl of a freshly prepared 1:1 (v/v) mixture of 40% paraformaldehyde and glacial acetic acid. Alternatively, multiple coverslips can be placed in a 9-cm culture dish and grown in 10 ml of media. Addition of 2.2 ml of a 1:1 (v/v) mixture of 40% paraformaldehyde and acetic acid directly to the media fixes the cells. Once the cells are fixed and washed as described, they can by stored in the 70% ethanol in which they were permeabilized at 4° for at least several months before use.

Choosing a Probe

Detection of the small RNAs made from the pol III expression cassettes can be challenging as the small size of transcripts limits the choice of probes. The probe may often be complementary to the entire transcript, so that the GC content reflects that of the target sequence and cannot be changed. In addition, it may not be possible to exclude hairpins from the probe. siRNAs that have been expressed as hairpins containing long duplexes are particularly difficult to detect because the target RNA in the cell, and possibly the probe, will have to be denatured for complementary base pairing between target and probe to occur. In most cellular RNAs, regions of base pairing alternate with single-stranded regions. Duplexes found in natural RNAs are usually shorter than the 19 uninterrupted base pairs found in siRNA and often contain mismatched bases or loops that decrease their stability. Despite the presence of many short duplexes, these endogenous RNAs are readily detected with DNA oligonucleotide probes (Fig. 3) that have fluorophores attached at the 5′ end. Labeled oligonucleotide probes can be purchased from commercial vendors, including Invitrogen (Carlsbad, CA) and Operon (Alameda, CA). Like endogenous RNAs, the hairpin ribozyme has short regions of base pairing alternating with open loops (Fig. 4B), making it easy to detect in fixed cells with DNA probes labeled at the 5′ terminus (Fig. 3). In contrast, the long base-paired region found in siRNA cannot be readily detected with a DNA probe. Use of a 2′-O-methyl-substituted RNA oligonucleotide instead of a DNA probe provides stable binding to the siRNA in the cell, allowing detection of the siRNA sequence in the cells. The increased stability of the duplex due to the modified backbone favors formation of hybrids with the probe. Signal from 2′-O-methyl-substituted RNA oligonucleotides labeled at the 5′ end shows that siRNA expressed from the U6+27 cassette is found in the nucleoplasm (Fig. 3). In contrast, the signal from a 5′-end-labeled DNA probe could not be detected (data not shown) even though hybridization conditions were the same as those used to detect the endogenous small RNAs and the hairpin ribozyme expressed from the U6+27 cassette (Fig. 3). Trilink Biotechnologies (San Diego, CA) can provide 2′-O-methyl oligonucleotides that are labeled at the 5′ end with a fluorophore for localizing siRNA.

The length of the probe will often be dictated by the size of the insert sequence. Extra bases, such as those from the flanking restriction sites or the endogenous RNA from which the cassette sequence was derived, should not be included in the probe so that cross hybridization with other RNAs in the cell can be avoided. When detecting siRNAs in which each strand is expressed individually, the probe will consist of the entire length of one strand of the siRNA. When detecting siRNA expressed as a hairpin, the probe sequence can extend through the loop; however, the full length of the hairpin is not included to prevent the probe oligonucleotide from folding back on itself. Probes identical in sequence to the target in the message will recognize the complementary strand of the siRNA, but not the mRNA, simplifying interpretation of the results.

The choice of fluorophore for labeling each probe will depend on the microscopy equipment available, because appropriate filters are required to detect individual fluorophores and the appropriate combination of filters is required for the use of multiple fluorophores without overlap of signals.

Sensitive detection of small amounts of RNA depends on a fluorescent dye that generates a strong signal and does not undergo rapid photobleaching. A large variety of fluorophores are available commercially and on request oligonucleotide probes can be commercially synthesized with a fluorophore attached to the 5' end. End labeling of probes is usually sufficient for these short probes, because incorporation of multiple fluorophores can result in quenching of the signal if two dye molecules are in close proximity.

At least two separate fluorophores are required to establish the distribution pattern in the cell, one for the siRNA probe and another for an RNA that serves as an endogenous marker for which the subcellular location is known. For example, subcellular distribution of the cassette inserts can be confirmed by comparing the signal from oligonucleotide probes labeled with Cy3 to that of a probe for an endogenous RNA that is labeled with another fluorophore, such as Oregon Green. Alternatively, their location can be determined by comparison with the signal from nuclei stained with DAPI (Fig. 3). To be certain that there is clear separation of the two signals, control slides labeled with only one fluorophore should be included in experiments that use multiple labels.

Hybridization Conditions

In the cell, siRNA is likely to be base paired to the complementary siRNA strand or the target mRNA and is also likely to be complexed with proteins, such as those found in the RISC. During incubation with the probe, conditions must allow disruption of enough of these interactions so that the probe can bind to its target, but at the same time the structure of the cell must be maintained for accurate localization. The same conditions that are used for hybridization of DNA probes to less structured cellular RNAs, while keeping the organization of the cell intact, allow hybridization of 2'-O-methyl RNA oligonucleotide probes to the long duplexes in siRNA. We are using conditions that are minor modifications of published protocols. The hybridization solution contains 10% dextran sulfate, 2 mM vanadyl ribonucleoside complex (New England BioLabs, Beverly, MA), 0.02% bovine serum albumin (BSA), 1 mg/ml yeast tRNA, 2× SSC, and 50% formamide (http://singerlab.aecom.yu.edu/protocols). The amount of probe used for each slide is best determined empirically, usually varying from 6 to 30 ng. A higher amount of probe may be required if the signal is weak, whereas reducing the amount of probe can decrease background fluorescence. Sufficient hybridization solution is made from stock solutions (Table I) for each experiment so that 40 μl of the solution can be used for each slide. A sample should be removed from the master mix before the probe is added so that a control slide to check for endogenous cellular fluorescence can be examined side-by-side with the slides with probe. This prevents fluorescence due to cellular components from being mistaken for hybridization signal. This background fluorescence can vary with growth conditions and fixation and so should be checked in each experiment. After removal of this aliquot, the probe is added to the master mix and then both are brought to the appropriate volume with water.

Adherent cells that have been grown on coverslips are incubated with hybridization solution by placing a 40-μl aliquot on each coverslip and then placing the coverslip on a glass microscope slide. Forceps are used to gently lower the coverslip at an angle to the slide so that bubbles are not trapped underneath it. The coverslips are sealed to the slides with rubber cement and placed in a humid environment for incubation at 37° overnight. Commercially available slide incubators such as the Boekel Scientific (Feasterville, PA) slide moat provide the humid environment needed. Alternatively, slides can be incubated on moist paper towel in a sealed plastic container that is placed in an available laboratory incubator.

After incubation with the probe is complete, coverslips are removed from the slides and are conveniently washed in six-well tissue culture plates. Cells are washed two times for 30 min each time in 2× SSC, 50% formamide at 37° and then once in PBS for 15 min at room temperature. Following the washes, the cells can be stained with DAPI for the identification of nuclei by incubating the coverslips for 5 min in 2 ng/μl DAPI in PBS. The incubation can be done on glass microscope slides or between two tightly stretched layers of parafilm (Chartrand et al., 2000), minimizing the

TABLE I

HYBRIDIZATION SOLUTION FOR THE INCUBATION OF FLUORESCENTLY
TAGGED PROBES WITH CELLS FOR IN SITU HYBRIDIZATION

Stock solution	One slide (μl)	Ten slides (μl)
Dextran sulfate, 50%	8.0	80.0
Vanadyl ribonucleoside complex, 200 mM	0.4	4.0
Bovine serum albumin, 10 mg/ml	0.8	8.0
Yeast tRNA, 10 mg/ml	4.0	40.0
20 × SSC	4.0	40.0
Formamide, 50%	20.0	200.0
Probe (5 ng/μl)	1.2	12.0[a]
dH$_2$O	1.6	16.0[a]
Total volume	40.0	400.0[a]

[a] Volumes must be adjusted to prepare the control sample that lacks probe as described in the text.

volume of solution needed. Coverslips are then placed back in the six-well plates and washed three times with PBS for 5 min each time. Finally, coverslips are mounted for viewing on clean microscope slides, using ProLong™ mounting medium (Molecular Probes, Eugene, OR). During all manipulations, care should be taken to minimize the exposure of the probe to light because of the sensitivity of the fluorophores.

Microscopy and Data Collection

A fluorescence microscope with the appropriate filters and a camera is required. Filter sets must be designed to prevent overlap of the signals from multiple fluorophores. A Nikon (Melville, NY) Eclipse E800 microscope and a Hamamatsu (Bridgewater, NJ) Orca II digital camera were used to prepare the photographs presented here (Fig. 3). Filter sets were designed by Chroma Technology Corporation (Rockingham, VT) to allow double and triple labeling. Clear images can be obtained by eliminating fluorescence originating from the different planes of the three-dimensional cell. Confocal microscopes record the fluorescence found in a single plane of the cell. Comparable images can be obtained through deconvolution of a set of images taken at regular intervals through one axis of the cell (Z-stack). Using deconvolution software, signal is returned to the plane from which it originated and eliminated from other planes (Shaw, 1995). Deconvolution of images in Fig. 3 was performed with ISEE Analytical Imaging Software (Inovision Corporation, Raleigh, NC).

Experimental Design and Interpretation of Results

Subcellular location of the RNA inserts is determined by comparison with markers for which the distribution is known. The locations of the hairpin ribozyme and siRNA expressed from the U6, 5S, and 7 SL cassettes are shown relative to the location of the cell nuclei (Fig. 3). DAPI stains the DNA, providing clear identification of the nucleoplasm in contrast to the nucleoli and cytoplasm, regions of the cell that are not detectably stained. Small RNAs expressed from each cassette are distributed in cells in a pattern similar to that of the corresponding endogenous RNA. Endogenous U6 transcripts are found in speckles in the nucleoplasm (Fig. 3). Hairpin ribozyme and siRNA inserts expressed from the U6+27 cassette are distributed in an identical pattern in the cell (Fig. 3). Endogenous 5S rRNA and a hairpin ribozyme insert expressed from the 5S cassette are found in the nucleoli and cytoplasm (Fig. 3). A similar distribution is seen when endogenous 7 SL or a hairpin ribozyme expressed from the 7 SL cassette are localized (Fig. 3). Colocalization of two RNAs can be demonstrated by attaching different labels to probes that recognize each of the RNAs and then incubating a slide with both probes simultaneously.

To establish the cellular distribution of an RNA, it is necessary to include several controls in each experiment. Fluorescence due to endogenous cellular components can be identified by examining slides that have not been probed with labeled oligonucleotides. Cross hybridization of the probe to other cellular nucleic acids or binding of the probe to other cellular components can be identified if signal occurs in cells that are not expressing the small RNA. Untransfected cells or cells that have been transfected with a cassette with no insert may be used for this purpose. In multilabel experiments, slides that are hybridized to each probe separately can be compared to a slide that is probed with all the labels together, allowing identification of bleed-through from one signal to another.

When analyzing the experimental results, a large number of cells should be examined on each slide so that the subcellular distribution of the small RNA can be firmly established and the selection of cells that are not representative avoided. For example, cell components can be redistributed during cell division when the nuclear membrane breaks down and siRNA is found throughout the cell at this stage of the cell cycle. Dividing cells can be identified by the presence of clearly visible chromosomes when the cells are stained with DAPI. Photographs taken at low magnification, so that they include many cells, are useful in establishing the most common distribution of the RNA.

Conclusions

Identification of sites of siRNA accumulation in cells will help unravel the overlapping pathways by which small RNAs provide the specificity for gene silencing. The ability of a small RNA to enter a particular pathway is dependent on its availability for interaction with specific cellular components. This requires colocalization of the siRNA and these factors in the cell. Cassettes based on the U6 snRNA promoter provide for accumulation of small RNAs in the nucleus. The expression of siRNA from U6-based cassettes results in downregulation of gene expression, whereas expression from cassettes that provide for nucleolar and cytoplasmic accumulation of the insert does not. The development of several pol III expression cassettes that send the small RNAs to different subcellular locations provides options for expression of small RNAs for a variety of experimental and therapeutic purposes. The subcellular location of siRNAs can be found through *in situ* hybridization with a fluorescently tagged 2'-*O*-methyl oligonucleotide. The modified backbone of the probe allows it to form stable hybrids with its target and to successfully compete with the complementary strand of the siRNA.

Acknowledgments

I thank Dave Engelke and Paul Good for their contributions to this work.

References

Bredow, S., Kleinert, H., and Benecke, B.-J. (1990a). Sequence and factor requirements for faithful *in vitro* transcription of human 7SL DNA. *Gene* **86,** 217–225.

Bredow, S., Sürig, D., Müller, J., Kleinert, H., and Benecke, B.-J. (1990b). Activating-transcription-factor (ATF) regulates human 7SL RNA transcription by RNA polymerase III *in vivo* and *in vitro*. *Nucleic Acids Res.* **18,** 6779–6784.

Brummelkamp, T. R., Bernards, R., and Agami, R. (2002). A system for stable expression of short interfering RNAs in mammalian cells. *Science* **296,** 550–553.

Carmo-Fonseca, M., Pepperkok, R., Carvalho, M. T., and Lamond, A. I. (1992). Transcription-dependent colocalization of the U1, U2, U4/U6, and U5 snRNPs in coiled bodies. *J. Cell Biol.* **117,** 1–14.

Cerruti, H. (2003). RNA interference: Traveling in the cell and gaining functions. *Trends Genet.* **19,** 39–46.

Chartrand, P., Bertrand, E., Singer, R. H., and Long, R. M. (2000). Sensitive and high-resolution detection of RNA in situ. *Methods Enzymol.* **318,** 493–506.

Danzeiser, D. A, Urso, O., and Kunkel, G.R. (1993). Functional characterization of elements in a human U6 small nuclear RNA gene distal control region. *Mol. Cell. Biol.* **13,** 4670–4678.

Denli, A. M., and Hannon, G. J. (2003). RNAi: An ever-growing puzzle. *Trends Biochem. Sci.* **28,** 196–201.

Dorsett, Y., and Tuschl, T. (2004). siRNAs: Application in functional genomics and potential as therapeutics. *Nat. Rev. Drug Discov.* **3,** 318–329.

Dykxhoorn, D. M., Novina, D. D., and Sharp, P. A. (2003). Killing the messenger: Short RNAs that silence gene expression. *Nat. Rev. Mol. Cell. Biol.* **4,** 457–467.

Elbashir, S. M., Harborth, J., Lendeckel, W., Yalcin, A., Weber, K., and Tuschl, T. (2001). Duplexes of 21-nucleotide RNAs mediate RNA interference in cultured mammalian cells. *Nature* **411,** 494–498.

Elgin, S. C. R., and Grewel, S. I. S. (2003). Heterochromatin: Silence is golden. *Curr. Biol.* **13,** R895–R898.

Good, P. D., Krikos, A. J., Li, S. X. L., Bertrand, E., Lee, N. S., Giver, L., Ellington, A., Zaia, J. A., Rossi, J. J., and Engelke, D. R. (1997). Expression of small, therapeutic RNAs in human cell nuclei. *Gene Ther.* **4,** 45–54.

Gupta, S., Schoer, R. A., Egan, J. E., Hannon, G. J., and Mittal, V. (2004). Inducible, reversible, and stable RNA interference in mammalian cells. *Proc. Natl. Acad. Sci. USA* **101,** 1927–1932.

Hallenberg, C., and Frederiksen, S. (2001). Effect of mutations in the upstream promoter on the transcription of human 5S rRNA genes. *Biochim. Biophys. Acta* **1520,** 169–173.

He, X.-P., Bataillé, N., and Fried, H. M. (1994). Nuclear export of signal recognition particle RNA is a facilitated process that involves the Alu sequence domain. *J. Cell Sci.* **107,** 903–912.

Hutvagner, G., and Zamore, P. D. (2002). RNA: Nature abhors a double-strand. *Curr. Opin. Genes Dev.* **12,** 225–232.

Jacobson, M. R., and Pederson, T. (1998). Localization of signal recognition particle RNA in the nucleolus of mammalian cells. *Proc. Natl. Acad. Sci. USA* **95,** 7981–7986.

Kasim, V., Miyagishi, M., and Taira, K. (2004). Control of siRNA expression using the Cre-lox P recombination system. *Nucleic Acids Res.* **32,** e66.

Kawasaki, H., and Taira, K. (2003). Short hairpin type of dsRNAs that are controlled by tRNA$^{(Val)}$ promoter significantly induce RNAi-mediated gene silencing in the cytoplasm of human cells. *Nucleic Acids Res.* **31,** 700–707.

Kleinert, H., Gladen, A., Geisler, M., and Benecke, B.-J. (1988). Differential regulation of transcription of human 7SK and 7SL RNA genes. *J. Biol. Chem.* **263,** 11511–11515.

Kunkel, G. R., Maser, R. L., Calvet, J. P., and Pederson, T. (1986). U6 small nuclear RNA is transcribed by RNA polymerase III. *Proc. Natl. Acad. Sci. USA* **83,** 8575–8579.

Kunkel, G. R., and Pederson, T. (1989). Transcription of a human U6 small nuclear RNA gene *in vivo* withstands deletion of intragenic sequences but not of an upstream TATATA box. *Nucleic Acids Res.* **17,** 7371–7379.

Lee, N. S., Dohjima, T., Bauer, G., Li, H., Li, M.-J., Ehsani, A., Salvaterra, P., and Rossi, J. (2002). Expression of small interfering RNAs tarteted against HIV-1 rev transcripts in human cells. *Nat. Biotechnol.* **20,** 500–505.

Matsukara, S., Jones, P. A., and Takai, D. (2003). Establishment of conditional vectors for hairpin siRNA knockdowns. *Nucleic Acids Res.* **31,** e77.

McManus, M., Petersen, C. P., Haines, B.B., Chen, J., and Sharp, P. A. (2002). Gene silencing using micro-RNA designed hairpins. *RNA* **8,** 842–850.

McManus, M. T., and Sharp, P. A. (2002). Gene silencing in mammals by small interfering RNAs. *Nat. Rev. Genet.* **3,** 737–747.

Miyagishi, M., and Taira, K. (2002). U6 promoter-driven siRNAs with four uridine 3′ overhangs efficiently suppress targeted gene expression in mammalian cells. *Nat. Biotechnol.* **20,** 497–500.

Nelson, P., Kiriakidou, M., Sharma, A., Maniataki, E., and Mourelatos, Z. (2003). The microRNA world: Small is mighty. *Trends Biochem. Sci.* **28,** 534–540.

Nielsen, J. N, Hallenberg, C., Frederiksen, S., Sorensen, P. D., and Lomholt, B. (1993). Transcription of human 5S rRNA genes is influenced by an upstream DNA sequence. *Nucleic Acids Res.* **21,** 3631–3636.

Noonberg, S. B., Scott, G. K., Garovoy, M. R., Benz, C. C., and Hunt, C. A. (1994). *In vivo* generation of highly abundant sequence-specific oligonucleotides for antisense and triplex gene regulation. *Nucleic Acids Res.* **22,** 2830–2836.

Paddison, P. J., Caudy, A. A., Bernstein, E., Hannon, G. J., and Conklin, D. S. (2002). Stable suppression of gene expression by RNAi in mammalian cells. Short hairpin RNAs (shRNA) induce sequence-specific silencing in mammalian cells. *Genes Dev.* **16,** 948–958.

Paul, C. P., Good, P. D., Li, S. X. L., Kleihauer, A., Rossi, J. J., and Engelke, D. R. (2003). Localized expression of small RNA inhibitors in human cell. *Mol. Ther.* **7,** 237–247.

Paul, C. P., Good, P. D., Winer, I., and Engelke, D. R. (2002). Effective expression of small interfering RNA in human cells. *Nat. Biotechnol.* **20,** 505–508.

Pringle, J. R., Preston, R. A., Adams, A. E., Stearns, T., Drubin, D. G., Haarer, B. K., and Jones, E. W. (1989). Fluorescence microscopy methods for yeast. *Mol. Cell Biol.* **31,** 357–435.

Shaw, P. J. (1995). Comparison of wide-field deconvolution and confocal microscopy for 3D imaging. *In* "Handbook of Biological Confocal Microscopy" (J. B. Pawley, ed.), pp. 373–387. Plenum Press, New York.

Sui, G., Soohoo, C., Affar, E. B., Gay, F., Shi, Y., Forrester, W. C., and Shi, Y. (2002). A DNA vector-based RNAi technology to suppress gene expression in mammalian cells. *Proc. Natl. Acad. Sci. USA* **99,** 5515–5520.

Strub, K., Moss, J., and Walter, P. (1991). Binding sites of the 9- and 14-kilodalton heterodimeric protein subunit of the signal recognition particle (SRP) are contained exclusively in the Alu domain of SRP RNA and contain a sequence motif that is conserved in evolution. *Mol. Cell. Biol.* **11,** 3949–3959.

Ullu, E., and Weiner, A. M. (1985). Upstream sequences modulate the internal promoter of human 7SL RNA. *Nature* **318,** 371–374.

Weichenrieder, O., Kapp, U., Cusack, S., and Strub, K. (1997). Identification of a minimal Alu RNA folding domain that specifically binds SRP 9/14. *RNA* **3,** 1262–1274.

Weichenrieder, O., Wild, K., Strub, K., and Cusack, S. (2000). Structure and assembly of the Alu domain of the mammalian signal recognition particle. *Nature* **408,** 167–173.

White, R. J. (1998). "RNA Polymerase III Transcription." R.G. Landis, Georgetown.

Xia, H., Mao, O., Paulson, T. C., and Davidson, B. L. (2002). siRNA-induced gene silencing *in vitro* and *in vivo*. *Nat. Biotechnol.* **20,** 1006–1010.

Yu, J.-Y., DeRuiter, S. L., and Turner, D. L. (2002). RNA interference by expression of short-interfering RNAs and hairpin RNAs in mammalian cells. *Proc. Natl. Acad. Sci. USA* **99,** 6047–6052.

Zamore, P. D. (2002). Ancient pathways programmed by small RNAs. *Science* **296,** 1265–1269.

Zeng, Y., and Cullen, B. R. (2002). RNA interference in human cells is restricted to the cytoplasm. *RNA* **8,** 855–860.

[9] Viral Delivery of Recombinant Short Hairpin RNAs

By BEVERLY L. DAVIDSON and SCOTT Q. HARPER

Abstract

Recent work demonstrates that RNA interference (RNAi) can coordinate protein expression. Inhibitory RNAs are expressed naturally in cells as microRNAs (miRNAs) or introduced into cells as small interfering RNAs (siRNAs). Both types of small RNAs can be used at the bench to silence mRNA expression. For many researchers, transfection of siRNAs synthesized *in vitro* or purchased from commercial sources is impractical for the cellular system under study. As an alternative to transfection-based methods, we provide a practical approach to accomplish siRNA-mediated gene silencing through the generation and introduction of recombinant viral vectors expressing short hairpin RNAs (shRNAs). shRNAs are subsequently processed to siRNAs *in vivo*, leading to efficient, and, in some cases, long-term silencing.

Introduction

Recent work has demonstrated that RNA interference (RNAi) helps coordinate the flow of information from transcription to protein expression, complicating tremendously our former understanding of how protein expression is regulated (Davidson and Paulson, 2004; Hannon, 2002). Inhibitory RNAs can be expressed naturally in cells as microRNAs (miR-NAs) or introduced into cells as small interfering RNAs (siRNAs). Both miRNAs and siRNAs inhibit gene expression, but in general miRNAs inhibit translation and siRNAs most often degrade target mRNA. Both small RNAs can be used at the bench to silence mRNA expression. In this chapter, we provide a practical approach to accomplish siRNA-mediated gene silencing through the generation and introduction of recombinant viral vectors expressing short hairpin RNAs (shRNAs), which are subsequently processed to siRNAs *in vivo*.

Gene silencing can be accomplished in many tissue culture cell lines, using a variety of transfection methods and starting materials (see Chapters 6, 7, 11, and 12). However many primary cell culture systems and cells *in vivo* are refractory to, or very inefficiently, transfected. Recombinant viral vectors provide an alternative for difficult-to-transfect systems.

Viral Vector Systems

Overview

Several viral vector systems are in widespread use for delivery of genetic material, and they have recently been co-opted for the delivery of shRNAs. Three systems are discussed: adenovirus, adeno-associated virus, and lentivirus-based systems. Adenoviruses (Ads) are encapsidated and contain a linear double-stranded DNA genome that remains episomal after infection (Horwitz, 1996; Shenk, 1996). Recombinant Ads (rAds) containing expression cassettes of interest are readily generated and can be purified to very high titers [up to 10^{13} infectious units (IU)/ml; Anderson *et al.*, 2000]. To date, rAd5-based vectors have been used most often for gene transfer to cells, animals, and humans. More recently, "gutless Ads," which refer to rAds deleted of all native viral genes, have been used for long-term expression in liver and brain tissues (Schiedner *et al.*, 1998; Thomas *et al.*, 2000). Vectors based on current generations of rAds remain very useful for addressing short-term questions, largely because they infect such a broad range of cell types.

Lentiviruses are the most complex of the retrovirus family (Narayan and Clements, 1990). They have RNA genomes that are reverse transcribed by virally encoded reverse transcriptase. Lentiviruses, and vectors derived from them, have the capacity to integrate into the host genome in the absence of cell division (Naldini, 1996). This feature sets the lentiviruses apart from murine oncoretroviruses [e.g., murine Moloney leukemia virus (MMLV)], another retrovirus commonly used as a vector for gene transfer. In the case of MMLV, cell division is required for nuclear entry and integration (Coffin, 1990). Vectors developed from feline immunodeficiency virus (rFIV) and human immunodeficiency virus (rHIV) infect nondividing cells as well, and transduce a variety of cell types and are also easily generated in the laboratory.

Adeno-associated viruses (AAVs) are small encapsidated viruses with a single-stranded DNA genome. AAVs are called dependoviruses because they require helper viruses for productive infections—they are not themselves pathogenic (Berns, 1990). Wild-type AAV2 integrates into a specific site in chromosome 19; however, removal of *rep* genes reduces integration site specificity as well as the propensity to integrate (Duan *et al.*, 1997; Kotin *et al.*, 1991, 1992; Weitzman *et al.*, 1994; Wu *et al.*, 1998). AAV does not transduce many cell types *in vitro* with high efficiency. However, it is very useful for long-term expression in liver, muscle, and the central nervous system (CNS). At least eight serotypes of AAV are under study, with demonstrable differences in tropism of cell types transduced (Alisky *et al.*, 2000; Davidson *et al.*, 2000).

Designing the Short Hairpin RNA (shRNA) Expression Cassette

Viral vectors are generated from shuttle vectors, containing the expression cassette of interest and plasmids required for vector production (see Fig. 2). PCR-based methodology allows construction of a universal shRNA expression cassette for eventual expression from lentivirus, AAV, or Ad.

Principle. We use a one-step PCR method for generating the shRNA expression cassette. In addition to commonly used antibiotic resistance genes and *cis*-acting elements required for DNA replication, all shRNA expression plasmids contain three key features: (1) a promoter, (2) the shRNA "cDNA," and (3) a transcription termination signal.

PROMOTER. Various promoters directing both RNA polymerases II and III (pol II and III) are used to transcribe shRNA from plasmids in mammalian cells, including H1, U6, tRNAval, and the cytomegalovirus I/E promoter (CMV; Elbashir *et al.*, 2001; Paul *et al.*, 2002; Xia *et al.*, 2002). Current data suggest that the production of functional shRNA requires proper positioning at or near the transcription start site, particularly for pol II-based promoters driving shRNA expression (Xia *et al.*, 2002). miRNAs as carriers for siRNAs overcome this limitation (Zeng and Cullen, 2003). shRNAs cloned in our laboratory contain no more than 6 bp of sequence

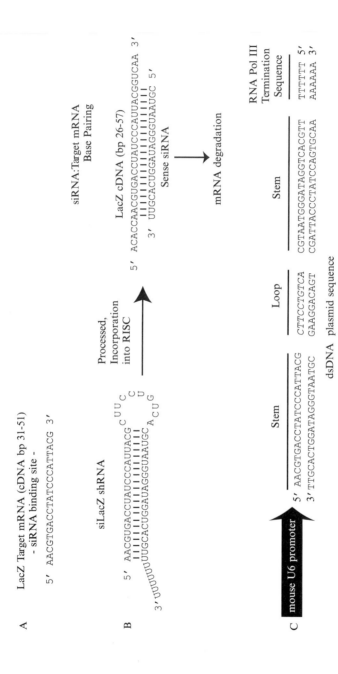

Fig. 1. General outline for generating plasmids for shRNA expression, using shRNAs targeting LacZ as an example. (A) The site within the LacZ cDNA to which the shRNA is targeted is depicted. (B) Representation of shRNA processing to provide a LacZ-specific siRNA. (C) Sequence of LacZ shRNA, with mouse U6 promoter to drive hairpin expression, and pol III termination signal.

(i.e., one common restriction enzyme recognition site) between the promoter's transcription start site and the 5' end of the DNA sequence giving rise to the shRNA. Following this general rule, the construction methods described here can be applied to any promoter of interest.

shRNA-ENCODING DNA. This DNA sequence serves as the transcription template that gives rise to shRNA (Fig. 1). An shRNA directed against *E. coli* β-galactosidase serves as an example of the complementarity between the guide strand of the processed shRNA (siRNA guide strand) and the cDNA target. The functional shRNA has a predicted stem-loop structure and leads to target mRNA degradation in a sequence-specific manner (Fig. 1). Therefore, production of an effective shRNA requires sequence specificity with a target mRNA of interest and base pairing that allows for RNA hairpin formation (Fig. 1). There are three key elements to consider when constructing DNA sequences from which shRNAs arise:

1. *Target site sequence.* One half of the stem encodes 19–22 nt of sequence derived from the mRNA target (Fig. 1). Our laboratory generally uses a 21-nt sequence for shRNA construction. Because no free and reliable siRNA prediction algorithms have been published to date, we generally select a 21-nt target mRNA sequence randomly but keeping in mind established criteria for proper guide strand loading (Khvora *et al.*, 2003; Zeng and Cullen, 2003).

2. *siRNA sequence.* The siRNA sequence is the perfect reverse complement of the mRNA target site and constitutes the other half of the stem (Fig. 1B).

3. *Loop sequence.* Various publications have described loops of varying sizes and sequence derivations, including loops of palindromic restriction enzyme sites. It is now established that the loop sequence need be >6 nucleotides long and nonpalindromic. The loop sequence is placed between the two halves of the stem sequence (Fig. 1B, C).

TRANSCRIPTION TERMINATION SIGNAL. The transcription termination signal should be cloned at or near the 3' end of the DNA sequence encoding the shRNA (Fig. 1C). shRNAs cloned in our laboratory contain no more than 6 bp of sequence (i.e., one common restriction enzyme recognition site) between the transcription termination signal and the 3' end of the sequence encoding the shRNA. The sequence of the termination signal will depend on the promoter used to drive shRNA expression. For promoters directing pol III activity, the DNA sequence should be five or six thymidines (T). For promoters directing pol II activity, the DNA sequence should be AATAAA.

Method. The PCR method for generating shRNA expression plasmids does not significantly differ from a standard PCR reaction (Fig. 2A). The

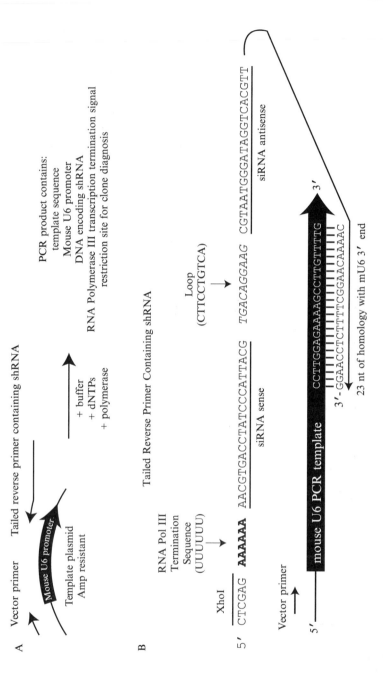

Fig. 2. General schematic for generating shRNAs by a PCR-based reaction. (A) Outline of PCR strategy and contents of PCR product. (B) Full characterization of the key features of the reverse primer for hairpin generation.

reaction requires a DNA template containing the promoter of interest, a DNA oligonucleotide primer pair, a polymerase, standard buffers, and dNTPs. The most outstanding feature of this PCR reaction is the length of the reverse primer that encodes the shRNA transcription template (Fig. 2B). Again, the shRNA specific for *E. coli* β-galactosidase is shown for illustrative purposes only. Only about one third of this primer actually binds the 3' end of the promoter, but the remaining tailed end of the primer encoding the shRNA becomes incorporated into the product (Fig. 2). Subsequent amplification produces a distinct PCR product containing the entire promoter followed by the shRNA transcription template DNA. Following amplification, the product is cloned directly into a PCR cloning vector, transformed into bacteria, and clones are selected and analyzed by restriction digestion and DNA sequencing. The entire process from the PCR reaction to sequence verification can take as little as four total days. However, the total hands-on time required to generate an shRNA plasmid can be less than 5 h over the course of those 4 days.

In our example, an ampicillin-resistant plasmid containing the mouse U6 promoter is used as a PCR template (Fig. 2A). Once amplified, the product is cloned directly into the pCR-Blunt II-TOPO vector (Invitrogen, Carlsbad, CA), which confers kanamycin resistance. Selection of transformed bacteria on kanamycin plates allows for selection of colonies containing only the shRNA expression cassette.

The forward PCR primer should encompass the 5' end of the promoter of interest and ideally satisfy standard criteria for PCR primer design [e.g., 50–60° melting temperature (T_m); 40–60% GC content; 17–25 nt in length]. The forward primer used in this example is located in Ad5 sequence 85 bp upstream of the mU6 promoter. The reverse primer used to generate the shRNA is 87 nt in length (Fig. 2B). The 3' end of the primer contains 23 nt of sequence that hybridize the top strand of the mU6 promoter. In the first few PCR cycles, only these sequences participate during the annealing. Thus, one should choose the minimum region of homology for a primer T_m of >50°. The next 58 nt encode the shRNA transcription template. The final 6 nt encode an *Xho*I restriction site that allows for selection of fully extended clones. Inclusion of a restriction site eliminates the requirements for a PAGE or HPLC purified primer. This strategy therefore provides a cheap and quick way to screen recombinants without broad-scale sequencing and costly purification of long primers.

1. Amplify using the following thermocycling parameters: initial denaturation 1 cycle, 3 min, 94°; 30 cycles, 94° for 30 sec, 50° for 30 sec, 72° for 30 sec; final extension 7 min, 72°.

2. Following thermocycling, load 4 μl of the PCR reaction and verify by agarose gel electrophoresis that a single band was produced.

3. As per the manufacturer's protocol, add 4 μl of PCR product directly to a tube containing 1 μl pCR-Blunt II-TOPO vector (Invitrogen) and 1 μl salt solution (1.2 M NaCl, 0.06 M MgCl$_2$; provided with TOPO® Kit). Incubate for 30 min at room temperature.

4. Transform 2–3 μl of TOPO® cloning reaction into chemically competent *E. coli* by standard procedures.

5. Spread one-tenth volume of transformed bacteria onto a kanamycin-selective plate.

6. Centrifuge the remaining transformation at low speed (e.g., ~2000 rpm/0.4 rcf for 1 min in an Eppendorf centrifuge, Model 5415) until bacteria form a pellet. Remove supernatant. Resuspend pellet in 50–100 ml of bacterial growth medium (e.g., LB, SOC). Spread remaining cells on a second plate.

7. Incubate plates overnight at 37°. A typical cloning reaction yields hundreds of colonies.

8. Pick 10 colonies into a culture tube containing 3 ml of LB medium and 50 μg/ml kanamycin. Grow overnight at 37°.

9. Isolate plasmid DNA by using standard procedures or a commercial miniprep kit. There are numerous methods to purify bacterial DNA. If the primer contains, for example, the *Xho*I site for insert verification, set up a standard restriction digest to confirm the presence of the insert before sequencing.

10. Sequence positive clones. An automated sequencer that uses Sanger-based fluorescent chemistry will greatly expedite this procedure. The University of Iowa DNA Facility uses Perkin Elmer-Applied Biosystems sequencers (Models 3100 and 3700) and ABI Prism® Big Dye™ Terminator Cycle Sequencing reagents. If automated sequencing is not available, manual sequencing that uses similar dideoxy terminator chemistry (e.g., ThermoSequenase cycle sequencing kit, Amersham Biosciences, Piscataway, NJ) will suffice but will add time and labor involved. Our initial attempts at sequencing shRNA plasmids by using standard chemistry and cycling conditions resulted in a high incidence of truncated or failed sequencing reads, probably due to the inability of the polymerase to accurately read through the shRNA hairpin. In collaboration with the University of Iowa DNA Facility, we modified the standard protocol to produce more reliable and consistent sequencing of shRNA hairpins. Cycle parameters are as follows: initial denaturation 1 cycle, 2 min, 98°; 30 cycles of 96° for 1 min, 50° for 5 sec, 60° for 4 min.

11. Move the shRNAs into the vector shuttles. The *Xho*I site in the oligomer for PCR of hairpins, together with the restriction sites in TOPO, enables directed cloning into Ad, AAV, or FIV shuttle plasmids (Fig. 3). Moving the hairpins into the shuttles entails standard cut-and-paste

reactions. The *Mfe*I site in FIV is compatible with *Eco*RI restricted ends. Studies in my laboratory indicate that placement of the hairpin in this region (just downstream of gag) of the FIV packaging transcript is important for maintaining vector titer (S. Harper and B. Davidson, unpublished observations).

Materials
PCR AMPLIFICATION AND TOPO CLONING REACTIONS.
For a 50 ml PCR reaction:
5 ml of 10× *Pfu* reaction buffer
4 ml of 2.5 m*M* dNTP mix
1 ml forward primer at 100 ng/ml
1 ml reverse primer at 100 ng/ml
0.5 ml *Pfu* polymerase at 2.5 units/ml
1 ml pAd5mU6 template plasmid at 10 ng/ml
37.5 ml water
CYCLE SEQUENCING REACTIONS
250 ng plasmid DNA template
3 pmoles T7 or SP6 primer
1 μl DMSO
8 μl Perkin Elmer dGTP sequencing dye terminator
Water to 20 μl

Recombinant Feline Immunodeficiency Virus (FIV) Production

Principle. Figure 3 depicts the general method for generating recombinant lentivirus vectors. There are no changes from this general production method, which has been reported previously (Stein and Davidson, 2002), except that the RNA to be packaged expresses a shRNA directed against the gene of interest. Because methods to detect cells expressing the shRNAs are laborious, we include reporters in the vector systems. Thus, reporter-positive cells are presumed to be also expressing the shRNA. In this example, an expression cassette for humanized *Renilla* green fluorescent protein (hrGFP) is encoded.

The FIV packaging construct (pCFIVΔorf2Δvif; Johnston *et al.*, 1999) was derived from the FIV molecular clone p34TF10. The packaging construct retains full-length *gag* and *pol*, and *rev*, but contains a deletion in the *env* gene and mutations in the *vif* and *orf2* genes. The vif and orf2 accessory proteins have been determined to be dispensable both for recombinant particle production and for transduction *in vivo*. The third accessory gene *rev* is retained because it is necessary for efficient nuclear export of long and full-length transcripts. The native 5′ long terminal repeat

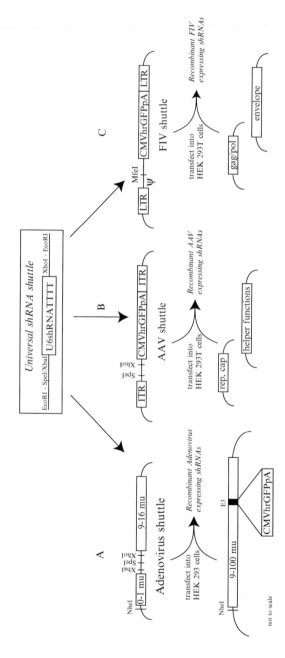

Fig. 3. Cloning of the shRNA expression cassettes into shuttle plasmids for generating (A) adenovirus, (B) adeno-associated virus (AAV), and (C) lentivirus.

(LTR) has been replaced by the human CMV immediate early promoter/ enhancer, and the 3' LTR has been replaced with the simian virus 40 polyadenylation signal. The vector construct (pVETLCβ) (Johnston et al., 1999) carries the shRNA expression cassette and the hrGFP reporter driven by an internal promoter. The 5' U3 of FIV has been replaced with the CMV promoter. pVETLCβ is deleted of all viral coding regions with the exception of a 5' portion of gag (which contains part of the packaging signal). Lastly, a third plasmid, the envelope construct, provides the envelope protein in trans. In this example, a plasmid encoding the glycoprotein from the vesicular stomatitus virus (VSV-G) envelope glycoprotein, expressed from the CMV promoter, is used (Yee et al., 1994). Other envelopes may be better for transduction of specific cell types of interest, as discussed in more detail later.

293T cells (Shen et al., 1995) are maintained in exponential growth in DMEM-10, and grow as monolayers. They are easily lifted with trypsin for subculturing.

Methods. We routinely produce recombinant FIV particles by using the triple plasmid system and constructs described by Johnston et al. (1999) (Fig. 1; see "Notes on FIV," Point 1). This system of particle production involves concurrent transfection of 293T cells with three plasmids, followed by harvest of particle-containing culture medium and concentration of particles. Steps for preparing 3 ml of concentrated vector particles (usually 108–109 transduction units/ml), from an 18-plate (150-mm diameter) transfection are given next.

FIV PRODUCTION

1. Seed 293T cells into eighteen 150-mm-diameter flat-bottom tissue culture dishes at a density of 107 cells per dish.
2. The next day, add 34 ml of Hepes-buffered saline (HBS) to two 50-ml conical tubes. HBS should be at room temperature.
3. Add 225 μg of the packaging plasmid, 337.5 μg of the vector plasmid, and 112.5 μg of the envelope plasmid to each HBS-containing tube and vortex well.
4. Slowly add 1.7 ml of 2.5 M CaCl$_2$ to each tube while slowly vortexing or shaking the HBS–plasmid mixture. CaCl$_2$ should be at room temperature.
5. Let the solution stand for 25 min to allow precipitate formation. The solution should appear slightly translucent or cloudy.
6. Add both tubes of precipitate directly to 200 ml of DMEM. Briefly mix.
7. Aspirate off the medium from the cells (nine plates at a time).

8. Gently pipette the transfection solution onto the cells (15 ml per dish), and return the cells to the incubator.

9. Four to six hours after transfection, aspirate off the medium and provide 15 ml of fresh DMEM-10 per dish.

10. Collect the medium (containing vector particles) at 24, 36, and 72 h, each time replacing this medium with fresh DMEM-10. At each collection, filter the medium through a 0.45-μm filter (Nalgene PES, low-protein-binding 500-ml bottle top filter) and store short term at 4° or long term in 50-ml aliquots at −80° (see "Notes on FIV," Point 2).

11. Just before intended use, concentrate the particles by centrifuging the collected medium at 4° for 16 h at 7400g (7000 rpm in the SLA 1500 rotor, Sorvall Centrifuge, 275-ml-capacity tubes). Carefully pour off the supernatants and resuspend particles in lactose buffer (see "Notes on FIV," Point 3). We typically resuspend the particles produced from an 18-plate transfection into total volume of 3 ml.

DETERMINING TITER BY LIMITING DILUTION AND ASSAY OF TRANSGENE EXPRESSION. Standardized methods of determining the concentration of FIV-based lentivirus particles and the concentration of transduction-competent particles within lentiviral preparations have not yet been established. An ELISA assay can be used to measure the FIV p24 nucleocapsid antigen (Johnston et al., 1999; Mastromarino, 1987; Tilton, 1990) as an indicator of particle concentration. For determination of transduction titers, we transduce HT-1080 cells with serially diluted particles, followed by quantification of transduced cells either by staining and counting the transgene-expressing cells or by quantitative PCR detection of vector sequences (see "Notes on FIV," Points 4 and 5).

1. One day before transduction, seed a six-well flat-bottom plate with 2 million HT-1080 cells per well in DMEM-10.

2. For transduction, make a 10-fold dilution series of concentrated FIV as follows. Add 1.584 ml of DMEM-2/polybrene in the first tube and 1.35 ml in tubes 2 through 6. Add 15 μl of virus to the first tube and vortex. Transfer 150 μl from the first to the second tube and vortex, and so on for the remaining tubes.

3. Remove culture media from wells. Add 1 ml of each dilution to separate wells. For wells 1 through 6, the dilution factors will thus be 10^3, 10^4, 10^5, 10^6, 10^7, and 10^8, respectively. Return the cells to the incubator.

4. Incubate the HT-1080 cells for 72 h and then feed with 1 ml of DMEM-10.

5. Incubate a further 24 h. Rinse monolayers with PBS.

6. Using an inverted fluorescent microscope, visualize and count the number of hrGFP-expressing cells in each well. The first two wells will often have too many positive cells to count. Doublets or small clusters of cells are counted as one, as they likely originated by division of a single transduced cell.

7. For each well, multiply the total blue cell count by the dilution volume (1 ml) and by the dilution factor. Determine the mean of all the wells. This number represents the transducing units per milliliter (TU/ml) of concentrated virus. By using this method, our concentrated FIVhrGFP preparations typically contain 10^8–10^9 TU/ml.

TITERING ASSAY OF FIV VECTOR BY USING REAL-TIME PCR (SEE "NOTES ON FIV," POINT 5). Concentrated FIV vector is diluted 100-fold and 1000-fold and applied to ~1 million HT1080 cells in triplicate. Four days later, the cells are harvested and genomic DNA is purified. Real-time PCR is performed using the TaqMan™ Universal PCR Master Mix and an ABI thermocycler to quantify FIV vector copies per genome. For each titering PCR run, standard curves are generated for both FIV vector and genomic GAPDH, using 10^8, 10^7, 10^6, 10^5, and 10^4 copies of purified plasmid. Based on the quantitative PCR method, our concentrated FIV preparations typically contain 10^8 to 5×10^9 TU/ml. The cycle conditions are 1 cycle of 50° for 2 min and 95° for 10 min, followed by 40 cycles of 95° for 15 sec and 60° for 1 min.

Pseudotyping with Alternative Envelopes. We have described here the production of a lentivirus vector pseudotyped with the vesticular stomatitis virus G (VSV-G) envelope protein. VSV-G lends enhanced structural stability to vector particles, allowing for concentration by ultracentrifugation with minimal loss of infectivity (Burns *et al.*, 1993). In addition, VSV-G mediates viral entry by interaction with a membrane phospholipid component (DePolo *et al.*, 2000) rather than with a specific cell surface receptor protein, and this imparts VSV-G-pseudotyped vectors with an extremely broad host-cell range, including cells of nonmammalian species (Burns *et al.*, 1993). These features have resulted in the common use of VSV-G for pseudotyping murine leukemia virus (MLV) and lentiviral gene transfer vectors.

However, there can be disadvantages to using the VSV-G envelope. Despite the wide tropism, VSV-G-pseudotyped vectors do not always mediate efficient transduction. For example, VSV-G-pseudotyped FIV was unable to transduce polarized airway epithelial cells when applied to the apical surface (Wang *et al.*, 1999). Furthermore, the observations that VSV-G can be toxic to cells (Burns *et al.*, 1993) and that VSV-G-pseudotyped vectors are inactivated on exposure to human serum (Page *et al.*, 1990) may limit the clinical application of VSV-G-pseudotyped vectors. Lastly, particularly for direct *in vivo* applications, widespread tropism can be a drawback when it is desirable to restrict transduction to specific cell types or tissues.

Several reports describe successful pseudotyping of HIV-based lentiviral vectors with alternative envelope proteins. Early HIV-1 vectors were pseudotyped with the MLV amphotropic envelope (Landau *et al.*, 1991; Reiser *et al.*, 1996), human T-cell leukemia virus envelope (Chan *et al.*, 2000), as well as the VSV-G envelope glycoprotein (Landau *et al.*, 1991). Recently, HIV vectors incorporating envelope glycoproteins from Marburg or Ebola viruses have been produced and show a wide range of infectivity (Zeilfelder and Bosch, 2001).

Attempts at pseudotyping can also meet with failure. Efficient incorporation of envelope proteins into particles depends on appropriate interaction of cytoplasmic envelope sequences with encapsidated genomes. Maedi-visna virus envelope constructs were unable to pseudotype MLV- or HIV-derived vector particles (Stitz *et al.*, 2000). Pseudotyping of an HIV vector with the gibbon ape leukemia virus (GaLV) envelope glycoprotein was similarly unsuccessful (Miletic *et al.*, 1999). However, use of a chimeric construct substituting the cytoplasmic tail of the GaLV envelope with that of the MLV amphotropic envelope protein resulted in the formation of infective GaLV-pseudotyped HIV vectors (Miletic *et al.*, 1999).

Because FIV and HIV are related lentiviruses, successful pseudotyping of HIV with the aforementioned envelope proteins should predict similar results with FIV vectors. Preliminary studies in our laboratory indicate that FIV vectors can be pseudotyped with unmodified envelope proteins from amphotropic MLV, Marburg virus, and Ross River virus (Kang *et al.*, 2002). Also worthy of mention are a few alternative envelopes, including the lymphocytic choriomeningitis virus envelope. This envelope efficiently pseudotyped MLV, allowing generation of structurally stable particles (Spiegel *et al.*, 1998). Pseudotyping of MLV with the Sendai virus fusion protein imparted a restricted tropism for asialoglycoprotein receptor-bearing cells, which could be useful for hepatocyte-directed shRNA delivery (Taylor and Wolfe, 1994).

The successful production of FIV-based vectors pseudotyped with an alternative envelope will involve following the steps outlined previously, with a plasmid encoding high-level expression of the alternative envelope protein in place of the VSV-G envelope plasmid. High titer production of particles that display appropriate cell tropism would suggest successful pseudotyping.

Materials

TRANSIENT TRANSFECTION FOR VECTOR PRODUCTION

Incubator at 37° with 5% CO_2

Hepes buffered saline (HBS):

5.0 g Hepes

8.0 g NaCl

0.37 g KCl

0.188 g $Na_2HPO_4 \cdot 7H_2O$

1.0 g glucose

Bring to 1 L in ddH_2O, adjust pH to 7.1 with concentrated NaOH, filter sterilize, and store at 4°.

2.5 *M* $CaCl_2$

DMEM: Dulbecco's modified Eagle's medium

DMEM-10: DMEM with 10% fetal calf serum, 100 U/ml penicillin, and 100 μg/ml streptomycin

DMEM-2: DMEM with 2% fetal bovine serum and 100 U/ml penicillin and 100 μg/ml streptomycin.

Lactose buffer: phosphate-buffered saline (PBS), pH 7.4 (Sigma P-3813), with 40 mg/ml lactose, filter sterilized

Sorvall Centrifuge RC 26 Plus, with SLA 1500 rotor

Centrifugation bottles (250-ml capacity)

PCR FOR TITERING VIRAL GENOMES/UNIT VOLUME

Genomic DNA purification kit (Promega, No. A2361)

TaqMan™ Universal PCR Master Mix (ABI, No. 4304437)

ABI thermocycler (ABI7000)

FIV vector upstream primer: 5'-AGCAGAACTCCTGCTGACC-TAAA-3'

FIV vector downstream primer: 5'-TCGAGTCTGCTTCACTAGA-GATACTC-3'

FIV vector probe: 5'FAM-ACTGTTAGCAGCGTCTGCTACTGCT TCCCT-TAMRA3'

Genomic GAPDH upstream primer: 5'-GGTTTACATGTTCCAA-TATGATTCCA-3'

Genomic GAPDH downstream primer: 5'-ATGGGATTTCCATT-GATGACAAG-3'

Genomic GAPDH probe: 5'JOE-ATGGCACCGTCAAGGCTGA-GAACG-TAMRA3'

TITER BY TRANSGENE EXPRESSION: HUMANIZED *RENILLA* GREEN FLUORESCENT PROTEIN (HRGFP) FLUORESCENCE. HT-1080 cells (ATCC CRL-121) maintained in exponential growth in DMEM-10. These cells are derived from human fibrosarcoma and grow as monolayers. They are easily lifted with trypsin for subculturing.

Incubator at 37° with 5% CO_2
Six-well tissue culture dishes
DMEM-2 and DMEM-10 (see previously)
Polybrene stock: 8 mg/ml in ddH_2O, filter sterilized
DMEM-2/polybrene: On the day of use, dilute the polybrene stock
    1:2000 in DMEM-2 for a final polybrene concentration of 4 μg/ml
Dilution tubes (3.5-ml polystyrene sterile tubes)

Notes on FIV

1. We have found that transfections work best when plasmids have been purified by CsCl banding.

2. Lentivirus vector-containing culture medium suffers minimal loss in transduction titer when stored at −80°, whereas centrifuge-concentrated vector loses approximately one log in titer after freezing. Substantial loss of titer of concentrated preparations is also observed within 24 h of storage at 4°. Thus, we routinely concentrate the virus immediately before use.

3. We routinely suspend our concentrated particles in PBS/lactose as this buffer is physiological and acceptable for *in vivo* use and the lactose has a stabilizing effect. However, other buffers maintaining similar pH and salt concentrations may be suitable alternatives. These include saline, TNE (50 mM Tris-HCl, pH 7.8, 130 mM NaCl, 1 mM EDTA), and culture medium.

4. The presence of shRNA in the vector may reduce titers from 5- to 50-fold. (The packaged transcript is a substrate for the shRNA, which may be expressed efficiently in the packaging cell lines if plasmid ratios are not optimal.)

5. Titering of transduction-competent particles by staining and counting of transgene-expressing cells can be applied only when a suitable staining method is available. X-gal staining as described previously is a simple and reliable method for detecting β-galactosidase-expressing cells. For other transgene products, immunostaining methods can be applied, assuming there is a suitable antibody available and the HT-1080 cells have low to no endogenous expression of the antigen. X-gal staining or immunohistochemistry is useful for titer comparison between preparations of the same FIV vector. However, it is not recommended for comparisons between vector preparations carrying different transgenes, as the sensitivities of staining procedures for different transgene products may vary. Titering by PCR requires access to a quantitative PCR system and an

effort to optimize the procedure, but once this system is established it can be applied to FIV vectors encoding any transgene. Both titering methods are based on transduction of HT-1080 cells. HT-1080 cell transduction readily occurs with VSV-G-pseudotyped FIV. However, HT-1080 cell transduction may be less efficient if the FIV vector is pseudotyped with a different envelope. Most envelope proteins exhibit a restricted cellular tropism, and it may be necessary to find or generate a more suitable cell line for determining transduction titers of alternative pseudotypes.

Generation of Recombinant Adeno-Associated Virus (AAV)

Principle. Inverted terminal repeats (ITRs) serve as the viral origin of replication and are required in *cis* for packaging. The viral Rep proteins are required in *trans* (Fig. 3). For a number of serotypes the ITRs and Rep proteins are sufficiently homologous to allow for cross complementation and the production of pseudotyped particles, for example, an AAV2 genome in an AAV4 capsid (Chiorini *et al.*, 1997). Although the Rep proteins of AAV2 and AAV5 bind to either a type 2 ITR or a type 5 ITR, efficient genome replication occurs only when type 2 Rep replicates the type 2 ITR and type 5 Rep replicates the type 5 ITR. This specificity is the result of a difference in DNA cleavage specificity of the two Reps, which is necessary for replication (Chiorini *et al.*, 1999). AAV5 Rep cleaves at CGGTGTGA and AAV2 Rep cleaves at CGGTTGAG. Therefore, to package an AAV5 genome, the helper plasmid must supply AAV5 Rep functions. It is possible to make a pseudotyped AAV5 particle, for example, an AAV2 genome in an AAV5 capsid, but the helper plasmid must contain the AAV2 *rep* gene. For the production of AAV4-pseudotyped particles, an AAV2 genome vector with AAV2 ITRs can be replicated and packaged by AAV4 Rep in an AAV4 capsid (see "Notes on AAV Production," Points 1 and 2).

The original method of producing recombinant AAV involved transfection with two plasmids: (1) a vector plasmid containing the gene flanked by ITRs, and (2) a helper plasmid to supply the appropriate *rep* and *cap* genes. Virus production is then induced by transduction with either wt or a deleted Ad. The resulting cell lysate would contain recombinant AAV particles and adenoviral particles. The two viruses can be separated by heating the lysate to 56° for 30 min to inactivate the Ad, followed by density gradient centrifugation to physically separate the two particles. Although this system is effective, there are a number of drawbacks. Depending on overlapping sequence homology between the AAV helper plasmid and vector plasmid, replication-competent AAV (rcAAV) could also be produced with this system. Although wt AAV is not considered pathogenic, the presence of rcAAV and expression of the AAV *rep* genes can affect the stability of the recombinant genome and alter gene expression within the transduced cell. This system also results in lower yields compared with the amount of virus produced during a wt AAV infection.

To improve this system, the helper plasmids were modified in a variety of ways to increase their expression of Rep and Cap proteins in the cell and thus the amount of recombinant virus produced (see "Notes on AAV Production," Points 1 and 2). Adenoviral terminal repeats were added to the helper plasmids, which enhanced the expression of the AAV genes when the Ad was added to the cells (pAAV/Ad). Substitution of the natural viral p5 or p40 promoters with other viral promoters and changes in the translation efficiency of the viral messages have also improved the yield of virus (Flotte *et al.*, 1995; Grimm *et al.*, 1998; Li *et al.*, 1997; Vincent *et al.*, 1997). Origins of replications from other virus such as SV40 were added to the helper plasmid and could replicate the helper plasmid to high copy number when T-antigen was provided by the packaging cell such as Cos cells (Chiorini *et al.*, 1995; Inoue and Russell, 1998). Although these modifications improved the yield of recombinant AAV and removal of homologous sequences in the vector and helper plasmid reduced the level of replication-competent AAV that can form, large amounts of Ad were still present in the cell lysate. The development of a third plasmid for this system, the adenovirus helper plasmid, resulted in the production of a cell lysate free of adenovirus contamination. Several examples of this plasmid exist and are commercially available. In general, all contain the VA RNA, E1, E2, and E4 genes of Ad (Alisky *et al.*, 2000; Grimm *et al.*, 1998; Xiao *et al.*, 1998; see "Notes on AAV Production," Points 3 and 4). Some of the helper functions can also be supplied by the packaging cells. HEK 293 cells contain the E1 gene of Ad type 5 and have been used extensively to provide helper function for producing AAV. More recently, Kotin and colleagues developed a baculovirus production system (Urabe *et al.*, 2002). Baculoviruses expressing the construct to be packaged, along with baculoviruses expressing the necessary Cap and Rep proteins, are used to infect insect cells. AAV is then harvested from the cells and concentrated or purified by standard methods.

As mentioned previously, several pieces of evidence suggest that distinct mechanisms of uptake exist for each of the different serotypes of AAV. This presents a problem for determining a biological titer for the different viruses. One serotype may transduce a cell line very efficiently, but this same cell may be nonpermissive for another serotype. We have tested a number of different cell lines and have found that AAV serotypes

1–6 are able to transduce Cos cells but with different efficiencies. For example, AAV4 and AAV5 transduce Cos cells 2- to 4- and 50-fold less efficiently, respectively, than AAV2. Whereas a biological titer is useful for comparing different preparations of the same serotype of AAV, it is not a meaningful assay for directly comparing relative transduction efficiencies of different AAV serotypes. A more appropriate method is to determine DNase-resistant genomes packaged in the AAV capsids. Originally, this assay was done by DNA dot blots. However, the development of quantitative PCR assays has greatly simplified the measurement of packaged AAV DNA. A number of different systems as well as programs for designing the primers and probes are currently available. Users should follow the manufacturer's instructions for the system of their choice. The methods for PCR-based titering are similar to those described for FIV, except that particles are treated with DNase, followed by heat inactivation of the DNase, and then PCR. We generally use the standard curve method, described in detail in the ABI technical notes, to quantify AAV genomes.

Methods
RECOMBINANT VIRUS PREPARATION

1. *Transfections and virus harvesting.* Split one confluent 150-mm plate of 293T cells 1:10 into 10- × 150-mm^2 plates 24 h before transfection. The plates should be 30% confluent for transfection. Prepare transfection solutions A and B as described previously. Place a 2-ml pipette in a tube of solution A and gently bubble. Then add solution B dropwise while continuing to bubble. Incubate mixture for 20 min at room temperature and allow precipitant to form. Mix gently and add 2 ml of the transfection mixture dropwise to each of the plates, swirling intermittently while adding the transfection mixture to the plates. Supplement the cultures with 10 ml of fresh medium 24 h after transfection. Incubate for 48 h at 37°, 5% CO$_2$. To harvest the virus, remove the medium and quickly rinse the plates twice with 5 ml of citric saline solution. After the second rinse incubate the cells for 5 min at room temperature, and then collect the cells by adding 10 ml of TD buffer to the plate. Then pellet the cells by centrifugation at 2000g for 5 min.

2. *Virus concentration.* For every 10 plates, resuspend the cell pellet in 5 ml of TD buffer and store at −70°. Thaw the cell pellet at 37° and add the benzonase to a final concentration of 20 U/ml. Then add sodium deoxycholate to a final concentration of 0.5% and incubate for 1 h at 37°. Once the lysate has cleared, centrifuge at 3000g for 20 min to pellet the cellular debris. Carefully remove the supernatant to a fresh tube and add 0.55 g of CsCl per 1.0 ml of supernatant. This should result in a final density of 1.4 g/cm^3. After confirming the density with a refractometer, transfer the

lysate to SW40.1 polyallomer ultracentrifuge tubes and adjust to a final volume of 11.8 ml with the CsCl/TD solution. The meniscus should be approximately 3 mm from the top of the tube. Centrifuge at 38,000 rpm for 65 h at 20°. Following centrifugation, remove the tubes and place in a ring stand holder. Using one end of a clamped butterfly needle, puncture the side of the tube approximately 2 cm from the bottom. Slowly open the clamp and begin collecting 300- to 400-μl fractions. For AAV2 and AAV5, pool fractions with a refractive index of 1.373–1.371. AAV4 bands a slightly higher density of 1.376–1.374 because of its different protein:DNA ratio. For a number of applications this level of purity is sufficient. However, to increase the titer and purity of the virus the fractions can be combined and centrifuged again by using an SW50.1 rotor. Centrifuge at 38,000 rpm for 24 h at 20°. As an alternative to CsCl gradients, iodixonal gradients can be employed. These allow effective separation of empty and DNA-containing capsids, but again serve to concentrate the virus only.

3. *Dialysis and storage.* The virus is best stored long term in CsCl as small aliquots. Repeated freeze–thaw cycles will result in a loss of activity. Once an aliquot is thawed, it can be stored for at least 1 week at 4° in CsCl. For *in vivo* applications, the CsCl must be removed by dialysis. We have found that AAV is not stable in PBS; therefore, its use is not recommended. We routinely dialyze against isotonic saline before transduction, but other buffers may be preferred. For any buffer, it is recommended that its effects on virus stability and transduction properties be formally tested. We routinely dialyze small volumes by using mini collodion membranes (No. 25300, Schleicher & Schuell, Keene, NH) as per the manufacturer's instructions. Other devices may also be used. The dialyzed virus should be used immediately.

TITERING AAV. Encapsidated viral genomes can be quantified by using an Applied Biosystems 7700 and a primer set specific for the viral genome. If a number of genes are being studied in the laboratory, development of a primer set specific for a common promoter may be more versatile. For the RSV promoter, the following primer set has been used successfully: forward 5'-GATGAGTTAGCAACATGCCTTACAA and reverse 5'-TCGTACCA CCTTA-CTTCCA-CCAA. Viral samples for quantitative PCR (QPCR) need to be treated to remove any contaminating plasmid DNA from the transfection and the capsid digested for maximum efficiency during cycling. Samples are first treated with benzonase (6 U/μl) and incubated at 37° for 30 min, followed by digestion with protease K (40 μg/ml) for 1 h. The DNA is extracted with phenol–chloroform, precipitated, and resuspended in water. The standard reaction consists of 300 nM of each primer; 1 μl of viral DNA sample, and 1× SYBR green master mix in a total volume of 25 μl. Following a 10 min activation step at 95°, 40 cycles of a two-step PCR cycle are run: 95° for 15 sec and 6° for 1 min. Additional primer sets can be developed to test for Ad and replication-competent AAV.

Materials
GENERATION OF RECOMBINANT VIRUS LYSATE
  15-cm plates
  50-ml polystyrene tubes
  10× HBS: 40.9 g Hepes free acid, 29.79 g NaCl, final volume 250 ml
  2 *M* calcium solution: 147.02 g $CaCl_2$ dihydrate, final volume 500 ml
  140 m*M* phosphate solution: add monobasic to dibasic solution to
    final pH of 6.8 (3.48 g dibasic $NaPO_4$, final volume 200 ml; monobasic
    3.36 g $NaPO_4$ final volume 200 ml)
  120 m*M* dextrose solution (10×): 2.16 g dextrose in 100 ml H_2O
  Calcium phosphate solution A: For five plates combine 3.915 ml H_2O,
    0.51 ml 120 m*M* dextrose, 0.5 ml 10× HBS, 0.075 ml 1 *N* NaOH,
    and 0.1 ml 140 m*M* $NaPO_4$ (pH 6.8). Final volume 5.1 ml, adjust pH
    to 7.01–7.05, and sterile filter.
  Calcium phosphate solution B: For five plates combine 125 μg of
    DNA (30.5 μg vector plasmid, 30.5 μg of helper plasmid, 64 μg of
    Ad plasmid) in a final volume of 4.4 ml of H_2O. For the last step,
    add 0.6 ml 2 *M* $CaCl_2$.
VIRUS ISOLATION AND GRADIENT PURIFICATION
  10× citric saline: 50 g KCl, 22 g sodium citrate, final volume 500 ml
  10× TD buffer: 1.4 *M* NaCl, 50 m*M* KCl, 7 m*M* K_2HPO_4, 250 m*M*
    Tris, pH 7.4
  Benzonase (Sigma, St. Louis, MO)
  10% Sodium deoxycholic acid in H_2O (Sigma)
  CsCl
  CsCl/TD solution (TD buffer adjusted to 1.4 g/ml with solid CsCl)
BIOLOGICAL TITER
  Refractometer (Sigma)
  96-Well plates (Falcon™)
  Wild-type Ad stocks: ATCC
  Cos 7 cells: ATCC
PARTICLE QUANTITATIVE PCR (QPCR)
  Optical plates (Applied Biosystems, Foster City, CA)
  Optical caps (Applied Biosystems)
  QPCR machine (Applied Biosystems)
  Probe/primer pairs (Applied Biosystems)
  Proteinase K: 20 mg/ml in H_2O

Benzonase
Phenol–chloroform–isoamylalcohol in a 1:1:24 ratio can be purchased
   from many suppliers
95% Ethanol
70% Ethanol
3 M sodium acetate (pH 5.2)
200-μl PCR tubes (Marsh Lab Supplies)

Notes on AAV Production

1. Significant homology exists between the ITRs and Rep proteins of
AAV4 and AAV2. As a result, it is possible to pseudotype these viruses by
using a plasmid that contains AAV2 ITRs flanking your gene of interest
and a helper plasmid that contains the AAV4 *rep* and *cap* genes. Thus,
standard AAV2 vector plasmids such as PAV$_2$ and psub201 can be used
(Li *et al.*, 1997). The resulting pseudotyped viral particles will have an
AAV2 genome packaged in an AAV4 capsid. Two helper plasmids have
been described for producing recombinant AAV4. Both pAAV4SV40ori
and p4RC contain the AAV4 *rep* and *cap* genes, but pAAV4SV40ori also
contains an SV40 for high-level expression in T-antigen-expressing cells
(Alisky *et al.*, 2000; Chiorini *et al.*, 1997).

2. AAV5 Rep and ITRs are distinct from those of AAV2, and it is not
possible to produce pseudotyped particles as described previously. For
AAV5, the same Rep and ITR serotype must be used. For example, to
package an AAV5 genome, AAV5 *rep* genes must be present on the
helper plasmid. One vector plasmid for AAV5 is pAAV5LacZ. This
plasmid contains an RSV promoter and a nuclear-localized β-galactosidase
gene with an SV40 poly A signal. The cassette can be removed by digestion
with *Bgl*II and *Bsm*I, leaving only the AAV5 ITRs or *Not*I and *Bsm*I to
remove the nuclear-localized β-galactosidase gene and SV40 poly A signal.
The AAV5 helper plasmids for use with this vector plasmid are
p5RepCapB or p5RC. Both plasmids contain the AAV5 *rep* and *cap*
genes and p5RepCapB also contains an SV40 origin of replication SV40
for high-level expression in T-antigen-expressing cells (Alisky *et al.*, 2000;
Chiorini *et al.*, 1999). Other laboratories have developed methods for
pseudotyping AAV2 within AAV5 capsids (Duan *et al.*, 2001; Rabinowitz
et al., 2002). For this to work, a universal shuttle was created containing
AAV2 ITRs and the transgene. The packaging construct contained AAV2
rep and AAV5 *cap*, AAV2 and AAV4 *cap*, etc. In this manner, virions
differing only in capsid proteins could be produced.

3. A number of helper and vector plasmids have been described for
AAV2 (Chiorini *et al.*, 1999; Xiao *et al.*, 1998). If the purpose of the
experiment is to compare the relative transduction efficiency of different

serotypes of AAV for a given tissue, it is important to use vector plasmids with similar expression cassettes. The plasmid pAAV2LacZ contains the same expression cassette as pAAV5LacZ (an RSV promoter and a nuclear-localized β-galactosidase gene with an SV40 poly A signal), but possesses AAV2 instead of AAV5 ITRs. Because of the similarity of the ITRs and Rep ORFs, pAAV2LacZ can be used for producing recombinant AAV4.

4. As discussed previously, the development of a plasmid to provide the necessary helper functions resulted in the production of cell lysates free of Ad contamination. Several examples of this plasmid exist and are commercially available. In general, all contain the VA RNA, E1, E2, and E4 genes of Ad (Alisky *et al.*, 2000; Grimm *et al.*, 1998; Xiao *et al.*, 1998).

Production of Recombinant Adenovirus

Principle. Recombinant Ad vectors are derived from human Ads, non-enveloped, encapsidated linear, double-stranded DNA viruses that commonly cause respiratory and gastrointestinal infections. A total of 43 different human Ad serotypes have been characterized (Horwitz, 1990). Ad5 replication-impaired vectors contain deletions in the E1 and E3 region of the Ad genome, and transgenes are driven by a variety of promoters, including viral and cell- or tissue-specific promoters (Millecamps *et al.*, 1999). Recently, fiber-modified and 'gutless' Ad vectors have been developed for improved use *in vivo* (Chillon and Kremer 2001; Chillon *et al.*, 1999; Davidson and Bohn, 1997; Heistad and Faraci, 1996; Soudais *et al.*, 2001; Umana *et al.*, 2001; Xia *et al.*, 2000).

First-generation viruses have deletions in *E1*, a gene required for replication of the virus in most mammalian cells. Most rAds also retain a small deletion in E3 because dl309 (Jones and Shenk, 1978), also known as sub360, was used to generate the first recombinants. dl309 backbones are used for more than 95% of applications; of ease recombination between dl309 and shuttle plasmids is very efficient, and this backbone supports high titer production. With appropriate controls, first-generation Ad vectors provide investigators with the tools required to answer important biological questions *in vitro* and *in vivo*.

The shuttle plasmid generally contains convenient restriction sites for cloning of expression cassettes, including those driving shRNA expression from pol II- (Xia *et al.*, 2002) and pol III- based promoters (Shen *et al.*, 2003). To our knowledge, tandem siRNA expression cassettes have not been generated in Ads (see "Notes on AAV Production," Point 1).

The backbone plasmid contains sequences homologous to a portion of the shuttle plasmid, or a recombination site (Fig. 3). Many backbone plasmids now also contain expression cassettes in the E3 region. This is advantageous for vectors expressing shRNAs so that transduced cells can be monitored by simultaneous expression of reporter genes such as GFP, DsRed, or *E. coli* β-galactosidase (Davidson, 2003; Xia *et al.*, 2002).

Methods. There are several commercially available kits for generation of Ads in the laboratory and academic vector cores and for-profit entities that provide fee-for-service generation of Ad vectors. Examples of facilities include, but are not limited to, http://www.uiowa.edu/%7Egene//, http://www.med.umich.edu/vcore/, http://www3.mdanderson.org/depts/genetherapy/vectorrevised.html, http://www.viraquest.com/, http://www.merlincustomservices.com/v-adenovirus.shtml, and http://www.adenovirus.com/products/custom/. More recently, several companies have capitalized on the ease of adenovirus production and now provide kits for generating shRNA expressing adenoviruses (http://www.imgenex.com/; http://www.ambion.com/; see "Notes on AAV Production," Point 2).

Details of Ad production, concentration, titering, and storage are provided at http://www.uiowa.edu/%7Egene//, but are as outlined in Fig. 3. In brief, shuttle plasmids containing the shRNA expression cassettes are transfected into HEK 293 cells along with the desired backbone. In 10 days, the Ad has amplified to the extent that the cells will lift off the plate. Cells are harvested, lysed, and the virus in the cell lysate used to infect more HEK 293 cells. The overall process outlined in Fig. 3 takes on average 2–3 weeks from transfection to final banding of high-titer shRNA-expressing rAd.

Materials. The materials for Ad production, titering, storage, and concentration are available at http://www.uiowa.edu/%7Egene//.

Notes

1. If two shRNA expression cassettes are placed in tandem, care should be taken to avoid using the same promoter to drive the sense and antisense strands because recombination between the promoters will lead to rearrangement and deletion.
2. Hairpins for pol II-based systems should range from 19 to 21 bp, with a 6- to 8-bp loop between the sense and antisense strands. Hairpins for pol III-based systems have ranged from 19 to 27 bp. The inclusion of a loop from a microRNA sequence improves their export and silencing capabilities in some systems (Kawasaki and Taira, 2003).

Applying the shRNA Expression Vectors to Cells and Tissues

For all viruses, it is imperative to optimize viral infection parameters. Viruses expressing reporter constructs are incubated with the target cell at varying dilutions, generally in serum-free medium. After a 4 h to an overnight incubation, the virus-containing media is removed and replaced with complete media. Cells are then monitored for evidence of gene transfer 24–72 h later. For shRNA studies, the extent of silencing of protein will depend on the target protein half-life. For this reason, monitoring of silencing by evaluation of target RNA levels may be warranted. All vector systems have been used *in vivo*. [For sample methods for CNS application see Alisky and Davidson (2004), Davidson and Chiorini (2003), and Stein and Davidson (2002).]

Conclusions

The protocols provided should allow the researcher to develop Ad-, AAV-, and FIV-based silencing tools. Moreover, the shuttle plasmids used to generate the various viruses are themselves excellent expression plasmids for shRNA expression in tissue culture cell lines.

References

Alisky, J., Hughes, S., Sauter, S., Jolly, D., Dubensky, T., Staber, P., Chiorini, J., and Davidson, B. (2000). Transduction of murine cerebellar neurons with recombinant FIV and AAV5 vectors. *NeuroRep.* **11,** 2669–2673.

Alisky, J. M., and Davidson, B. L. (2004). Gene transfer to brain and spinal cord using recombinant adenoviral vectors. *Methods Mol. Biol.* **246,** 91–120.

Anderson, R. D., Haskell, R. E., Xia, H., Roessler, B. J., and Davidson, B. L. (2000). A simple method for the rapid generation of recombinant adenovirus vectors. *Gene Ther.* **7,** 1034–1038.

Berns, K. I. (1990). Parvoviridae and their replication. *In* "Virology" (B. N. Fields and D. Knipe, eds.), pp. 1743–1763. Raven Press, New York.

Burns, J. C., Friedmann, T., Driever, W., Burrascano, M., and Yee, J. K. (1993). Vesicular stomatitis virus G glycoprotein pseudotyped retroviral vectors: Concentration to very high titer and efficient gene transfer into mammalian and nonmammalian cells. *Proc. Natl. Acad. Sci. USA* **90,** 8033–8037.

Chan, S. Y., Speck, R. F., and Ma, M. C. (2000). Distinct mechanisms of entry by envelope glycoproteins of Marburg and Ebola (Zaire) viruses. *J. Virol.* **74,** 49233–49237.

Chillon, M., Bosch, A., Zabner, J., Law, L., Armentano, D., Welsh, M. J., and Davidson, B. L. (1999). Group D adenoviruses infect primary central nervous system cells more efficiently than those from Group C. *J. Virol.* **73**(3), 2537–2540.

Chillon, M., and Kremer, E. J. (2001). Trafficking and propagation of canine adenovirus vectors lacking a known integrin-interacting motif. *Hum. Gene Ther.* **12**(14), 1815–1823.

Chiorini, J. A., Afione, S., and Kotin, R. M. (1999). Adeno-associated virus (AAV) type 5 Rep protein cleaves a unique terminal resolution site compared with other AAV serotypes. *J. Virol.* **73,** 4293–4298.

Chiorini, J. A., Wendtner, C. M., Urcelay, E., Safer, B., Hallek, M., and Kotin, R. M. (1995). High-efficiency transfer of the T cell co-stimulatory molecule B7-2 to lymphoid cells using high-titer recombinant adeno-associated virus vectors. *Hum. Gene Ther.* **6,** 1531–1541.

Chiorini, J. A., Yang, L., Liu, Y., Safer, B., and Kotin, R. M. (1997). Cloning of adeno-associated virus type 4 (AAV4) and generation of recombinant AAV4 particles. *J. Virol.* **71,** 6823–6833.

Coffin, J. M. (1990). Retroviridae and their replication. *In* "Virology" (B. N. Fields and D. M. Knipe, eds.), pp. 1437–1500. Raven Press, New York.

Davidson, B. L. (2003). Generating adenoviruses for shRNA or siRNA delivery. *In* "RNA Interference (RNAi). Nuts and Bolts of RNAi Technology" (D. R. Engelke, ed.), pp. 205–216. DNA Press, Eagleville, PA.

Davidson, B. L., and Bohn, M. C. (1997). Recombinant adenovirus: A gene transfer vector for study and treatment of CNS diseases. *Exp. Neurol.* **144**(1), 125–130.

Davidson, B. L., and Chiorini, J. A. (2002). Recombinant adeno-associated virus vector types 4 and 5: Preparation and application for CNS gene transfer. *In* "Meth. Mol. Med." (C. A. Machida, ed.), Vol. 76, pp. 269–283. Humana Press, Totowa, NJ.

Davidson, B. L., and Paulson, H. L. (2004). Molecular medicine for the brain: Silencing disease genes with RNA interference. *Lancet Neurol.* **3,** 145–149.

Davidson, B. L., Stein, C. S., Heth, J. A., Martins, I., Kotin, R. M., Derksen, T. A., Zabner, J., Ghodsi, A., and Chiorini, J. A. (2000). Recombinant adeno-associated type 2, 4 and 5 vectors: Transduction of variant cell types and regions in the mammalian CNS. *Proc. Natl, Acad. Sci. USA* **97**(7), 3428–3432.

DePolo, N. J., Reed, J. D., Sheridan, P. L., Townsend, K., Sauter, S. L., Jolly, D. J., and Dubensky, T. W. J. (2000). VSV-G pseudotyped lentiviral vector particles produced in human cells are inactivated by human serum. *Mol. Ther.* **2,** 218–222.

Duan, D., Fisher, K., Burda, J., and Engelhardt, J. F. (1997). Structural and functional heterogeneity of integrated recombinant AAV genomes. *Virus Res.* **48,** 41–56.

Duan, D., Yan, Z., Yue, Y., Ding, W., and Engelhardt, J. F. (2001). Enhancement of muscle gene delivery with pseudotyped adeno-associated virus type 5 correlates with myoblast differentiation. *J. Virol.* **75**(16), 7662–7671.

Elbashir, S. M., Harborth, J., Lendeckel, W., Yalcin, A., Weber, K., and Tuschl, T. (2001). Duplexes of 21-nucleotide RNAs mediate RNA interference in cultured mammalian cells. *Nature* **411,** 494–498.

Flotte, T. R., Barraza-Ortiz, X., Solow, R., Afione, S. A., Carter, B. J., and Guggino, W. B. (1995). An improved system for packaging recombinant adeno-associated virus vectors capable of *in vivo* transduction. *Gene Ther.* **2,** 29–37.

Grimm, D., Kern, A., Rittner, K., and Kleinschmidt, J. A. (1998). Novel tools for production and purification of recombinant adeno-associated virus vectors. *Hum. Gene Ther.* **9,** 2745–2760.

Hannon, G. J. (2002). RNA interference. *Nature* **418,** 244–251.

Heistad, D. D., and Faraci, F. M. (1996). Gene therapy for cerebral vascular disease. *Stroke* **27**(9), 1688–1693.

Horwitz, M. S. (1990). Adenoviruses. *In* "Fields Virology" (B. N. Fields and D. M. Knipe, eds.), Vol. 2, pp. 1723–1740. Raven Press, New York.

Horwitz, M. S. (1996). Adenoviruses. *In* "Fields Virology" (A. B. Rickinson, E. Kieff, B. N. Fields, D. M. Knipe, and P. M. Howley, eds.), pp. 2149–2171. Lippincott-Raven, Philadelphia.

Inoue, N., and Russell, D. W. (1998). Packaging cells based on inducible gene amplification for the production of adeno-associated virus vectors. *J. Virol.* **72,** 7024–7031.

Johnston, J. C., Gasmi, M., Lim, L. E., Elder, J. H., Yee, J. K., Jolly, D. J., Campbell, K. P., Davidson, B. L., and Sauter, S. L. (1999). Minimum requirements for efficient transduction of dividing and nondividing cells by feline immunodeficiency virus vectors. *J. Virol.* **73**, 4991–5000.

Jones, N., and Shenk, T. (1978). Isolation of deletion and substitution mutants of adenovirus type 5. *Cell* **13**, 181–188.

Kang, Y., Stein, C. S., Heth, J. A., Sinn, P. L., Penisten, A. K., Staber, P. D., Shen, H., Barker, C. K., Martins, I., Sharkey, C. M., Sanders, D. A., McCray, P. B. J., and Davidson, B. L. (2002). *In vivo* gene transfer using a nonprimate lentiviral vector pseudotyped with Ross River virus glycoproteins. *J. Virol.* **76**, 9378–9788.

Kawasaki, H., and Taira, K. (2003). Short hairpin type of dsRNAs that are controlled by tRNA(Val) promoter significantly induce RNAi-mediated gene silencing in the cytoplasm of human cells. *Nucleic Acids Res.* **31**, 700–707.

Khvorova, A., Reynolds, A., and Jayasena, S.D. (2003). Functional siRNAs and miRNAs exhibit strand bias. *Cell* **115**, 209–216.

Kotin, R. M., Linden, R. M., and Berns, K. I. (1992). Characterization of a preferred site on human chromosome 19q for integration of adeno-associated virus DNA by non-homologous recombination. *EMBO J.* **11**, 5071–5078.

Kotin, R. M., Menninger, J. C., Ward, D. C., and Berns, K. I. (1991). Mapping and direct visualization of a region-specific viral DNA integration site on chromosome 19q13-qter. *Genomics* **10**, 831–834.

Landau, N. R., Page, K. A., and Littman, D. R. (1991). Pseudotyping with human T-cell leukemia virus type I broadens the human immunodeficiency virus host range. *J. Virol.* **65**, 162–169.

Li, J., Samulski, R. J., and Xiao, X. (1997). Role for highly regulated rep gene expression in adeno-associated virus vector production. *J. Virol.* **71**, 5236–5243.

Mastromarino, P., Conti, C., Goldoni, P., Hauttecoeur, B., and Orsi, N. (1987). Characterization of membrane components of the erythrocyte involved in vesicular stomatitis virus attachment and fusion at acidic pH. *J. Gen. Virol.* **68**, 2359–2369.

Miletic, H., Bruns, M., Tsiakas, K., Vogt, B., Rezai, R., Baum, C., Kuhlke, K., Cosset, F. L., Ostertag, W., Lother, H., and von Laer, D. (1999). Retroviral vectors pseudotyped with lymphocytic choriomeningitis virus. *J. Virol.* **73**, 6114–6116.

Millecamps, S., Kiefer, H., Navarro, V., Geoffroy, M. C., Robert, J. J., Finiels, F., Mallet, J., and Barkats, M. (1999). Neuron-restrictive silencer elements mediate neuron specificity of adenoviral gene expression. *Nat. Biotechnol.* **17**, 865–869.

Naldini, L., Blomer, U., Gage, G. H., Trono, D., and Verma, I. M. (1996). Efficient transfer, integration, and sustained long-term expression of the transgene in adult rat brains injected with a lentiviral vector. *Proc. Natl. Acad. Sci. USA* **93**, 11382–11388.

Narayan, O., and Clements, J. E. (1990). Lentiviruses. *In* "Virology" (B. N. Fields and D. M. Knipe, eds.), pp. 1571–1592. Raven Press, New York.

Page, K. A., Landau, N. R., and Littman, D. R. (1990). Construction and use of a human immunodeficiency virus vector for analysis of virus infectivity. *J. Virol.* **64**, 5270–5276.

Paul, C. P., Good, P. D., Winer, I., and Engelke, D. R. (2002). Effective expression of small interfering RNA in human cells. *Nat. Biotechnol.* **20**, 505–508.

Rabinowitz, J. E., Rolling, F., Li, C., Conrath, H., Xiao, W., Xiao, X., and Samulski, R. J. (2002). Cross-packaging of a single adeno-associated virus (AAV) type 2 vector genome into multiple AAV serotypes enables transduction with broad specificity. *J. Virol.* **76**, 791–801.

Reiser, J., Harmison, G., Kluepfel-Stahl, S., Brady, R. O., Karlsson, S., and Schubert, M. (1996). Transduction of nondividing cells using pseudotyped defective high-titer HIV type 1 particles. *Proc. Natl. Acad. Sci. USA* **93**, 15266–15271.

Schiedner, G., Morral, N., Parks, R. J., Wu, Y., Koopmans, S. C., Langston, C., Graham, F. L., Beaudet, A. L., and Kochanek, S. (1998). Genomic DNA transfer with a high-capacity adenovirus vector results in improved *in vivo* gene expression and decreased toxicity. *Nat. Genet* **18**, 180–183.

Shen, C., Buck, A. K., Liu, X., Winkler, M., and Reske, S. N. (2003). Gene silencing by adenovirus-delivered siRNA. *FEBS Lett.* **539**, 111–114.

Shen, E. S., Cooke, G. M., and Horlick, R. A. (1995). Improved expression cloning using reporter genes and Epstein-Barr virus ori-containing vectors. *Gene* **156**, 235–239.

Shenk, T. (1996). Adenoviridae: The viruses and their replication. *In* "Fields Virology" (A. B. Rickinson, E. Kieff, B. N. Fields, D. M. Knipe, and P. M. Howley, eds.), pp. 2111–2148. Lippincott-Raven, Philadelphia.

Soudais, C., Laplace-Builhe, C., Kissa, K., and Kremer, E. J. (2001). Preferential transduction of neurons by canine adenovirus vectors and their efficient retrograde transport *in vivo. FASEB J.* **15**, 2283–2285.

Spiegel, M., Bitzer, M., Schenk, A., Rossmann, H., Neubert, W. J., Seidler, U., Gregor, M., and Lauer, U. (1998). Pseudotype formation of Moloney murine leukemia virus with Sendai virus glycoprotein F. *J. Virol.* **72**, 5296–5302.

Stein, C. S., and Davidson, B. L. (2002). Gene transfer to the brain using feline immunodeficiency virus-based lentivirus vectors. *In* "Methods in Enzymology" (M. I. Phillips, ed.), Vol. 346, pp. 433–454. Academic Press, San Diego, CA.

Stitz, J., Buchholz, C. J., Engelstadter, M., Uckeret, W., Bloemer, U., Schmitt, I., and Cichutek, K. (2000). Lentiviral vectors pseudotyped with envelope glycoproteins derived from gibbon ape leukemia virus and murine leukemia virus 10A1. *Virology* **273**, 16–20.

Taylor, R. M., and Wolfe, J. H. (1994). Cross-correction of β-glucuronidase deficiency by retroviral vector-mediated gene transfer. *Exp. Cell Res.* **214**, 606–613.

Thomas, C. E., Schiedner, G., Kochanek, S., Castro, M. G., and Lowenstein, P. R. (2000). Peripheral infection with adenovirus causes unexpected long-term brain inflammation in animals injected intracranially with first-generation, but not with high-capacity, adenovirus vectors: Toward realistic long-term neurological gene therapy for chronic diseases. *Proc. Natl. Acad. Sci. USA* **97**(13), 7482–7487.

Tilton, G. K., O'Connor, T. P. J., Seymour, C. L., Lawrence, K. L., Cohen, N. D., Andersen, P. R., and Tonelli, Q. J. (1990). Immunoassay for detection of feline immunodeficiency virus core antigen. *J. Clin. Microbiol.* **28**(5), 898–904.

Umana, P., Gerdes, C. A., Stone, D., Davis, J. R., Ward, D., Castro, M. G., and Lowenstein, P. R. (2001). Efficient FLPe recombinase enables scalable production helper-dependent adenoviral vectors with negligible helper-virus contamination. *Nat. Biotechnol.* **19**, 582–585.

Urabe, M., Ding, C., and Kotin, R. M. (2002). Insect cells as a factory to produce adeno-associated virus type 2 vectors. *Hum. Gene Ther.* **13**, 1935–1943.

Vincent, K. A., Piraino, S. T., and Wadsworth, S. C. (1997). Analysis of recombinant adeno-associated virus packaging and requirements for rep and cap gene products. *J. Virol.* **71**, 1897–1905.

Wang, G., Slepushkin, V., Zabner, J., Keshavjee, S., Johnston, J. C., Sauter, S. L., Jolly, D. J., Dubensky, T., Davidson, B. L., and McCray, P. B. J. (1999). Feline immunodeficiency virus vectors persistently transduce nondividing airway epithelia and correct the cystic fibrosis defect. *J. Clin. Invest* **104**, R55–R62.

Weitzman, M. D., Kyostio, S. R., Kotin, R. M., and Owens, R. A. (1994). Adeno-associated virus (AAV) Rep proteins mediate complex formation between AAV DNA and its integration site in human DNA. *Proc. Nal. Acad. Sci. USA* **91**, 5808–5812.

Wu, P., Phillips, M. I., Bui, J., and Terwilliger, E. F. (1998). Adeno-associated virus vector-mediated transgene integration into neurons and other nondividing cell targets. *J. Virol.* **72**(7), 5919–5926.

Xia, H., Anderson, B., Mao, Q., and Davidson, B. L. (2000). Recombinant human adenovirus: Targeting to the human transferrin receptor improves gene transfer to brain microcapillary endothelium. *J. Virol.* **74**(23), 11359–11366.

Xia, H., Mao, Q., Paulson, H. L., and Davidson, B. L. (2002). siRNA-mediated gene silencing *in vitro* and *in vivo*. *Nat. Biotechnol.* **20**(10), 1006–1010.

Xiao, X., Li, J., and Samulski, R. J. (1998). Production of high-titer recombinant adeno-associated virus vectors in the absence of helper adenovirus. *J. Virol.* **72**, 2224–2232.

Yee, J.-K., Friedmann, T., and Burns, J. C. (1994). Generation of high-titer pseudotyped retroviral vectors with very broad host range. *In* "Methods in Cell Biology" Vol. 43, pp. 99–112. Academic Press, San Diego, CA.

Zeilfelder, U., and Bosch, V. (2001). Properties of wild-type, C-terminally truncated, and chimeric maedi-visna virus glycoprotein and putative pseudotyping of retroviral vector particles. *J. Virol.* **75**, 548–555.

Zeng, Y., and Cullen, B. R. (2003). Sequence requirements for micro RNA processing and function in human cells. *RNA* **9**, 112–123.

[10] Targeting Cellular Genes with PCR Cassettes Expressing Short Interfering RNAs

By Daniela Castanotto and Lisa Scherer

Abstract

The synthesis and transfection of PCR short interfering/short hairpin RNA (si/shRNA) expression cassettes described in this chapter can be used to rapidly test siRNA targeting and function in cells. One critical element in the design of effective siRNAs is the selection of siRNA–target sequence combinations that yield the best inhibitory activity. This can be accomplished by using synthetic siRNAs and transfection procedures, but these can be costly and time consuming. By using the PCR strategy, it is possible to create several expression cassettes and simultaneously screen for the best target sites on any given mRNA. This PCR strategy allows a rapid and inexpensive approach for intracellular expression of si/shRNAs and subsequent testing of target site sensitivity to downregulation by RNA interference (RNAi).

Introduction

At present, RNA interference (RNAi) is perhaps the most powerful genetic tool for target-specific knockdown in mammalian cell (Hannon, 2002; Tuschl, 2001). siRNAs and the related short hairpin RNAs (shRNAs) can be expressed *in vivo* against human, viral, and cellular targets. The most popular method for targeting a gene is based on transfection of chemically synthesized siRNA duplexes or intracellular expression of siRNAs using RNA polymerase III and polymerase II (pol III and pol II) cassettes (Elbashir *et al.*, 2001a,b; Tuschl, 2001, 2002). Despite the great potential of siRNAs, identifying optimal siRNAs target sites can be problematic, which is critical for siRNAs function (Holen *et al.*, 2002; Lee *et al.*, 2002) and specificity. Although these molecules are generally highly specific, in high concentration they can elicit nonspecific off-target effects. Therefore, it is critical to identify the best target site – siRNA combination that provides the highest knockdown at the lowest possible concentration of siRNAs. Moreover, at high concentrations, it is possible that some siRNAs can activate the interferon response pathway, which could lead to nonspecific degradation of cellular RNA (Bridge *et al.*, 2003; Sledz *et al.*, 2003).

Several Web sites contain programs that use algorithms to find the best siRNA design for a specific target. Combining these programs with new rules for the thermodynamic properties of the sense and antisense siRNAs (Khvorova *et al.*, 2003; Schwarz *et al.*, 2003) is helpful for designing effective interfering/short hairpin RNAs (si/shRNAs), but the best target site selection can only be found by systematically and empirically testing a variety of potential targets (Vickers *et al.*, 2003). However, constructing several vectors expressing si/shRNAs to test their downregulation activity *in vivo* is considerably time consuming. One possible way of overcoming this limitation is to have a simple screening procedure that allows the testing of several different sites along a messenger RNA (mRNA) for sensitivity to siRNA. The PCR-based method previously described (Castanotto *et al.*, 2002) and detailed here has proven to be both straightforward and effective. The method involves creation of pol III transcription units by PCR and direct transfection and testing of the PCR products for siRNA function in cell culture.

Design of Small Interfering RNA (siRNA)-Expressing PCR Cassettes

The expressed siRNA may consist of two separate, annealed single strands of 21 nt, in which the terminal two 3' nucleotides are unpaired (3' overhang) or it may be in the form of a single stem-loop, often referred to as a shRNA. siRNAs exogenously introduced into mammalian cells usually, but not always (Sledz *et al.*, 2003), bypass the Dicer step and directly activate homologous mRNA degradation without initiating the interferon response (Caplen *et al.*, 2002; Elbashir *et al.*, 2001a; Harborth *et al.*, 2001), whereas double-stranded RNA (dsRNA) longer than 30 nt triggers the nonspecific interferon pathway. Activation of the interferon pathway can lead to global downregulation of translation as well as global RNA degradation. Thus, it is advisable to design the siRNA sense and antisense strands to be shorter than 30 nt. Duplexes of 21 nt are usually sufficient to obtain full activity of expressed si/shRNAs. However, slightly longer (24–27) si/shRNAs, which undergo processing to 21-nt molecules by the Dicer enzyme, could be more efficiently incorporated into the RNA-induced silencing complex (RISC) and thus more effective in downregulating their target.

With the exception of the two nucleotides that form the 3' overhang, the antisense sequence should be completely complementary to the target to direct its cleavage. However, some mismatches can be tolerated [Amarzguioui et al., 2003; Bernstein et al., 2001; Brummelkamp et al., 2002; Chiu and Rana, 2002; Dohjima et al., 2003; Elbashir et al., 2001; Holen et al., 2002; Randall and Rice, 2001; Yu et al., 2002; Hutvagner and Zamore, 2002] and this should be considered when designing inactive, mutant forms of the siRNAs for use as negative controls. In general, we find that four consecutive nucleotide mismatches in the center of the antisense strand work for the majority of siRNA–target combinations. Standard computer searching programs (e.g., BLAST) should also be used to exclude the presence of long stretches of siRNA complementarity with nontargeted cellular genes. Because U6 initiates transcription with a G, it will be incorporated as the first base of the sense or antisense strand. For this reason, in the design of siRNAs we prefer placing the sense strand preceding the antisense strand, because the active strand (the antisense) will not obligatorily contain a 5' G. If an extra G is included at the 5' end of the sense strand, it is not necessary to include a corresponding C at the 3' end of the antisense strand because the G can form a G–U pair with one of the 3' terminal Us that serve as a termination signal for RNA pol III; it is also important that the si/shRNAs does not include more than three Us in a row. Thus, the selected target site (and the sense strand) should not contain more then three contiguous As.

Construction of si/shRNA PCR Cassettes

PCR Oligo Design

The procedure for the PCR approach employs a universal primer complementary to the 5' end of the U6 promoter (other pol III and possibly pol II promoters can also be used) along with a unique primer complementary to the 3' end of the promoter. The 3' primers can be designed to include complementary sequences to either the sense or antisense siRNAs or to include both the sense and antisense sequences linked by a short loop. The first design is used when two separate PCR cassettes independently expressing the sense and antisense siRNA are constructed (Fig. 1A). The sense or

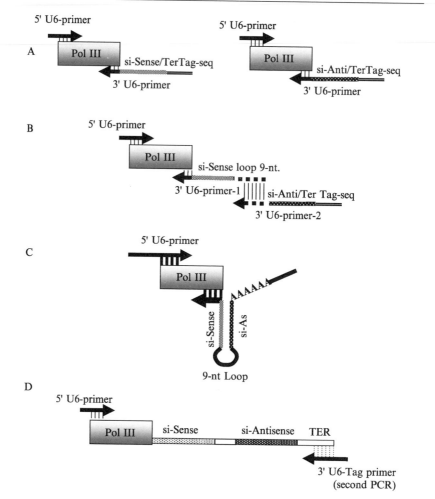

Fig. 1. Schematic representation of the PCR strategy used to generate U6 transcription cassettes expressing siRNAs. The 5′ PCR primer is complementary to the 5′ end of the U6 promoter and is standard for all PCR reactions. (A) The 3′ PCR primer is complementary to sequences at the 3′ end of the U6 promoter followed by either the sense or antisense sequences, a stretch of six deoxyadenosines (Ter), and an additional stuffer-tag sequence, which contains a restriction site and six additional random nucleotides. The As are complementary to the termination signal for the U6 pol III promoter; therefore, any sequence added after this signal will not be transcribed by the pol III polymerase and will not be part of the siRNA. (B) The first 3′ PCR primer is complementary to sequences at the 3′ end of the U6 promoter followed by the sense sequences and a 9-nt loop. The second 3′ PCR primer contains a sequence complementary to the 9-nt loop followed by the antisense sequences, a stretch of four to six deoxyadenosines (Ter), and the additional stuffer-tag sequence. The sense and antisense sequences are inserted in the cassette by a two-step PCR reaction (see text). (C) The sense and antisense sequences, linked by a 9-nt loop and followed by the stretch

the antisense sequences are followed by a stretch of six deoxyadenosines (Ter) and by a short additional sequence, which we term the stuffer-tag sequence, which includes a restriction site for possible cloning at a later stage as well as six random nucleotides. Thus, the resulting PCR products include the U6 promoter sequence, the sense or antisense, a terminator sequence, and the selected restriction sites at the 5′ and 3′ termini of the product. The two PCR cassettes making the sense or the antisense siRNA strands should be simultaneously cotransfected in cells to test their activity. The best combination among the sets of cassettes tested can be later cloned into a single expression vector for long-term studies. To maximize intracellular expression and avoid promoter interference, it is advisable to clone the two U6 cassettes expressing the sense or antisense siRNA strand in the same orientation and separated by a short spacer. If the selected vector backbone does not contain suitable restriction sites, it is possible to include the spacer in the design of the 5′ PCR oligonucleotide by starting with a restriction site at the 5′ end of the oligo and proceeding with the spacer and with the nucleotides complementary to U6 promoter sequences.

A preferable design is to include both the sense and antisense sequences within the same cassette (Fig. 1B, C). In this approach, the 3′ primer contains sequences that encode the complete shRNA, a pol III terminator sequence (T_6), a restriction site for possible cloning at a later stage, as well as four to six random nucleotides. The sense and antisense siRNA sequences contained in the 3′ primer are linked with a short 9-nt loop (UUUGUGUAG), which in our experience has proven to be an effective loop with several different targets. However, a number of other loop sequences are suitable for siRNA function. When selecting a loop sequence it is advisable to base the selection on comparison among various microRNA sequences, because naturally evolved sequences are likely to be processed more efficiently. If the UUUGUGUAG loop sequence is selected for the shRNA design, it is important that the siRNA sense strand does not contain a U at its 3′ terminus as this would create a stretch of four Us that could serve as a pol III terminator element.

To construct the U6-shRNA cassette, two 3′ primers or a single 3′ primer can be used. When two 3′ primers are used, the first PCR reaction employs the 5′ U6 universal primer and a 3′ primer complementary to 25 nt of the U6 promoter, followed by sequences complementary to the sense and the 9-nt loop (Fig. 1B). One microliter of this first reaction is reamplified in a second PCR reaction that employs the same 5′ U6 primer and

of As and the stuffer-tag sequences, are included in a single 3′ primer. (D) Complete PCR expression cassette obtained by the PCR reaction. To increase the yield of the PCR product shown in (B), an additional PCR step can be performed using the universal 5′ U6 primer and a 3′ primer complementary to the Tag sequence, as indicated in the figure and in the text.

a 3′ primer harboring sequences complementary to the 9-nt loop appended to the antisense strand, Ter, and restriction site sequence (stuffer-tag sequence, Fig. 1B). It is possible to add a few additional nucleotides complementary to the sense strand after the 9-nt loop to increase the T_m of this second PCR primer; however, it is important to use caution in designing the extended complementarities because this will cause snap-back at the 3′ end of the PCR primer and create difficulties in the extension step. The resulting PCR products include the U6 promoter, the sense and antisense coding sequences, followed by the pol III terminator sequence and the restriction site (Fig. 1D).

When a single 3′ primer is used, the procedure consists of a one-step PCR reaction with a 3′ primer containing sequences complementary to the 3′ end of the promoter and sequences complementary to the sense, the loop, and the antisense siRNA gene followed by a stretch of six deoxyadenosines (Ter) and by the short sequence that includes the restriction site and six random nucleotides (Fig. 1C). The resulting PCR product includes the U6 promoter, the sense and antisense siRNAs in the form of a stem loop, the terminator sequence, and the restriction sites (Fig. 1D). This second approach employs a considerably long and structured 3′ PCR primer that with a few sequences may cause difficulties in the amplification reaction, in which case it is helpful to slightly modify the standard amplification conditions (see next section). However, this strategy minimizes the possibility of inserting any polymerase-induced mutations in the siRNA sequence during the amplification reaction. In addition, the RNA expressed intracellularly by this PCR cassette is already in the form of a stem loop, which should have greater stability than when the sense and antisense siRNAs are expressed from two separate cassettes as single-stranded RNAs. A number of commercial DNA synthesis facilities are capable of synthesizing the resulting 3′ primers, which are typically 85–95 nt. If required, there are several options for primer purification. HPLC purification is efficient, but not recommended because it has poor resolution for primers of this length. Urea-PAGE purification has higher resolution, but yields are often low and if not done carefully can introduce contaminants that interfere with subsequent PCR reactions. However, it is a procedure that can be easily performed in most laboratories. In practice, however, simple desalting should be sufficient. Generally, most failure sequences will reflect incomplete products from the last rounds of synthesis of the 5′ end of the primer, which encodes the 3′ end of the PCR product. Designing the primer with sufficient 3′ noncoding sequence, along with purification of the final PCR product (see later), should minimize this potential problem.

For direct transfection and testing of the PCR-amplified siRNA genes, the 5′ termini of the PCR primers must be phosphorylated by using DNA polynucleotide kinase and nonradioactive ATP. This modification results in enhanced function of the PCR products, perhaps by stabilizing them intracellularly or promoting enzymatic ligation and multimerization of the PCR cassettes. PCR primers are phosphorylated before the PCR reactions, using a 600-pmole primer, 1× buffer (New England Biolabs, Beverly, MA), 1 mM ATP, and 20 units T4 polynucleotide kinase in a total volume of 100 μl. After a 30-min incubation at 37°, reactions are adjusted to 2.0 M ammonium acetate, extracted once with phenol:chloroform:isoamyl alcohol (25:24:1), once with chloroform:isoamyl alcohol (24:1), and precipitated with 2.5 volumes of ethanol and glycogen carrier (Roche, Switzerland). After collecting precipitates by centrifugation, the resulting pellets are washed in 70% ethanol, dried briefly, and resuspended at 50 pmoles/ml. Alternatively, some companies that synthesize oligos provide the option of phosphorylating them as well, which may save both time and money when it is anticipated that an oligo (such as the universal 5′ oligo) is to be used repeatedly.

PCR Reaction Conditions

PCR reactions use a hot-start protocol, beginning with a mix of 1× Vent polymerase buffer (New England Biolabs), 40 nmoles of each dNTP, and 100 pmoles of each of the universal U6 5′ and unique 3′ primers in a final volume of 25 μl in a PCR tube containing an Ampliwax pellet, which is then incubated at 85° for 10 min, followed by a brief cooling to harden the wax layer. The top mix consists of 75 μl of 1× Vent buffer, 2 units Vent polymerase, and 1 μg of a template plasmid containing the desired promoter, preferably linearized immediately downstream of the transcription start site. Although linearizing of the template in this manner is not required, it can increase the efficiency of amplification. In most cases, a simple PCR reaction that cycles 1 min at 95°, 1 min at 55°, and 1 min at 72° is sufficient to obtain enough product for functional tests in mammalian cells. When specific sequences present difficulties in the amplification reaction, it is preferable to perform a ramp PCR, in which the cycling parameters are typically 1 min at 95°, 3–5 min slow cool (ramping down) to the annealing temperature, 1 min at the final annealing temperature followed by 3–4 min ramping up to the extension temperature of 82°, and 1 min at 82°. The final annealing temperature is determined by the 3′ primer and should be at least 10° below the theoretical T_m for the 3′ primer and the U6+1 template; 15° below the theoretical T_m can increase yield. These reactions give very low yield under typical PCR conditions, in which the annealing temperatures are usually 5° below the theoretical T_m for the primer–template pair with the lowest annealing temperature. We postulate that the 3′ primer snaps back into the hairpin conformation very efficiently at a relatively high temperature, which may interfere with annealing to the template. For this reason, we recommend at least 22–24 nt of complementarity with the template promoter sequence. (As an aside, snap-back formation may have the favorable effect of reducing the formation of primer dimers.) The occasional 3′ primer may have a higher, narrow-range optimum annealing temperature, but, in general, ramping down between the denaturation and annealing steps allows efficient annealing of all primers without time-consuming optimization. The synthesis step is carried out at 82°, 10° higher than is usual for Vent polymerase, to allow for more efficient read-through of the stem loop of the 3′ primer. Vent has close to maximum activity at this temperature, whereas the enzyme half-life is not significantly affected (personal communication, technical services, New England Biolabs). The ramp-up between the final annealing and synthesis steps is to allow for some extension of the primers, stabilizing interaction with the template and preventing premature dissociation from the template. These reactions are often not as efficient as standard PCR reactions and may require more cycles than theoretically necessary; typically, we use 30 cycles of amplification.

For all PCR reactions we primarily use Vent polymerase, because of its high fidelity relative to Taq polymerase. PfuTurbo (Stratagene, La Jolla, CA) can sometimes improve PCR of highly structured templates, which may be useful for difficult shRNA templates. The temperature of the synthesis step with PfuTurbo can also be increased to the 74–78° range. Because PfuTurbo has very little activity between 40 and 50°, the hot-start protocol and slow ramp-up between the annealing and synthesis temperatures is probably unnecessary. Also, this enzyme tolerates addition of DNA denaturants that can improve results (see Stratagene protocols for details); we have used 1% high-quality DMSO with success. However, in the few cases we found difficulties in amplifying the template with Vent or PfuTurbo® polymerases, the Taq polymerase was effective in yielding sufficient amounts of shRNA PCR cassettes, which were subsequently confirmed to be devoid of mutations. When using Taq polymerase, we have performed synthesis at temperatures up to 75°. Finally, a 3′ primer designed to be complementary to the short sequence located at the 3′ end of the PCR cassette (stuffer-tag sequence) can be used together with the U6 universal primer to increase the production of poorly amplified PCR cassettes (Fig. 1D). The 3′ primer includes the pol III terminator (six Ts), the restriction site, and the six additional random nucleotides that follow the restriction site, for a total of 18 nt (Castanotto et al., 2002). This reamplification step should only be performed if all other strategies to improve the yield of the PCR product have failed.

When the PCR products appear homogeneous, they can be purified directly using Qiaquick® PCR columns. These column-purified products can be transfected into cells after assessing their concentration with a spectrophotometer. It is not unusual to see an additional band migrating just below the main product band. The proportion of this secondary product relative to the main product is typical of a given construct. The nature of this band is unknown; possibilities include alternative conformers that migrate differently (e.g., a cruciform structure of the shRNA region) or incomplete PCR products. Careful isolation of the upper band does not result in reappearance of the lower band on reanalysis. PCR products are isolated from a 1.6% agarose gel and full-length bands purified by Qiaquick columns (Qiagen), employing the extra washes recommended for microinjection. Recovery is quantified by gel electrophoresis and ethidium bromide staining relative to an internal standard.

Transfection Conditions

Transfection conditions depend on the cell system being used. Once the PCR reaction is completed and the products are column purified, they can be applied to cells by using cationic liposomes, calcium phosphate, or electroporation, depending on the cells in question. PCR cassettes become localized to the nucleus, where they will transcribe the shRNA or siRNA molecules capable of downregulating their cellular target (Fig. 2). The purified pol III-PCR cassettes can be either cotransfected with a plasmid expressing a specific target or directed against an endogenous cellular target. With 25–100 ng of the PCR cassette expressing the si/shRNA, 250 ng of the target plasmid can be cotransfected. As little as 25 ng of the PCR product can be effective in producing siRNAs. However, it is advisable to transfect at least 50 ng of PCR product.

If co-transfections of the target gene on a plasmid and the PCR products are to be utilized for experiments, the ratio between the PCR products and the plasmid expressing the target should be calculated to fall into molar ratio ranging from 1:1 to 1:5. To facilitate transfection of small amounts of PCR-amplified DNA, 400 ng to 1 μg of a plasmid such as BlueScript® (Stratagene) should be added to each reaction to serve as carrier. We have had poor success using chromosomal DNAs as carrier. It is important that the total amount of DNA be the same for each transfection. The BlueScript plasmid can be used in each case to achieve the desired amount of DNA recommended for each transfection procedure. For readily transfectable cell lines, transfections can be performed in six-well plates, using Lipofectamine Plus™ (Life Technologies, GibcoBRL, Gaithersburg, MD) or other transfection reagents as described by the manufacturer. Transfection reagents specifically developed for transfection of PCR cassettes in cell cultures are commercially available (e.g., siPortX-1, Ambion, Austin, TX).

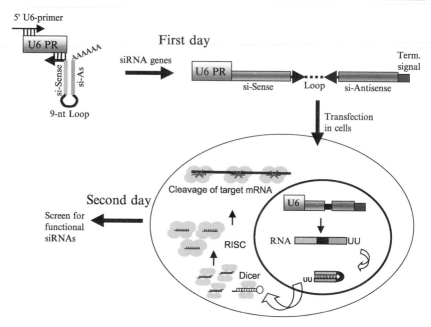

FIG. 2. Schematic representation of the steps required to inhibit cellular gene expression with pol III-driven shRNA PCR cassettes. A template containing the U6 pol III promoter is amplified with a 3' primer containing the shRNA sequence as described in the text. Within a couple of hours the PCR reaction generates a transcription unit capable of producing functional siRNA. The purified cassette is directly transfected into selected cell lines to target an expressed cellular gene. If the experimental cell line does not produce the shRNA target, the PCR cassettes can be cotransfected with a plasmid expressing the target of choice (see text). The RNAs transcribed from the PCR cassettes fold into the double-stranded shRNA structure, are transported to the cytoplasm where they are processed into siRNAs, and enter RISC, resulting in degradation of their mRNA target. The various siRNAs can be screened for function 24–48 h after transfection. (See color insert.)

Strong and specific downregulation of the target gene by the siRNAs should be detected 36–48 h post-transfection. One example of this PCR methodology is shown in Fig. 3. In this experiment, a PCR cassette was used to express a shRNA directed against the HIV-1 rev target gene, which is fused to a gene encoding for the green fluorescence protein. The gene fusion, which can be potentially created for any targeted gene, allows a quick screen for the efficacy of the si/shRNA through monitoring inhibition of fluorescence. A plasmid expressing the red fluorescence protein is co-transfected with the target and the PCR cassettes as a control for transfection efficiency. A control for transfection efficiency should always be included in all transfection experiments, especially if different si/shRNA

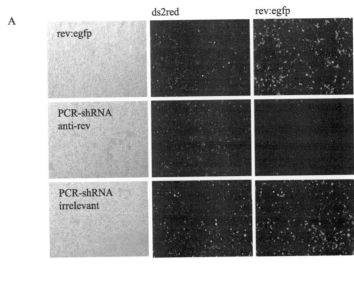

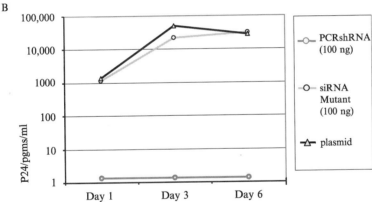

Fig. 3. (A) Downregulation of a rev:egfp gene fusion by an anti-rev shRNA-PCR cassette in 293 cells. The right panels show the cells under a bright field. The middle panel shows the red fluorescence generated by the ds2Red protein-expressing plasmid that was cotransfected with the experimental samples and used to determine the efficiency of transfection for each construct. The left panels show the green fluorescence generated by the intracellular expression of the rev:egfp fusion gene. (*Top*) The rev:egfp target was cotransfected with the ds2Red plasmid. (*Middle*) The rev:egfp target was cotransfected with the ds2Red plasmid and with 50 ng of a functional anti-rev shRNA PCR cassette. (*Bottom*) The rev:egfp plasmid was cotransfected with the ds2Red plasmid and with 50 ng of a shRNA PCR cassette with an irrelevant (nontargeted) sequence. (B) Inhibition of HIV-1 p24 antigen production by the anti-rev shRNA PCR cassette. A total of 100 ng of the anti-rev shRNA cassette was cotransfected with pNL4-3 HIV proviral DNA. A nonfunctional anti-rev shRNA-PCR cassette (siRNA mutant) containing four mutations in the middle of the shRNA sequence was

cassettes are tested to assess their relative ability to downregulate the target. FACS analyses can be used to obtain the exact levels of green fluorescence protein, which can be normalized to the efficiency of transfection of the various PCR cassettes by using the level of red fluorescence protein detected in each experimental sample (also obtained by FACS). The anti-rev shRNA that proved to be effective in cell cultures against the HIV-1 rev:egfp fusion target (Fig. 3A) was equally effective when used in bona fide HIV-challenge assays (Fig. 3B) and after it was cloned in an expression vector (data not shown). The anti-HIV rev PCR cassette used in cotransfection assays with HIV-1 proviral DNA was able to yield 1000-fold inhibition of p24 antigen production (Fig. 3B). For these assays, we typically use 0.5 μg proviral DNA, 100–200 ng of U6 + 1 shRNA PCR product, and enough pBluescript to bring the total amount of DNA up to 1 μg and transfect by using Lipofectamine Plus (Invitrogen, Carlsbad, CA) in Opti-MEM, according to the manufacturer's instructions. With a good target, as little as 25 ng of PCR product can inhibit HIV-1 replication, measured by p24 antigen production.

Conclusions

The direct transfection of PCR cassettes allows a simple and rapid screening of many genes and many sites within the same target gene. Using the si/shRNA PCR approach and a 96-well plate for cells transfection, it is potentially feasible to simultaneously test for the accessibility and siRNA sensitivity of 96 sites present in the target gene in a single transfection experiment within a short time frame (Fig. 2), thereby saving time and effort.

By using the PCR cassettes, we have observed target inhibition for at least six days. A nonfunctional mutant siRNA or a target with silent codon changes in the region of the siRNA base pairing should always be used as a control for nonspecific effects. Once the PCR product that works best for a given target is identified, it can be easily cloned into a plasmid or viral vector for transfection and transduction into primary cells and to conduct long-term studies of gene knockdown.

Acknowledgments

This work was supported by NIH grants A129329, AI 42552, and HL074704.

---

used as control for nonspecific effects. The pTZ U6 + 1 empty vector (plasmid) was used as a control to exclude nonspecific effects caused by the U6 promoter. The anti-rev shRNA PCR cassette yielded a 1000-fold inhibition of p24 antigen production and inhibited p24 antigen production by HIV-1 for up to 6 days post-transfection. (See color insert.)

References

Amarzguioui, M., Holen, T., Babaie, E., and Prydz, H. (2003). Tolerance for mutations and chemical modifications in a siRNA. *Nucleic Acids Res.* **31,** 589–595.

Bernstein, E., Denli, A. M., and Hannon, G. J. (2001). The rest is silence. *RNA* **7,** 1509–1521.

Bridge, A. J., Pebernard, S., Ducraux, A., Nicoulaz, A. L., and Iggo, R. (2003). Induction of an interferon response by RNAi vectors in mammalian cells. *Nat. Genet.* **34,** 263–264.

Brummelkamp, T. R., Bernards, R., and Agami, R. (2002). A system for stable expression of short interfering RNAs in mammalian cells. *Science* **296,** 550–553.

Caplen, N. J., Taylor, J. P., Statham, V. S., Tanaka, F., Fire, A., and Morgan, R. A. (2002). Rescue of polyglutamine-mediated cytotoxicity by double-stranded RNA-mediated RNA interference. *Hum. Mol. Genet.* **11,** 175–184.

Castanotto, D., Li, H., and Rossi, J. J. (2002). Functional siRNA expression from transfected PCR products. *RNA* **8,** 1454–1460.

Chiu, Y. L., and Rana, T. M. (2002). RNAi in human cells: Basic structural and functional features of small interfering RNA. *Mol. Cell* **10,** 549–561.

Doghjima, T., Lee, N. S., Li, H., Ohno, T., and Rossi, J. J. (2003). Small interfering RNAs expressed from a Pol III promoter suppress the EWS/Fli-1 transcript in an Ewing sarcoma cell line. *Mol. Ther.* **7,** 811–816.

Elbashir, S. M., Harborth, J., Lendeckel, W., Yalcin, A., Weber, K., and Tuschl, T. (2001a). Duplexes of 21-nucleotide RNAs mediate RNA interference in cultured mammalian cells. *Nature* **411,** 494–498.

Elbashir, S. M., Lendeckel, W., and Tuschl, T. (2001b). RNA interference is mediated by 21- and 22-nucleotide RNAs. *Genes Dev.* **15,** 188–200.

Hannon, G. J. (2002). RNA interference. *Nature* **418,** 244–251.

Harborth, J., Elbashir, S. M., Bechert, K., Tuschl, T., and Weber, K. (2001). Identification of essential genes in cultured mammalian cells using small interfering RNAs. *J. Cell Sci.* **114,** 4557–4565.

Holen, T., Amarzguioui, M., Wiiger, M. T., Babaie, E., and Prydz, H. (2002). Positional effects of short interfering RNAs targeting the human coagulation trigger tissue factor. *Nucleic Acids Res.* **30,** 1757–1766.

Hutvagner, G., and Zamore, P. D. (2002). A microRNA in a multiple-turnover RNAi enzyme complex. *Science* **297,** 2056–2060.

Khvorova, A., Reynolds, A., and Jayasena, S. D. (2003). Functional siRNAs and miRNAs exhibit strand bias. *Cell* **115,** 209–216.

Lee, N. S., Dohjima, T., Bauer, G., Li, H., Li, M. J., Ehsani, A., Salvaterra, P., and Rossi, J. (2002). Expression of small interfering RNAs targeted against HIV-1 rev transcripts in human cells. *Nat. Biotechnol.* **20,** 500–505.

Randall, G., and Rice, C. M. (2001). Hepatitis C virus cell culture replication systems: Their potential use for the development of antiviral therapies. *Curr. Opin. Infect. Dis.* **14,** 743–747.

Schwarz, D. S., Hutvagner, G., Du, T., Xu, Z., Aronin, N., and Zamore, P. D. (2003). Asymmetry in the assembly of the RNAi enzyme complex. *Cell* **115,** 199–208.

Sledz, C. A., Holko, M., de Veer, M. J., Silverman, R. H., and Williams, B. R. (2003). Activation of the interferon system by short-interfering RNAs. *Nat. Cell Biol.* **5,** 834–839.

Tuschl, T. (2001). RNA interference and small interfering RNAs. *Chembiochem* **2,** 239–245.

Tuschl, T. (2002). Expanding small RNA interference. *Nat. Biotechnol.* **20,** 446–448.

Vickers, T. A., Koo, S., Bennett, C. F., Crooke, S. T., Dean, N. M., and Baker, B. F. (2003). Efficient reduction of target RNAs by small interfering RNA and RNase H-dependent antisense agents. A Comparative Analysis. *J. Biol. Chem.* **278,** 7108–7118.

Yu, J. Y., DeRuiter, S. L., and Turner, D. L. (2002). RNA interference by expression of short-interfering RNAs and hairpin RNAs in mammalian cells. *Proc. Natl. Acad. Sci. USA* **99,** 6047–6052.

[11] Use of Short Hairpin RNA Expression Vectors to Study Mammalian Neural Development

By Jenn-Yah Yu, Tsu-Wei Wang, Anne B. Vojtek,
Jack M. Parent, and David L. Turner

Abstract

The use of RNA interference (RNAi) in mammalian cells has become a powerful tool for the analysis of gene function. Here we discuss the use of DNA vectors to produce short hairpin RNAs (shRNAs) and inhibit gene expression in mammalian neural progenitors and neurons. Protocols are presented for introducing shRNA vectors into mouse P19 cells differentiated as neurons *in vitro* and for electroporation of shRNA vectors into primary neural progenitors from the embryonic mouse dorsal telencephalon (prospective cerebral cortex). Transfected primary cortical progenitors can be differentiated *in vitro* either in dissociated culture or organotypic slice culture. The use of shRNA vectors for RNAi provides a versatile approach to understand gene function during mammalian neural development.

Introduction

RNA interference (RNAi) has proven to be a powerful tool for studying gene function in animals (Dorsett and Tuschl, 2004; McManus and Sharp, 2002; Shi, 2003). In 2001, Tuschl and colleagues observed that synthetic short interfering RNAs (siRNAs) mediate effective sequence-specific RNAi in mammalian cells (Elbashir *et al.*, 2001). Subsequently, we and others have developed expression vectors that transcribe either pairs of siRNAs or short hairpin RNAs (shRNAs) in mammalian cells, permitting vector-based RNAi (Brummelkamp *et al.*, 2002; Lee *et al.*, 2002; McManus *et al.*, 2002; Miyagishi and Taira, 2002; Paddison *et al.*, 2002; Paul *et al.*, 2002; Sui *et al.*, 2002; Yu *et al.*, 2002; Zeng *et al.*, 2002). We are using shRNA vectors to study neurogenesis and neuronal differentiation during mammalian development. shRNA vectors can be used to inhibit the expression of individual proteins during neuronal differentiation (Yu *et al.*, 2002), and combinations of shRNA vectors can be used to inhibit expression of multiple proteins simultaneously, facilitating analysis of genes with overlapping functions (Yu *et al.*, 2003). We have used combinatorial inhibition with shRNAs to demonstrate a redundant role for Akt1 and Akt2 kinases in regulating transcription during neuronal differentiation (Vojtek

METHODS IN ENZYMOLOGY, VOL. 392

et al., 2003). shRNA vectors have also been used to analyze the roles for a number of other genes during neural development and differentiation, both *in vitro* and *in vivo* (Bai *et al.*, 2003; Gaudilliere *et al.*, 2002, 2004; Konishi *et al.*, 2004; Krichevsky and Kosik, 2002; Matsuda and Cepko, 2004). Here we discuss the design and testing of shRNAs for RNAi. We also present a protocol for introducing shRNA vectors into the mouse P19 cell line, as well as protocols for electroporation of shRNA vectors into neural progenitors of the embryonic mouse cerebral cortex, followed by *in vitro* neuronal differentiation in dissociated culture or organotypic slice culture.

Short Hairpin RNA (shRNA) Vectors

shRNA transcripts are constructed by connecting the sense and anti-sense strands of an siRNA duplex with a loop sequence, allowing a single transcript to fold back into a duplex structure on being transcribed. After transcription, shRNAs are processed into siRNAs by the Dicer enzyme (Brummelkamp *et al.*, 2002; McManus *et al.*, 2002; Paddison *et al.*, 2002). RNA polymerase III promoters have been used to synthesize shRNAs with defined 5′ and 3′ ends (Good *et al.*, 1997; Noonberg *et al.*, 1994). Importantly, RNA polymerase III terminates transcription at stretches of four consecutive thymidine residues in its DNA template, allowing precise control of the location of the 3′ end of the shRNA (Booth and Pugh, 1997; Tazi *et al.*, 1993). The human or mouse RNA polymerase III promoters from the U6 small nuclear RNA (Reddy, 1988) and the ribonuclease P H1 subunit genes (Baer *et al.*, 1990) are most commonly used for shRNA synthesis. Both promoters are located entirely 5′ to their transcripts, without *cis*-acting regulatory elements in their transcribed regions, and both initiate transcription at a single position. A human tRNA^{Val}-derived promoter has also been used for shRNA synthesis, and this promoter has been shown to facilitate nuclear export of shRNA sequences after synthesis (Kawasaki and Taira, 2003). Inducible RNA polymerase III promoter-driven shRNA vectors have also been described (Czauderna *et al.*, 2003; Matsukura *et al.*, 2003; van de Wetering *et al.*, 2003). Potential advantages for shRNA expression-vector-based RNAi in comparison with chemically synthesized siRNAs are that more delivery methods are available for DNA-based vectors, cell lines stably expressing shRNAs can be established for stable inhibition of expression, and shRNA vectors can be delivered by using viral vectors.

Design of shRNAs

An example of an shRNA transcribed from the mouse U6 promoter is shown in Fig. 1. The shRNA sequence starts with the first nucleotide of the

U6 snRNA transcript (G). After the shRNA sequence, five consecutive Ts are included in the DNA template to terminate the transcription of RNA polymerase III. The strategy for inserting shRNA templates into the mU6pro vector used here has been described previously (Yu *et al.*, 2002). In brief, two synthetic DNA oligonucleotides encoding the entire shRNA and terminator are annealed to form a DNA duplex that is inserted into the vector by standard molecular biology techniques. (Additional information on the mU6pro vector is available at http://sitemaker.umich.edu/dlturner. vectors.)

Either the first (5′) or the second (3′) strand of an shRNA can be complementary to its target sequence. In the example shown in Fig. 1, the strand complementary to the target (the antisense strand) comes first (5′). Highly-effective shRNAs can have duplex lengths between 19 and 29 nt, but duplexes at the longer end of this range may increase the chance of effective inhibition (Paddison *et al.*, 2002; Yu *et al.*, 2003), possibly by increasing the probability of processing by Dicer or by allowing the generation of several overlapping siRNAs (Dykxhoorn *et al.*, 2003). We usually select target

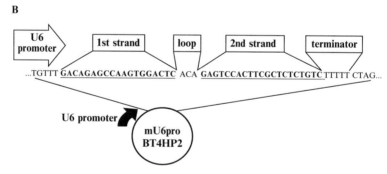

FIG. 1. An example of a short hairpin RNA (shRNA) expressed from the mU6pro vector. (A) The expected structure of an shRNA directed against the neuronal class III β-tubulin is shown (Yu *et al.*, 2002). (B) The first strand sequence is complementary to neuronal β-tubulin mRNA. (This strand is also referred to as the antisense strand because it is antisense with respect to the target mRNA.) The second strand shRNA sequence is complementary to the functional shRNA strand and corresponds to the target mRNA sense strand. A mismatched nucleotide is included in the second strand (see text). A three-nucleotide loop connects the two strands of the shRNA, and a transcription terminator (5 Ts) is present after the shRNA sequences. The DNA template for this shRNA is inserted in the mU6pro vector (see text).

sequences within the coding sequence or 3' untranslated region of an mRNA. Target sequences within the 3' untranslated region can facilitate rescue of an RNAi phenotype with an expression vector for the coding region (see later). We also try to avoid target sequences with high GC content (>70%). Note that the first nucleotide of an shRNA may be constrained by the choice of promoter (G for U6 promoter, A or G for H1 promoter). In the example shown in Fig. 1, the strand of the shRNA that is not complementary to the target mRNA has one nucleotide sequence alteration to create a mismatch within the shRNA duplex. Such mismatches do not alter the ability of the shRNA to mediate RNAi (Yu *et al.*, 2002), but they can facilitate DNA sequence analysis of the shRNA template.

Recently, it has been shown that the two strands of a siRNA duplex can function asymmetrically (Khvorova *et al.*, 2003; Schwarz *et al.*, 2003). The strand with its 5' end at the less stable end of the siRNA duplex is preferentially incorporated into the protein complex that triggers RNAi, whereas the other strand is apparently degraded. This observation can be used to improve siRNA and shRNA effectiveness by choosing target sequences with preferentially A or U for the four nucleotides at their 3' end (which basepair with the 5' end of the shRNA antisense strand) and preferentially G or C at the other end. However, it may be more difficult to apply these criteria when using long (28–29 nt) shRNAs, because the precise locations of the ends of the mature siRNAs (derived from the shRNAs by Dicer processing) are uncertain. Several Web sites also offer convenient searches and sequences that have been used in published results. However, most of the rules used for design remain empirical, and not all shRNAs prove equally effective for RNAi. Therefore, it is best to design and test several shRNAs for their ability to inhibit a specific target mRNA.

The two shRNA strands are joined by a short loop sequence. We have generally used a three-base extension of the shRNA strand that is complementary to the target as the loop (Yu *et al.*, 2002, 2003). Other groups have used a variety of loop sequences and sizes (up to 23 nucleotides) (Brummelkamp *et al.*, 2002; Lee *et al.*, 2002; Miyagishi and Taira, 2002; Paddison *et al.*, 2002; Paul *et al.*, 2002; Sui *et al.*, 2002). However, recent observations suggest that a microRNA-based loop may in some cases increase the efficiency of shRNA-mediated RNAi (Kawasaki and Taira, 2003).

Testing the Functionality of shRNAs

Not all shRNAs (or siRNAs) are equally effective for RNAi; therefore, it is important to confirm that an shRNA inhibits the expression of its target. Ideally, reduced expression of the protein encoded by the target mRNA or reduction of the mRNA level is measured directly (e.g., by Western or Northern blot analysis, immunohistochemistry, or RT-PCR). For transient transfections of cultured cells with shRNAs, transfection efficiency is often much less than 100%. Expression of the target RNA or protein in the untransfected cells can interfere with the assessment of shRNA inhibition in the transfected cells. (For example, at 50% transfection efficiency, the maximum possible reduction of expression by a transfected shRNA is only twofold.) One method to circumvent this problem is to purify cells transiently transfected with the shRNA vector, using a cotransfected puromycin-resistance gene (Vara et al., 1986) and high-dose puromycin selection (Yu et al., 2003). Alternatively, fluorescence-activated cell sorting can be used to purify shRNA vector transfected cells cotransfected with a green fluorescent protein (GFP) expression vector (Kojima et al., 2004). RNAs or proteins from the purified cells can be extracted and analyzed to assess shRNA inhibition. In some cell types (e.g., HEK293 cells), transfection efficiencies may be sufficient to permit evaluation of shRNA target inhibition without purification of the transfected cells.

Another approach is to assess shRNA inhibition by using a cotransfected reporter (Paddison et al., 2002; Yu et al., 2002). For example, shRNA target sequences can be inserted downstream of a luciferase reporter in an expression vector. On cotransfection of the shRNA vector and the luciferase reporter–target construct, the level of luciferase activity will be reduced by the shRNA. Because the shRNA and luciferase reporter are introduced into the same cells by cotransfection, the untransfected cells do not interfere in this assay. This method provides an easy and rapid means for evaluating shRNA inhibition in a quantitative manner; therefore, we often use this method for screening candidate shRNAs. In general, effective shRNAs should reduce the target expression by 10-fold or more. [The amount of inhibition depends in part on how the luciferase target expression is regulated; see Yu et al. (2003) for additional information.] Although reporter inhibition is a good indication of effective shRNA function, a few shRNAs that reduce reporter expression substantially are not as effective at reducing expression of the endogenous target, perhaps because of target accessibility issues and a difference in protein stability between luciferase and the endogenous protein. Therefore, it is important to confirm target inhibition directly.

Specificity Controls for RNA Interference (RNAi) Experiments

It is important to include appropriate controls to show the specificity of shRNA-mediated RNAi. Unrelated sequences or mismatched bases in shRNA sequences have been commonly used as controls to show sequence specificity of shRNA effects. Demonstrating that the phenotypic effect of an shRNA is caused by specific inhibition of its target is also essential. One method for demonstrating that an effect reflects specific inhibition of a target is to compare the effects of two (or more) functional shRNA vectors that are directed against distinct sequences within that target. A similar effect should be observed with both shRNAs, although the degree of effect may vary depending on the efficiency of the shRNAs to inhibit gene expression. Alternatively, if shRNA sequences are directed against the 3' untranslated region of a target gene, forced expression of the coding sequence for the target protein from an expression vector can be used to reverse the effect of the shRNAs (Lassus et al., 2002). However, such rescue experiments may be difficult to perform in some cases, because forced expression of the target gene may not allow proper spatial or temporal expression and it may not provide an appropriate level of expression.

Delivery of DNA Vectors into Neurons Differentiated from P19 Cells

P19 cells are a multipotential mouse teratocarcinoma cell line that can differentiate into neurons and glia after aggregation and retinoic acid (RA) treatment, as well as additional cell fates under other conditions (McBurney et al., 1982). We have previously shown that transient expression of neural basic helix-loop-helix (bHLH) transcription factors can differentiate P19 cells into neurons in monolayer cultures without RA treatment (Farah et al., 2000). This method provides an efficient in vitro system to study gene function during a relatively synchronous neuronal differentiation process. Transfected cells can be fixed at different times after transfection to examine cell cycle withdrawal, neuron-specific gene expression, and morphological differentiation (Farah et al., 2000). By co-transfection of U6 vectors expressing shRNAs and neural bHLH expression vectors, we can inhibit expression of specific genes during neuronal differentiation of P19 cells and examine the phenotypic consequences on neuronal differentiation (Vojtek et al., 2003; Yu et al., 2002).

Protocol for Delivery of DNA Vectors into Neurons Differentiated from P19 Cells

1. Maintain P19 cells undifferentiated in MEMα (Invitrogen, Carlsbad, CA) with 10% serum (7.5% calf serum, 2.5% fetal bovine serum), and 100 units/ml penicillin plus 100 μg/ml streptomycin (Invitrogen). Typically, passage cells every 2–3 days. Maintain cell density at 80% confluence or lower. Higher densities can lead to spontaneous differentiation into

alternative cell fates and thus reduce neuronal differentiation. Culture cells on untreated tissue-culture-grade plastics (Corning, Corning, NY).

2. Plate cells onto laminin-coated dishes 1 day before transfection, with a seeding density of 1×10^5 to 1.5×10^5 cells per 35-mm dish. (To minimize toxicity, 1.5×10^5 cells per 35-mm dish are recommended for Lipofectamine™ 2000 transfections.) For coating, resuspend natural mouse laminin (Invitrogen) at 2 μg/ml in phosphate-buffered saline (PBS) and filter sterilize. Incubate dishes with laminin solution for at least 2 h to overnight at room temperature. Rinse dishes once with PBS before use.

3. Change medium into antibiotic-free MEMα with 10% serum 30 min before transfection.

4. Transfect cells using either Lipofectamine 2000 (Invitrogen) or FuGENE 6 (Roche, Switzerland), according to the manufacturer's instructions. Typically, we use 1 μg of neural bHLH expression vectors, 1 μg of GFP expression vectors, and 1 μg of mU6pro constructs per 35-mm dish of P19 cells. Both DNA and Lipofectamine 2000 are diluted in Opti-MEM1 medium (Invitrogen). Cotransfection efficiency is higher than 90%.

5. Change medium to reduced serum medium Opti-MEM1 (with 1% fetal bovine serum, 100 units/ml penicillin, and 100 μg/ml streptomycin) the day after transfection. Replace half of the medium with fresh medium every other day thereafter.

6. The transfection efficiency for P19 cells is generally between 30 and 50% with either Lipofectamine 2000 or FuGENE 6. Two days after transfection, 30–50% of the transfected cells (GFP-positive cells) withdraw from the cell cycle and express neuronal β-tubulin. Transfected cells (GFP-positive cells) that fail to differentiate into neurons may continue to divide and thus reduce the percentage of the transfected cells that differentiate into neurons.

7. Fix cells with 3.7% formaldehyde or 4% paraformaldehyde in PBS; analyze differentiation (e.g., by indirect immunofluorescent detection of neuronal antigens) (Farah et al., 2000).

Electroporation of shRNA Vectors into Primary Cortical Progenitors

The efficiency of delivering DNA or RNA into primary neural cultures is relatively low when using common transfection methods such as calcium phosphate coprecipitation, cationic liposome-mediated transfection, or polyethylenimine transfection. In contrast, electroporation can provide a higher DNA delivery efficiency into primary neurons or neural progenitor cells, despite the fact that it may kill a certain proportion of electroporated cells (Miyasaka et al., 1999). Several electroporation protocols have been developed to deliver DNA into neural tubes of early embryonic stage

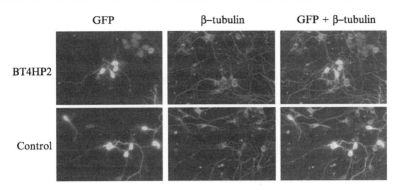

GFP   β–tubulin   GFP + β–tubulin

BT4HP2

Control

FIG. 2. Inhibition of neuron-specific β-tubulin in mouse primary cortical neurons by vector-derived shRNAs. Mouse cortical progenitors from E13.5 embryos were electroporated with a U6 vector expressing either an shRNA against neuronal β-tubulin (BT4HP2) or a control shRNA; a vector expressing GFP was coelectroporated to label the shRNA-transfected cells (Yu et al., 2002). After electroporation, cells were dissociated and cultured in vitro. Cells were fixed 3 days after transfection and both neuronal β-tubulin and green fluorescent protein (GFP) were detected by indirect immunofluorescence. Cells that received BT4HP2 differentiated as neurons based on their morphology, but failed to express significant levels of neuronal β-tubulin. (See color insert.)

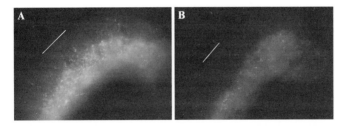

FIG. 3. shRNA-mediated inhibition of the Akt1 and Akt2 kinases disrupts radial migration in a slice culture from the embryonic mouse telencephalon. The dorsal telencephalon of an E14.5 mouse embryo was coelectroporated with either a control U6 shRNA vector (A) or two U6 vectors that express shRNAs against Akt1 and Akt2 (B) (Vojtek et al., 2003). The CS2 + GFP expression vector was coelectroporated to identify the shRNA-expressing cells (Yu et al., 2002). After electroporation, organotypic slices were prepared and cultured for 3 days. An apoptosis inhibitor was included in the culture medium to reduce cell death from Akt inhibition, as previously described (Vojtek et al., 2003). After fixation, anti-GFP antibody and an Alexa-488 conjugated secondary antibody were used to enhance detection of GFP in the electroporated cells. In the control, a subset of control cells migrated radially into the cortical layers, whereas most Akt1/2 shRNA-transfected cells failed to migrate and remained in the ventricular zone without elaborating processes. The white lines indicate the outer surface of the cortex.

embryos both inside and outside the uterus (Bai *et al.*, 2003; Fukuchi-Shimogori and Grove, 2001; Hatanaka and Murakami, 2002; Inoue *et al.*, 2001; Matsuda and Cepko, 2004; Miyasaka *et al.*, 1999; Takahashi *et al.*, 2002). Embryos can then either be maintained inside the pregnant mother or cultured as whole embryos or explants. We describe a protocol for electroporation of mouse telencephalon/cerebral cortex from embryonic days 13–15. In this protocol, the majority of the electroporated cells are neural progenitor cells that line the ventricles. Cotransfection of a GFP expression vector permits identification of the transfected cells as the cotransfection efficiency of the electroporation procedure is at least 90% (unpublished observations). The tissue can be dissociated and cultured in monolayer, allowing differentiation of the transfected cortical progenitors (Fig. 2). Alternatively, the cortex can be cultured as an organotypic slice culture (Stuhmer *et al.*, 2002). Slice culture allows the observation of the migration and differentiation of transfected cortical cells in an intact three-dimensional environment (Fig. 3).

Protocol for Electroporation

1. Dissect mouse embryos from the uterus of a timed-pregnant female at embryonic days 13–15, and keep embryos on ice in sterile PBS.
2. Mix DNA plasmids (GFP and mU6pro) and dilute to 1–2 $\mu g/\mu l$ with water. Add trypan blue (0.04%) as tracer to visualize the injection and monitor the amount of solution injected into the ventricles.
3. Using a stereomicroscope, inject DNA solution (1–2 μl) into the telencephalic ventricles, using pulled glass capillary tubes connected to a micropipettor (Gilson, P200).
4. Incubate the embryos in ice-cold PBS for 2 min to allow the DNA to diffuse.
5. Fill the chamber of the petri dish platinum bar electrode (CUY-523 from Protech International, San Antonio, TX) with PBS. Put an intact embryo between the electrodes and orient it such that the dorsal part of the telencephalon is toward the positive side of the electrode.
6. Electroporate with the following parameters: pulse, 85–100 V, pulse duration 10 msec, interpulse interval 1 sec, 10–15 pulses (ECM 830 Electroporator from BTX).

Protocol for Dissociated Primary Cultures after Electroporation

1. Continue from previous electroporation protocol. Dissect dorsal part of the telencephalon from embryos in sterile PBS. Culture the

telencephalons as explants overnight in culture plate inserts (Millicell-CM, 0.4-μm culture plate insert, Millipore, Billerica, MA) in L15 medium with N2 supplement, B27 supplement, 30 mM glucose, 26 mM NaHCO$_3$, 100 units/ml penicillin, 100 μg/ml streptomycin, and 20 ng/ml FGF2 (Bonni *et al.*, 1997). Mix medium components freshly. The explant culture step allows the cells to recover from electroporation and improves cell survival significantly. Dissociation immediately after electroporation leads to substantial cell death.

2. Transfer the explants into tubes containing sterile HBSS buffer without Ca^{2+} and Mg^{2+} (Invitrogen). Spin down the explants at 4° at 1600 rpm for 5 min. Wash the explants once with HBSS without Ca^{2+} and Mg^{2+} to remove Ca^{2+} thoroughly, which facilitates the dissociation. Spin down the explants again at 4° at 1600 rpm for 5 min.

3. Dissociate the explants in ice-cold L15 medium by trituration gently with fire-polished Pasteur pipettes. Use trypan blue exclusion to identify and count live cells. The expected number of cells per E14.5 embryo is 3 $\times$ 10^5 to 5 $\times$ 10^5 after the electroporation, explant culture, and dissociation processes. The transfection efficiency is generally 5–10%. The electroporation of DNA appears limited to the cells lining the ventricular surfaces; therefore, most electroporated cells are progenitor cells instead of differentiating or differentiated neurons.

4. Plate cells at 5 $\times$ 10^5 to 10^6 cells/well in 12-well plates coated with poly-L-lysine (10 μg/ml, 3–5 h) and laminin (2 μg/ml, overnight). Use the same medium as described in Step 1.

5. Replace half the medium with fresh medium without FGF2 every other day. Removal of the FGF2 increases the frequency of neuronal differentiation. Cells can be fixed 4 days after dissociation and stained with different neuronal markers. To improve fluorescent detection of transfected cells, slices can be processed for indirect immunofluorescent detection of the GFP with a rabbit antibody to the GFP protein (Molecular Probes, Eugene, OR) and an Alexa-488 conjugated goat anti-rabbit secondary antibody (Molecular Probes).

Protocol for Organotypic Slice Culture after Electroporation

1. Continue from previous electroporation protocol. Carefully dissect the whole telencephalon with diencephalon together from the electroporated embryo.

2. Transfer the tissues into a mold containing 4% low-melting agarose in PBS, which is kept at 42° until use.

3. Incubate embedded tissues at 4° and allow the agarose to solidify thoroughly.

4. Remove the agarose cube from the mold and transfer to the vibratome.

5. Cut coronal sections of 250-μm thickness through the brain tissue with a vibratome (Leica VT1000s), speed 3, high vibration frequency (7–8), in sterile HBSS without Ca^{2+} and Mg^{2+}.

6. Using a sterile laboratory spatula, transfer the slices carefully onto polycarbonate membranes (0.4-μm pore size, Whatman nucleopore poly-carbonate Track-etch membrane) floating on Neurobasal™ medium (Invitrogen) with B27 supplement, penicillin/streptomycin, and GlutMAX (Invitrogen; Stuhmer *et al.*, 2002). For each embryo, collect three to five slices containing the middle part of the telencephalon. Most electroporated cells are within this region.

7. Replace one half of the medium with fresh medium without FGF2 every other day. Culture for 3 days before fixation. (The ventricle may become filled in because of cell proliferation after culturing for more than 3 days.)

8. Fix the slices with 4% paraformaldehyde in PBS for 2–3 h at room temperature.

9. To improve fluorescent detection of transfected cells, slices can be processed for indirect immunofluorescent detection of the GFP with a rabbit antibody to the GFP protein (Molecular Probes) and an Alexa-488 conjugated goat anti-rabbit secondary antibody (Molecular Probes). Block slices in 10% goat serum, 0.4% Triton X-100, 3% BSA, 1% glycine in PBS for at least 1 h at room temperature. Dilute the primary antibody 1:1000 in the blocking solution and incubate with the slices overnight at 4°. Wash slices in PBS three times, each wash for at least 1 h. Dilute the secondary antibody 1:1000 in the blocking solution and incubate with the slices overnight at 4°. Wash slices again in PBS three times, each wash for at least 1 h, and store in PBS.

Slices also can be resectioned with a cryostat or vibratome into thinner sections for further analysis.

Acknowledgments

This work was supported by NIH NS38698 (DLT) and Research Scholar Grant RSG-01-177-01-MGO from the American Cancer Society (ABV).

References

Baer, M., Nilsen, T. W., Costigan, C., and Altman, S. (1990). Structure and transcription of a human gene for H1 RNA, the RNA component of human RNase P. *Nucleic Acids Res.* **18**, 97–103.

Bai, J., Ramos, R. L., Ackman, J. B., Thomas, A. M., Lee, R. V., and LoTurco, J. J. (2003). RNAi reveals doublecortin is required for radial migration in rat neocortex. *Nat. Neurosci.* **6,** 1277–1283.

Bonni, A., Sun, Y., Nadal-Vicens, M., Bhatt, A., Frank, D. A., Rozovsky, I., Stahl, N., Yancopoulos, G. D., and Greenberg, M. E. (1997). Regulation of gliogenesis in the central nervous system by the JAK-STAT signaling pathway. *Science* **278,** 477–483.

Booth, B. L., Jr., and Pugh, B. F. (1997). Identification and characterization of a nuclease specific for the 3′ end of the U6 small nuclear RNA. *J. Biol. Chem.* **272,** 984–991.

Brummelkamp, T. R., Bernards, R., and Agami, R. (2002). A system for stable expression of short interfering RNAs in mammalian cells. *Science* **296,** 550–553.

Czauderna, F., Santel, A., Hinz, M., Fechtner, M., Durieux, B., Fisch, G., Leenders, F., Arnold, W., Giese, K., Klippel, A., and Kaufmann, J. (2003). Inducible shRNA expression for application in a prostate cancer mouse model. *Nucleic Acids Res.* **31,** e127.

Dorsett, Y., and Tuschl, T. (2004). siRNAs: Applications in functional genomics and potential as therapeutics. *Nat. Rev. Drug Discov.* **3,** 318–329.

Dykxhoorn, D. M., Novina, C. D., and Sharp, P. A. (2003). Killing the messenger: Short RNAs that silence gene expression. *Nat. Rev. Mol. Cell Biol.* **4,** 457–467.

Elbashir, S. M., Harborth, J., Lendeckel, W., Yalcin, A., Weber, K., and Tuschl, T. (2001). Duplexes of 21-nucleotide RNAs mediate RNA interference in cultured mammalian cells. *Nature* **411,** 494–498.

Farah, M. H., Olson, J. M., Sucic, H. B., Hume, R. I., Tapscott, S. J., and Turner, D. L. (2000). Generation of neurons by transient expression of neural bHLH proteins in mammalian cells. *Development* **127,** 693–702.

Fukuchi-Shimogori, T., and Grove, E. A. (2001). Neocortex patterning by the secreted signaling molecule FGF8. *Science* **294,** 1071–1074.

Gaudilliere, B., Konishi, Y., de la Iglesia, N., Yao, G., and Bonni, A. (2004). A CaMKII-neuroD signaling pathway specifies dendritic morphogenesis. *Neuron* **41,** 229–241.

Gaudilliere, B., Shi, Y., and Bonni, A. (2002). RNA interference reveals a requirement for myocyte enhancer factor 2A in activity-dependent neuronal survival. *J. Biol. Chem.* **277,** 46442–46446.

Good, P. D., Krikos, A. J., Li, S. X., Bertrand, E., Lee, N. S., Giver, L., Ellington, A., Zaia, J. A., Rossi, J. J., and Engelke, D. R. (1997). Expression of small, therapeutic RNAs in human cell nuclei. *Gene Ther.* **4,** 45–54.

Hatanaka, Y., and Murakami, F. (2002). *In vitro* analysis of the origin, migratory behavior, and maturation of cortical pyramidal cells. *J. Comp. Neurol.* **454,** 1–14.

Inoue, T., Tanaka, T., Takeichi, M., Chisaka, O., Nakamura, S., and Osumi, N. (2001). Role of cadherins in maintaining the compartment boundary between the cortex and striatum during development. *Development* **128,** 561–569.

Kawasaki, H., and Taira, K. (2003). Short hairpin type of dsRNAs that are controlled by tRNA(Val) promoter significantly induce RNAi-mediated gene silencing in the cytoplasm of human cells. *Nucleic Acids Res.* **31,** 700–707.

Khvorova, A., Reynolds, A., and Jayasena, S. D. (2003). Functional siRNAs and miRNAs exhibit strand bias. *Cell* **115,** 209–216.

Kojima, S., Vignjevic, D., and Borisy, G. G. (2004). Improved silencing vector co-expressing GFP and small hairpin RNA. *Biotechniques* **36,** 74–79.

Konishi, Y., Stegmuller, J., Matsuda, T., Bonni, S., and Bonni, A. (2004). Cdh1-APC controls axonal growth and patterning in the mammalian brain. *Science* **303,** 1026–1030.

Krichevsky, A. M., and Kosik, K. S. (2002). RNAi functions in cultured mammalian neurons. *Proc. Natl. Acad. Sci. USA* **99,** 11926–11929.

Lassus, P., Rodriguez, J., and Lazebnik, Y. (2002). Confirming specificity of RNAi in mammalian cells. *Sci STKE* **2002,** PL13.

Lee, N. S., Dohjima, T., Bauer, G., Li, H., Li, M. J., Ehsani, A., Salvaterra, P., and Rossi, J. (2002). Expression of small interfering RNAs targeted against HIV-1 rev transcripts in human cells. *Nat. Biotechnol.* **20,** 500–505.

Matsuda, T., and Cepko, C. L. (2004). Electroporation and RNA interference in the rodent retina *in vivo* and *in vitro*. *Proc. Natl. Acad. Sci. USA* **101,** 16–22.

Matsukura, S., Jones, P. A., and Takai, D. (2003). Establishment of conditional vectors for hairpin siRNA knockdowns. *Nucleic Acids Res.* **31,** e77.

McBurney, M. W., Jones-Villeneuve, E. M., Edwards, M. K., and Anderson, P. J. (1982). Control of muscle and neuronal differentiation in a cultured embryonal carcinoma cell line. *Nature* **299,** 165–167.

McManus, M. T., Petersen, C. P., Haines, B. B., Chen, J., and Sharp, P. A. (2002). Gene silencing using micro-RNA designed hairpins. *RNA* **8,** 842–850.

McManus, M. T., and Sharp, P. A. (2002). Gene silencing in mammals by small interfering RNAs. *Nat. Rev. Genet.* **3,** 737–747.

Miyagishi, M., and Taira, K. (2002). U6 promoter-driven siRNAs with four uridine 3′ overhangs efficiently suppress targeted gene expression in mammalian cells. *Nat. Biotechnol.* **20,** 497–500.

Miyasaka, N., Arimatsu, Y., and Takiguchihayashi, K. (1999). Foreign gene expression in an organotypic culture of cortical anlage after *in vivo* electroporation. *Neuroreport* **10,** 2319–2323.

Noonberg, S. B., Scott, G. K., Garovoy, M. R., Benz, C. C., and Hunt, C. A. (1994). *In vivo* generation of highly abundant sequence-specific oligonucleotides for antisense and triplex gene regulation. *Nucleic Acids Res.* **22,** 2830–2836.

Paddison, P. J., Caudy, A. A., Bernstein, E., Hannon, G. J., and Conklin, D. S. (2002). Short hairpin RNAs (shRNAs) induce sequence-specific silencing in mammalian cells. *Genes Dev.* **16,** 948–958.

Paul, C. P., Good, P. D., Winer, I., and Engelke, D. R. (2002). Effective expression of small interfering RNA in human cells. *Nat. Biotechnol.* **20,** 505–508.

Reddy, R. (1988). Transcription of a U6 small nuclear RNA gene *in vitro*. Transcription of a mouse U6 small nuclear RNA gene *in vitro* by RNA polymerase III is dependent on transcription factor(s) different from transcription factors IIIA, IIIB, and IIIC. *J. Biol. Chem.* **263,** 15980–15984.

Schwarz, D. S., Hutvagner, G., Du, T., Xu, Z., Aronin, N., and Zamore, P. D. (2003). Asymmetry in the assembly of the RNAi enzyme complex. *Cell* **115,** 199–208.

Shi, Y. (2003). Mammalian RNAi for the masses. *Trends Genet.* **19,** 9–12.

Stuhmer, T., Anderson, S. A., Ekker, M., and Rubenstein, J. L. (2002). Ectopic expression of the Dlx genes induces glutamic acid decarboxylase and Dlx expression. *Development* **129,** 245–252.

Sui, G., Soohoo, C., Affar el, B., Gay, F., Shi, Y., and Forrester, W. C. (2002). A DNA vector-based RNAi technology to suppress gene expression in mammalian cells. *Proc. Natl. Acad. Sci. USA* **99,** 5515–5520.

Takahashi, M., Sato, K., Nomura, T., and Osumi, N. (2002). Manipulating gene expressions by electroporation in the developing brain of mammalian embryos. *Differentiation* **70,** 155–162.

Tazi, J., Forne, T., Jeanteur, P., Cathala, G., and Brunel, C. (1993). Mammalian U6 small nuclear RNA undergoes 3′ end modifications within the spliceosome. *Mol. Cell Biol.* **13,** 1641–1650.

van de Wetering, M., Oving, I., Muncan, V., Pon Fong, M. T., Brantjes, H., van Leenen, D., Holstege, F. C., Brummelkamp, T. R., Agami, R., and Clevers, H. (2003). Specific inhibition of gene expression using a stably integrated, inducible small-interfering-RNA vector. *EMBO Rep.* **4,** 609–615.

Vara, J. A., Portela, A., Ortin, J., and Jimenez, A. (1986). Expression in mammalian cells of a gene from *Streptomyces alboniger* conferring puromycin resistance. *Nucleic Acids Res.* **14,** 4617–4624.

Vojtek, A. B., Taylor, J., DeRuiter, S. L., Yu, J. Y., Figueroa, C., Kwok, R. P., and Turner, D. L. (2003). Akt regulates basic helix-loop-helix transcription factor-coactivator complex formation and activity during neuronal differentiation. *Mol. Cell Biol.* **23,** 4417–4427.

Yu, J. Y., DeRuiter, S. L., and Turner, D. L. (2002). RNA interference by expression of short-interfering RNAs and hairpin RNAs in mammalian cells. *Proc. Natl. Acad. Sci. USA* **99,** 6047–6052.

Yu, J. Y., Taylor, J., DeRuiter, S. L., Vojtek, A. B., and Turner, D. L. (2003). Simultaneous inhibition of GSK3alpha and GSK3beta using hairpin siRNA expression vectors. *Mol. Ther.* **7,** 228–236.

Zeng, Y., Wagner, E. J., and Cullen, B. R. (2002). Both natural and designed micro RNAs can inhibit the expression of cognate mRNAs when expressed in human cells. *Mol. Cell* **9,** 1327–1333.

[12] Analysis of T-Cell Development by Using Short Interfering RNA to Knock Down Protein Expression

By GABRIELA HERNÁNDEZ-HOYOS and JOSÉ ALBEROLA-ILA

Abstract

We have applied RNA interference (RNAi) technology to the analysis of genes involved in T-cell development, combining a reaggregate fetal thymic organ culture (rFTOC) system with retroviral delivery of short interfering RNA (siRNA) hairpins. The process involves the isolation of murine fetal liver or fetal thymocytes, infection with retroviral particles carrying the construct of interest, followed by reaggregation of the trans-duced precursors with fetal thymic stroma into lobes. Subsequently, individual lobes are harvested and analyzed for development at various time points. These reaggregate cultures recapitulate most features of T-cell development *in vivo*, including pre-TCR selection and expansion, positive selection of CD4 and CD8 T cells, and negative selection. In our hands, the combination of retroviral delivery of RNAi and rFTOCs is a quick alternative to conventional knockouts for the analysis of gene function during T-cell development. This chapter describes the methods we have developed to knock down gene expression in T-cell precursors, using retroviral delivery of siRNA hairpins.

Introduction

Over the past years, the field of posttranscriptional gene silencing or RNA interference (RNAi) has undergone an explosion with the discovery that 25-nt antisense RNA fragments were involved in sequence-specific gene silencing in plants and flies (Agrawal *et al.*, 2003; McManus and Sharp, 2002). RNAi is a conserved eukaryote system that might have evolved as a genome defense and gene regulation mechanism. Furthermore, RNAi has been exploited as a powerful tool to knock down gene expression in many species, including worms, flies, and mammals (Agrawal *et al.*, 2003; McManus and Sharp, 2002). This chapter describes the methods we have developed to knock down gene expression in T-cell precursors, using retroviral delivery of short interfering RNA (siRNA) hairpins.

Cumulative work from various groups has contributed to a two-step mechanistic model of RNAi (Agrawal *et al.*, 2003; Hannon, 2002). The first step involved the degradation of a large double-stranded RNA (dsRNA) into discrete double-stranded 21- to 25-nt duplexes known as siRNA. In the second step, the siRNAs are incorporated into RNA-induced silencing complexes (RISCs). These complexes unwind the siRNA and use the single-stranded fragments to mediate suppression of homologous mRNAs by targeted degradation. In addition to their size (21–25 nt), siRNAs have symmetric 3' two-nucleotide overhangs. Sequence-specific RNAi can be triggered in many eukaryote species by introducing or expressing long dsRNA. However, expression of long dsRNA (>30 nt) in mammalian tissues other than embryos triggers interferon production, generalized inhibition of protein synthesis, and may result in cell death (reviewed in McManus and Sharp, 2002). The direct introduction of siRNAs of 21–23 nt into mammalian cells by transfection, which bypasses this alarm response, has been extremely useful to knock down gene expression in *in vitro* cell assay systems (Caplen *et al.*, 2001; Elbashir *et al.*, 2001). Subsequently, the use of polymerase III (pol III) promoters for siRNA expression (Agrawal *et al.*, 2003; McManus and Sharp, 2002) combined with retroviral or lenti-viral vector delivery have been major breakthroughs, enabling the use of RNAi technology for analysis of gene function *in vitro*, *in vivo* in mammalian developmental systems, and in transgenic mice (Brummelkamp *et al.*, 2002a; Hernandez-Hoyos *et al.*, 2003; Michienzi *et al.*, 2003; Paddison *et al.*, 2004; Qin *et al.*, 2003; Rubinson *et al.*, 2003; Scherr *et al.*, 2003b). RNAi technology complements traditional gene knockout experiments. It is suitable for fast screening of large numbers of genes either in mammalian systems *in vivo* or *in vitro* and allows dose–response studies (Berns *et al.*, 2004; Hemann *et al.*, 2003; Paddison *et al.*, 2004). It can also be used to simultaneously knock down closely related genes. A note of caution for this approach are recent results suggesting that in some cases siRNAs may trigger nonspecific effects on gene expression (Bridge *et al.*, 2003; Persengiev *et al.*, 2004; Sledz *et al.*, 2003). We have applied RNAi technology to the analysis of genes involved in T-cell development, combining a reaggregate fetal thymic organ culture (rFTOC) system with retroviral delivery of siRNA hairpins. The process

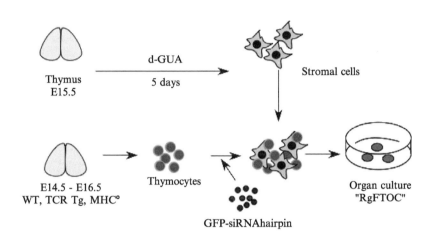

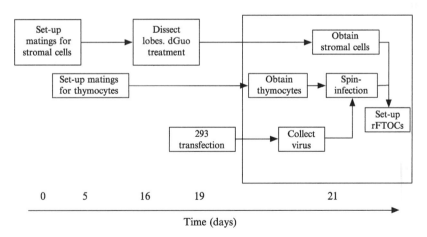

Fig. 1. Overall experimental design. Stromal cells are derived from E15.5 thymic lobes treated for 5 days with 2'-deoxyguanosine. T-cell precursors are obtained from E14.5 to E16.5 thymic lobes and infected with retroviral supernantant. The stromal cells and infected precursors are mixed to generate reaggregate fetal thymic organ cultures (rFTOCs). The timing of the procedures is indicated at the bottom of the panel. Timed matings are set on the late afternoon and plugs checked next morning; we consider the morning of plug as Day 0.5.

involves the isolation of murine fetal liver or fetal thymocytes, infection with retroviral particles carrying the construct of interest, followed by reaggregation of the transduced precursors with fetal thymic stroma into lobes (Fig. 1). Subsequently, individual lobes are harvested and analyzed for development at various time points. These reaggregate cultures recapitulate most features of T-cell development *in vivo*, including pre-TCR selection and expansion, positive selection of CD4 and CD8 T cells, and negative selection (Jenkinson and Anderson, 1994). In our hands, the combination of retroviral delivery and rFTOCs is a reliable and consistent system to analyze gene function by overexpression of wild-type or mutant gene alleles (Anderson *et al.*, 2002; Hernandez-Hoyos *et al.*, 2003). Together with conventional overexpression, the use of RNAi now allows us to do both gain-of-function and loss-of-function experiments for a complete analysis of gene function in the same developmental system. So far, our results from using RNAi in rFTOCs to analyze gene function during T-cell development have been consistent with those obtained with classical gene disruption technology. For example, using this system, we showed that the transcription factor GATA-3 is necessary for the development of CD4 but not CD8 T cells, a result that was confirmed later by another group that used inducible knockout technology (Hernandez-Hoyos *et al.*, 2003; Pai *et al.*, 2003).

There are a few general considerations to keep in mind when using siRNA. For every particular gene targeted, the extent to which its expression can be knocked down will vary depending on the hairpin sequence. To ensure efficient targeting of a specific sequence, it is recommended to generate more than one hairpin per gene and to determine the level of mRNA or protein reduction with each particular hairpin. Thus, it is important to have an assay to test the efficacy of the various hairpins. Efficient transduction of cells with retroviruses and good levels of expression are essential for rFTOCs. Different groups have developed different vectors and cloning or shuttling strategies for introducing the hairpins. We use a retroviral system designed by John Rossi's group, which yields efficient infection of murine lymphoid cells and cell lines. We describe the methods that we have optimized and developed for our system. Figure 1 shows a diagram of the overall experimental design.

Generation of Short Interfering RNA (siRNA)/Retroviral Constructs to Target Genes

Retroviral Vector

Retroviral vectors based on the Moloney murine leukemia virus (MMLV) and lentiviruses are commonly used for stable gene transfer and efficient gene expression in hematopoietic cells. Various replication-defective vectors with the same basic characteristics have been designed for siRNA expression (Brummelkamp *et al.*, 2002a; Hemann *et al.*, 2003; Michienzi *et al.*, 2003; Paddison *et al.*, 2004; Qin *et al.*, 2003; Rubinson *et al.*, 2003; Scherr *et al.*, 2003b). In these systems, siRNA expression is driven from the U6 or H1 pol III promoters, whereas reporter gene expression such as the green fluorescent protein (GFP) or Neo is driven either from the long terminal repeat (LTR) or from another internal promoter if the 3′ LTR is mutated (self-inactivating). There are two variations to siRNA expression systems: expression of sense and antisense siRNA strands from two independent pol III promoters or expression of a short hairpin RNA (shRNA) from a single pol III promoter in which the sense and antisense strands are connected by a short loop. In the latter case, the hairpin is processed into an siRNA. The majority of the retroviral and lentiviral systems applied employ the latter approach.

We have successfully used the MMLV-derived retroviral vector Banshee-GFP, in which shRNA expression is driven from the human U6 promoter and GFP from the LTR (Fig. 2A; Hernandez-Hoyos *et al.*, 2003). The vector is derived from the pMND-neo vector, an MMLV vector engineered in Donald Kohn's laboratory for improved expression frequency and decreased methylation in transduced hematopoietic stem cells (Halene *et al.*, 1999; Robbins *et al.*, 1998). pMND-neo was subsequently modified in John Rossi's laboratory for siRNA expression by engineering a multiple cloning site within the U5 region of the 3′ LTR to clone U6-siRNA sequences (pMND-Banshee). In this way, a copy of the U6-shRNA is introduced into the 5′ LTR during the first round of viral replication, doubling the levels of siRNA expression. We replaced the neo cassette for GFP to facilitate monitoring of the transduced cells by flow cytometry (Fig. 2A). Banshee-GFP shows high levels of expression in murine fetal thymocytes, fetal liver, and lymphoid cell lines.

The parental construct Banshee-GFP yields higher viral titers and higher levels of GFP expression than Banshee-GFP plus any shRNA constructs. The better the hairpin inhibits expression of its target gene, the lower are the titer and GFP expression levels. Presumably, this results from the shRNA targeting the LTR-derived genomic size transcript for degradation because Banshee-GFP is not self-inactivating. To overcome this problem, we introduce two interspersed U:G wobbles, with less stable pairing, in the sense arm of the coding strand of every hairpin sequence, as illustrated in Fig. 2B. This should prevent the shRNA from targeting the full-length genomic LTR-derived transcript. In practice, this modification restores both viral titer and levels of GFP expression without affecting the ability to knock down expression of its target (Fig. 2B). The use of pol III

A

Banshee-GFP and shRNA

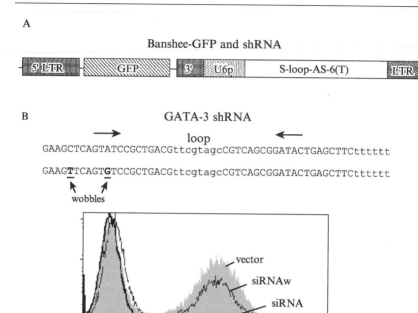

B   GATA-3 shRNA

→
loop   ←
GAAGCTCAGTATCCGCTGACGttcgtagcCGTCAGCGGATACTGAGCTTCtttttt

GAAGTTCAGTGTCCGCTGACGttcgtagcCGTCAGCGGATACTGAGCTTCtttttt

wobbles

GFP

Fig. 2. (A) Structure of the Banshee-GFP retroviral vector. The multiple cloning site for shRNA cloning lies within the 3′ LTR. (B) Sequence of a short hairpin RNA (shRNA) used to target GATA-3. Introduction of the indicated T:G wobbles in the sense arm of the hairpin restores levels of green fluorescent protein (GFP) expression and viral titer as shown in the histogram. The histogram shows the levels of GFP expression in a T-cell line infected 5 days earlier with the vector only (vector), the original GATA-3 shRNA (shRNA), and the modified construct (shRNAw). The mutations did not affect the level of protein silencing (not shown).

promoters U6 or H1 yields similar RNAi efficiency in other systems (Paddison *et al.*, 2004). We have not tested other pol II promoters in the Banshee-GFP vector.

siRNA Hairpin Design: Factors to Consider

Because the mechanisms that determine the efficiency of target sequence degradation by RNAi are not completely understood, various shRNA sequences should be tested. The length and sequence of the hairpin are important factors to consider when designing hairpins to target transcripts. Sequences of 19–29 nt have been successfully used for RNAi, with optimal size varying depending on the system (Caplen and Mousses, 2003; Elbashir *et al.*, 2002; Paddison *et al.*, 2004). We have been using hairpins with the following characteristics: 21-mer of sense (s) sequence, 8 nt of loop sequence, 21-mer of antisense (as) sequence tttttt (termination signal; Fig. 2A). The first nucleotide of the sense sequence needs to be a guanine because it is the first base transcribed by the pol III polymerase. Thus, stretches of 21 nt that begin with a guanine, with approximately 50% GC content and with no internal complementarity are selected for generating hairpins. The six thymidines at the end provide a termination of transcription signal; the transcript will end at or after the first thymidine and thus provide the 3′ overhang. Most groups include two thymidines as the first bases in the loop so that both arms of the hairpin sequence have symmetric 3′ thymidine overhang ends. Loops with random sequence ranging from 3 to 23 nt have been used with different results, depending on the system (reviewed in McManus and Sharp, 2002); changes in loop size may (Brummelkamp et al., 2002b) or may not (Paddison et al., 2002) have an effect. In our system, changing the loop size from 4 to 8 nt marginally improved the efficiency of RNAi (Hernandez-Hoyos et al., 2003). There are freely available software packages that can be used to select the target sequences and design the hairpins (see reagents and specifications). It is important to blast the chosen sequence to verify that it is not homologous to any other RNA. Two recent studies demonstrated that it is possible to favor the incorporation of either siRNA arm into RISC complexes by generating a less-stable interaction on either end of the duplex (Khvorova et al., 2003; Schwarz et al., 2003). This should contribute to the design of shRNAs in which incorporation of the antisense strand is preferred over that of the sense strand to improve targeting efficiency. Given the recent reports that dsRNA of <30 bp can trigger nonspecific effects on gene expression (Bridge et al., 2003; Persengiev et al., 2004; Sledz et al., 2003), it is advisable to generate control shRNAs constructs, for example, by introducing point mutations in the middle of the antisense arm.

Cloning siRNA Hairpins in the Retroviral Vector (Banshee-GFP)

We use PCR amplification to place the desired shRNA sequence in frame immediately downstream of the human U6 promoter and to introduce restriction sites at both ends. The amplicon is digested and cloned in Banshee-GFP. The PCR amplification requires a template containing the human U6 promoter (pTZ U6+1; Lee et al., 2002) and forward and reverse oligonucleotides (oligos). The reverse oligo includes the siRNA hairpin sequence. The siRNA hairpin sequence must begin exactly at the last guanine at the end of the promoter. The reverse 82-mer oligos need to be purified before PCR amplification to exclude incomplete synthesis products. The PCR product is separated on 2% agarose gels. Amplicons of >300 bp are purified by standard protocols, digested, and cloned into Banshee-GFP, previously cut and dephosphorylated. We have found gel purification necessary because smaller-size products (<300 bp) that contain the U6 promoter but exclude the hairpin are generated during the PCR amplification. This probably results from the reverse oligo adopting a hairpin conformation during amplification. To minimize recombination during plasmid amplification, we use RecA-electrocompetent bacteria (TOP 10, GIBCO, Langley, OK), and grow all cultures at 30°. Plasmid is prepared from maxicultures by standard polyethylene glycol precipitation or by commercial methods.

Stepwise Generation of shRNA Hairpin Construct

1. Search for sequences 5′-G(N18) to 5′-G(N20) with 50% G/C content in the cDNA of interest (make sure introns are excluded); others recommend avoiding 5′ and 3′ untranslated region if possible (Elbashir *et al.*, 2002).

2. Blast-search the sequence and verify that there is no homology with other genes.

3. Generate the sequence 5′–G(N20)–TT(N_6)–(N'20)C-(T_6)–3′, where N' is the reverse complement for N. N'20C is the antisense arm of the hairpin and is responsible for silencing.

4. Introduce one or two interspersed T:G wobbles in the sense strand (not the antisense) of the hairpin, as follows:

A → G or C → T so that A : T → G : T or C : G → T : G (mispair is T : G)

5. Blast-search the wobbled sequence to verify that there is no homology with other genes.

6. Because this will be the reverse oligo for PCR amplification, add the 3′ end of U6 promoter sequence at the 5′ end of sequence; at the 3′ end add the restriction site and two extra nucleotides for enzyme restriction and cloning. For example,

5′GGAAAGGACGAAACACC/G(N20)/<u>TTCGTAGC</u>/
(N'20)C/<u>TTTTTT</u>AAGCTTGGG

3′ end of U6 promoter sense *8nt* antisense T(6) *Hind*III

7. Reverse complement the sequence and order the oligo. This is the reverse oligo for PCR amplification of the hairpin; make sure it is purified to exclude incomplete shorter-size oligos.

8. PCR amplify fragments by using U6 promoter template, forward and reverse oligos. Set 2 to 5 × PCR reactions to have enough product for subsequent manipulations.

Forward oligo : 5′-GA<u>AGATCT</u>GGATCCAAGGTCGGGCAG-3′

                *Bgl*II 5′   end of U6

PCR reaction	Volume (μl)	Total	Source
DyNAzyme ext 10× buffer	5		Finnzyme
Primer 5′	1	20 pm	
Primer 3′	1	20 pm	
U6 template (PTZ plasmid)	2.5	25 pg	
dNTPs (100 m*M* stock)	1		Cat. No. 2795, Sigma, St. Louis, MO
DyNAzyme polymerase	2	(2U)	Finnzyme, Finland
H₂O	37.5		
	50		

PCR thermocycler conditions: (1) 95°, 5 min; (2) 95°, 1 min; (3) 60°, 1 min; (4) 72°, 45 sec; (5) go to (2) for a total of 30/31 cycles; (6) 72° 10.

9. Purify fragment on a 2% agarose gel; fragment should be >300 bp.

10. Digest the fragment and vector for cloning (*Bgl*II and *Hin*dIII in the example), dephosphorylate vector, purify both fragments, and clone. Additional unique sites are available for cloning the hairpin in Banshee-GFP.

11. Grow in recombination-deficient strain of bacteria (e.g., TOP 10) at 30°.

Preparation of Retroviral Supernatant and Infection of Cells

To package the virus we use a high-titer murine stem cell virus (MSCV) system (Naviaux *et al.*, 1996). Because our work involves murine cells, we have chosen to use an ecotropic coat to minimize the biosafety measures involved in the experimental work. The use of amphotropic (VSV-G) coated virus does not yield better results in our system. For preparing viral supernatant, human embryonic kidney cells (293) are cotransfected with Banshee-GFP and the packaging-helper plasmid PCL/Eco that contains *gag*, *pol*, and *env* genes (Naviaux *et al.*, 1996), using standard calcium phosphate precipitation as follows.

Transfections (See Reagents and Specifications)

1. Twenty-four hours before transfection, plate 0.4 × 10⁶ to 0.8 × 10⁶ exponentially growing 293 cells per 6-cm petri dish in 4 ml of Dulbecco's

modified Eagle medium (DMEM) 10% FBS Pen/Strep or 1×10^6 to 2×10^6 cells per 10-cm dish in 8 ml.

2. For the transfection, have 2× HBS and 2 M CaCl$_2$ solutions and sterile water at room temperature. Mix reagents as follows in polypropylene tubes for each 6-cm dish; double the mix for a 10-cm dish: 438 μl H$_2$O, 62 μl 2 M CaCl$_2$, 4 μg retroviral construct, and 4 μg packaging vector(s). Add dropwise to 500 μl 2× HBS while vortexing or bubbling air into the mix with a pipettor. Immediately add the mix to the 293s, distributing it dropwise throughout the plate, and then swirl the dish gently.

3. Incubate cells with the transfection mix for 16–20 h; after this, rinse the plates once with PBS or medium and feed with 3 ml of fresh medium. At this stage, the cells are slightly detached and rinses should be done gently to avoid stripping them off.

4. Collect supernatant 24 and 48 h after the rinse. We prepare fresh supernatant for every experiment, but snap-frozen supernatant ($-70°$) can be used with a 50% reduction in viral titer.

Infection with Retroviral Supernatant

Most primary hematopoietic and lymphoid cells show reduced viability when cultured for long periods of time with retroviral supernatant and other infection reagents such as lipids or polybrene. We use spin infection as our preferred method for retroviral transduction because it minimizes the time for which the cells are exposed to these reagents. In general, the method involves plating cells in 24-well plates with 2 ml of fresh warm viral supernatant and 20 μg/ml Lipofectamine™ (Invitrogen, Carlsbad, CA) or 5 μg/ml polybrene (Sigma), centrifuging cells at 20° for 1–1.5 h at 460g, culturing cells for 1 h at 37°, and then replacing the viral supernatant with medium; additional details are provided later. This approach yields good efficiency of infection: up to 90% with lymphoid lines and up to 60% with fetal liver and thymocytes.

Analysis of Protein or mRNA Knockdown

Not all hairpins generated against a particular protein work efficiently. Therefore, it is necessary to test their efficacy in knocking down protein expression or mRNA levels before using them for experiments. To test siRNA hairpin function, we express the hairpins in lymphoid cell lines that express the protein to be targeted. Downregulation of protein expression can be analyzed by flow cytometry on gated GFP$^+$ cells, or by Western blot or mRNA analysis on isolated GFP$^+$ cells. We find analysis by flow cytometry the most convenient method. However, it requires the use of a highly specific antibody that gives low background staining. Analysis by Western blot or mRNA analysis requires very high transduction efficiency or isolation of GFP$^+$ cells. In some cases, gene silencing results in poor survival or expansion, making isolation of GFP$^+$ cells essential. We use the following methods for infection of cells and analysis and intracellular staining for flow cytometric analysis.

Infection of Lymphoid Cell Lines

1. Resuspend cells from a culture in exponential growth at 5×10^5 per 100 μl of fresh medium. Mix each 100 μl of cells with 2 ml of fresh viral supernatant, 5 μg/ml polybrene, and dispense in one well in 24-well plates.
2. Centrifuge plate for 1 h at room temperature at 450g.
3. Incubate cells for 1 h in the incubator (37°, 5% CO_2). Replace 1.5 ml of medium in each well with fresh medium. Later, feed and split cells as necessary.

If a cell line is refractory to infection by using an ecotropic envelope, try an amphotropic coat.

Flow Cytometric Analysis of Intracellular Proteins and the Green Fluorescent Protein (GFP)

The following method works well for some proteins but should be tested for each particular case.

1. Resuspend up to 2×10^6 cells in 200 μl of PBS + 4% FBS.
2. Add paraformaldehyde at a final concentration of 2% and incubate for 10 min at 37°. Use fresh or frozen aliquots of paraformaldehyde only (Cat. No. 2650, Sigma). Concentrations of paraformaldehyde lower than 2% or shorter incubation times will result in loss of GFP from the cells.
3. Add 1.8 ml of ice-cold methanol while vortexing and incubate 30 min on ice. Spin (4 min, 450g) and wash with 2 ml of PBS + 4% FBS.
4. Resuspend in 100 μl of primary antibody for 20 min at room temperature.
5. Wash by adding 2 ml of PBS + 4% FBS.
6. Resuspend in secondary antibody as done for the primary, incubate, and wash. Cells can be run immediately or stored at 4° for flow cytometric analysis.

Addition of a biotinylated secondary and use of an avidin-conjugated fluorochrome improve the staining. The ideal concentration of antibodies needs to be determined to avoid high background.

Note that efficient knockdown of protein expression does not happen immediately after initiation of transcription from the retrovirus—it requires a few days. The delay most likely results from the time required for expression and incorporation of the siRNA into active RISC and the turnover time for the particular protein to be assayed. Presumably, more stable proteins require more time for efficient reduction in expression. This is an important consideration when determining how much time to give before assaying reduction of protein expression as well as analyzing function. This is a clear disadvantage over traditional gene knockouts, in which protein expression is completely eliminated from the beginning. The use of siRNA hairpins in transgenic mice or transduction of early precursors such as hematopoietic stem cells can overcome this problem (Hemann *et al.*, 2003; Rubinson *et al.*, 2003; Scherr *et al.*, 2003a). However, the disadvantage in the latter case is the inability to drive hairpin expression in a tissue-specific way. To bypass this difficulty in a developmental system such as the rFTOC, we use lymphoid precursors that will require a few days (e.g., 2 days) before reaching the desired developmental stage, thus allowing enough time for efficient knockdown of protein expression.

Analysis of T-Cell Development in Reaggregate Fetal Thymic Organ Culture (rFTOC)

For serially testing retroviral constructs, we have optimized an rFTOC system derived from the basic protocol reported by Anderson and colleagues (Jenkinson and Anderson, 1994). In our system, we isolate T-cell precursors, infect them with retrovirus, and reaggregate them with fetal thymic stromal cells. The stromal cells are derived from fetal thymic lobes treated with 2′-deoxyguanosine (d-Gua). The overall experimental design is outlined in Fig. 1. In our experience, reaggregates yield very consistent results, superior to those obtained with de-Gua-treated lobes recolonized with lymphoid precursors. Typically, we analyze three reaggregates per construct and time point in each experiment. Normally, two to three time points are analyzed per construct (e.g., Days 4, 8, and 12 of culture). This provides sufficient information to determine whether the construct has a consistent effect on a particular developmental stage in all the triplicates and over time. In addition to $\alpha\beta$ T-cells, $\gamma\delta$ T cells routinely develop in these cultures. The lobes are cultured on 0.4-μm filters that allow addition of cytokines or antibodies if necessary.

T-Cell Precursors: Calculating the Number of Embryos Required

T-cell precursors can be isolated from fetal thymus of different embryonic stages, depending on the thymocyte subpopulations required. T-cell development in the embryonic thymus is similar to that of the adult. Initially $CD4^-CD8^-$ double-negative (DN) precursors enter the thymus. These undergo proliferation and differentiation from DN1 to DN3, at which stage they undergo pre-TCR selection. The latter triggers substantial proliferation and differentiation into DN4 cells and finally into $CD4^+CD8^+$ double-positive cells. The DP cells undergo positive selection to differentiate into mature $CD4^+CD8^-$ and $CD4^-CD8^+$ single-positive cells. E13.5 thymic lobes contain only DN thymocytes, E14.5 also includes some cells undergoing pre-TCR selection, whereas E16.5 has a significant proportion of DP thymocytes, some of which are undergoing positive selection. (We consider the morning of plug as Day 0.5.) On average, each reaggregate is formed with approximately 0.25×10^6 stromal and up to 0.5×10^6 donor E16.5 thymocytes. The fraction of thymocytes undergoing proliferation decreases as development progresses; if unfractionated donor thymocytes are used, then fewer (approximately 0.1×10^6) E14.5 thymocytes can be used per reaggregate because these will proliferate more than E16.5 thymocytes. It is best to use embryos from an inbred strain to minimize genetic variation that can considerably alter T-cell development. The number of embryos required depends on the yield of thymocytes/lobe, which varies depending on the age and strain of the embryos. The following are approximate values for C57BL/6 embryos (B6).

Age	Embryos required	Yield/lobe (cells)
E14.5	~50	0.1×10^6
E15.5	~30	0.25×10^6
E16.5	~10	0.5×10^6

When earlier developmental stages are required, fetal liver is a very good source of T-cell precursors (Anderson *et al.*, 2002).

Stromal Cells: Calculating the Number of Embryos Required

Thymic stromal cells are obtained from lobes treated for 5 days with d-Gua (Jenkinson *et al.*, 1982). This treatment very efficiently depletes the endogenous lymphoid population, leaving the stromal cells, including dendritic cells, intact. We use B6D2F1 ([C57BL/6 × DBA/2]F1) × B6D2F1 timed matings because they mate efficiently and litter size ranges from 6 to 12 embryos. The genetic background of the F2 embryos varies, but stromal cells from all lobes are pooled, and, as a result, reaggregates are homogenous. Ideally, lobes should be obtained from E14.5 to E15.5 embryos. Treatment of lobes from older embryos with d-Gua does not efficiently deplete the lymphoid population, as does treatment of the lobes for less than 5 days. Treatment of the lobes for longer periods can affect the stromal cells. Approximately 100–160 lobes (50–80 embryos, 6–9 pregnant females) are required for one experiment that will test three to four constructs, including controls, at 6–10 reaggregates per construct, sufficient for 2–3 time points. About 100–160 lobes should yield between 3×10^6 and 9×10^6 stromal cells.

Detailed Protocols for Infection of Lymphoid Precursors and Generation of rFTOCs (See Reagents and Specifications)

Deoxy-Guanosine Treatment of Lobes

E15.5 thymic lobes are easier and faster to isolate and clean up than E14.5 lobes; they also contain more stromal cells.

1. Dissect the thymic lobes from E14.5 to E15.5 embryos in 10-cm petri dishes with approximately 6 ml of PBS. Use Dumont forceps #5. Do four pregnant females at a time so that each batch of embryos is dealt with within 1–2 h.
2. Rinse lobes in d-Gua/DMEM once; pipette them with a P1000 into a 6-cm petri dish containing 5 ml of d-Gua/DMEM.
3. Transfer lobes onto insert filters and distribute evenly or lobes will fuse. Use six-well plates with insert filters and 1 ml of d-Gua/DMEM per well under each filter. Up to 40–50 lobes can fit per insert filter.
4. Culture for 5 days. Replace the d-Gua/DMEM medium on the third day.

Isolation of Donor Thymocytes

1. Dissect thymic lobes from E14.5 to E16.5 embryos, as detailed previously.
2. Transfer into Eppendorf with 500–1000 μl of 2 mg/ml collagenase solution, and incubate at 37° for 30 min to 1 h. At 30 min, gently resuspend with a P1000 and then with a P200. Terminate incubation when lobes dissociate into a single-cell suspension. Rushing this process will rip apart the cells.
3. Add 500 μl of complete medium and spin (see centrifugation specifications later).
4. Wash twice with 1 ml of complete medium and count.

Retroviral Transduction

1. Plate 1.5×10^6 to 2×10^6 cells/well in six-well plates in 2 ml of retroviral supernatant plus Lipofectamine (GIBCO) at 20 μg/ml. Spin for 1–1.5 h at 450g at room temperature.

2. Wash cells by aspirating most of the supernatant without disturbing the layer of cells and replace with fresh medium.
3. Culture for 2–4 h.

Isolation of Stromal Cells

1. Transfer lobes into Eppendorf tube with 1 ml of complete medium and wash thrice with 1 ml of medium over a period of 3 h or more. (Begin once the donor lobes are placed in collagenase.) Place in incubator in between each wash. Place 50 to 100 lobes per tube.
2. Rinse thrice with PBS to remove protein.
3. Add 500 μl of warm trypsin-EDTA solution and place at 37° and incubate for 20 min. At 10 min, pipette gently with a P1000 and put back at 37°. At 20 min, pipette first with a P1000 and then with a P200 until a single-cell suspension is generated. Stop the reaction by adding 500 μl of complete medium. If a single-cell suspension is not generated, transfer the supernatant into another tube with complete medium, add more trypsin-EDTA to the original tube, and incubate for another 5 min. If the trypsin treatment is overdone, most of the cells will be lysed, DNA will be released, and the rest of the cells will be entangled into a huge clump.
4. Discard any clumps, and wash cells at least three times in Eppendorf or larger tubes with complete medium. Count and spin into approximately 500 μl of medium. Use cells immediately or they will reaggregate as they start pelleting in the tube.

Generating the Reaggregates

1. Harvest thymocytes into Eppendorf tubes and wash thrice with complete medium.
2. Mix in one Eppendorf enough stromal cells and thymocytes to generate the desired number of reaggregates: for example, 0.25×10^6 stromal cells/lobe $\times$ 10 lobes $= 2.5 \times 10^6$ cells $+ 0.25 \times 10^6$ thymocytes/lobe $\times$ 10 lobes $= 2.5 \times 10^6$ cells.
3. Spin and discard most of the supernatant, and then spin briefly in a microfuge ($450g$ for 20 sec) to concentrate the pellet and discard the rest of the supernatant.
4. Resuspend cells in a volume equivalent to the number of lobes $\times$ 0.7μl.
5. Place insert filters in six-well plate wells with 1 ml of complete medium under each filter. Dispense 0.7-μl drops onto insert filters placed in six-well plates. Place in incubator as described previously.

Analysis

We stain lobes individually and minimize cell losses by performing all steps in 96-well plates as follows.

1. Harvest the lobes at the required times (e.g., Days 3, 6 and 9), using Dumont forceps with folded tips to avoid breaking the lobes or filters. Place individual lobes in round-bottom 96-well plates (1 lobe/well) with 150 μl/ml collagenase solution (2 mg/ml) in each well. Avoid using tissue-culture-treated plates because cells tend to stick to the sides of the wells instead of forming a pellet.
2. Cover the plate and incubate at 37° for 30 min to 1 h.
3. Resuspend the lobes with a P200. Stain with antibodies in the 96-well plates.

Reagents and Specifications

Web Sites for siRNA Design

http://www.mekentosj.com/irnai/index.html
http://katahdin.cshl.org:9331/RNAi/html/rnai.html
http://www.ambion.com/techlib/misc/siRNA_finder.html

Solutions for Transfection

2 *M* CaCl$_2$: Filter through a 0.2-μm filter and store aliquots at $-20°$.

2× HBS: 50 m*M* HEPES, pH 7.05, 10 m*M* KCl, 12 m*M* dextrose, 280 m*M* NaCl, 1.5 m*M* Na$_2$HPO$_4$. Adjust pH of 2× HBS at the end to exactly 7.05 $\pm$ 0.05 and then filter sterilize (0.2-μm filter). Store aliquots at $-20°$. Solution deteriorates after 6 months, even when stored at $-20°$. We normally prepare three batches of 2× HBSS with very slight changes in pH and test them. We find that good batches yield consistent and better transfection efficiency than any of the lipids we have tested so far.

DMEM/FTOC Culture Medium

DMEM (Cat. No. 11995–065, GIBCO)
10% FBS (heat inactivated)
100 U/100 μg/ml Pen/Strep (we usually use one fifth of this concentration)
50 μ*M* 2-ME
20 m*M* Hepes
100 μ*M* Nonessential amino acids
2 m*M* L-Glu

2'-Deoxyguanosine (d-Gua; Cat. No. D-0901, Sigma)

Prepare a 13.5 mM (10×) stock solution in DMEM/HEPES without FBS, (100 mg d-Gua in 27.7 ml DMEM). Warm up to 37° to dissolve, and then filter sterilize with a 0.25-μm filter. Freeze aliquots at −20°. Thaw 10× d-Gua stocks at 37° and dilute 1:10 in complete DMEM. In our experience, frozen aliquots last up to 6 months. After this, the d-Gua forms precipitates that do not dissolve and become toxic to the FTOCs.

Collagenase

Prepare 2 mg/ml collagenase Type IV (Worthington, Freehold, NJ, or Boehringer, Germany) in RPMI/HEPES. Filter sterilize and store aliquots at −20°.

Trypsin-EDTA (Cat. No. 25300–054, GIBCO)

Use 0.05%; trypsin, 0.53 mM EDTA · 4Na.

Insert Filters

Use 0.4-μm insert filters (Millicell-CM PICM 030 50); they fit in six-well plates.

Forceps

Use Dumont dissection forceps #5 and Dumont forceps with angled tips.

Culture Conditions

Use a humidified incubator at 37°, 7% CO_2. All the culture plates are set inside Tupperware™ with just enough distilled sterile water to cover the bottom of the Tupperware box, Culture plates are set on top of a petri dish or plate lid on top of the water. The gas inside the Tupperware is allowed to equilibrate for a few hours or overnight, and then the lid is sealed to minimize changes in humidity, temperature, or CO_2.

Transferring Lobes

Transfer lobes by a mouth pipette. For some manipulations, a P1000 can be used, although the lobes can stick to the inside of the tip and some may be lost. Alternatively, use Dumont forceps with angled tips.

Centrifugations

When spinning cells in Eppendorf tubes, use bucket rotors or cells will smear along the side of the tubes and will be lost during all the washes. Spin at 450g (1.4 KRPM in a standard benchtop centrifuge), without brake, at room temperature unless specified. Stromal cells have a lower density and need to be spun for longer than thymocytes.

Acknowledgments

G. H. H. is Special Fellow, Leukemia & Lymphoma Society. J. A. I is Investigator, Cancer Research Institute. The work was supported by grants to J. A. I. from NIH (AI45072) and CRI.

References

Agrawal, N., Dasaradhi, P. V., Mohmmed, A., Malhotra, P., Bhatnagar, R. K., and Mukherjee, S. K. (2003). RNA interference: Biology, mechanism, and applications. *Microbiol. Mol. Biol. Rev.* **67,** 657–685.

Anderson, M. K., Hernandez-Hoyos, G., Dionne, C. J., Arias, A. M., Chen, D., and Rothenberg, E. V. (2002). Definition of regulatory network elements for T cell development by perturbation analysis with PU.1 and GATA-3. *Dev. Biol.* **246,** 103–121.

Berns, K., Hijmans, E. M., Mullenders, J., Brummelkamp, T. R., Velds, A., Heimerikx, M., Kerkhoven, R. M., Madiredjo, M., Nijkamp, W., Weigelt, B., Agami, R., Ge, W., Cavet, G., Linsley, P. S., Beijersbergen, R. L., and Bernards, R. (2004). A large-scale RNAi screen in human cells identifies new components of the p53 pathway. *Nature* **428,** 431–437.

Bridge, A. J., Pebernard, S., Ducraux, A., Nicoulaz, A. L., and Iggo, R. (2003). Induction of an interferon response by RNAi vectors in mammalian cells. *Nat. Genet.* **34,** 263–264.

Brummelkamp, T. R., Bernards, R., and Agami, R. (2002a). Stable suppression of tumorigenicity by virus-mediated RNA interference. *Cancer Cell* **2,** 243–247.

Brummelkamp, T. R., Bernards, R., and Agami, R. (2002b). A system for stable expression of short interfering RNAs in mammalian cells. *Science* **296,** 550–553.

Caplen, N. J., and Mousses, S. (2003). Short interfering RNA (siRNA)-mediated RNA interference (RNAi) in human cells. *Ann. N. Y. Acad. Sci.* **1002,** 56–62.

Caplen, N. J., Parrish, S., Imani, F., Fire, A., and Morgan, R. A. (2001). Specific inhibition of gene expression by small double-stranded RNAs in invertebrate and vertebrate systems. *Proc. Natl. Acad. Sci. USA* **98,** 9742–9747.

Elbashir, S. M., Harborth, J., Lendeckel, W., Yalcin, A., Weber, K., and Tuschl, T. (2001). Duplexes of 21-nucleotide RNAs mediate RNA interference in cultured mammalian cells. *Nature* **411,** 494–498.

Elbashir, S. M., Harborth, J., Weber, K., and Tuschl, T. (2002). Analysis of gene function in somatic mammalian cells using small interfering RNAs. *Methods* **26,** 199–213.

Halene, S., Wang, L., Cooper, R. M., Bockstoce, D. C., Robbins, P. B., and Kohn, D. B. (1999). Improved expression in hematopoietic and lymphoid cells in mice after transplantation of bone marrow transduced with a modified retroviral vector. *Blood* **94,** 3349–3357.

Hannon, G. J. (2002). RNA interference. *Nature* **418,** 244–251.

Hemann, M. T., Fridman, J. S., Zilfou, J. T., Hernando, E., Paddison, P. J., Cordon-Cardo, C., Hannon, G. J., and Lowe, S. W. (2003). An epi-allelic series of p53 hypomorphs created by stable RNAi produces distinct tumor phenotypes *in vivo. Nat. Genet.* **33,** 396–400.

Hernandez-Hoyos, G., Anderson, M. K., Wang, C., Rothenberg, E. V., and Alberola-Ila, J. (2003). GATA-3 expression is controlled by TCR signals and regulates CD4/CD8 differentiation. *Immunity* **19,** 83–94.

Jenkinson, E. J., and Anderson, G. (1994). Fetal thymic organ cultures. *Curr. Opin. Immunol.* **6**, 293–297.

Jenkinson, E. J., Franchi, L. L., Kingston, R., and Owen, J. J. (1982). Effect of deoxyguanosine on lymphopoiesis in the developing thymus rudiment *in vitro*: Application in the production of chimeric thymus rudiments. *Eur. J. Immunol.* **12**, 583–587.

Khvorova, A., Reynolds, A., and Jayasena, S. D. (2003). Functional siRNAs and miRNAs exhibit strand bias. *Cell* **115**, 209–216.

Lee, N. S., Dohjima, T., Bauer, G., Li, H., Li, M. J., Ehsani, A., Salvaterra, P., and Rossi, J. (2002). Expression of small interfering RNAs targeted against HIV-1 rev transcripts in human cells. *Nat. Biotechnol.* **20**, 500–505.

McManus, M. T., and Sharp, P. A. (2002). Gene silencing in mammals by small interfering RNAs. *Nat. Rev. Genet.* **3**, 737–747.

Michienzi, A., Castanotto, D., Lee, N., Li, S., Zaia, J. A., and Rossi, J. J. (2003). RNA-mediated inhibition of HIV in a gene therapy setting. *Ann. N. Y. Acad. Sci.* **1002**, 63–71.

Naviaux, R. K., Costanzi, E., Haas, M., and Verma, I. M. (1996). The pCL vector system: Rapid production of helper-free, high-titer, recombinant retroviruses. *J. Virol.* **70**, 5701–5705.

Paddison, P. J., Caudy, A. A., and Hannon, G. J. (2002). Stable suppression of gene expression by RNAi in mammalian cells. *Proc. Natl. Acad. Sci. USA* **99**, 1443–1448.

Paddison, P. J., Silva, J. M., Conklin, D. S., Schlabach, M., Li, M., Aruleba, S., Balija, V., O'Shaughnessy, A., Gnoj, L., Scobie, K., Chang, K., Westbrook, T., Cleary, M., Sachidanandam, R., McCombie, W. R., Elledge, S. J., and Hannon, G. J. (2004). A resource for large-scale RNA-interference-based screens in mammals. *Nature* **428**, 427–431.

Pai, S. Y., Truitt, M. L., Ting, C. N., Leiden, J. M., Glimcher, L. H., and Ho, I. C. (2003). Critical roles for transcription factor GATA-3 in thymocyte development. *Immunity* **19**, 863–875.

Persengiev, S. P., Zhu, X., and Green, M. R. (2004). Nonspecific, concentration-dependent stimulation and repression of mammalian gene expression by small interfering RNAs (siRNAs). *RNA* **10**, 12–18.

Qin, X. F., An, D. S., Chen, I. S., and Baltimore, D. (2003). Inhibiting HIV-1 infection in human T cells by lentiviral-mediated delivery of small interfering RNA against CCR5. *Proc. Natl. Acad. Sci. USA* **100**, 183–188.

Robbins, P. B., Skelton, D. C., Yu, X. J., Halene, S., Leonard, E. H., and Kohn, D. B. (1998). Consistent, persistent expression from modified retroviral vectors in murine hematopoietic stem cells. *Proc. Natl. Acad. Sci. USA* **95**, 10182–10187.

Rubinson, D. A., Dillon, C. P., Kwiatkowski, A. V., Sievers, C., Yang, L., Kopinja, J., Rooney, D. L., Ihrig, M. M., McManus, M. T., Gertler, F. B., Scott, M. L., and Van Parijs, L. (2003). A lentivirus-based system to functionally silence genes in primary mammalian cells, stem cells and transgenic mice by RNA interference. *Nat. Genet.* **33**, 401–406.

Scherr, M., Battmer, K., Dallmann, I., Ganser, A., and Eder, M. (2003a). Inhibition of GM-CSF receptor function by stable RNA interference in a NOD/SCID mouse hematopoietic stem cell transplantation model. *Oligonucleotides* **13**, 353–363.

Scherr, M., Battmer, K., Ganser, A., and Eder, M. (2003b). Modulation of gene expression by lentiviral-mediated delivery of small interfering RNA. *Cell Cycle* **2**, 251–257.

Schwarz, D. S., Hutvagner, G., Du, T., Xu, Z., Aronin, N., and Zamore, P. D. (2003). Asymmetry in the assembly of the RNAi enzyme complex. *Cell* **115**, 199–208.

Sledz, C. A., Holko, M., de Veer, M. J., Silverman, R. H., and Williams, B. R. (2003). Activation of the interferon system by short-interfering RNAs. *Nat. Cell Biol.* **5**, 834–839.

[13] Lentiviral Vector Delivery of Recombinant Small Interfering RNA Expression Cassettes

By MING-JIE LI and JOHN J. ROSSI

Abstract

Lentiviral vectors are able to transduce nondividing cells and maintain sustained long-term expression of transgenes. Many cells types, including brain, liver, muscle and hematopoietic stem cells, have been successfully transduced with lentiviral vectors carrying a variety of genes. These properties make lentiviral vectors attractive vehicles for delivering small interfering RNA (siRNA) genes into mammalian cells. RNA polymerase III (pol III) promoters are most commonly used for expressing siRNAs from lentiviral vectors. Pol III promoters are relatively small, have high activity, and use simple termination signals of short stretches of Us. It is possible to include several pol III expression cassettes in a single lentiviral vector backbone to express different siRNAs or to combine siRNAs with other transgenes. This chapter describes the delivery of pol III-promoted siRNAs by human immunodeficiency virus (HIV)-based lentiviral vectors and covers vector design, production, and verification of siRNA expression and function. This chapter should be useful for establishing a lentiviral vector-based delivery of siRNAs in experiments that require long-term gene knockdown or developing siRNA-based approaches for gene therapy applications.

Introduction

Small interfering RNAs (siRNAs) mediate RNA interference (RNAi), providing a powerful new tool for gene silencing in a sequence-specific manner. Not only are they an attractive approach for studying gene function and regulation, but they also have potential for gene therapy applications. However, synthetic siRNAs transfected into target cells can only be retained for a few days in cells and will be diluted after several generations of cell division. In many cases, a long-term effect of RNAi is required. Retroviral vectors derived from murine leukemia virus (MuLV) have long been favored in gene delivery for their efficient integration into the genome of the target cells and accompanying expression of the transgenes. However, these vectors require cell division for efficient gene transfer (Miller *et al.*, 1990; Verma, 1994). In contrast, lentiviral vectors are able

0076-6879/05 $35.00

to transduce nondividing cells with sustained long-term expression of the genes. The majority of target cell types for gene therapy are nondividing or slowly dividing. These include brain, liver, muscle, and hematopoietic stem cells. Each of these cell types has been efficiently transduced by lentivirus vectors (Kafri *et al.*, 1997; Naldini *et al.*, 1996; Uchida *et al.*, 1998). We and others (Li *et al.*, 2003; Qin *et al.*, 2003; Scherr *et al.*, 2003) have reported that lentiviral vectors can efficiently deliver siRNA expression cassettes into a variety of cells with sustained expression and potent function of the siR-NAs.

The issue of biosafety has been a concern for human immunodeficiency virus (HIV) based lentiviral vectors. The current third-generation replication-defective and self-inactivating (SIN) lentiviral vectors have minimized the potential risk of generating replication-competent helper virus. We used phIV-7-GFP, a typical third-generation replication-defective SIN vector as the transfer vector for siRNAs. As illustrated in Fig. 1, this vector contains a hybrid 5′ long terminal repeat (LTR), in which the U3 region is replaced with the cytomegalovirus (CMV) promoter and enhancer sequence. This strong promoter makes the transcription of the vector

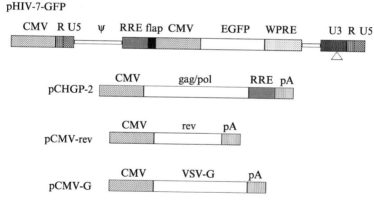

FIG. 1. Constructs of the lentiviral vector and the packaging plasmids. The transfer vector, phIV-7-GFP, contains a hybrid 5′ long terminal repeat (LTR) in which the U3 region is replaced with the cytomegalovirus (CMV) promoter, the packaging signal (Ψ), the rev-responsive element (RRE) sequence, the flap sequence, the enhanced green fluorescent protein (EGFP) gene driven by CMV promoter, the woodchuck posttranscriptional regulatory element (WPRE) and the 3′ LTR in which the *cis* regulatory sequences are completely removed from the U3 region. The genes of interest along with a pol III promoter can be inserted directly upstream of the CMV promoter of EGFP in the phIV-7-GFP vector. pCHGP-2 contains the *gag* and *pol* genes and RRE sequence from HIV-1 under the control of the CMV promoter. pCMV-rev contains the coding sequence of rev driven by the CMV promoter. pCMV-G contains the VSV-G protein gene under the control of the CMV promoter. PA indicates the polyadenylation signal from the human β-globin gene.

sequence independent of HIV Tat function, which is normally required for HIV gene expression (Arya *et al.*, 1985). The packaging signal (Ψ) is essential for encapsidation, and the Rev-responsive element (RRE) is required for producing high-titer vectors. The flap sequence or polypurine tract (cPPT) and the central termination sequence (CTS) are important for nuclear import of the vector DNA, a feature required for transducing nondividing cells (Sirven *et al.*, 2000). The enhanced green fluorescent protein (EGFP) reporter gene is driven by an internal CMV promoter. This is useful for vector titration, detection of transduction efficiency, and selection of transduced cells. The woodchuck hepatitis virus posttranscription regulation element (WPRE; Zufferey *et al.*, 1999) following the EGFP sequence improves the expression of the reporter gene by promoting RNA nuclear export and polyadenylation, or both. In the 3' LTR, the *cis*-regulatory sequences were completely removed from the U3 region. This deletion is copied to 5' LTR after reverse transcription, resulting in transcriptional inactivation of both LTRs. To produce the packaged vectors, the vector DNA, along with three other plasmids (Fig. 1), is cotransfected into 293T cells. The Gag/Pol proteins are supplied by the pCHGP-2 plasmid, pCMV-Rev encodes Rev which binds to the RRE for efficient RNA export from the nucleus, and pCMV-G encodes the vesicular stomatitis virus glycoprotein (VSV-G) that replaces HIV-1 Env. VSV-G expands the tropism of the vectors and allows concentration by ultracentrifugation. The siRNA sequence(s) along with a pol III promoter are inserted directly upstream of the CMV-EGFP sequence.

Construction of Lentiviral Vector Plasmid DNA Carrying Small Interfering RNA (siRNA) Sequences

The human U6 small nuclear RNA promoter and human H1 promoter are among the common pol III promoters used for expressing siRNAs. These promoters have relatively small size and the transcription is conveniently terminated within a stretch of four or more uridines. In our experience, at least three pol III expression cassettes can be delivered by a single lentiviral vector backbone. The sense and antisense sequences of siRNAs can be expressed from separate promoters or from a single promoter directing a short hairpin (shRNA) structure with sense and antisense connected by a loop sequence. In our early work, we demonstrated that anti-rev siRNAs expressed from separate promoters showed marked target downregulation as well as anti-HIV-1 activity (Lee *et al.*, 2002). We subsequently found that the shRNAs exhibited even more potent RNAi activity than the separately expressed siRNAs designed to inhibit the same target sequences (Li *et al.*, 2003). In addition, shRNA expression cassettes are

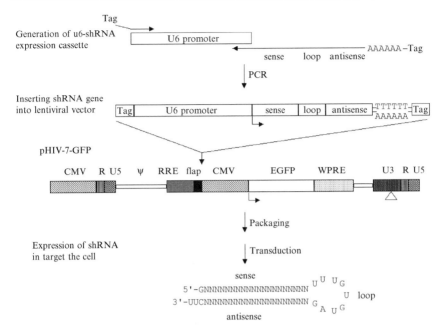

FIG. 2. Construction of lentiviral vector expressing short hairpin RNA (shRNAs). The U6-shRNA expression cassette is constructed by PCR with the primers shown. The PCR product contains restriction sites at both ends for cloning into the lentiviral vector. The shRNA expression cassette, including the U6 pol III promoter, the sense and antisense sequence of the shRNA separated by a 9-base loop, and a terminator composed of six thymidines is inserted directly upstream of the CMV promoter of EGFP in the pHIV-7-GFP vector. Arrows indicate the orientation of transcription for a given gene. The putative shRNA structure is shown at the bottom.

easier to construct than the siRNA counterparts. Therefore, this chapter focuses on shRNA constructs in the lentiviral vector delivery system incorporating the human U6 pol III promoter to direct shRNA expression.

1. PCR produce U6shRNA genes by amplification of a plasmid containing the human U6 promoter sequence, such as pTZ U6+1, with one primer complementary to the sequence upstream of the U6 promoter and another primer complementary to the 3' end of the U6 promoter and the sense, loop, antisense, and terminator sequences of the shRNA (Fig. 2; see Chapter 10). The first nucleotide of the U6 transcripts should be guanine (G) to ensure efficient transcription. If the first base in the target sequence is not a G, the extra G at the 5' end of the sense strand of the shRNA does not affect the function of the shRNA. Inclusion of a restriction site at the 5' end of both primers is useful for ligating the PCR

products to the corresponding or complementary sites in the vector. Alternatively, the restriction sites in the primers can be omitted by ligating the PCR products in the PCR cloning vector, pCR2.1, using the TA Cloning® Kit (Invitrogen, Carlsbad, CA).

2. Digest the PCR product with corresponding restriction enzyme(s) or excise the fragment containing the U6 promotor and shRNA gene from the polylinker in the pCR2.1 vector and ligate the insert into the multiple cloning sites located between the flap and CMV-EGFP in the lentiviral vector backbone.

3. Verify the ligated product by restriction analysis and DNA sequencing.

4. Amplify and purify the vector along with the other plasmids required for packaging, using a Plasmid Maxi Kit (Qiagen, Valencia, CA).

Production of Lentiviral Vectors

Lentiviral vectors are commonly produced by transient cotransfection of the transfer vector and other plasmids required for packaging into 293T cells, an easily transformed human embryonic kidney cell line. The cells are maintained in Dulbecco's modified Eagle's medium (DMEM) with high glucose (4500 mg/l), supplemented with 10% fetal bovine serum (FBS), 100 units/ml penicillin, and 100 μg/ml streptomycin in a 37° incubator with 10% CO_2.

1. Plate the 293T cells in 100-mm tissue culture dishes 24 h before transduction. The cell density should be 30–40% confluent when seeding and will be about 80% confluent for transfection. We usually prepare at least five dishes per vector to obtain reasonable amounts of packaged vector for multiple transductions.

2. Change the culture medium with 10 ml of fresh medium 5 h before transfection.

3. Prepare 1 ml of calcium phosphate-DNA suspension for each 100-mm plate of cells as follows:
   a. Set up two sterile tubes for transfection of one plate. Label the tubes 1 and 2.
   b. Add 0.5 ml of 2× HBS (0.05 M HEPES, 0.28 M NaCl, 1.5 mM Na$_2$HPO$_4$, pH 7.12) to Tube 1.
   c. Add TE 79/10 (1 mM Tris-HCl, 0.1 mM EDTA, pH 7.9) to Tube 2. The volume of TE 79/10 = 440 μl – the volume of DNA.
   d. Add 15 μg of the transfer vector containing the transgenes, 15 μg pCHGP-2, 5 μg pCMV-rev, and 5 μg pCMV-G to Tube 2 and mix.
   e. Add 10 μl of 2 M CaCl$_2$ solution to Tube 2 and gently mix.

   f. Add 50 μl of 2 M CaCl$_2$ solution to Tube 2 and gently mix.

   g. Transfer the contents from Tube 2 to Tube 1 dropwise while gently mixing.

   h. Allow the suspension to sit at room temperature for at least 30 min.

4. Mix the precipitate well by pipetting or vortexing.

5. Add 1 ml of suspension to a 100-mm plate containing cells. The suspension must be added dropwise and slowly while gently swirling the medium in the plate. Return the plates to the incubator and leave the precipitation for 5 to 6 h.

6. Replace the medium with 6 ml complete medium. Add 60 μl of 0.6 M butyric acid. Return to incubator.

7. After 24 h of incubator culturing, collect supernatant and freeze at $-80°$. Add 6 ml complete medium to each plate. Add 60 μl of 0.6 M butyric acid. Return to incubator.

8. After 12 h of incubator culture, collect supernatant. Freeze at $-80°$ or proceed to next step.

9. Centrifuge the freshly collected or thawed supernatant at 2000 rpm for 10 min to remove any cell debris in the supernatant. Filter the supernatant with a 0.2-μm syringe filter. It can also be filtered with a 0.2-μm cellulose acetate bottle-top filter for preparing large quantities and reducing loss of the vector.

10. Concentrate the supernatant by ultracentrifugation at 24,500 rpm and 4° for 1.5 h, using a Beckman SW28 swinging bucket rotor.

11. Remove the supernatant and resuspend the pellet in an appropriate amount of culture medium, for example, 150 μl for 30 ml of original supernatant if a 200-fold concentration is desired.

12. Divide the concentrated vector into 10- to 50-μl aliquots and store at $-80°$ until use. Avoid freeze–thaw cycles.

Titration of the Vectors

1. Seed 1×10^5 of HT1080 cells (a human fibrosarcoma cell line) per well in six-well plates in DMEM medium supplemented with 10% FBS and incubate overnight.

2. Add serial-diluted vector stock and 4 μl/ml polybrene to the cultured cells. Continue incubation for 48 h.

3. Trypsinize the cells. Following centrifugation, remove the supernatant and resuspend the pellet in 300 μl of 3.7% formaldehyde in PBS.

4. Determine the percentage of EGFP-positive cells by fluorescence-activated cell sorting (FACS) analysis. The titer will be represented as transduction units per milliliter concentrated vector (TU/ml).

$$\text{Titer} = \frac{\text{Cell number} \times \text{Percentage of EGFP Positive cells} \times \text{Dilution}}{\text{vector volume (ml)} \times 100}$$

In this formula, the cell number stands for the cell count when the vector was added. If a transfer vector does not contain a reporter gene, the vector titer may be determined as the number of vector DNA molecules/ml by real-time PCR, using primers complementary to the psi element of the vector (Sastry et al., 2002).

Assay for Replication-Competant Lentivirus (RCL)

To ensure that the packaged vector stock is free from replication-competent lentivirus, a p24 antigen assay or PCR assay for the HIV pol gene should be carried out. Transduce cells such as H9 or CEM with the vectors to be tested. Keep passaging the culture for up to 1 month. Collect the supernatant weekly for the p24 antigen assay. The genomic DNA will be extracted from the transduced cells for PCR analyses. The primer sequences for amplifying the pol gene are 5'-CCAGCACACAAAG-GAATTGG-3' and 5'-GTATGCTGTTTCTTGCCCTG-3'.

Transduction of Lentiviral Vectors to Target Cells

For monolayer-cultured cells, seed cells in culture plates 24 h before transduction. At the time of transduction, the cell density should be 30–40% confluent. Add the vector at an appropriate multiplicity of infection (MOI) and polybrene at a final concentration of 4 μg/ml and return the cells to the incubator. After overnight culture, replace the culture medium. For many monolayer-cultured cell lines, an MOI of 5 can achieve more than 90% transduction efficiency. For a new cell line to be used, a series of different MOIs should be tested to find a minimum effective MOI. To transduce suspension-cultured cell lines, seed 2×10^5 cells/well into a 24-well plate. Add the appropriate amount of vector and 4 μg/ml polybrene. After overnight incubation, replace the medium. For cells such as K562, the transduction efficiency is close to 100% at an MOI of 10. For some suspension-cultured cells such as CEM (a human T-cell line), centrifugation can remarkably enhance transduction efficiency. Place 2×10^5 cells in 1 ml culture medium in a 15-ml centrifuge tube. Add 4 μg/ml polybrene. Centrifuge at 2000 rpm at 20° for 30 min. Resuspend the cells with a pipette and transfer the cells to a 24-well culture plate. After overnight incubation, replace the medium. The transduction efficiency can be determined by FACS analysis 48 h after transduction. For transduction of hematopoietic stem cells, CD34$^+$ cells are enriched from umbilical cord blood or bone marrow by anti-CD34 antibody-coupled magnetic beads (Miltenyi Biotech, Aubum, CA). Fourty-eight hours before transduction, the CD34$^+$ cells are cultured in Iscove's modified Dulbecco's medium (IMDM) supplemented with 20% BIT9500 (Stem Cell Technology, Vancouver, Canada), 40 μg/ml human low-density lipoproteins, 10^{-4} M 2-mercaptoethanol, 100 ng/ml SCF, 100 ng/ml flt3-ligand, 10 ng/ml TPO (PeproTech, Rocky Hill, NJ), 20 ng/ml IL-3, and 20 ng/ml IL-6. The lentiviral vector stock is adjusted to an MOI of 40 in 200 μl culture medium and loaded onto a 24-well plate coated with RetroNectin™ (Takara Mirus Bio, Inc., Madison, WI). After incubation at 32° for 4 h, remove the vector supernatant and wash the wells with PBS. Add the prestimulated CD34$^+$ cells to the well at 5 × 10^4/ml in the growth medium.

Detection of the Expression of siRNAs Delivered by Lentiviral Vectors

Northern blotting is among the most reliable methods to verify whether the cloned siRNA or shRNA sequence can be expressed in the target cells. Extract total RNA from transduced cells with STAT-60 reagent (Tel-Test, Inc., Friendswood, TX) according to the manufacturer's protocol. Prepare a polyacrylamide gel containing 7 M urea. A 15% polyacrylamide gel is commonly used for detecting siRNAs, but an 8% polyacrylamide gel can resolve the siRNAs as well and is more efficient for transfer to nitrocellulose. In addition, an 8% polyacrylamide gel allows siRNAs and other larger RNAs up to several hundred nucleotides in length to be detected in a single blot. Mix 15 μg of total RNA solution with an equal volume of loading buffer (95% deionized formamide, 0.025% bromophenol blue, 0.025% xylene cyanol, 0.5 mM EDTA, 0.025% SDS). Heat the samples at 95° for 4 min and then place in ice. After electrophoresis, transfer the RNA to Hybond™-N nylon membrane (Amersham, Arlington Heights, IL) by electroblotting in 0.5 × TBE buffer. After UV cross-linking, prehybridize the membrane in buffer containing 6 × SSPE (20 × SSPE stock: 3 M NaCl, 0.2 M NaH$_2$PO$_4$, 0.02 M EDTA, pH 7.4), 5 × Denhart's reagent, 0.5% SDS, and carrier DNA for 2 h. Add the appropriate γ^{32}P-labeled oligonucleotide probe complementary to the antisense sequence of the siRNA and hybridize overnight at 37°. Wash the membrane with 6 × SSPE and 0.1% SDS at 37° for 10 min and then with 2 × SSPE and 0.1% SDS twice at 37° for 10 min each and expose to an X-ray film.

Determinization of the Efficacy of RNA Cleavage by siRNAs in Lentiviral Vector-Transduced Cells

For detecting the cleavage of endogenous mRNAs by siRNAs, RT-PCR, real-time PCR, and Northern blotting are among the choices. To determine protein-level knockdown, Western blotting and immunofluorescence are most common approaches. For functional assays, in the case of anti-HIV siRNAs, assaying for HIV-1 p24 antigen and reverse transcriptase are used.

For siRNAs targeting a fusion mRNA such as in certain leukemic cells, cell proliferation and apoptosis assays are good choices.

Acknowledgments

We are very grateful to Dr. Jiing-Kaun Yee for providing pHIV-7-GFP and the packaging plasmids. This work was supported by NIH grants AI29329 and AI42552 and HL074704 to J. J. R.

References

Arya, S. K., Guo, C., Josephs, S. F., and Wong-Staal, F. (1985). Trans-activator gene of human T-lymphotropic virus type III (HTLV-III). *Science* **229,** 69–73.

Kafri, T., Blomer, U., Peterson, D. A., Gage, F. H., and Verma, I. M. (1997). Sustained expression of genes delivered directly into liver and muscle by lentiviral vectors. *Nat. Genet.* **17,** 314–317.

Lee, N. S., Dohjima, T., Bauer, G., Li, H., Li, M. J., Ehsani, A., Salvaterra, P., and Rossi, J. (2002). Expression of small interfering RNAs targeted against HIV-1 rev transcripts in human cells. *Nat. Biotechnol.* **20,** 500–505.

Li, M.-J., Bauer, G., Michienzi, A., Yee, J.-K., Lee, N.-S., Kim, J., Li, S., Castanotto, D., Zaia, J., and Rossi, J. J. (2003). Inhibition of HIV-1 infection by lentiviral vectors Expressing Pol III promoted anti-HIV RNAs. *Mol. Ther.* **8,** 196–206.

Miller, D. G., Adam, M. A., and Miller, A. D. (1990). Gene transfer by retrovirus vectors occurs only in cells that are actively replicating at the time of infection. *Mol. Cell. Biol.* **10,** 4239–4242.

Naldini, L., Blomer, U., Gage, F. H., Trono, D., and Verma, I. M. (1996). Efficient transfer, integration, and sustained long-term expression of the transgene in adult rat brains injected with a lentiviral vector. *Proc. Natl. Acad. Sci. USA* **93,** 11382–11388.

Qin, X. F., An, D. S., Chen, I. S., and Baltimore, D. (2003). Inhibiting HIV-1 infection in human T cells by lentiviral-mediated delivery of small interfering RNA against CCR5. *Proc. Natl. Acad. Sci. USA* **100,** 183–188.

Sastry, L., Johnson, T., Hobson, M. J., Smucker, B., and Cornetta, K. (2002). Titering lentiviral vectors: Comparison of DNA, RNA and marker expression methods. *Gene. Ther.* **9,** 1155–1162.

Scherr, M., Battmer, K., Ganser, A., and Eder, M. (2003). Modulation of gene expression by lentiviral-mediated delivery of small interfering RNA. *Cell Cycle* **2,** 251–257.

Sirven, A., Pflumio, F., Zennou, V., Titeux, M., Vainchenker, W., Coulombel, L., Dubart-Kupperschmitt, A., and Charneau, P. (2000). The human immunodeficiency virus type-1 central DNA flap is a crucial determinant for lentiviral vector nuclear import and gene transduction of human hematopoietic stem cells. *Blood* **96,** 4103–4110.

Uchida, N., Sutton, R. E., Friera, A. M., He, D., Reitsma, M. J., Chang, W. C., Veres, G., Scollay, R., and Weissman, I. L. (1998). HIV, but not murine leukemia virus, vectors mediate high efficiency gene transfer into freshly isolated G0/G1 human hematopoietic stem cells. *Proc. Natl. Acad. Sci. USA* **95,** 11939–11944.

Verma, I. M. (1994). Gene therapy: Hopes, hypes, and hurdles. *Mol. Med.* **1,** 2–3.

Zufferey, R., Donello, J. E., Trono, D., and Hope, T. J. (1999). Woodchuck hepatitis virus posttranscriptional regulatory element enhances expression of transgenes delivered by retroviral vectors. *J. Virol.* **73,** 2886–2892.

[14] Paradigms for Conditional Expression of RNA Interference Molecules for Use Against Viral Targets

By DAVID S. STRAYER, MARK FEITELSON, BILL SUN, and
ALEXEY A. MATSKEVICH

Abstract

The rapid increase in the study of small interfering RNA (siRNA) as a means to decrease expression of targeted genes has led to concerns about possible unexpected consequences of constitutive siRNA expression. We therefore devised a conditional siRNA expression system in which siRNA targeting hepatitis C virus (HCV) would be produced in response to HCV. We found that HCV acts via NFκB to stimulate the HIV long terminal repeat (LTR) as a promoter. We exploited this observation by designing conditional siRNA transcription constructs to be triggered by HCV-induced activation of NFκB. These were delivered by using highly efficient recombinant *Tag*-deleted SV40-derived vectors. Conditional activation of HIV-LTR and consequent siRNA synthesis in cells expressing HCV were observed. HCV-specific RNAi decreased HCV RNA greatly within 4 days, using transient transfection of the whole HCV genome as a model of acute HCV entry into transduced cells. We then tested the effectiveness of rSV40-delivered anti-HCV siRNA in cells stably transfected with the whole HCV genome to simulate hepatocytes chronically infected with HCV. There is considerable need for regulated production of siRNAs activated by a particular set of conditions (HCV in this case) but quiescent otherwise. Approaches described here may serve as a paradigm for such conditional siRNA expression.

Introduction

Cellular mechanisms of gene silencing by targeting RNAs exist in plants and animals, and the molecular machinery seems to be ancient and highly conserved (Bosher and Labouesse, 2000; Hannon, 2002). In *Drosophila*, larger double-stranded RNAs are processed to short 22-mers by an RNase III-like enzyme, dicer (Bernstein *et al.*, 2001). These small interfering RNAs (siRNAs) become unwound and associate with an activated RNA-induced silencing complex (Nykanen *et al.*, 2001). The single-stranded siRNA then guides substrate selection by this complex, leading to cleavage by dicer of a homologous target RNA molecule (Hammond *et al.*,

2000). The siRNA complex is not altered by the cleavage process: after one mRNA molecule is destroyed, the complex survives to target others.

Although initial studies of RNA interference (RNAi) focused on cellular mRNA targets such as transposons, evidence suggests that RNAi may also be used to target viral RNAs. Posttranscriptional gene silencing is well documented as an antiviral defense mechanism in plants (Al-Kaff et al., 1998; Dougherty et al., 1994; Ruiz et al., 1998), and RNAi has recently been shown to have antiviral functions in animal cells (Li et al., 2002). The importance of RNAi in protecting hosts from viruses is suggested by the finding of inhibitors of RNAi in several plant viruses (Brigneti et al., 1998; Li and Ding 2001; Voinnet et al., 2000) and in an animal virus, the flock house virus (Li et al., 2002; Lindenbach and Rice, 2002). It has also been shown that transfected siRNAs can efficiently eliminate replicating hepatitis C virus (HCV) RNAs from human hepatoma cells (Kapadia et al., 2003; Randall et al., 2003; Wilson et al., 2003).

However, recent reports that RNAi recognition sequences may allow mismatches in target binding (Hamada et al., 2002; Saxena et al., 2003) raise concerns as to the safety and specificity of these species in destroying only their intended target RNAs. Thus far there has been little information concerning potential side effects of RNAi. Among the mechanisms that may be used to increase specificity and safety of siRNA expression is the use of conditional promoters, which are responsive to a stimulus associated with the target RNA. Employing conditional RNAi transcription to focus its expression to target cells has not received much attention.

We present here a paradigm of conditional RNAi expression that may be useful for investigators targeting viral and other transcripts. The ability of the HIV-LTR to act in hepatocytes as a conditional promoter responsive to hepatitis B virus (HBV) is well documented (Gómez-Gonzalo et al., 2001; Haviv et al., 1995; Lin et al., 1997). By using reporter gene constructs, we have recently documented the effectiveness of the HIV-1 LTR in hepatocytes as a conditional promoter responsive to hepatitis C virus (Matskevich et al., in press). HIV-LTR activation by both HBV and HCV is largely mediated through its NFκB binding sites (Lin et al., 2003; Matskevich, et al., in press). Specifically, the NS5a HCV protein activates NFκB, which in turn activates the promoter function of HIV-LTR (Gong et al., 2001; Matskevich et al., 2003). We exploit these findings to illustrate the potential applicability of such conditional expression approaches to drive transcription of inhibitory RNAs. The experimental target of the siRNA in this illustration is HCV mRNA. The siRNA used was delivered with highly efficient recombinant Tag-deleted SV40-derived vectors (Sun et al., in press). We propose that this approach may represent a paradigm for designing conditionally expressed RNAi constructs targeting viral RNAs.

Materials and Methods

Cell Lines

The handling of all cell lines has been described in detail (Lin *et al.*, 1997; Matskevich *et al.*, in press; Sun *et al.*, in press) and is not repeated here: the human hepatoblastoma cell line, HepG2; HepG2 cells stably expressing HCV full genome (HepG2-HCV; Sun *et al.*, in press); and COS-7 cells, used to package rSV40 vectors.

Plasmids and rSV40s

rSV40 vectors are made by cloning the siRNA-coding cDNAs (with or without specific promoter) into a plasmid pT7ΔΔ.1pa (Fig. 1A). This plasmid carries a SV40 genome in which the *Tag* gene is replaced by a polylinker. Transcription from the SV40 early promoter (which overlaps the ori, and so cannot be deleted) was blocked by multiple tandem polyadenylation signals. Two HIV-1 LTR constructs were cloned into this vector: one was wild-type HIV-1$_{NL4-3}$ LTR (wtHIV-LTR; Jayan *et al.*, 2001) and the other was the same LTR doubly mutated to delete its two NFκB binding sites (muHIV-LTR). A generic map of this rSV40 genome is shown in Fig. 1B. An siRNA cDNA specific for the C region of HCV genome was cloned into these plasmids, so that their expression would be driven by either wtHIV-LTR or muHIV-LTR. All structures were verified by automated DNA sequencing (PE Applied Biosystems, Inc., Foster City, CA; Kimmel Cancer Center, Thomas Jefferson University, Philadelphia, PA).

Construction of recombinant SV40 derivative viruses (rSV40) for gene transfer has been described previously (Strayer *et al.*, 2001). In brief, rSV40 genomes were excised from the modified pT7blue (Novagen, Madison, WI) carrier plasmid, gel purified, recircularized, and transfected into COS-7 cells. These cells supply all packaging functions *in trans*. Replication-incompetent vectors are isolated from COS-7 cell lysates, purified by ultracentrifugation, and titered by *in situ* PCR as described. Typical infectious titers for vectors prepared in this manner are between 10^{11} and 10^{12} infectious units (IU)/ml. rSV40 vectors were named according to the promoter (in brackets, [ ], wt or mutant HIV-LTR) with the transgene following in parentheses. Thus, SV[wtHIV-LTR](siHCV) carries wt HIV-LTR + siRNA-coding cDNA against HCV, the vector carrying mutant HIV-LTR + siRNA is SV[muHIV-LTR](siHCV) (Fig. 1). SVsiLAM carries a cDNA, encoding an siRNA against lamin, driven by the constitutive adenoviral pol III promoter VA1 (Cordelier *et al.*, 2003). A cDNA form of HCV, strain 1b, complete virus genome was used as the plasmid, pRC/CMV-HCV (Sun *et al.*, in press).

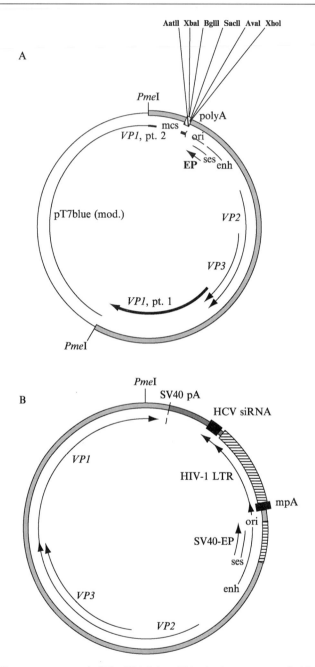

FIG. 1. Key constructs used. (A) pT7ΔΔ.1pa. This cloning target, carried in a modified pT7blue (Novagen) plasmid, contains a wtSV40 genome from which the *Tag* and *tag* genes have been replaced with a polylinker. Transcription from the SV40 early promoter (EP) is

Detection of HCV RNA by Nested RT-PCR

Primers from the NS5A coding region of the HCV genome were selected for RT-PCR (Watanabe *et al.*, 2001):

5'-TGGATGGAGTGCGGTTGCACAGGTA as the outer sense strand primer

5'-TCTTTCTCCGTGGAGGTGGTATTGC as the outer antisense strand primer

5'-CAGGTACGCTCCGGCGTGCA as the inner sense strand primer

5'-GGGGCCTTGGTAGGTGGCAA as the inner antisense primer

The expected PCR products are 617 bp for the outer set and 571 bp for the inner set.

The reverse transcription (RT) reaction was performed on 35 ng of extracted DNase I-treated RNA, using the RNeasy Mini Kit (Qiagen, Valencia, CA). The reaction mixture contained RT buffer (single-step RT-PCR kit, Invitrogen, Carlsbad, CA; 50 mM Tris-HCl, 75 mM KCl, 1.5 mM MgCl$_2$) containing 10 mM dithiothreitol, 200 U of SuperScript™ II reverse transcriptase (Invitrogen), 40 U of RNase inhibitor (RNAsin, Pharmacia, Piscataway, NJ), 300 μM each dGTP, dATP, dCTP, 5 U of Platinum™ Taq polymerase (Invitrogen) and 0.2 μM of each primer. The extracted RNA (5 μl) was added to the mix. The first round of PCR was done by using the outer set of primers for the first round (Matskevich *et al.*, 2003). The first round began with incubation for 30 min at 58°; followed by 15 cycles of 1 min at 94°, 1 min at 60°, and 1 min at 72°; and finally 7 min at 72°. The first-round product (5 μl) was added to 45 μl of PCR mix

---

blocked by polyadenylation signals (pA). SV40 encapsidation sequences (ses), enhancer (enh), and ori are intact, as are the SV40 capsid genes, *VP1*, *VP2*, and *VP3*. Thus, siRNA-encoding DNAs, together with the promoter of choice, can be cloned into the polylinker (mcs), whereupon this rSV40-containing plasmid could be used to express siRNAs driven by an array of conditional promoters of choice. To make virus from derivatives of this plasmid, the rSV40 genome is excised from the pT7blue backbone with *Pme*I (sites indicated), recircularized, and transfected into COS-7 cells, as described in Strayer *et al.* (2001) and in materials and methods. (B) SV[HIV-LTR](siHCV). The viral genome map for SV[HIV-LTR](siHCV), which was derived from the above cloning target by cloning the HIV-1$_{NL4-3}$ long terminal repeat (LTR), plus the siHCV RNAi, and then excising the virus genome from the carrier plasmid and recircularizing it. Either the wt LTR or a LTR that was mutated to destroy the NFκB binding sites was used. Construction and cloning of the siRNA used here are described in materials and methods. The structure of SV(siLAM) is similar, except that the siRNA in question targets lamin and that the adenoviral VA promoter was used instead of the HIV-LTR.

containing 1× buffer, 300 μ*M* each dNTP, 2.5 U of Platinum Taq polymerase, and 0.2 μ*M* of the inner set of primers. Thermocycling was performed for 25 cycles, each consisting of 1 min at 94°, 1 min at 60°, and 1 min at 72°.

Detection of HCV RNA by In Situ RT-PCR

HepG2 cells were cultured on chamber slides, washed in sterile saline, fixed in ice-cold 10% buffered formaldehyde solution and kept for 2 h at 37° and treated with proteinase K at 37° for 10 min, washed again, and treated with 40 U of DNase (Invitrogen) for 4 h at 37°. Control cells were then treated with 40 U of RNase. The permeabilized HepG2 cells were then washed in PBS and treated with RT buffer (single-step RT-PCR kit, Invitrogen; 50 m*M* Tris-HCl, 75 m*M* KCl, 3 m*M* MgCl$_2$) containing 10 m*M* dithiothreitol, 200 U of SuperScript II reverse transcriptase (Invitrogen), 40 U of RNase inhibitor (RNAsin, Pharmacia), 0.2 μ*M* each of dGTP, dATP, and dCTP, and 0.15 m*M* dTTP and 50 n*M* of dUTP-FITC; Platinum Taq polymerase (IU/100 ng template) and 50 pmol of the outer oligonucleotide primers of HCV. The tubes were incubated at 58° for 45 min and 94° for 5 min to inactivate residual RT activity. The RT step was followed by first-round PCR: 35 cycles of 1 min each at 94°, 58°, and 72°, with the final extension at 72° for 10 min. At the end of RT-PCR, slides were incubated in PBS. Slides were washed in 0.1× SSC at 37° for 15 min and then analyzed under an inverted epifluorescence microscope (Olympus).

Transduction and Transfection Experiments

For transduction with SV40-derived virus, HepG2 or HepG2-HCV cells were treated once at an MOI of 100, as described. Three days later, HepG2 cells were transfected with HCV-cDNA-containing plasmid (Lipofectamine™ 2000, Invitrogen), according to manufacturer's instructions, and then cultured for 2 days.

Northern Blotting Analysis

Total RNA from 10^6 HepG2 cells SV[HIV-LTR]siHCV transduced, mock transduced, or SVsiLAM transduced and subsequently transfected with HCV (if mentioned) was extracted by the RNeasy Mini Kit (Qiagen). Samples of 15 μg of total RNA were electrophoresed on a 1% agarose/formaldehyde gel, transferred on a nylon filter (Nytran Super Charge, Schleicher & Schuell, Germany), UV cross-linked with Stratalinker® oven

(Stratagene, La Jolla, CA), and baked for 2 h at 80° in a vacuum oven. Filters were prehybridized in 50% formamide, 5× SSPE, 20 mg/ml denatured salmon sperm DNA, 5× Denhardt's solution, 0.1% SDS at 42° for 8 h. Filters were subsequently hybridized under the same conditions with a region NS5A HCV cDNA probe that had been labeled with α32P-dCTP by using a random priming labeling kit (Gibco BRL, Gaithersburg, MD). Hybridization was performed at 42° in 2× SSC overnight. After hybridization, filters were washed under high stringency conditions in 0.1× SSC 0.1% SDS at 37° for 30 min and signals were visualized by a Strom 840 PhosphoImager (Molecular Dynamics). To assess loading of various lanes, the same filters were stripped and reprobed with a radiolabeled cDNA for human β-actin.

RNA Interference (RNAi) Design

siRNAs were designed as follows, and for each the sense-strand sequence is described (a complementary oligonucleotide synthesized for each): lamin A/C siRNA (siLAM), 5'-aacuggacuuccagaagaacaTT; HCV siRNA (siHCV), 5'-aacctcaaagaaaaaccaaacTT. Chemically synthesized cDNAs coding RNA oligonucleotides were annealed and cloned into pT7[HIVLTR] plasmids. The HCV-specific siRNA described previously was shown to be effective at silencing HCV expression in stably transfected cells (Sun *et al.*, in press).

Primer 1:

Cap – *Sac*II(sense orientation) – siDNA(sense) – TTT–

siDNA(antisense) – *Aat*II (antisense orientation) – Cap

Primer 2:

Cap – *Sac*II (antisense orientation) – siDNA (sense)–

TTT – siDNA (antisense) – *Aat*II (antisense orientation) – Cap

Strategy Used

Two primers coding siRNA [two tandem reverse-oriented (sense and antisense) siDNA separated by a small loop sequence] flanked by *Sac*II and *Aat*II restriction sites and protective caps were self-annealed at 95° for 5 min. Resulting double-stranded DNAs were cut with *Sac*II and *Aat*II and gel purified by using the QuiaEXII gel extraction kit (fragment about 50 bp) and then cloned into pT7[HIVLTR] that had been opened with *Sac*II and *Aat*II.

Effectiveness of the Techniques

Vector Delivery of HCVRNAi Decreases HCV RNA Species Following Transient Transfection of HCV cDNA

Because HCV is very difficult to grow *in vitro*, we tested the effectiveness of the delivered anti-HCV siRNA and the control constructs by transient transfection of full-length genomic HCV cDNA in HepG2 cell culture. In addition, we used an established HepG2 cell line that had been stably transfected with HCV and that expressed HCV transcripts continuously (Sun *et al.*, in press).

We tested the potential of such an SV40-derived viral vector to silence HCV RNAs in HepG2 cells. Cells were transduced with SV[HIVLTR]-siHCV or, as controls, SVsiLAM (which carries an siRNA against lamin) and SV[muHIVLTR](siHCV), which carries the same siRNA but under the control of an HIVLTR from which the NFκB binding sites were deleted. Three days after transduction, cells were transfected with full-length genomic HCV cDNA. SV[wtHIVLTR]siHCV greatly reduced HCV RNA (Fig. 2) compared with SVsiLAM and SV[muHIVLTR]siHCV.

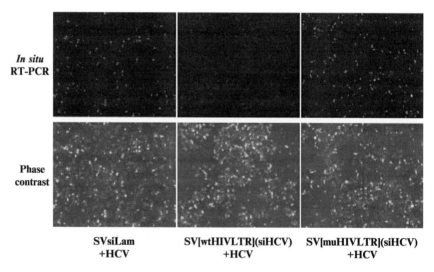

FIG. 2. Hepatitis C virus (HCV)-specific silencing by siRNA delivered by rSV40 vector and driven by HIVLTR. HepG2 cells were transduced with SV[wtHIVLTR]siHCV, SV[mu-HIVLTR](siHCV), or SVsiLam, and 3 days later transfected with HCV full genomic cDNA. Two days thereafter, the cells were washed and fixed and their HCV RNA was visualized by *in situ* RT-PCR; that is, following DNase treatment, the HCV RNA was reverse transcribed *in situ* with HCV-NS5A-specific primers, then amplified *in situ* by PCR by using fluoresceinated nucleotides, as described in materials and methods.

SV[wtHIVLTR](siHCV) Decreases HCV RNA as Measured by RT-PCR

The effectiveness of rSV40-delivered, conditionally expressed anti-HCV siRNA was further tested by semiquantitative RT-PCR. HepG2 cells were transduced with SV[wtHIVLTR](siHCV), SV[muHIVLTR](siHCV), or SVsiLam and then transfected with HCV 2 days later, as described previously. Negative-control HepG2 cultures were mock transfected and mock transduced. Six days after transfection, equal amounts of RNA (35 ng) from each group were subjected to nested RT-PCR as described in materials and methods. PCR products were visualized under UV light following agarose gel electrophoresis and ethidium bromide staining. Transduction with SV[wtHIVLTR](siHCV) greatly reduced levels of the targeted HCV transcript as compared with control-transduced cells (Fig. 3).

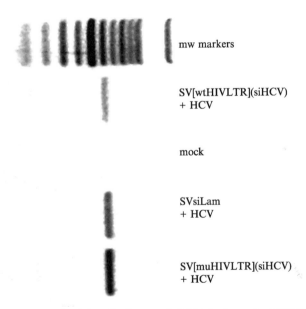

mw markers

SV[wtHIVLTR](siHCV)
+ HCV

mock

SVsiLam
+ HCV

SV[muHIVLTR](siHCV)
+ HCV

Fig. 3. RT-PCR analysis of the effectiveness of siRNA in decreasing HCV RNA. HepG2 cells were transduced with SV[wtHIVLTR](siHCV), SV[muHIVLTR](siHCV), or SVsiLam. Two days later, they were transfected with HCV. One control group (mock) was both mock transduced and mock transfected. Six days after transfection, RNA was harvested. A total of 35 ng of whole-cell RNA was subjected to nested RT-PCR as described in materials and methods. PCR products were visualized by electrophoresis on ethidium bromide-containing agarose gels. An RNA molecular size ladder is shown at the left. The image is inverted for clarity.

HCV RNAi Decreases HCV RNA in Cells Stably Transfected with HCV cDNA

To further assess the potential of this approach to using conditionally expressed RNAi to target viral RNAs, we tested whether SV[HIVLTR] (siHCV) could alter HCV RNA levels in HepG2 cells that were stably transfected with the HCV genome (HepG2-HCV). Thus, HepG2-HCV cells were transduced with the several SVsiRNA constructs. Expression of HCV was assayed 3 days later by *in situ* RT-PCR (Fig. 4). In cells transduced with SV[wtHIVLTR]siHCV, expression of HCV RNAs was reduced to virtually undetectable levels in >98% of cells. Specifically, in cultures treated with SV[wtHIVLTR]siHCV, HCV transcript was detected in <2% of cells. The remaining fluorescence, indicating detectable HCV RNA, in a small percentage of cells is consistent with the ≥95% transduction efficiency of SV40 in unselected cells (Matskevich *et al.*, in press). SV[muHIVLTR]siHCV and SVsiLAM had no discernable effect on HCV transcript levels. Thus, HCV-specific siRNA delivered by rSV40 vectors and driven by the HIVLTR as a conditional promoter greatly decreases levels of HCV RNA in >98% of hepatocytes.

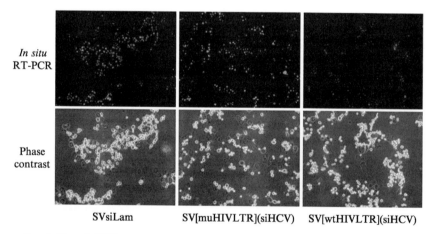

In situ
RT-PCR

Phase
contrast

SVsiLam   SV[muHIVLTR](siHCV)   SV[wtHIVLTR](siHCV)

Fɪɢ. 4. Use of rSV40 vectors carrying siRNA to target HCV RNA in HepG2 cells that carry a replicating HCV genome. HepG2-HCV cells were transduced with SV[wtHIVLTR]-siHCV, SV[muHIVLTR]siHCV, or SVsiLam and maintained for 3 days. These cells were then washed, fixed in 2% paraformaldehyde, and subjected to *in situ* RT-PCR , as described in materials and methods. Slides were mounted and HCV RNAs visualized by fluorescence microscopy.

Conclusions

These data illustrate the use of siRNAs, delivered by rSV40 vectors and expressed using the HIVLTR as a promoter, to inhibit a viral RNA. Cells harboring actively replicating HCV were sensitive to siRNA targeting of viral (HCV) transcripts. It has been reported that both HIV and poliovirus are sensitive to RNAi in cell culture (Coburn and Cullen, 2002; Gitlin et al., 2002; Jacque et al., 2002; Lee et al., 2002; Novina et al., 2002). These findings, together with our data, lend credence to the potential utility of siRNAs as general antiviral agents.

Recently, several approaches have been described for generating loss-of-function phenotypes in mammalian systems by using RNAi (Brummelkamp et al., 2002; McCaffrey et al., 2002; Mercer et al., 2001). However, these approaches have limited applications and are not especially applicable to long-term silencing in vivo, such as would be necessary to target HCV in people who are chronically infected. To address these issues, long-term RNAi expression in the liver is necessary. In this context, recombinant SV40-derived vectors appear to have the desired characteristics. These vectors are produced at very high titers, sufficient to transduce an organ the size of a human liver. They efficiently transduce both hepatocyte cell lines, as shown here, and primary hepatocytes (Lin et al., 1997; Matakevich et al., in press; Sauter et al., 2000). Gene delivery to both resting and dividing cells is equally efficient. Most importantly, rSV40s elicit no detectable neutralizing immune response. They can thus be readministered if repeat dosing is needed, and they can be used in sequence to deliver different transgenes, that is, combination genetic therapy (Kondo et al., 1998; McKee and Strayer, 2002).

Several groups have described the use of small nuclear RNA promoters (H1 and U6) to express siRNAs in mammalian cells (Brummelkamp et al., 2002; Sui et al., 2002; Yu et al., 2002). We have found that constitutive pol III promoters, such as adenovirus VA1 promoter, can also be effective in driving siRNAs in rSV40 vectors (Cordelier et al., 2003). We describe here a paradigm for virus-responsive conditional expression of siRNA, in which the HIV-1 long terminal repeat (HIVLTR) is used as a promoter for siHCV. We previously reported that HCV-induced activation of NFκB (Gong et al., 2001) stimulates HIVLTR promoter activity, and that this effect can be exploited to deliver transgene expression that responds to the presence of HCV (Matskevich et al., 2003). Therefore, regulated expression of siRNAs in rSV40 vectors could be useful for reducing any potential toxicities that may affect constitutive siRNA expression.

The fact that the HIVLTR is activated by HCV, through NFκB, makes it a potentially useful conditional promoter. The HIV-1NL4-3 LTR is not specific for HCV in its current form, however; that is, other stimuli, such as HBV (Gómez-Gonzalo *et al.*, 2001; Haviv *et al.*, 1995; Lin *et al.*, 1997) and HIV, may activate the HIVLTR. Additional work is needed to increase the HCV specificity of this promoter.

The specific silencing of a particular transcript by RNAi without induction of nonspecific IFN responses is attractive from a therapeutic standpoint. However, the challenge of silencing distinct HCV genotypes highlights the possibility of reduced efficacy against other genotypes or of selecting for viral escape mutants. It has been shown that target sequences that differed by three nucleotides from the siRNA could not be silenced. In another study, poliovirus escape mutants rapidly emerge after siRNA treatment (Gitlin *et al.*, 2002).

The HCV sequences targeted in this study are in the coding region and are not completely conserved in different HCV genotypes. Given the demonstration of principle presented here, a next step could be to identify accessible and highly conserved HCV nucleotide sequences as potential targets. Use of multiple siRNAs targeting evolutionarily conserved HCV RNA sequences, driven by a conditional promoter responsive to the presence of HCV, may both limit the emergence of escape mutants and minimize potential risks associated with constitutive siRNA expression. We have devised rSV40s capable of delivering two or more siRNAs simultaneously and documented improved effectiveness of such constructs compared with those delivering one moiety alone (P. Cordelier and coworkers, unpublished).

The specific HCV challenge systems used here are imperfect, because the virus replicates and infects cultured cells poorly, if at all. Nonetheless, transfected whole-genomic HCV cDNA is transcribed and therefore presents both a reasonable marker for virus activity and a good therapeutic target. SV[HIVLTR](siHCV) was a powerful inhibitor of HCV by this assay. Inhibition of HCV RNAs was demonstrated by *in situ* RT-PCR, which because it involves considerable amplification allows visualization of the number of cells with HCV RNA with great sensitivity. These data were further supported by standard RT-PCR. Although the latter technique is semiquantitative, it demonstrated very large differences in HCV RNA between SV[wtHIVLTR](siHCV)-transduced cells and control-transduced cells.

The combination of SV40 vectors with conditionally expressed siRNAs to target viral transcripts shows great promise, but the technology requires further development and study. Such an approach may represent an attractive approach to therapy for severe chronic HCV infection that could avoid the side effects of systemic treatment regimens.

Acknowledgments

The authors acknowledge the advice of and provision of materials by Drs. Janet S. Butel, Henry Chu, J. Roy Chowdhury, Pierre Cordelier, Jan Hoek, Geetha Jayan, Aleem Siddiqui, and Jay Schneider. This work was supported by NIH grants AI48244 and RR13156.

References

Al-Kaff, N. S., Covey, S. N., Kreike, M. M., Page, A. M., Pinder, R., and Dale, P. J. (1998). Transcriptional and posttranscriptional plant gene silencing in response to a pathogen. *Science* **279,** 2113–2115.

Bernstein, E., Caudy, A. A., Hammond, S. M., and Hannon, G. J. (2001). Role for a bidentate ribonuclease in the initiation step of RNA interference. *Nature* **409,** 363–366.

Bosher, J. M., and Labouesse, M. (2000). RNA interference: Genetic wand and genetic watchdog. *Nat. Cell Biol.* **2,** E31–E36.

Brigneti, G., Voinnet, O., Li, W. X., Ji, L. H., Ding, S. W., and Baulcombe, D. C. (1998). Viral pathogenicity determinants are suppressors of transgene silencing in *Nicotiana benthamiana. EMBO J.* **17,** 6739–6746.

Brummelkamp, T. R., Bernards, R., and Agami, R. (2002). A system for stable expression of short interfering RNAs in mammalian cells. *Science* **296,** 550–553.

Coburn, G. A., and Cullen, B. R. (2002). Potent and specific inhibition of human immunodeficiency virus type 1 replication by RNA interference. *J. Virol.* **76,** 9225–9231.

Cordelier, P., Morse, B., and Strayer, D. S. (2003). Targeting CCR5 with siRNAs: Using recombinant SV40-derived vectors to protect macrophages and microglia from R5-tropic HIV. *Oligonucleotides* **13,** 281–294.

Dougherty, W. G., Lindbo, J. A., Smith, H. A., Parks, T. D., Swaney, S., and Proebsting, W. M. (1994). RNA-mediated virus resistance in transgenic plants: Exploitation of a cellular pathway possibly involved in RNA degradation. *Mol. Plant Microbe Interact.* **7,** 544–552.

Gitlin, L., Karelsky, S., and Andino, R. (2002). Short interfering RNA confers intracellular antiviral immunity in human cells. *Nature* **418,** 430–434.

Gómez-Gonzalo, M., Carretero, M., Rullas, J., Lara-Pezzi, E., Aramburu, J., Berkhout, B., Alcamí, J., and López-Cabrera, M. (2001). The hepatitis B virus X protein induces HIV-1 replication and transcription in synergy with T-cell activation signals: Functional roles of NF-κB/NF-AT and SP1-binding sites in the HIV-1 long terminal repeat promoter. *J. Biol. Chem.* **276,** 35435–354343.

Gong, G., Waris, G., Tanveer, R., and Siddiqui, A. (2001). Human hepatitis C virus NS5A protein alters intracellular calcium levels, induces oxidative stress, and activates STAT-3 and NFκB. *Proc. Natl. Acad. Sci. USA* **98,** 9599–9604.

Hamada, M., Ohtsuka, T., Kawaida, R., Koizumi, M., Morita, K., Furukawa, H., Imanishi, T., Miyagishi, M., and Taira, K. (2002). Effects on RNA interference in gene expression (RNAi) in cultured mammalian cells of mismatches and the introduction of chemical modifications at the 3′-ends of siRNAs. *Antisense Nucleic Acid Drug Dev.* **12,** 301–309.

Hammond, S. M., Bernstein, E., Beach, D., and Hannon, G. J. (2000). An RNA-directed nuclease mediates post-transcriptional gene silencing in *Drosophila* cells. *Nature* **404,** 293–296.

Hannon, G. J. (2002). RNA interference. *Nature* **418,** 244–251.

Haviv, I., Vaizel, D., and Shaul, Y. (1995). The X protein of hepatitis B virus coactivates potent activation domains. *Mol. Cell. Biol.* **15,** 1079–1085.

Jacque, J. M., Triques, K., and Stevenson, M. (2002). Modulation of HIV-1 replication by RNA interference. *Nature* **418**, 435–438.

Jayan, G. C., Cordelier, P., Patel, C., BouHamdan, M., Johnson, R. P., Lisziewicz, J., Pomerantz, R. J., and Strayer, D. S. (2001). SV40-derived vectors provide effective transgene expression and inhibition of HIV-1 using constitutive, conditional, and pol III promoters. *Gene Ther.* **8**, 1033–1042.

Kapadia, S. B., Brideau-Anderson, A., and Chisari, F. V. (2003). Interference with hepatitis C virus RNA replication by short interfering RNAs. *Proc. Natl. Acad. Sci. USA* **100**, 2014–2018.

Kondo, R., Feitelson, M. A., and Strayer, D. S. (1998). Use of SV40 to immunize against hepatitis B surface antigen: Implications for the use of SV40 for gene transduction and its use as an immunizing agent. *Gene Ther.* **5**, 575–582.

Lee, N. S., Dohjima, T., Bauer, G., Li, H., Li, M. J., Ehsani, A., Salvaterra, P., and Rossi, J. (2002). Expression of small interfering RNAs targeted against HIV-1 rev transcripts in human cells. *Nat. Biotechnol.* **20**, 500–505.

Li, H., Li, W. X., and Ding, S. W. (2002). Induction and suppression of RNA silencing by an animal virus. *Science* **296**, 1319–1321.

Li, W. X., and Ding, S. W. (2001). Viral suppressors of RNA silencing. *Curr. Opin. Biotechnol.* **12**, 150–154.

Lin, Y., Tang, H., Nomura, T., Dorjsuren, D., Hayashi, N., Wei, W., Ohta, T., Roeder, R., and Murakami, S. (1997). Hepatitis B virus X protein is a transcriptional modulator that communicates with transcription factor IIB and the RNA polymerase II subunit 5. *J. Biol. Chem.* **272**, 7132–7139.

Lindenbach, B. D., and Rice, C. M. (2002). RNAi targeting an animal virus: News from the front. *Mol. Cell* **9**, 925–927.

Matskevich, A., Cordelier, P., Strayer, D. S. (in press). Conditional expression of interferons α and activated by HBV as genetic therapy for hepatitis B. *J. Interferon Cytokine Res.*

Matskevich, A. A., and Strayer, D. S. (2003). Exploiting hepatitis C virus activation of NFκB to deliver HCV-responsive expression of interferons α and γ. *Gene Ther.* **10**, 1861–1873.

McCaffrey, A. P., Meuse, L., Pham, T. T., Conklin, D. S., Hannon, G. J., and Kay, M. A. (2002). Determinants of hepatitis C translational initiation *in vitro*, in cultured cells and mice. *Nature* **418**, 38–39.

McKee, H. J., and Strayer, D. S. (2002). Immune response against SIV envelope glycoprotein, using recombinant SV40 as a vaccine delivery vector. *Vaccine* **20**, 3613–3625.

Mercer, D. F., Schiller, D. E., Elliott, J. F., Douglas, D. N., Hao, C., Rinfret, A., Addison, W. R., Fischer, K. P., Churchill, T. A., Lakey, J. R. T., Tyrell, D. L. J., and Kneteman, N. M. (2001). Hepatitis C virus replication in mice with chimeric human livers. *Nat. Med.* **7**, 927–933.

Novina, C. D., Murray, M. F., Dykxhoorn, D. M., Beresford, P. J., Riess, J., Lee, S. K., Collman, R. G., Lieberman, J., Shankar, P., and Sharp, P. A. (2002). siRNA-directed inhibition of HIV-1 infection. *Nat. Med.* **8**, 681–686.

Nykanen, A., Haley, B., and Zamore, P. D. (2001). ATP requirements and small interfering RNA structure in the RNA interference pathway. *Cell* **107**, 309–321.

Randall, G., Grakoui, A., and Rice, C. M. (2003). Clearance of replicating hepatitis C virus replicon RNAs in cell culture by small interfering RNAs. *Proc. Natl. Acad. Sci. USA* **100**, 235–240.

Ruiz, M. T., Voinnet, O., and Baulcombe, D. C. (1998). Initiation and maintenance of virus-induced gene silencing. *Plant Cell* **10**, 937–946.

Sauter, B. V., Parashar, B., Chowdhury, N. R., Kadakol, A., Ilan, Y., Singh, H., Milano, J., Strayer, D. S., and Chowdhury, J. R. (2000). Gene transfer to the liver using a replication-deficient recombinant SV40 vector results in long-term amelioration of jaundice in Gunn rats. *Gastroenterology* **119,** 1348–1357.

Saxena, S., Jonsson, Z. O., and Dutta, A. (2003). Small RNAs with imperfect match to endogenous mRNA repress translation. Implications for off-target activity of small inhibitory RNA in mammalian cells. *J. Biol. Chem.* **278,** 44312–44319.

Strayer, D. S., Lamothe, M., Wei, D., Milano, J., and Kondo, R. (2001). Generation of recombinant SV40 vectors for gene transfer. *In* "Methods in Molecular Biology, Vol. 165: SV40 Protocols" (L. Raptis, ed.). Humana Press, Totowa, NJ.

Strayer, D. S., Zern, M. A., and Chowdhury, J. R. (2002). What can SV40-derived vectors do for gene therapy? *Curr. Opin. Mol. Ther.* **4,** 313–323.

Sui, G., Soohoo, C., Affar el, B., Gay, F., Shi, Y., and Forrester, W. C. (2002). A DNA vector-based RNAi technology to suppress gene expression in mammalian cells. *Proc. Natl. Acad. Sci.* **99,** 5515–5520.

Sun, B. S., Pan, J., Clayton, M. M., Liu, J., Yan, X., Matskevich, A. A., Strayer, D. S., Gerber, M., Feitelson, M. A. (in press). Hepatitis C virus replication in stably transfected HepG2 cells promotes hepatocellular growth and tumorigenesis. *J. Cell Physiol.*

Voinnet, O., Lederer, C., and Baulcombe, D. C. (2000). A viral movement protein prevents spread of the gene silencing signal in *Nicotiana benthamiana*. *Cell* **103,** 157–167.

Watanabe, H., Enomoto, N., Nagayama, K., Izumi, N., Marumo, F., Sato, C., and Watanabe, M. (2001). Number and position of mutations in the interferon (IFN) sensitivity-determining region of the gene for nonstructural protein 5A correlate with IFN efficacy in hepatitis C virus genotype 1b infection. *J. Infect. Dis.* **183,** 1195–1203.

Wilson, J. A., Jayasena, S. S., Khvorova, A., Sabatinos, S., Rodrigue-Gervais, I. G., Arya, S., Sarangi, F., Harris-Brandts, M., Beaulieu, S., and Richardson, C. D. (2003). RNA interference blocks gene expression from hepatitis C replicons propagated in human liver cells. *Proc. Natl. Acad. Sci. USA* **100,** 2783–2788.

Yu, J. Y., DeRuiter, S. L., and Turner, D. L. (2002). RNA interference by expression of short-interfering RNAs and hairpin RNAs in mammalian cells. *Proc. Natl. Acad. Sci. USA* **99,** 6047–6052.

[15] High-Throughput RNA Interference Strategies for
Target Discovery and Validation by Using
Synthetic Short Interfering Rnas:
Functional Genomics Investigations of
Biological Pathways

By Christoph Sachse, Eberhard Krausz, Andrea Krönke, Michael
Hannus, Andrew Walsh, Anne Grabner, Dmitriy Ovcharenko,
David Dorris, Claude Trudel, Birte Sönnichsen, and
Christophe J. Echeverri

Abstract

During the past five years, RNA interference (RNAi) has emerged as
arguably the best functional genomics tool available to date, providing
direct, causal links between individual genes and loss-of-function phenotypes through robust, broadly applicable, and readily upscalable methodologies. Originally applied experimentally in *C. elegans* and *Drosophila*,
RNAi is now widely used in mammalian cell systems also. The development
of commercially available libraries of short interfering RNAs (siRNAs) and
other RNAi silencing reagents targeting entire classes of human genes
provide the opportunity to carry out genome-scale screens to discover and
characterize gene functions directly in human cells. A key challenge of these
studies, also faced by earlier genomics or proteomics approaches, resides in
reaching an optimal balance between the necessarily high throughput and
the desire to achieve the same level of detailed analysis that is routine in
conventional small-scale studies. This chapter discusses technical aspects
of how to perform such screens, what parameters to monitor, and which
readouts to apply. Examples of homogenous assays and multiplexed
high-content microscopy-based screens are demonstrated.

Introduction: The Advent of RNA Interference
(RNAi)-Based Genomics

The completion of major genome sequencing projects, from those of
C. elegans, *Drosophila melanogaster*, mouse, and rat to the human genome
project itself, have marked major milestones in recent biomedical research.
The enormous amounts of data that have emerged from these projects now
offer huge potential for significantly advancing both basic academic research and pharmaceutical drug development. Thus, there has been the

0076-6879/05 $35.00

advent of so-called functional genomics technologies to efficiently exploit those mountains of genome sequence data, extracting new insights into the functional relevance of individual genes.

These technologies have most notably included microarray-based expression profiling, high-throughput (HT) bioinformatic tools for *in silico* data mining, and large-scale two-hybrid screens for generating protein–protein interaction maps. Although these and other first-generation functional genomics technologies have accelerated our understanding of a broad range of biological phenomena, their inherent limitations are now becoming evident. In particular, these approaches typically yield indirect evidence, at best, of a gene product's function or possible interactions with other cellular components. For instance, correlations between gene expression levels and a given disease state, as generated by comparative transcriptional profiling of diseased versus normal tissue biopsies, cannot distinguish between therapeutically relevant causes and irrelevant consequences of the diseased state. Based on this, one can argue that this approach is better adapted to identify new diagnostic biomarkers than new therapeutic drug targets. Evidence from purely *in vitro* or *in silico* analyses also tend to offer questionable pathophysiological significance and high rates of false positives, making the resulting data difficult to prioritize for eventual follow-up, especially when its sheer volume is already far beyond the scale usually faced by traditional research laboratories.

RNA-mediated interference (RNAi) has emerged over the past few years as perhaps the best way of overcoming many of these obstacles, making it arguably the most powerful second-generation functional genomics technology available to date (Carpenter and Sabatini, 2004). Its ability to induce the destruction of individual mRNAs in a targeted way with high efficacy and specificity enables the generation of direct relationships between a gene's expression level and its functional role in any biological process being studied. Furthermore, RNAi-based methodologies in several key experimental systems from *C. elegans* to cultured human cells have proven robust enough to be applied in a high-throughput manner. Combined with the availability of full genome sequences and associated gene structure predictions, these advances have enabled the systematic, genome-scale application of this technology, thus heralding the advent of RNAi-based genomics.

The development of large-scale RNAi screening, though clearly exciting, has not been without its own challenges and, like all other experimental techniques, its own limitations and caveats have emerged steadily. The feasibility of applying RNAi as a systematic genome-scale screening method was first demonstrated rather early on in *C. elegans* (Fraser *et al.*, 2000;

Gönczy *et al.*, 2000; Kamath *et al.*, 2003), although the delivery methods and readout assays chosen in these studies only allowed for low to medium throughputs. Significantly higher throughputs were only achieved once these methods were successfully implemented in cultured cell systems, in which more highly parallelized experimentation is possible, starting with *Drosophila* cells (Clemens *et al.*, 2000; Kiger *et al.*, 2003; Lum *et al.*, 2003). Although at that time many were already dreaming of high-throughput genomewide screens in cultured human cells, the experimental applicability of RNAi in vertebrate systems remained in doubt due to the uncertainty of how to trigger RNAi without also triggering interferon response in those organisms. Thankfully, the first broadly applicable solution to this problem was demonstrated in 2001 by Tuschl and colleagues (Elbashir *et al.*, 2001), whose success came from the use of chemically synthesized double-stranded RNA (dsRNA) molecules designed to mimic the size and structure of so-called short interfering RNAs (siRNAs), which had been identified in plants as apparent intermediates in the RNAi pathway (Hamilton and Baulcombe, 1999). The result of this breakthrough, not only for the RNAi field itself but also more broadly for an ever-growing range of biomedical research overall, has been nothing short of a revolution. Although most researchers began to experience success with small-scale applications of RNAi focusing on handfuls of their favorite genes, those aiming for genome-scale applications quickly began to recognize new choices and challenges inherent to this pursuit, beyond those already faced with *C. elegans* or *Drosophila*. Chief among these were the following:

1. *Building a library of silencing reagents:*
   - *Choice of source target sequences:* Using cDNA libraries or predicted gene sequences from genomic databases.
   - *Choice of reagent type:* Chemically synthesized siRNAs or vector-based constructs expressing short hairpin RNAs (shRNAs).
   - Automated selection of optimal siRNA/shRNA sequences to achieve genomewide coverage with reproducibly high silencing efficacy and specificity.
2. *Delivering the silencing reagents:* The efficient delivery of siRNAs into cultured cells, with minimal associated cytotoxicity or delivery-associated side effects.
3. Controlling the experiments adequately to ensure that not only is the RNAi treatment effective, but, most importantly, that the observed RNAi-induced phenotypes truly are target specific rather than reagent specific.
4. Designing a screening strategy to maximize cost efficiency while maintaining scientific integrity and depth of the experiment.

This chapter reviews our progression in exploring and addressing these issues, as we have developed our genomewide RNAi screening program in human cells, after having done so in *Drosophila* cells and *C. elegans*. Having made the strategic choice early on of focusing on the use of chemically synthesized siRNAs rather than exploring alternative approaches such as vector-based expression of short hairpin RNA constructs, the scope of this chapter will necessarily be restricted to the former, whereas the latter is covered in more depth in other chapters in this volume. We show basic concepts, necessities, and technical considerations. We also discuss the range of applications that are at present considered feasible and being pursued by our group and others not only in basic research but also in the more applied areas of discovery and development of new therapeutic drugs.

Building a Genome-Scale Short Interfering RNA (siRNA) Library

First Choices: Source Sequences and Reagent Types

A number of strategies either focusing on speed or comprehensive screening coverage have emerged for building genome-scale libraries of RNAi silencing reagents, involving choices for both the source sequences and the type of reagent to be used. We have opted to aim for maximally comprehensive genome coverage, believing strongly that this represents an important opportunity afforded by the combination of currently available resources, that is, fully sequenced genomes and a targeted method of silencing, which can and should be realized. To achieve this aim, we have had to focus on the use of chemically synthesized siRNAs as the best-characterized, best-performing gold standard reagents in the field to date, custom designed from available genome sequence data to target all predicted genes.

This approach has therefore required that we develop new bioinformatics tools (discussed later) to systematically apply to all known available selection criteria for effective siRNA designs in an automated manner over the entire genome sequence. Thus, the primary pitfall of this strategy is its dependence on the science of gene structure predictions, which remains in rapid evolution at present. As a result, many genes predictions, that is, the locations of intron–exon boundaries, are still constantly evolving, thus requiring the concomitant updating of RNAi libraries to keep up with the new annotations. However, the alternative—the use of cDNA libraries as source material—avoids this issue by ensuring that all silencing reagents necessarily target expressed gene sequences. However, this precludes the achievement of anything close to genomewide coverage and biases such libraries heavily toward only those genes that showed good expression and therefore good representation in the cDNA library. This bias can be used to significant advantage in those cases in which the scope of the screen is meant to be focused on a subset of genes that may be preferentially enriched in the said cDNA library.

Chemically synthesized siRNAs and vector-encoded shRNAs are at present the most broadly used reagents for all scales of RNAi experimentation in human cells. The use of viral vector-mediated expression of shRNAs (Arts *et al.*, 2003; Rubinson *et al.*, 2003) has the primary advantage of generating RNAi-based silencing in a more sustained manner, beyond the 5- to 7-day transient effect afforded by siRNAs. The viral delivery approach also facilitates studies in certain cell types that are otherwise difficult to transfect. Nonetheless, for our purposes, these benefits have so far been outweighed by the accompanying disadvantages, which include the need for much more specialized, dedicated laboratory infrastructure (especially if one aims to carry out this work at the genome scale), the high complexity and fallibility of required large-scale cloning strategies, and the much higher variability in silencing performance exhibited by shRNA than by siRNAs. Importantly, the inability to control the effective dose of shRNA generated inside the expressing cell also emerges as a significant shortcoming of the shRNA approach, especially in view of published reports that indicate an increased risk of off-target and nonspecific effects (including the activation of interferon response genes) when using excessive concentrations of RNAi silencing reagents (Bridge *et al.*, 2003; Persengiev *et al.*, 2004; Sledz *et al.*, 2003). Although these issues clearly do not preclude the overall use of vector-based shRNA libraries for RNAi screening [as has already been successfully demonstrated, e.g., Arts *et al.* (2003), Berns *et al.* (2004), and Paddison *et al.* (2004)], they limit their breadth of applicability, making them best adapted for groups who are unconcerned with the issue of being comprehensive and who accept that some phenotypes may be missed.

Automated Design of siRNAs

In model organisms such as *C. elegans* and *Drosophila melanogaster*, libraries of relatively long dsRNAs (usually ~500–1500 bp) have been successfully built here and elsewhere through *in vitro* transcription of RNA from DNA templates generated by PCR amplification of appropriate portions of purified chromosomal DNA. Customized software algorithms were developed to efficiently screen through available genomic sequence data and select target amplicons chosen primarily to maximize their content in exon sequences from a single targeted gene, to ensure their uniqueness during the PCR amplification process, and to otherwise avoid known problems that might hinder synthesis. These long dsRNAs proved to be highly specific and potent silencers, but the need to avoid the interferon response in mammalian cells required the development of new algorithms to select optimal 2-mer siRNA sequences.

siRNA design algorithms comprise a wide range of different criteria to maximize silencing efficacy, which in our case have included specific base compositions at defined positions along the 19 core siRNA base pairs, thermodynamic base-pairing profiles defining regional base compositions (GC content in particular), base composition of 3′ overhangs, positions along the targeted mRNA, and lack of variability of the targeted mRNA over the relevant site (avoiding known single nucleotide polymorphisms, etc.). These and other criteria were derived empirically from a large experimental dataset of siRNA-derived silencing efficacies, covering hundreds of genes. One of the most potent design features implemented in our algorithm, as well as algorithms of others, is the creation of a differential between base-pairing thermodynamics at either end of the siRNA, ensuring a significantly weaker pairing of the 5′ end of the anti-sense strand. As recently confirmed by Zamore and colleagues (Schwarz *et al.*, 2003), such asymmetry strongly favors the loading of the antisense strand over that of the sense strand into available RNA-induced silencing complexes (RISCs), thereby ensuring the recognition and subsequent destruction of the correct target mRNA.

Moreover, because multiple studies have reported the detection of complex sequence-dependent off-target effects at the mRNA level (Jackson *et al.*, 2003; Persengiev *et al.*, 2004), most siRNA design algorithms also include measures to maximize specificity by minimizing the risk of such off-target effects. Although this issue has led us and others to integrate design requirements that siRNAs should have minimal numbers of mismatches against any and all off-target gene sequences, the field's current understanding of these risks remains quite superficial, and most siRNAs commonly available at present are likely to have quite complex targeting footprints, at least when analyzed at the mRNA level. This is further discussed later, but it is important to point out here that this issue is readily neutralized by confirming the target-specific nature of any observed RNAi-induced phenotypes through the use of multiple, distinct siRNA designs against the same target gene.

Ultimately, no matter how good the *in silico* predictions look, the perfor-mance of the algorithm-designed siRNAs in silencing endogenously ex-pressed genes must be demonstrated experimentally in cultured human cells under strictly standardized conditions. Furthermore, it is our experience that such tests should integrate large enough numbers of genes so as to gen-erate a statistically relevant sample and that a suitably quantitative analysis

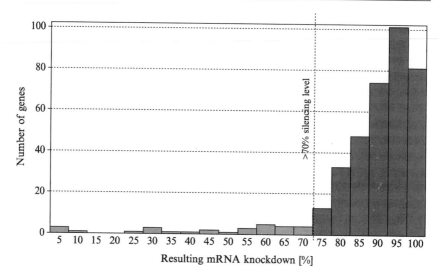

FIG. 1. Results of efficacy testing for short interfering RNAs (siRNAs) against 379 human genes. Among 379 genes (1106 siRNAs), more than 70% silencing was achieved for more than 93% of the genes (as measured by real-time RT-PCR at 48 h after transfection into HeLa cells), applying our proprietary siRNA design algorithm.

of target mRNA levels [e.g., real-time RT-PCR or quantitative (qRT-PCR)] should be applied. As an example, results of our qRT-PCR testing of silencing efficacies are shown in Fig. 1 for a set of more than 1100 siRNAs designed by using our proprietary algorithm to target 379 human genes.

Applying this type of algorithm on a genome- or transcriptomewide scale, we have completed the designs of multiple siRNAs targeting virtually every predicted gene of the human, mouse, and rat genomes. These pre-designed siRNAs now enable not only genomewide screens but essentially any more focused screening scope, for example, on subsets of genes corresponding to known families [e.g., kinases, G-protein-coupled receptors (GPCRs), or "druggable genome"] or custom sets defined by expression profiling studies.

Optimizing Large-Scale siRNA-Based Experiments

Delivery of Chemically Synthesized siRNAs

Large-scale RNAi experiments require a method for delivering siRNAs into cells with high efficiency, low toxicity, very high reproducibility, as well as cost efficiency. The manipulations involved in the delivery protocols must be simple and fast enough to be amenable to highly parallelized experimentation, usually in 96-well or even 384-well microplate formats. The most commonly used and best-characterized approaches are reviewed next.

Lipofection. For most commonly used cell lines, especially transformed cells, transfection with cationic liposomal reagents is a straightforward way of getting siRNAs into cells. Several siRNA lipofection reagents are available, which exhibit different profiles in terms of their range applicability to different cell lines, both in terms of transfection efficiency and in toxicity to the cells.

Because every cell line has its unique requirements, the type of lipofection reagent as well as the exact transfection amounts and conditions must be tested and adjusted for each cell line to determine optimal conditions. Normally, transfection optimization for a new cell line consists in testing of one or more transfection reagents and several siRNA:transfection reagent ratios. Other sensitive parameters to be optimized include the cell seeding density and the use (or omission) of serum and antibiotics during the transfection process.

To monitor delivery success, one can begin by visualizing the internalization of fluorescently labeled siRNAs. However, microscopy examinations of this type often reveal siRNAs accumulating in discrete membranous compartments (presumably endocytic vesicles or organelles) with little or no detectable signal in the cytosol. Thus, unless a high correlation can be demonstrated between the presence of labeled siRNAs in these compartments and the triggering of an RNAi response in those cells (which is not always the case), such signals should not be considered reliable predictors of silencing. Instead, we favor the direct monitoring of silencing either in individual cells or, more commonly, in screening projects over the entire cell population in each well.

Thus, two readouts are necessary: (1) monitoring for knockdown of the gene of interest, preferably by qRT-PCR, and (2) proper monitoring for cytotoxicity carried out in parallel, which is highly recommended [e.g., ToxiLight™ assay (Cambrex, Baltimore, MD), which is conveniently performed by using a sample of the culture's growth medium]. Optimal conditions are usually reached when the best compromise between transfection efficiency and toxicity is achieved. The sensitivity of the desired functional assay readouts should also be considered, as some are more highly sensitive to moderate toxicity than others. A transfection optimization example is illustrated in Fig. 2.

Electroporation. For cell types proving to be recalcitrant to lipofection-based protocols, electroporation can be a valuable alternative. Electroporation involves applying an electric field pulse to induce the formation of microscopic pores in the cell's plasma membrane, which then allow the

Protocol 1. Lipofection-Mediated siRNA Transfection of Human Cells (Oligofectamine™ Protocol for 96-Well Plates, Final siRNA Concentration 100 nM)

1. Twenty-four hours before transfection, seed cells at the appropriate density to each well of a 96-well plate (e.g., 20,000 cells/well for HepG2, 6000 cells/well for HeLa, and 13,000 cells/well for MCF-7).
2. Dilute siRNAs to a 10 μM working stock concentration.
3. Prepare siRNA mix: 1 μl of 10 μM siRNA + 16 μl Opti-MEM$^®$.
4. Prepare Oligofectamine™ mastermix: 0.4 μl Oligofectamine (Invitrogen, Carlsbad, CA) + 2.6 μl Opti-MEM. Incubate at room temperature for 10 min.
5. Combine solutions in (3) and (4) by gently pipetting up and down (do not vortex); incubate 20 min at room temperature.
6. Remove the culture medium from the cells (Optional: wash with 200 μl serum-free medium).
7. Serve cells (40–50% confluent) with 80 μl Dulbecco's modified Eagle medium (DMEM; without phenol red, serum, or antibiotics) per 96 wells.
8. Carefully add 20 μl transfection mix to the center of each well; shake the plate gently.
9. After 4 h, add 50 μl of medium (containing 3× serum and antibiotics) per well, to give a total culture volume of 150 μl.
10. Incubate for 48 or 72 h.

siRNA to traverse the membrane. Under specific pulse conditions, the pores reseal quickly and the electroporated cells recover and resume growth. A distinct advantage of electroporation is that it is not dependent on cell division, and RNAi-induced mRNA reduction can be detected just a few hours after delivery.

Most existing electroporation protocols were developed to deliver plasmid DNA to cell nuclei, and these protocols often suffer from high cell mortality. For transient RNAi experiments, siRNAs need to be delivered into the cytoplasm only, and therefore milder electroporation conditions can be used that minimize cellular mortality and trauma while ensuring highly efficient siRNA delivery.

Different primary and neuronal cell types require different electroporation parameters. In our experience, varying the number of pulses is the most influential parameter and a key determinant for the mortality or viability of cells (Fig. 3). For example, HUVEC cells require one electropulse (150 μS,

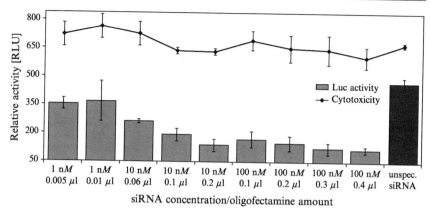

FIG. 2. Example of an siRNA transfection optimization by lipofection. An siRNA targeting luciferase was transfected into MCF-7 cells (pretransfected with a luciferase reporter plasmid) by using Oligofectamine™ (Invitrogen). Readouts were luciferase reporter activity (measured with Bright-Glo™, Promega) and cytotoxicity (detected using ToxiLight™, Cambrex). Best results were obtained with 100 nM siRNA/0.4 µl Oligofectamine.

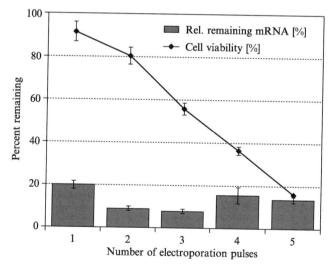

FIG. 3. Example of an siRNA transfection optimization by electroporation. An siRNA targeting GAPDH was electroporated into NHDF-Neo primary cells, using a varying number of electroporation pulses. In a 1-mm electroporation cuvette, 1.5 µg siRNA was transfected in 75 µl at 150 µS and 900 V, using a Gene Pulser Xcell™ (Bio-Rad). Twenty-four hours posttransfection, total RNA from cells was analyzed by real-time RT-PCR for target mRNA levels (normalized against 18S rRNA). Remaining mRNA was calculated as a percentage of mRNA compared with the negative control sample that was transfected with unspecific siRNA.

250 V; Ovcharenko *et al.*, 2004) whereas primary NHDF-Neo cells respond optimally to two pulses (70 μS, 900 V, 1-mm electroporation cuvette). Parameters that can further be optimized are the composition of electroporation buffer (which ideally facilitates rapid pore resealing) and the electro-pulse generator itself [e.g., Gene Pulser Xcell™ (Bio-Rad, Hercules, CA) or ECM® 830 (BTX, Holliston, MA)]. However, the biggest technical limitation of electroporation has been the lack of a commercially available 96-well electroporation device to enable high-throughput application with good well-to-well reproducibility. A number of prototypic devices are now appearing on the market, offering the hope that this may be overcome in the near future.

Validation of siRNA Efficacy

One crucial prerequisite for performing RNAi experiments is the need to link any observed phenotype to a demonstrable degree of siRNA-mediated knockdown of the target, either at the mRNA or the protein level. Despite any and all assurances offered by siRNA vendors, experimental validation of siRNA performance in the particular cell line of interest and with the transfection reagent of choice should always be part of any RNAi project. There are two scenarios:

1. *Low-throughput assays:* Monitoring RNAi-mediated knockdown for all siRNAs is feasible (e.g., preferably by qRT-PCR).
2. *High-throughput screens:* Monitoring knockdown for all siRNAs of a library of hundreds or thousands of siRNAs is not feasible. Instead, transfection optimization and siRNA validation should be done for a set of relevant control siRNAs, and this protocol be taken for the screen. Any positive hits coming out of the screen should then be tested for silencing efficacy as outlined previously.

Although in our experience protein and corresponding mRNA levels often mirror each other well, the choice of which to monitor (both is always preferable) depends primarily on the questions being asked: mRNA levels offer the most direct measure of success of the RNAi silencing and the protein level the most direct link to loss of function. In either case, the level of target mRNA or protein reduction can only offer an indirect and imperfect predictor of loss of function, as, in some cases, even as little as 5–10% remaining protein might be enough to maintain wild-type functions at levels that are not detectably different from controls. Conclusions drawn from RNAi experiments must therefore take this into account, which is why negative RNAi results are always very difficult to interpret.

siRNA Validation at the mRNA Level. For monitoring target mRNA levels, Northern blotting, branched-DNA (bDNA), or real-time RT-PCR (also known as qRT-PCR) are all valid assays. However, in view of its high sensitivity, upscalability, and reproducibility, qRT-PCR has become the method of choice (Fig. 4). It can be performed either with gene-specific primers (SybrGreen method), in which case monitoring melting curves represents an additional necessary quality control step, or with dual-labeled fluorescent probes (TaqMan® probes), which has the advantage that predesigned ready-to-run probes for most human, mouse, and rat genes are available commercially (Assays-on-Demand™, ABI, Foster City, CA).

Protocol 2. Validation of siRNA Efficacy in Human Cells by Real-Time RT-PCR (qRT-PCR) (SybrGreen Method, 384-Well Plate Setup, ABI-7900-HT Real-Time PCR Machine)

1. Perform siRNA transfection experiment according to Protocol 1.
2. At 48 h or 72 h after transfection, extract total RNA from cells (e.g., using RNAqueous™, Ambion, Austin, TX, or Invisorb® kits, Invitek, Germany), following the manufacturer's protocol.
3. Check quality and quantity of total RNA on an agarose gel.
4. Produce cDNA (e.g., using TaqMan RT reagents, ABI), following the manufacturer's instructions.
5. Check quality and quantity of cDNA on an agarose gel.
6. Run real-time qPCR with gene-specific primers:
   5.5 μl 2× SybrGreen PCR mix (e.g., from ABgene, Surrey, UK, or ABI)
   3.0 μl cDNA
   2.5 μl 2 μM F/R primers
   11 μl total
7. *Real-time qPCR program:* 50° for 2 min, 95° for 10 min, 45 cycles (95° for 15 sec and 60° for 1 min), 95° for 15 sec, 60° for 15 sec, 95° for 15 sec (melting curve).
8. *Normalization:* 18S rRNA or GAPDH as a housekeeper; run for each sample.
9. *Degree of knockdown:* This is calculated by comparing the amplification level of the gene of interest, normalized through the level of the housekeeper gene, between samples transfected with a specific siRNA and negative control samples (e.g., transfected with Negative 1 nonsense siRNA (Ambion), also at a final concentration of 100 nM).

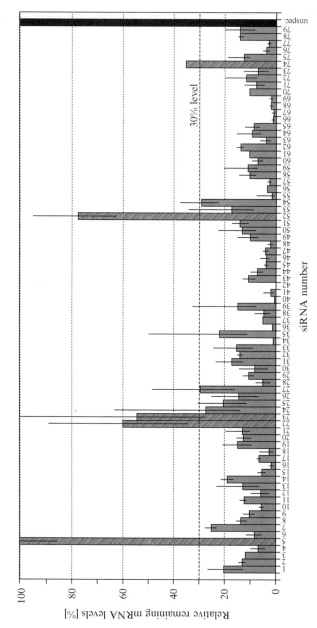

FIG. 4. Real-time RT-PCR-based validation of RNA interference (RNAi) efficacy. Transfection was performed in HeLa cells, with 79 individual siRNAs (each at 100 n*M* final concentration) against endogenously expressed human kinases. Forty-eight hours after transfection, total RNA was extracted from cells and cDNA was subjected to real-time RT-PCR (see also Protocol 2). siRNAs above the 30% level (i.e., below 70% silencing) are indicated by light bars. For 74 of 79 (94%) siRNAs, the mean remaining mRNA level was less than 30% compared with the unspecific control siRNA.

siRNA Validation at the Protein Level. To monitor RNAi knockdown at the protein level, several options are principally possible:

• Western blotting or ELISA-based methods are most commonly used for analyzing protein levels (Fig. 5). However, potential differences in the specificities of the antibody and the siRNAs should be taken into account as they can otherwise lead to misleading results.
• An enzyme activity assay can also give a direct answer about the amount of a particular protein or enzyme left after RNAi treatment.
• If a reporter gene construct is cotransfected, the readout for the degree of its depletion can be performed [e.g., luciferase reporter activity or green fluorescent protein (GFP) fluorescence]. On the other hand, extrapolation from such cotransfection assay results to the eventual performance of the same siRNA on the endogenously expressed target is not always reliable, especially when the cotransfection is in a cell type that is significantly different from the natural one.

However, the limitations common to most of these methods are that they are only applicable for monitoring the levels of a few or even only one protein at a time and that most are only applicable in a relatively low-throughput scale.

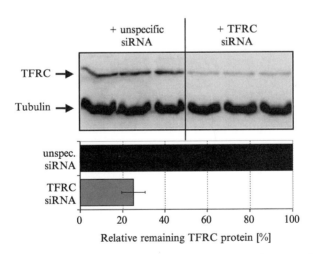

FIG. 5. RNAi-mediated knockdown of transferrin receptor (TFRC) measured by Western blot. Protein extracts from HeLa cells (24-well plates) were prepared 72 h after transfection; quantification was done on an ECL reader, with two different exposure times to account for the different abundance of the two proteins. TFRC expression levels were normalized against tubulin and compared to the unspecific control.

Kinetics of siRNA-Based Screens

The kinetics of the RNAi response derive from a complex multistep enzymatic process in which the target mRNA is specifically degraded, resulting in depletion of the target protein pool through natural turnover. Thus, the relative kinetics of degradation of the mRNA versus the protein should be taken into account when monitoring RNAi phenotypes. Whereas the mRNA degradation, as the first step of the RNAi mechanism, is believed to start almost immediately following delivery of siRNAs into cells, the reduction of the protein and accompanying loss of function completely depend on the protein's half-life, progressing more slowly over the hours and days that follow delivery of the siRNA. The half-life differs markedly from protein to protein, thus making the choice of time points for monitoring RNAi-induced phenotypes a particularly important issue. Although the ideal solution, that is, to collect data at multiple time points, can allow one to document the progression of increasingly severe hypomorphic phenotypes, this is rarely feasible in the context of large-scale screening projects. Thus, a single time point must usually be chosen, which inevitably represents a compromise to minimize the risk of leaving any phenotypes undetected.

In our experience, whether monitoring target mRNA levels by qRT-PCR or examining functional readouts, the time window of 36–72 h posttransfection (with 48 h being the most convenient) has yielded the most informative results in siRNA-based experiments.

In this context, also note that the relatively slow kinetics of the RNAi response represent a major difference with compound screening paradigms. As seen in Fig. 6, the RNAi effect needs 1–2 days to become detectable, but is then stable over several days after transfection. Although a second transfection could be an option to prolong that time (although many cells show adverse effects from such repeated transfection protocols), an assay window of up to 5 days is normally convenient and flexible enough for most assays. Assays requiring sustained gene silencing lasting longer than 5–7 days reach the limits of what is advisable with siRNAs and at present would be best addressed by using vector-based shRNA expression (discussed previously). Also, in pathway analysis screens, transduction of the signal that is triggered by a silenced constituent of the pathway can take time to effect in an actual signal (Fig. 7).

Assay Screenability and Evaluation of Hits

Assessing Assay Screenability. As a statistic value of assessing the quality of an assay, the screening window coefficient or Z-factor introduced by Zhang *et al.* (1999) can be used to describe the assay's suitability for

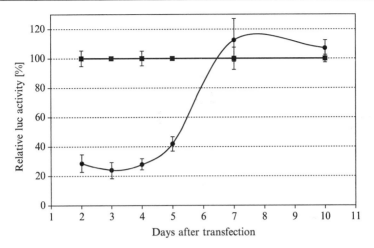

FIG. 6. Time course of siRNA-mediated RNAi knockdown. An siRNA against luciferase was transfected at 100 nM into MCF-7 cells (cotransfected with luciferase reporter plasmid). Luciferase activity readout (Bright-Glo™, Promega) was normalized by Wst activity (Wst reagent, Roche). Relative luciferase activity for samples treated with luciferase targeting siRNA is shown in circles, and samples treated with unspecific siRNAs are indicated by squares.

large-scale screening (see equation). It is a measure of the separation band between data variability of positive controls (for RNAi screens, samples transfected with a positive control siRNA) and of negative controls (for RNAi screens, samples transfected with an unspecific scrambled control siRNA).

$$Z = 1 - \frac{(3\times \text{SD}_{\text{treated}} + 3\times \text{SD}_{\text{unspecific}})}{I \text{ mean}_{\text{treated}} - \text{mean}_{\text{unspecific}} I}$$

The suggestions of Zhang and coworkers on how to interpret Z-factors are useful indicators:

- $Z = 1$ describes an ideal assay (either no variation or an unlimited distant signal)
- $1 > Z > 0.5$ reflects an excellent assay with a large separation band
- $0.5 > Z > 0$ reflects a double assay with a small separation band
- $Z = 0$ describes a yes/no assay (sample and control variations touch)
- $Z < 0$ does not allow proper screening, because sample and control variations overlap.

The same rules can also be applied to RNAi-based assays (Fig. 8). Therefore, before an RNAi assay is implemented as a large-scale screen, it is highly advisable to monitor its robustness by using the Z-factor method (positive

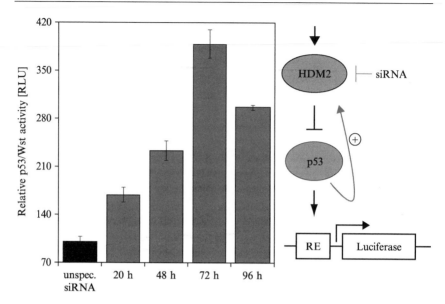

FIG. 7. Pathway analysis: time course of p53 activation after HDM2 silencing. Transfection of an siRNA against HDM2 (in MCF-7 cells cotransfected with p53 reporter plasmid) results in a downstream p53 activation, proving the applicability of RNAi for pathway dissection. The reporter construct consists of a TK basal promoter, a p53-responsive element (RE), and a firefly luciferase gene downstream. RNAi-based inhibition of HDM2 first activates p53, which is seen through the induction of luciferase expression. Later, a feedback loop (induction of endogenous HDM2 that also has a p53 response element in its promoter) results in the decrease of luciferase signal at 96 h.

versus negative controls). This is ideally done also to determine intraplate, interplate, interoperator, and interexperiment sources of variability.

Evaluation and Confirmation of Hits. To evaluate screening results from homogenous and microscopy-based RNAi screens, a threshold of three standard deviations (SDs) from the negative control (cells transfected with an unspecific siRNA) is generally considered to be the hit limit, equivalent to a 99.73% confidence limit to the sample data (Zhang *et al.*, 1999). Depending on the screen, this threshold can be adjusted to modulate the size of the resulting hit collection to be further examined in secondary screening.

Because the key goal of the primary screening pass is always to reduce the scale of the task to something more manageable, longer assay time points and higher siRNA concentrations are advisable to ensure maximal inclusiveness and minimal risk of missing positives (high detection sensitivity but low specificity or accuracy). Although this approach inevitably yields a high rate of false positives, these can be readily weeded out during

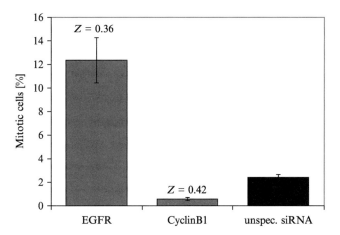

FIG. 8. Calculation of Z-factor. For a mitotic index assay in HeLa cells, the Z-factor was calculated for two positive controls, EGFR and cyclin B1. Both positive controls, for increase and decrease of mitotic index, gave Z-factors around 0.4, which means that this assay can be classified as being a *double assay* according to the terminology of Zhang *et al.* (1999) (see text).

the secondary screening pass, which we recommend to design for maximum specificity, and thereby to increase the accuracy and confidence in a putative hit (confirmatory pass). Therefore, to verify primary hits from an RNAi screen, the secondary screen should not only involve retransfection of the same siRNA against all putative hits but also the validation of those results through the use of at least one additional siRNA targeting the same gene. Ultimately, one must always link the observed phenotypes to the knockdown of the targeted gene, but this is usually not feasible during the primary screening pass, and therefore can and should be integrated in the secondary screen. Thus, in our view, a positive hit should be one that displays the same phenotype by using at least two distinct siRNAs in two separate experiments, showing a clear correlation with reduced expression levels.

Common Caveats of siRNA-Based Screens

Interpreting Unexpected Negatives

The most basic caveat of all RNAi-based screens is that negative results, that is, the absence of a detectably altered phenotype, is nearly always impossible to interpret conclusively. This is because of the inescapable fact that RNAi is inherently a knockdown rather than a knockout technology: siRNAs inevitably cause varying degrees of loss of function, and although the best siRNA design algorithms produce very efficient siRNAs, the level of silencing is rarely 100%. Thus, it is always expected that some level of the target protein will remain, and, depending on the type of protein (e.g., enzymes versus structural components), this may offer a possible justification for maintaining wild-type readings (a situation comparable to human genetics, in which the carrier of an autosomal recessive disorder does not exhibit the disease). Indeed, the threshold (i.e., the degree of silencing) required for achieving a detectable loss-of-function phenotype differs markedly between proteins and depends on the detection sensitivity of the assay. Conversely, if a detectable phenotype is generated, the ability of RNAi to titrate silencing by performing dose–response experiments is a significant strength of RNAi technology that can provide allelic series and often yields further insights into a gene's functions. These considerations underscore the importance of documenting the silencing efficacy in these experiments.

Specificity

As a clear reminder of how young the RNAi field is, a number of studies over the past year have documented instances of direct off-target effects caused by reagents, including siRNAs, designed to trigger RNAi responses (Jackson *et al.*, 2003; Scacheri *et al.*, 2004; Snove and Holen, 2004). Of course, these findings have challenged the initial hope that these reagents could exquisitely mediate gene-specific silencing, with clear implications on the design of such reagents, as well as the associated experimental strategies, screening paradigms, and data analysis criteria. Although the general understanding that emerges to date remains far from complete, it does suggest that many siRNAs in use at present not only recognize the intended, perfectly matched target mRNA but also direct the destruction of other imperfectly matched secondary target mRNAs, though usually to a significantly lesser extent. These sequence-dependent off-target effects are proving difficult to predict, as the underlying stretches of base-pair complementarity have not exhibited readily recognizable thresholds in size, composition, or other obvious patterns. Thus, the top BLAST hits for a given siRNA do not necessarily represent the most likely secondary targets. Although some have worried that imperfect siRNA-target pairings showing discrete mismatches may be triggering translational suppression, as natural miRNAs do, the finding that secondary target mRNA levels actually decrease in these cases suggests that these types of off-target effects are in fact mediated through RNAi-like modulation of mRNA stability. Nonetheless, the sequence dependence of this type of off-target effect can be used to completely and easily neutralize this problem. Because each individual siRNA's off-target footprint is defined by its sequence, one can quickly ascertain the target specificity of any observed phenotype simply by eliciting it by using multiple distinct siRNAs targeting the same gene of interest (Anon, 2003).

A second type of off-target effect that has been noted in RNAi experiments consist of the concentration-dependent modulation of the expression of nontargeted stress response genes, including factors of the interferon response pathway (Persengiev et al., 2004; Sledz et al., 2003). Although still poorly understood, these effects are thought to be dependent on such factors as cell type and delivery method, and, unlike the sequence-dependent off-target effects, their risk is known to be significantly increased by the use of excessively high concentrations of silencing reagents such as siRNAs (Persengiev et al., 2004). Thus, titration of the siRNA concentration (e.g., from 100 nM down to about 10 pM) has the potential of attenuating or even eliminating these effects. Having a powerful design algorithm that generates very efficient siRNAs (see Automated Design of siRNAs) allows transfection of siRNAs at concentrations lower than 100 nM without losing efficiency (which is, however, only recommended for confirmatory experiments). In addition, parallel qRT-PCR monitoring of the interferon response markers such as OAS1, OAS2, and STAT1 at each step of the dose–response experiment can provide a further valuable control. One or two scrambled or unspecific siRNAs, that is, siRNAs that do not exhibit significant matches to any gene of the targeted genome, are usually included in most RNAi experiments to control for this issue. This is a more appropriate control than a so-called mock control (with transfection reagent but without siRNA), because the latter often yields toxic effects that are irrelevant to the experimental samples. Moreover, untransfected samples, though necessary to control for transfection effects, are also inadequate negative controls if used on their own for RNAi experiments. A test of several unspecific siRNAs is advisable. In fact, in essentially all RNAi screens, the vast majority of tested genes yield negative results and therefore provide a huge abundance of baseline control values.

Beyond these basic precautions, the following controls may also be considered in the validation procedure for primary hits (one siRNA) or confirmed hits (at least two distinct siRNAs) coming out of an RNAi screen:

- mRNA/protein/phenotype controls: If an observed loss-of-function phenotype is observed with multiple distinct siRNAs against the same target, the ultimate proof of a valid RNAi experiment is to relate that phenotype to a significant knockdown of both the mRNA and protein for each of the effective siRNAs.
- Functional controls: The ultimate proof of validity of an RNAi experiment is to rescue expression of the target in an siRNA-refractory manner. Although these experiments currently remain too laborious and technically complex for wide application, the development of new tools in this area is expected to facilitate this very powerful approach.

Screening Automation and Data Handling

Screening Automation. The availability of liquid handling robots makes automation of various screening steps affordable. Nonetheless, these instruments do not necessarily solve all problems, and careful thought and planning should therefore be invested to determine whether individual tasks in a screening project would best benefit from automation, or perhaps only mechanization, or from being maintained strictly under human manual control. Although the gain in throughput from liquid handling is sometimes not so significant compared to manual work with multichannel pipettes, the gain in robustness, reproducibility, and the reduction of possible errors caused by manual intervention are the strongest arguments for automation. Furthermore, some processes associated with screening, namely, rearraying, reformatting, and hit-picking of siRNAs, cannot be performed reliably enough by applying a manual approach. For reasons of further improving reliability, avoiding contamination, and streamlining the screening process, the use of several individual robots that are dedicated for particular defined steps of the protocol is highly recommended. In our experience, all automated liquid handling systems (e.g., from Tecan, Switzerland; Perkin Elmer, Wellesley, MA; Hamilton, Reno, NV; Beckman, Fullerton, CA) have their own pros and cons, dependent on the particular project, and there is no general recommendation for the best system.

Lastly, to deploy the full power of automation, the establishment of a laboratory information management system (LIMS) to coordinate task scheduling, controlling, and monitoring of all the automated processes is highly recommended, as discussed next.

Data Handling. One key requirement that is widely underestimated for successfully carrying out large-scale or genomewide RNAi screens is the necessity for an appropriate LIMS and data-handling setup with a robust, well-structured database at its core, in addition to the use of robotic liquid handlers. Especially in academic environments, the scientific ambitions to perform genomewide screens often reach far higher than the bioinformatics and informatics infrastructure that are available to enable them. Grant funding review boards are slowly recognizing this reality now.

The absence of this biocomputing infrastructure can potentially cause serious problems at every stage of such screens, creating very significant risks of misattribution or even complete loss of large datasets, severely compromising the overall quality and comprehensiveness of the screen. Even if errors occur only 1% of the time, because of the repetitive, systematic nature of these projects, these few mishaps often bear the potential to cast doubt on huge, entire datasets, which risk being rendered useless.

Furthermore, those few mishaps can be extremely difficult and time consuming to track down. Thus, crucially important is putting in-depth thought into the design of a data flow tracking system, be it electronic form or based on a well-conceived paper records (or better yet both), before starting such a large scale project. The following are some important aspects:

- *Gene/siRNA information and screening results:* The handling of multiple siRNAs targeting thousands of genes requires a proper database system for storing the gene/siRNA information and linking it to the screening results.
- *Rearraying and reformatting reagent stocks and experimental plates:* Trying to coordinate these processes solely based on the use of Excel sheets is highly error prone and should be avoided. Worklists, output lists, and reformatted plate layouts cannot be generated or handled manually.
- Scheduling and monitoring screening plate processing steps are highly desirable features of a good LIMS/database system.
- *Data acquisition, mining, and storage:* Because of the amount and complexity of data and their file size, these issues require major IT support, especially before starting microscopy-based screens.

Although tight integration of the LIMS interfaces with laboratory hardware can deliver the most optimal streamlining of data flow management and lowest chances of error, we have found this to be a particularly time-consuming development task. In addition, this aspect of a LIMS significantly reduces the versatility of the system, locking it in with specific instrument choices, which only makes sense if these are to be extensively used in the long run. We therefore recommend approaching the issue of tight hardware integration carefully and selectively, focusing initially perhaps more on those developing instrument-independent functionalities that maximally streamline repetitive and error-prone tasks. For example, we have found the careful implementation of a well-thought-out barcoding system, covering not only the full range of tubes, plates, and other sample types involved throughout the experiment process but also essentially any other types of data that must be entered repetitively (such as technicians' names, if this is being tracked, which we also highly recommend). Data entries by using barcode scanners should be considered whenever possible to replace the much more error-prone keyboard entries. (Mild dyslexia is more prevalent than one might expect.)

Thus, the overall combination of laboratory and computing infrastructures, reagent libraries, and detailed expertise needed to effectively run such screens represents a very significant investment of money and in most cases at least 1–2 years of development time. It is not surprising then that most major research funding organizations have now recognized the wisdom of entrusting such studies to specialized core facilities and service providers, thereby minimizing unnecessary and very costly duplication of effort.

Readout Assays for siRNA-Based Screens

We describe some examples of commonly used homogenous and microscopy-based high-throughput assays. Although the choice may seem restricted to oncology-related assays, their underlying principles can be applied widely and adapted to other research fields of interest. Our guiding principle in designing such assays is, wherever it is feasible and appropriate, to strive toward implementing the richest possible readouts, such that the resulting dataset offers maximal depth and breadth in its phenotypic classification of screened genes.

Cell Proliferation Assays

One assay that is most broadly applied in cell cycle research and oncology is quantifying the number of living cells to measure growth arrest or induction of proliferation. Numerous colorimetric, luminescent, or fluorescent homogenous assay kits based on different assay principles are available:

- Quantification of mitochondrial reductase activity by tetrazolium salts (MTT, XTT), or WST-1.
- ELISA-based measurement of BrdU incorporation into chromosomal DNA during S-phase.
- Detection of ATP levels in cell lysates (e.g., ATPlite™, Perkin Elmer) (Fig. 9).
- Staining of nucleic acids (e.g., CyQUANT™, Molecular Probes, Eugene, OR).
- Quantification of the reduction potential of intact cell membranes (e.g., CytoLite™, Perkin Elmer).
- Live/dead staining assays.

The most notable differences in the performance of these assays come from the degree of linearity, sensitivity and stability, the signal-to-noise ratio, and the complexity of the protocol and its costs. For high-throughput screens, systems with sustained stable emission and with an easy-to-follow protocol make most sense.

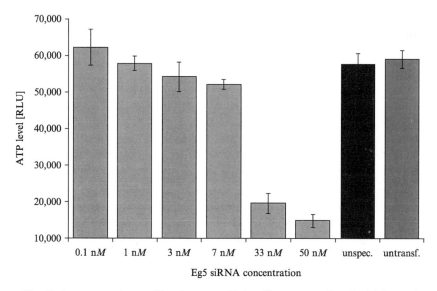

FIG. 9. Assay examples: proliferation assay. Hela cells were transfected with increasing concentrations of a siRNA targeting Eg5 (RefSeq name KIF11) in a mixture that was brought to 100 nM with unspecific negative control siRNA. Seventy-two hours after transfection, ATP levels were measured on a multilabel reader, using the ATPlite™ kit (Perkin Elmer).

When observing proliferation inhibition, one major limitation is that the underlying causes, namely, necrosis, apoptosis, or cell cycle deregulation, cannot be distinguished from each other by applying these assays on their own. This dilemma can be solved either by the combination of homogenous proliferation, apoptosis, and necrosis assays, or, more compellingly, through a microscopy-based approach whereby these multiple parameters can be detected simultaneously in so-called multiplexed assays (see later).

Mitotic Index Assay

Another example of an oncology-relevant assay is the mitotic index assay, which can reveal a gene's role in cell cycle progression by quantifying the proportion of cells undergoing mitosis at a given time point. The protocol, to be run as a microscopy assay, involves fixation of cells, followed by staining of chromosomes [6-diamidino-2-phenylindole (DAPI) or Hoechst], microtubules (anti-tubulin antibody) and with a mitosis-specific marker such as anti-phosphohistone H3 antibodies. In a large-scale, high-throughput scenario, data can then be acquired with an automated

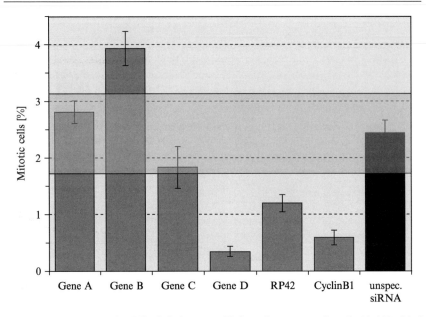

FIG. 10. Assay examples: Mitotic index assay. HeLa cells were transfected with 100 n*M* of validated siRNAs targeting several genes. Forty-eight hours after transfection, cells were fixed and stained with DAPI, anti-tubulin and anti-phosphohistone H3 (see text). The mitotic index (mitotic cells/total cells) was evaluated by fluorescence microscopy. The $\pm 3\times$ SD range of the negative control (hit limit) is indicated by a shaded box.

fluorescence microscope (e.g., the Discovery-1™ system, Molecular Devices, Sunnyvale, CA), using image processing software (e.g., MetaMorph®, Molecular Devices, or Cellenger®, Definiens, Germany).

In the example illustrated in Fig. 10, targets with a significantly increased or decreased mitotic index are shown, thus revealing their impact on cell cycle regulation. From the perspective of an oncology-focused target discovery screen, for example, cases in which silencing of a gene caused a decreased mitotic index (i.e., in which the gene is needed for passage through G1, S, and G2 phases) could point to a potential novel antiproliferative target. The combination of mitotic index screens with proliferation and apoptosis assays is highly advisable to derive even more compelling and comprehensive conclusions about potential antiproliferative targets.

Apoptosis Assays

Apoptosis, the multistep process of programmed cell death, is one of the most intensively studied processes in cell cycle research as well as in applied cancer research and drug development. Most notably from an assay

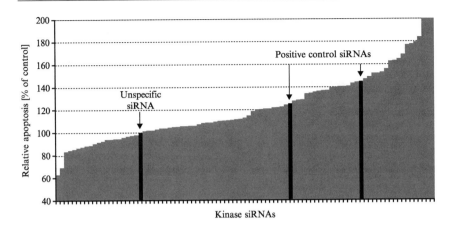

Fig. 11. Assay examples: apoptosis assay. Eighty-eight kinases were targeted, using individual siRNAs of validated efficiency (from the *Silencer™* Kinase siRNA Library, Ambion). HeLa cells were seeded in 96-well plates at 8000 cells/well. Twenty-four hours after seeding, siRNAs were transfected (100 n*M*, see Protocol 1). Forty-eight hours after transfection, cells were subjected to the apoptosis Apo-One assay (Promega), measured on a Victor-2 multilabel reader (Perkin Elmer). siRNAs against cyclin B1 and RP42 were used as positive controls; the Negative 1 siRNA (Ambion) was taken as the negative control. Arrows indicate negative and positive controls (black). Experiments were performed in triplicates, but for better visability graphs are shown without error bars.

point of view, apoptosis involves the activation of caspase cascades, loss of mitochondrial membrane potential, acidification, cell membrane permeability, chromosome condensation, DNA fragmentation, and the formation of apoptotic bodies.

Homogenous assays are established for different stages of apoptosis, roughly divided into early and late ones: very common assay principles with numerous commercially available kits are caspase 3/7 assays, as well as ELISA assays of cytochrome C release or DNA fragmentation (TUNEL assay). Figure 11 gives an example of an apoptosis screen through a large set of human kinases, using the ApoOne™ kit (Promega, Madison, WI).

Alternatively, antibodies against poly (ADP-ribose) polymerase (PARP), phosphorylated Akt-1, caspase-3, lamin A, cytokeratin 18, or cytochrome C are tools for use as part of microscopical high-content assays. An example is discussed in High-Content Microscopy Assays.

Reporter Gene Assays

Applying reporter gene technology is a powerful way of monitoring pathways of interest. The idea is simple and widely applied in small molecule compound screening: a target gene, a target's promoter, or a responsive element is positioned upstream of a reporter gene (e.g., luciferase) carried within an appropriate expression vector. This plasmid is either transiently transfected into the desired cell line (see Protocol 3) or a stable cell line is generated. The cells can then be used for large-scale projects, screening libraries of siRNAs, or chemical compounds. The adaptation of such an assay for RNAi is illustrated in Fig. 12. The strength of this kind of assay is the ease of the protocol, and the weakness the narrowness of the window afforded by this approach into the biology resulting from the experimental perturbation.

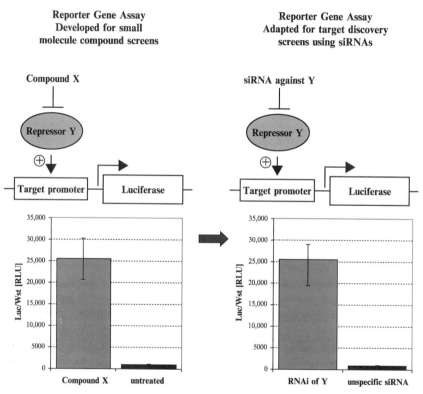

FIG. 12. Development of a typical RNAi reporter gene assay. Reporter gene assays are widely used for small molecule compound screening. The principle of converting a compound screen into an RNAi target discovery screen is illustrated. Instead of screening for small molecule inhibitors of a known target repressor Y, an siRNA-based target discovery screen can reveal novel repressors of a known target. Variants of this principle can come from different target promoters, responsive elements, or from screening for inducers rather than repressors (i.e., having an off-assay instead of an on-assay).

Although having the disadvantage of being more expensive than the plasmid-based reporter-gene assays, a version similar to reporter gene assays, in a wider sense, can be the direct monitoring of mRNA levels of two or more targets by qRT-PCR. For example, when monitoring for two or three key activity markers of a particular pathway, this strategy could reveal novel components of that pathway with high sensitivity.

Protocol 3. Cotransfection of Plasmid DNA and siRNA into Human Cells (Two Separate Transfections)

1. Twenty-four hours before transfection, seed cells at the appropriate density into each well of a six-well plate (e.g., 700,000 for MCF-7).
2. Next day, with 50–80% confluent cells, transfect 2 μg plasmid vector with 7 μl FuGene (Roche Diagnostics, Pleasanton, CA) per well, following the steps given next.
3. Per well, mix 7 μl FuGene with 93 μl serum-free medium and incubate for 5 min at room temperature.
4. Per well, prepare 10 μl plasmid vector at 0.2 μg/μl in sterile water and place into fresh tubes at 10 μl each.
5. Per tube, add dropwise the 100 μl of solution in Step 3 and gently tap the tube to mix. Incubate for 15 min at room temperature.
6. Remove the old medium from cells, and overlay with 2 ml fresh serum-containing growth medium.
7. Per well, add dropwise the 110 μl of solution in Step 5 onto the cells and gently shake the plate.
8. Incubate for 6 h, and then harvest and pool the cells from all transfected wells. Seed cells into a 96-well plate at the appropriate density (e.g., 13,000 cells/well for MCF-7). Two wells of a six-well plate are normally sufficient for one 96-well plate. Incubate cells overnight.
9. Next day, continue with Step 2 of the siRNA transfection method as outlined in Protocol 1.

High-Content Microscopy Assays

As previously noted, it is our view that the full value of cell-based screens in general, and of RNAi screens in particular, can best be realized by acquiring rich, multiparameter readouts. One strategy of doing that could be the combination of multiple homogenous assays, but an even more elegant approach is to carry out rich *in situ* analyses by establishing so-called high-content microscopy assays. Ideally, this effort to broaden our viewing window onto the biological consequences of a silencing experiment should integrate, whenever possible, both spatial and temporal information, as illustrated by the time-lapse microscopy readout used by our group and others in studying cell division genes in *C. elegans* embryos (Gönczy *et al.*, 2000; Sönnichsen *et al.*, 2004). However, such an endeavor inevitably poses a much more considerable challenge not only for the overall size and complexity of the study but most notably in terms of the analysis and annotation of resulting datasets. The development of automated image analysis algorithms and associated image handling software tools remains in its infancy, especially in cases in which kinetic data such as time-lapse recordings are concerned. Nonetheless, with the development of increasingly sophisticated automated microscopy instruments in recent years from such established vendors as Molecular Devices/Universal Imaging and Cellomics, these goals are becoming feasible and clearly well worth the extra effort. Indeed, by enabling more contextual analyses, genome-scale discovery efforts are becoming smarter, allowing for significantly more informative phenotypic classifications of genes and thereby reducing much of the guesswork that has so far been required in their follow-up.

As one very striking example, the screen illustrated in Fig. 13 revealed very complex data on proliferation, mitotic index, apoptosis, and necrosis: RNAi-mediated knockdown of both Eg5 (RefSeq name KIF11) and Cenix target X led to similar reductions in cell proliferation, as measured by the density of nuclei in imaged fields. Thus, the cellular response to these two siRNAs would be hard to distinguish by classical reader-based cell proliferation assays, which would require subsequent extra rounds of experimental follow-up to do so. In the present case though, our microscopy-based screen vastly expanded the available "discovery space" in a single round of screening, allowing us to further classify and interpret these cellular phenotypes with only modest extra effort invested. The silencing of Eg5 yielded a strong accumulation of mitotically arrested cells exhibiting condensed chromosomes and aberrant monopolar spindles (as reported previously by Mayer *et al.*, 1999). Many cells exit this arrest by entering apoptosis, which can be detected by increased levels of a caspase-cleaved fragment of PARP, a chromatin-associated protein used as an apoptotic marker. The unchanged interphase morphology of Eg5-silenced cells is consistent with the established function of Eg5 as a protein exclusively required during mitosis. The knockdown of target X, in contrast, did not cause a mitotic arrest and no increase in apoptosis was detected, thus initially suggesting that the observed reduction in cell proliferation might come from increased necrosis or a general, uniformly slowed progress

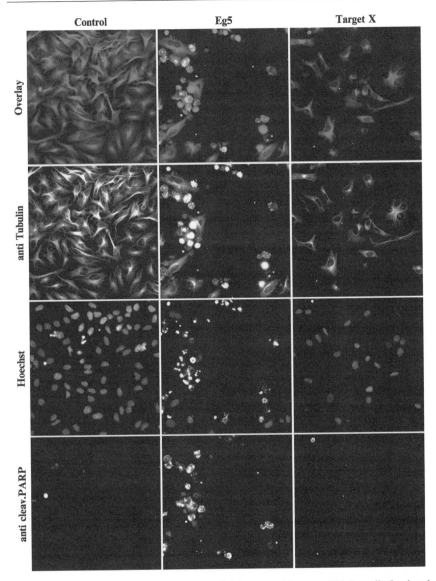

FIG. 13. High-content screening example. Triple-channel images of HeLa cells fixed and stained 48 h after treatment with three different siRNAs: unspecific control siRNA (left column), Eg5 (central column), and Cenix target X (right column). After fixation, cells were stained with specific antibodies against cleaved poly (ADP-ribose) polymerase (PARP; fourth row) and tubulin (second row). Nuclei were stained with Hoechst 33342 (third row). The three-channel overlays (first row) show cleaved PARP in red, tubulin in green, and Hoechst in blue. See text for a further discussion of this study. (See color insert.)

through all cell cycle stages (thus not affecting mitotic index). Our more detailed analysis of the same dataset also revealed a reproducible increase in the incidence of binucleated cells, suggesting a possible role of target X in the progression through cytokinesis, a deficiency that could explain the increased necrosis and lower proliferation without seeing an effect on mitotic index.

Thus, the use of high-content assays allows the identification of hits whose phenotypes may not represent drastic changes in a single parameter but the combination of subtle changes in several parameters. This is especially valuable in the context of RNAi screens, in which subtle phenotypes may result from partial gene silencing, and therefore a multiparameter approach can greatly improve not only the depth and breadth of the readout but also the overall sensitivity of the screen.

Case Study: RNAi Screening with a Kinase siRNA Library

The following case study serves as an example of a focused screen by using transient transfection of chemically synthesized siRNAs to illustrate the opportunities afforded by a multiplexed readout. The goal of this screen was to identify kinases that are involved in apoptosis, cell proliferation, and cell cycle control. The study itself was focused on 88 human kinases known to be expressed in HeLa cells, using prevalidated siRNAs known to trigger silencing of their target mRNAs by at least 70% (Cenix predesigned siRNAs, Ambion).

Forty-eight hours after transfection, cells were fixed and stained with DAPI, anti-tubulin and anti-phosphohistone H3 antibodies. Data were obtained by using an automated fluorescence microscope (Discovery-1, Molecular Devices) with automated image processing software (Meta-Morph), in combination with homogenous readouts obtained with a Victor-2 multilabel reader (Perkin Elmer). This enabled the simultaneous high-throughput acquisition of proliferation (cell numbers), mitotic index (percentage of cells in mitosis), apoptosis, and cytotoxicity, as well as cytoskeletal organization and overall cell morphologies. As a result (Fig. 14), several siRNAs were identified as causing dramatic increases or reductions in mitotic index, compared to the ~2.5% mitotic cells observed in control samples. First, this analysis, together with the proliferation data, gave indications on the underlying cell cycle regulation effects, namely, whether a higher mitotic index was elicited by mitotic arrest or a shorter interphase and whether a lower mitotic index points to an interphase arrest or a shorter mitosis. Second, it can clearly be distinguished between kinases whose inhibition leads to apoptosis and necrosis, or both. Third, the immunofluorescence staining of cell components (chromatin, microtubules)

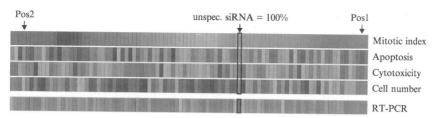

Fig. 14. Case study: mitotic index, apoptosis, necrosis, and proliferation. Multiple functional screening data illustrated for a subset of 88 kinases whose apoptosis data are shown in Fig. 11. The 88 kinases (plus two positive controls and one negative control) were targeted with individual validated siRNAs (part of the *Silencer* Kinase siRNA Library, Ambion). In 96-well plates, HeLa cells were transfected with individual siRNAs at a final concentration of 100 n*M* (Oligofectamine transfection reagent; see Protocol 1). Forty-eight hours after transfection, cells were subjected to different assays. To determine mitotic index (percentage of mitotic cells) and cell numbers, cells were fixed and stained with DAPI as well as with anti-tubulin and anti-phosphohistone H3 antibody, and data were acquired by automated fluorescence microscopy (Discovery-1 microscope, Molecular Devices). Other readouts were apoptosis (ApoOne™ kit) and cytotoxicity (ToxiLight kit), both measured on a Victor-2 multilabel reader (Perkin Elmer). The degree of RNAi silencing triggered by each siRNA, measured as the relative remaining mRNA level in real-time RT-PCR (see Protocol 2), is also depicted (always at the <30% level). Data are given in a heatmap format, illustrating the relative changes compared with the negative control (unspecific siRNA). Red indicates increasing values and green decreasing values. Data are ordered by the mitotic index (bright red to bright green). All data points show the average of a triplicate. (See color insert.)

allowed a direct evaluation of microscopy data to further classify mitotic arrest phenotypes. Taken together, the data lead to new, detailed insights into the function of a broad set of human kinases. The concept that was applied here as well as in an earlier study (Krönke *et al.*, 2004) serves as a proof of principle for the use of siRNA libraries and the value of combining large-scale siRNA-based screening with multiparametric readout assays.

Outlook: Reagents, Techniques, and Overall Strategies

In this chapter, we have reviewed and discussed the opportunities and challenges of genome-scale RNAi-based screens, using transient transfection of chemically synthesized siRNAs. We now look forward to the emergence of new reagents and techniques that will help further extend the applicability of RNAi-based functional genomic screens both horizontally, that is, using cell-based methods to study a broader range of biological processes, and vertically, applying RNAi screens to advance downstream applications in the development of new therapeutic drugs.

The latter will proceed through the discovery and characterization of gene functions that are not measurable in cell systems by using RNAi screening in animals. Furthermore, those RNAi screening methodologies currently being used in cells can and will be used, with minor adjustments, to gain new insights into the mechanisms of action, sources of toxicities, and side effects of therapeutic compounds currently under development.

One particularly elegant technique, pioneered by Sabatini and colleagues (Carpenter and Sabatini, 2004), and now known as retrotransfection or solid-phase optimized transfection (SPOT), is being further developed by several groups, allowing the creation of SPOT-RNAi arrays of siRNAs or even viruses, arranged as discrete spots on microscopically compatible growth supports (slides or plates) onto which cells can be seeded. Although it remains unclear whether, in practice, this will truly allow the use of smaller amounts of reagents, as is promised, this approach does offer the potential of greatly accelerating the throughput of compatible screening assays. The range of compatibility of screening assays for SPOT-RNAi will also depend on the method's overall experimental robustness, particularly in accommodating different cell types exhibiting varying levels of motility, as well as the numbers of cells required within each sample (i.e., over each siRNA spot) to yield statistically significant results from the desired assay.

Another up-and-coming development of note will likely be the emergence of new types of RNAi reagents, both for silencing and for better delivery. In the former category, we take particular note the discovery of so-called endonuclease-derived siRNAs, or esiRNAs, by Yang et al. (2002). These pools of siRNA-like molecules are generated through the controlled RNase III digestion of long dsRNA molecules (usually >200 bp) produced by in vitro transcription. The resulting pool of heterogeneous sequences, all targeting the same transcript, is thought to increase the potency of the reagent while diluting out the off-target effects of each individual molecule. In view of the method's inherently low production costs this sounds very promising, and therefore a more detailed characterization of esiRNA's actual experimental performance, especially in screening applications, is keenly awaited.

In the meantime, although siRNAs at present represent the gold quality standard, the associated cost factor is limiting its widespread adoption by academic groups for large-scale studies, some of whom are opting for vector-based screening strategies. Although renewal costs of shRNA vector libraries are thought to be generally lower than for siRNA libraries, the initial investments are considerable in both cases, the general quality and accuracy of vector libraries are notoriously variable (in large part due to the complexities of large-scale cloning), and the subsequent updating of vector libraries is much more cumbersome. It is most heartbreaking to note the acceptance by some of almost undoubtedly higher rates of false negatives (i.e., missed positives) afforded by shRNA vector libraries compared to what siRNAs could yield in the same experiments, as an acceptable trade-off, because the laboratory "will get a good number of genes to follow-up on anyway." Although this attitude reflects a commonly shared pragmatic wisdom necessary to keep successful research programs moving forward, because funding is unlikely to be granted for a second group to repeat the same screen, this is precisely the type of scenario whereby missed positive genes remain undiscovered by the research community for years or decades to follow. Thus, the obvious need to be pragmatic notwithstanding, we suggest that the motivation driving the design and implementation of these screens should, whenever possible, reach beyond the mere feeding of project pipelines of individual groups rather striving more explicitly toward the most comprehensive coverage possible of the targeted "discovery space." In this context then, a more defensible justification for choosing shRNA vectors for RNAi screening, in our view, is found in those cases in which delivery of siRNAs into the cells of interest is simply not feasible or when the silencing effect must be sustained for longer periods than what is afforded by siRNAs. Then, at least as RNAi technology stands at the moment, the choice of shRNAs makes clear sense. Ideally, the cost arguments should only offer stronger motivation to request and insist on higher grant funding levels for this type of research or to improve the availability of specialized outsourcing facilities or service providers, such that the scientific quality of the investigations is not compromised.

Finally, beyond these considerations, the chapter has hopefully illustrated the value of combining genome-scale siRNA-based screening with the application of rich, high-content readout assays. The potential of effectively generating complex, gene-specific phenotypic signatures, fingerprints, or perhaps phenoprints offers an important step forward in the evolution of functional genomics, advancing both systems biology and many more traditional cell, molecular and developmental investigations.

Acknowledgments

The authors thank Mary-Ann Grosse, Sindy Kluge, Moddasar Khan, Evelyn Oswald, and Corina Frenzel for their expert technical assistance in acquiring the data shown in this chapter. The work presented in this article was partially supported by BMBF grant 0314002.

References

Anon (2003). Whither RNAi? Editorial. *Nat. Cell Biol.* **5**, 489–490.

Arts, G. J., Langemeijer, E., Tissingh, R., Ma, L., Pavliska, H., Dokic, K., Dooijes, R., Mesic, E., Clasen, R., Michiels, F., van der Schueren, J., Lambrecht, M., Herman, S., Brys, R., Thys, K., Hoffmann, M., Tomme, P., and van Es, H. (2003). Adenoviral vectors expressing siRNAs for discovery and validation of gene function. *Genome Res.* **13**(10), 2325–2332.

Berns, K., Hijmans, E. M., Mullenders, J., Brummelkamp, T. R., Velds, A., Heimerikx, M., Kerkhoven, R. M, Madiredjo, M., Nijkamp, W., Weigelt, B., Agami, R., Ge, W., Cavet, G., Linsley, P. S., Beijersbergen, R. L., and Bernards, R. A. (2004). Large-scale RNAi screen in human cells identifies new components of the p53 pathway. *Nature* **428**(6981), 431–437.

Bridge, A. J., Pebernard, S., Ducraux, A., Nicoulaz, A. L., and Iggo, R. (2003). Induction of an interferon response by RNAi vectors in mammalian cells. *Nat. Genet.* **34**(3), 263–264.

Carpenter, A. E., and Sabatini, D. M. (2004). Systematic genome-wide screens of gene function. *Nat. Rev. Genet.* **5**(1), 11–22.

Clemens, J. C., Worby, C. A., Simonson-Leff, N., Muda, M., Maehama, T., Hemmings, B. A., and Dixon, J. E. (2000). Use of double-stranded RNA interference in *Drosophila* cell lines to dissect signal transduction pathways. *Proc. Natl. Acad. Sci. USA* **97**(12), 6499–6503.

Elbashir, S. M., Harborth, J., Lendeckel, W., Yalcin, A., Weber, K., and Tuschl, T. (2001). Duplexes of 21-nucleotide RNAs mediate RNA interference in cultured mammalian cells. *Nature* **411**(6836), 494–498.

Fraser, A. G., Kamath, R. S., Zipperlen, P., Martinez-Campos, M., Sohrmann, M., and Ahringer, J. (2000). Functional genomic analysis of *C. elegans* chromosome I by systematic RNA interference. *Nature* **408**(6810), 325–330.

Gönczy, P., Echeverri, C., Oegema, K., Coulson, A., Jones, S. J., Copley, R. R., Duperon, J., Oegema, J., Brehm, M., Cassin, E., Hannak, E., Kirkham, M., Pichler, S., Flohrs, K., Goessen, A., Leidel, S., Alleaume, A. M., Martin, C., Ozlu, N., Bork, P., and Hyman, A. A. (2000). Functional genomic analysis of cell division in *C. elegans* using RNAi of genes on chromosome III. *Nature* **408**(6810), 331–336.

Hamilton, A. J., and Baulcombe, D. C. (1999). A species of small antisense RNA in posttranscriptional gene silencing in plants. *Science* **286**(5441), 950–952.

Jackson, A. L., Bartz, S. R., Schelter, J., Kobayashi, S. V., Burchard, J., Mao, M., Li, B., Cavet, G., and Linsley, P. S. (2003). Expression profiling reveals off-target gene regulation by RNAi. *Nat. Biotechnol.* **21**(6), 635–637.

Kamath, R. S., Fraser, A. G., Dong, Y., Poulin, G., Durbin, R., Gotta, M., Kanapin, A., Le Bot, N., Moreno, S., Sohrmann, M., Welchman, D. P., Zipperlen, P., and Ahringer, J. (2003). Systematic functional analysis of the *Caenorhabditis elegans* genome using RNAi. *Nature* **421**, 231–237.

Kiger, A., Baum, B., Jones, S., Jones, M., Coulson, A., Echeverri, C., and Perrimon, N. A. (2003). Functional genomic analysis of cell morphology using RNA interference. *J. Biol.* **2**(4), 27.

Krönke, A., Grabner, A., Hannus, M., Sachse, C., Dorris, D., Echeverri, C. (Spring 2004). Using RNAi to identify and validate novel drug targets — Targeting human kinases with an siRNA library. *Drug Discovery World* **5**, 53–62.

Lum, L., Yao, S., Mozer, B., Rovescalli, A., Von Kessler, D., Nirenberg, M., and Beachy, P. (2003). A. Identification of hedgehog pathway components by RNAi in *Drosophila* cultured cells. *Science* **299**(5615), 2039–2045.

Mayer, T. U., Kapoor, T. M., Haggarty, S. J., King, R. W., Schreiber, S. L., and Mitchison, T. J. (1999). Small molecule inhibitor of mitotic spindle bipolarity identified in a phenotype-based screen. *Science* **286**(5441), 971–974.

Ovcharenko, D., Jarvis, R., Kelnar, K., and Brown, D. (2004). Delivering siRNAs to difficult cell types: Electroporation of primary, neuronal and other hard-to-transfect cells. *Ambion TechNotes* **11**(3), 14–15.

Paddison, P. J., Silva, J. M., Conklin, D. S., Schlabach, M., Li, M., Aruleba, S., Balija, V., O'Shaughnessy, A., Gnoj, L., Scobie, K., Chang, K., Westbrook, T., Cleary, M., Sachidanandam, R., McCombie, W. R., Elledge, S. J., and Hannon, G. J. (2004). A resource for large-scale RNA-interference-based screens in mammals. *Nature* **428**(6981), 427–431.

Persengiev, S. P., Zhu, X., and Green, M. R. (2004). Nonspecific, concentration-dependent stimulation and repression of mammalian gene expression by small interfering RNAs (siRNAs). *RNA* **10**(1), 12–18.

Rubinson, D. A., Dillon, C. P., Kwiatkowski, A. V., Sievers, C., Yang, L., Kopinja, J., Rooney, D. L., Ihrig, M. M., McManus, M. T., Gertler, F. B., Scott, M. L., and Van Parijs, L. (2003). A lentivirus-based system to functionally silence genes in primary mammalian cells, stem cells and transgenic mice by RNA interference. *Nat. Genet.* **33**(3), 401–406.

Scacheri, P. C., Rozenblatt-Rosen, O., Caplen, N. J., Wolfsberg, T. G., Umayam, L., Lee, J. C., Hughes, C. M., Shanmugam, K. S., Bhattacharjee, A., Meyerson, M., and Collins, F. S. (2004). Short interfering RNAs can induce unexpected and divergent changes in the levels of untargeted proteins in mammalian cells. *Proc. Natl. Acad. Sci. USA* **101**(7), 1892–1897.

Schwarz, D. S., Hutvagner, G., Du, T., Xu, Z., Aronin, N., and Zamore, P. D. (2003). Asymmetry in the assembly of the RNAi enzyme complex. *Cell* **115**(2), 199–220.

Sledz, C. A., Holko, M., de Veer, M. J., Silverman, R. H., and Williams, B. R. (2003). Activation of the interferon system by short-interfering RNAs. *Nat. Cell Biol.* **5**(9), 834–839.

Snove, O., Jr., and Holen, T. (2004). Many commonly used siRNAs risk off-target activity. *Biochem. Biophys. Res. Commun.* **319**(1), 256–263.

Sönnichsen, B., Koski, L., Walsh, A., Marshall, P., Neumann, F., Brehm, M., Alleaume, A.-M., Artelt, J., Bettencourt, P., Cassin, E., Hewitson, M., Holz, C., Khan, M., Lazik, S., Martin, C., Nitschke, B., Ruer, M., Stanford, J., Winzi, M., Heinkel, R., Röder, M., Finell, J., Häntsch, H., Jones, S., Jones, M., Coulson, A., Oegema, K., Gönczy, P., Hyman, A. A., and Echeverri, C. J. (submitted). Genome-wide screening by RNA interference identifies 661 genes required for the first cell division of *C. elegans*.

Yang, D., Buchholz, F., Huang, Z., Goga, A., Chen, C. Y., Brodsky, F. M., and Bishop, J. M. (2002). Short RNA duplexes produced by hydrolysis with *Escherichia coli* RNase III mediate effective RNA interference in mammalian cells. *Proc. Natl. Acad. Sci. USA* **99**(15), 9942–9947.

Zhang, J. H., Chung, T. D., and Oldenburg, K. R. (1999). A simple statistical parameter for use in evaluation and validation of high throughput screening assays. *J. Biomol. Screen* **4**(2), 67–73.

[16] RNA Interference *In Vivo*: Toward Synthetic Small Inhibitory RNA-Based Therapeutics

By Antonin de Fougerolles, Muthiah Manoharan,
Rachel Meyers, and Hans-Peter Vornlocher

Abstract

Small interfering RNA (siRNA) mediated inhibition of gene expression has rapidly become a major tool for *in vitro* analysis of protein function. *In vivo* gene silencing by siRNAs will play an important role for target validation and is the first step towards the development of siRNA-based therapeutics. This chapter reviews the early and intriguing successes in using siRNAs for *in vivo* gene silencing. The impact of chemical modification on siRNA efficacy *in vitro* and the potential for employing such modifications to alter the pharmacokinetic properties of siRNAs is also summarized. A protocol describing siRNA-based gene silencing in tumor models can serve as guide for the design of individual *in vivo* RNA interference experiments.

Introduction

RNA interference (RNAi) is a revolutionary discovery that is transforming biological research. Short double-stranded RNA (dsRNA) molecules are easily and rapidly available and can, in a sequence-specific manner, downregulate expression of any target mRNA molecule (Dorsett and Tuschl, 2004; Novina and Sharp, 2004). RNAi is now widely accepted as the *in vitro* method of choice for rapidly assessing biological mechanisms and pathways through loss-of-function analysis. More recently, RNAi libraries have been constructed, which allow functional phenotypic screening in mammalian cells (Berns *et al.*, 2004; Paddison *et al.*, 2004). Such libraries take advantage of the ability to express short hairpin structures from a variety of vector systems, which get correctly processed to generate the functional unit, a small inhibitory RNA (siRNA). Both short hairpin RNAs (shRNAs) and siRNAs can be used to downregulate mRNA expression in cells, but we largely restrict ourselves to discussing progress achieved with siRNAs.

siRNAs are synthetic RNA duplexes commonly composed of two 21-mer oligonucleotides with 19 nt of complementarity and a 2-nt single-stranded overhang at each 3′-end. It is clear that alternative structures of siRNAs can also function *in vitro* (Czauderna *et al.*, 2003), but we focus our remarks on the conventional structure, as all *in vivo* applications to date have used siRNAs of this type. For *in vitro* use, siRNA can be readily introduced into cell lines by a variety of transfection reagents and approaches. One of the major hurdles for using RNAi *in vivo* is efficient intracellular delivery of a highly charged and structurally rigid siRNA.

Substantial progress has been made in delivering siRNA *in vivo*, first through hydrodynamic methods and more recently through standard local and systemic administration. Delivery can be enhanced either by improving siRNA attributes directly through chemical modifications of the duplex or by using different formulations, such as cationic liposomes or polymers. Significant success has already been seen in using siRNA in many localized contexts such as the eye, the CNS, and the lung. In addition, encouraging data have been generated in a variety of other *in vivo* settings, including xenograft models and viral infections, demonstrating the potential for the systemic delivery of siRNA-based therapeutics. In this chapter, we outline the rapid progress made in the delivery of siRNA *in vivo* and describe the application of chemistry for the development of the next generation of siRNA for both local and systemic administration.

In Vivo Use of Synthetic siRNAs

Hydrodynamic Delivery

The first successful *in vivo* delivery of siRNA into cells used a hydrodynamic methodology that was initially developed to deliver gene therapy vectors into hepatocytes (Song *et al.*, 2002). Hydrodynamic delivery involves rapid large-volume intravenous injection equivalent to the total blood volume. In the mouse this involves injecting 2–3 ml over 10–15 sec. In the earliest applications of siRNA *in vivo*, coinjection of enhanced green fluorescent protein (EGFP; Lewis *et al.*, 2002) or luciferase (McCaffrey *et al.*, 2002) encoding plasmids together with the cognate-specific siRNA yielded significant downmodulation of target gene expression in the liver, suggesting that once delivered to hepatocytes, siRNAs were functional *in vivo* (Table I, hydrodynamic siRNA delivery). By this approach, several therapeutically relevant hepatic targets were specifically downregulated, including Fas (Song *et al.*, 2003) and caspase 8 (Zender *et al.*, 2003). In both these cases, the target genes were downregulated by 60–90%. In the case of Fas, the effect was sustained for 10 days and resulted in protection from fulminant hepatitis and liver failure. Similarly, hydrodynamically delivered siRNA directed toward hepatitis B-inhibited viral replication in mice, when coinjected with HBV-expressing plasmids (Giladi *et al.*, 2003). More recently, 2′-F modified siRNAs with enhanced serum stability showed efficacy in silencing luciferase expressed from a coinjected plasmid, demonstrating that chemically modified siRNAs are active *in vivo* (Layzer *et al.*, 2004). Although hydrodynamic injection provides an important proof-of-concept demonstration that RNAi can function *in vivo*, it does not represent a therapeutically viable method. As discussed later, chemical modification of siRNA will be crucial for the efficient delivery of siRNAs in a therapeutic modality.

Local Delivery

Local administration of siRNA involves direct introduction of siRNA into a particular tissue or organ. There are now multiple reports of siRNA being successfully introduced into the eye, lung, and brain (Table I, local siRNA administration).

The eye was one of the first sites to be extensively investigated by RNAi. There is strong precedence for utilization of oligonucleotide-based drugs in the eye as evidenced by both Vitravene®, an approved antisense oligonucleotide therapeutic for cytomegalovirus (CMV) retinitis, and Macugen®, an RNA aptamer specific for vascular endothelial growth factor that recently successfully completed phase III trials for age-related macular degeneration. In both instances, the oligonucleotides were chemically modified and delivered every 4–6 weeks by localized injections into the vitreal cavity of the eye.

Not surprisingly, given these precedents and the general clinical validation of the vascular endothelial growth factor (VEGF) pathway in humans, a significant amount of work has been done with siRNA targeting the VEGF pathway in the eye. siRNA targeting VEGF has been shown to be active in mouse and non-human primate models of choroidal neovascularization (Reich *et al.*, 2003; Tolentino *et al.*, 2004). In both model systems, siRNA was administered by intravitreal injection immediately before inducing choroidal neovascularization, and reduced vessel growth and leakage were seen over the course of the experiments (2 weeks for mouse and 6 weeks for non-human primates). In addition, other researchers have reported activity of siRNA targeting VEGF receptor 1 (VEGFR1) in inhibiting both choroidal neovascularization and corneal angiogenesis in rodents (Sirna press release, 2003).

The respiratory system is another example in which the direct RNAi approach has already been shown to be very promising. Intranasal delivery of an unmodified siRNA targeting heme oxygenase-1 (HO-1), a cytoprotective enzyme, has been shown to enhance ischemia-reperfusion-induced lung apoptosis (Zhang *et al.*, 2004). Prophylactic treatment with HO-1

TABLE I

In Vivo Application of Synthetic Small Interfering RNA

Target	Dosing and route of administration	Effect/indication	Reference
Hydrodynamic siRNA delivery			
Luciferase	Single HD IV; 40 μg siRNA	Coinjection of luciferase reporter plasmid; analysis with whole-body imaging; 70% inhibition of luciferase expression 72 h after coinjection	McCaffrey et al., 2002
Luciferase, sec. alkaline phosphatase, GFP	Single HD IV; 0.05–5 μg (luciferase), 5 μg (SEAP), 50 μg (GFP)	Coinjection of luciferase reporter plasmid; determination of luciferase activity in tissue extracts; 36–88% inhibition of luciferase activity; coinjection of SEAP reporter plasmid; determination of SEAP activity in serum; ≤83% inhibition 24 h postinjection; back to control level after 14 days; transgenic GFP-mice: inhibition in hepatocytes 48 h postinjection by fluorescence microscopy	Lewis et al., 2002
FAS	3 HD IV (0, 8, and 32 h) with 50 μg each	Endogenous gene targeting; 10-fold reduction of FAS-mRNA (RNase protection); FAS protein reduction to background (Western blot); stable reduction of mRNA and protein for 10 days and recovery from inhibition after 20 days; improved survival of FAS-siRNA treated animals fulminant hepatitis model	Song et al., 2003
Caspase 8, β-galactosidase	Single HD IV or HD into portal vein, 150–200 μg, (caspase 8), 150 μg (β-galactosidase)	Endogenous gene targeting; >3-fold reduction of caspase 8 mRNA (Northern) after IV; LacZ transgenic; 70% of hepatocytes show reduced LacZ expression and a three- to fourfold reduction of β-gal activity	Zender et al., 2003

(Continued)

TABLE I (Continued)

Target	Dosing and route of administration	Effect/indication	Reference
Hepatitis B virus	Single HD IV, 15 μg or 5–25.5 μg	Coinjection of HBV plasmid, assayed for HbsAg in serum; 98% inhibition 24 h postinjection diminishing to 60% at Day 3; 50% reduction of all three major HBV transcripts (Northern blot); reduction of HBV DNA in serum below detectable levels (Southern blot)	Giladi et al., 2003
Luciferase	Single HD IV, 10 μg	Coinjection of luciferase reporter analysis with whole-body imaging; >85% inhibition of luciferase expression 48 h after siRNA injection; no impact on magnitude or duration of inhibition with 2'-F modified siRNAs	Layzer et al., 2004
Local siRNA administration			
Unformulated siRNA			
AGRP	Single dose of 7 μg; stereotactic surgery, bilateral injection	50% Reduction of AGRP mRNA 24 h postinjection; increased metabolic rate (heat production and oxygen consumption) after AGRP-siRNA treatment	Makimura et al., 2002
HO-1	Single dose of ~50 μg; intranasal	Strong inhibition of heme oxygenase-1 protein expression (Western) in an ischemia/reperfusion model; detection of biotinylated siRNA in lung tissue	Zhang et al., 2004
P2X₃ cation channel	Continuous infusion for 6–7 days; 400 μg/day; intrathecal infusion through indwelling cannula attached to minipump	40% mRNA reduction in dorsal root ganglia; significant protein reduction (measured by immunohistochemistry); diminished mechanical hyperalgesia and tactile allodynia	Dorn et al., 2004

VEGF	Single dose of 70–350 μg; intravitreal (50-μl volume)	Inhibition in neovascular area by more than 50% in a choroidal neovascularization model	Tolentino et al., 2004
CSF-1 and CSF-1 receptor	Six injections; 10 μg/injection (every 3 days); intratumoral	Reduction of CSF-1 and CSF-1 receptor mRNA (RT-PCR) and protein (Western); tumor growth was inhibited ~40% by CSF-1 and CSF-1 receptor siRNAs	Aharinejad et al., 2004
Complexed siRNA			
VEGF EGFP	Single dose of 0.3 μg; subretinal, siRNA complexed with Transit TKO	Coinjection of hVEGF viral vector, assayed for VEGF protein and for neovascularization; 75% reduction of VEGF protein in eyes (ELISA) 60 h postinjection; significant reduction in choroidal neovascularization after VEGF siRNA treatment (14 days after laser treatment and 12.5 days after siRNA injection)	Reich et al., 2003
VEGF	Four injections 0.7-7 μg/injection (every 10 days); intratumoral, siRNA formulated with atelocollagen	85% Reduction of VEGF protein in tumor (ELISA) and up to 75% reduction in tumor volume at highest siRNA doses; reduced vessel density in tumors after VEGF-siRNA treatment; detection of siRNA in tumor 7 days postinjection	Takei et al., 2004
Luciferase and FGF-4	Single dose of 2.5 μg; intratumoral, siRNA formulated with atelocollagen	>75% Reduction of tumor growth at 21 days (bioluminescence) and ~50% reduction of FGF4 protein after 3 days (ELISA) with siRNAs targeting FGF-4	Minakuchi et al., 2004
Systemic siRNA administration			
Unformulated siRNA			
β-catenin	Single IV dose of ~3.5 μg	Significant reduction of β-catenin protein (Western) in xenograph tumors at 48–72 h postinjection	Verma et al., 2003
Luciferase	Single dose of 3 μg; IP, IV, SC, and intratumoral	50% Reduction of luciferase 72 h postinjection for IP, IV, SC; no luciferase inhibition after intratumoral injection	Filleur et al., 2003

(Continued)

TABLE I (*Continued*)

Target	Dosing and route of administration	Effect/indication	Reference
VEGF	16 daily IV injections of 3 μg/ injection	66% Tumor growth reduction 70% reduction of VEGF protein (ELISA from tumor samples) after 16 days	Duxbury et al., 2003
FAK	12 IV injections of ~4 μg (twice per week) in combination with gemcitabine	40% Reduction in tumor growth by FAK siRNA; ~75% reduction in tumor mass in combination with gemcitabine; combined treatment also yielded 50% reduction of FAK protein in tumor homogenate (Western), increased apoptotic fraction, and enhanced Caspase 3 activity	Duxbury et al., 2004
M2 subunit of ribonucleotide reductase (RRM2)	12 IV injections of ~4 μg (twice per week) in combination with gemcitabine	20% Reduction in pancreatic tumor growth after RRM2 siRNAs alone; >85% reduction in tumor mass in combination with gemcitabine; combined treatment also yielded complete inhibition of liver metastasis, increased apoptotic cells, and enhanced Caspase 3 activity	Duxbury et al., 2004
NF-κB p65	Multiple IV injections of ~4 μg (twice per week) in combination with CPT-11	Reduction of tumor growth only in combination with CPT-11; 50% reduction of p65 protein concentration in tumor samples	Guo et al., 2004
Formulated siRNA			
ß-Catenin	Multiple IP injections of ~4 μg; complexed with oligofectamine	Prolonged survival	Verma et al., 2003
Influenza virus	Single IV dose of 60 or 120 μg; formulated with polyethylenimine	Influenza virus specific siRNAs can reduce virus titer in lung when administered pre- or post-virus infection	Ge et al., 2004

siRNA, but not control siRNA, inhibited expression and ischemia-reperfusion-induced upregulation of HO-1 protein in the lung. The reduced HO-1 protein expression is restricted to the lung, as intranasal delivery of HO-1 siRNA has no effect on HO-1 expression in other organs such as the liver and kidney. As expected, inhibition of lung HO-1 gene expression by siRNA resulted in increased apoptosis within the lung in response to ischemia-reperfusion injury. Lastly, the presence of HO-1 siRNA was detected in numerous cells in the lung airway and parenchyma.

Progress has also been made in the local delivery of siRNA to the central nervous system (CNS). The first report of local siRNA injection into the CNS involved downregulation of the agouti-related peptide (AGRP; Makimura et al., 2002). Injection of unmodified siRNA targeting AGRP into the hypothalamus reduced AGRP mRNA expression by 50% after 24 h, whereas a nonspecific siRNA had no effect. Consistent with the role of AGRP in regulating metabolic function, siRNA-mediated reduction in AGRP levels was associated with increased metabolic rate and reduced body weight. The efficacy seen in this initial study may have been hampered by the lack of sustained siRNA exposure, but it provides an encouraging proof of concept for siRNA activity in the CNS. Dorn et al. (2004) clearly demonstrate that local intrathecal infusion of a chemically modified siRNA targeting the pain-related cation channel $P2X_3$ diminished pain responses in models of agonist-evoked pain and chronic neuropathic pain. Exposure of up to 400 μg/day of siRNA (delivered into the spinal fluid through an indwelling cannula attached to an osmotic minipump) resulted in significant reductions in pain with no signs of neurotoxicity. Molecular analysis of the surrounding neural tissue revealed inhibition of $P2X_3$ mRNA and protein in the dorsal root ganglion and spinal cord. These studies highlight the potential clinical utility of local siRNA infusion to treat CNS diseases, especially as this route of administration is commonly used clinically to treat severe pain and neurodegenerative disorders.

The relatively rapid success with localized delivery of siRNA is presumably due to a variety of factors. First, in many cases such as the eye and lung, there exist inherent host defense and clearance mechanisms that may promote cellular uptake of siRNA. For instance, retinal pigment epithelial cells readily take up siRNA (Reich et al., 2003); these cells are naturally phagocytic as they underlie the retinal photoreceptor cells and act to digest shed rod and cone segments. A second and more practical reason for the rapid progress on local delivery of siRNA is that the amount of material required is small and thus more amenable to widespread in vivo experimentation. Some reports detailing localized delivery of siRNA use delivery or complexation agents, but it is clear from many of the studies that these may not be necessary, though it will be interesting to determine whether these additional agents improve uptake of siRNA.

Systemic RNAi Mediated by siRNAs

As most target tissues or organs cannot be addressed by local administration of potential siRNA therapeutics, a systemic route of delivery is the ultimate goal in developing siRNA drugs. In addition to the practical considerations of delivering therapeutics to internal organs, there are scenarios in which inhibition of gene expression in multiple tissues is desirable. For example, treatment of highly metastatic tumors will require multiple organ targeting. As is the case for local siRNA administration, recent success in systemic delivery has been achieved by using both naked and complexed siRNAs.

The inhibitory potency of systemically administered siRNAs in tumor xenograft models has been extensively studied. Multiple injection routes and doses were tested for their ability to reduce target gene expression in tumor tissue and to inhibit tumor growth. (Table I, systemic siRNA administration). To assess the effect of siRNA-mediated downmodulation of β-catenin expression on tumor growth, Verma et al. (2003) introduced oligofectamine-formulated siRNAs by intraperitoneal (IP) injection. Multiple injections of siRNA ($\sim$140 μg/kg) over a period of 25 days resulted in a prolonged survival of animals treated with the β-catenin-specific siRNAs compared with control siRNA-treated animals. A single intravenous (IV) injection of "naked" siRNA yielded a significant reduction of the β-catenin protein.

To address whether tumor tissue can be reached by different routes of administration, subcutaneous tumors from a fibrosarcoma cell line constitutively expressing luciferase served as a model system (Filleur et al., 2003). A single dose of 3 μg luciferase-specific siRNA ($\sim$125 μg/kg) was administered IV, IP, subcutaneously (SC), or intratumorally. All three systemic routes resulted in 40–50% inhibition of luciferase activity, whereas intratumoral injection did not result in any downmodulation. Intraperitoneal injections of a VEGF-specific siRNA (125 μg/kg daily for 16 days) to mice bearing SC tumors resulted in a 66% reduction of tumor growth and a 70% reduction in VEGF protein compared with control animals receiving a luciferase-specific siRNA (Filleur et al., 2003). Why intratumor injection did not work in this system while other routes of administration did is unclear. In contrast to this report, several groups have demonstrated that intratumor injection of siRNA is effective. Direct intratumor injection of unmodified and uncomplexed siRNA to CSF-1 (400 μg/kg every 3 days for

15 days) resulted in a 50% reduction in both CSF-1 protein and tumor volume (Aharinejad *et al.*, 2004). Lastly, two recent reports have shown that complexation of siRNA with atelocollagen provided a beneficial effect in stabilizing and delivering siRNA to the tumors (Minakuchi *et al.*, 2004; Takei *et al.*, 2004). In these reports, intratumor injection of atelocollagen-complexed siRNA resulted in dramatic decreases in tumor volume and target protein expression.

An antitumor strategy that has proven valuable *in vitro* is to sensitize cells toward established chemotherapeutics by siRNA-mediated reduction of a protein that counteracts the effects of the given drug (Siegmund *et al.*, 2002; Wacheck *et al.*, 2003). A similar approach was taken by Duxbury *et al.* (2003, 2004) in an orthotopic xenograft model using pancreatic adenocarcinoma cells. siRNA-mediated reduction of the M2 subunit of ribonucleotide reductase and focal adhesion kinase dramatically sensitized tumor growth toward treatment with gemcitabine. Inhibition of ribonucleotide reductase by siRNA in combination with gemcitabine treatment also inhibited tumor metastases completely (Duxbury *et al.*, 2004). For both targets, a 50% reduction in protein level was observed, demonstrating that a complete inhibition of target protein is not required to sensitize cells toward chemotherapy. The same strategy was applied when systemically administered siRNA targeting NFκB was combined with the small-molecule topoisomerase inhibitor CPT-11 (Guo *et al.*, 2004). Once again, though siRNA treatment alone was successful in modulating target protein levels, only the combined therapy was successful in significantly reducing tumor growth. Such combined therapies may allow for reduced dosing of the chemotherapeutic and therefore eliminate undesired side effects.

More recently, systemically administered siRNA complexed with polyethyleneimine (PEI) was effective in mediating RNAi in the lung (Ge *et al.*, 2004). PEI-complexed siRNA delivered systemically was shown to promote specific uptake of siRNA to the lung with more than 90% inhibition of a reporter gene. The IV administration of a single dose (2.4 mg/kg) of an siRNA targeting the influenza virus nucleocapsid was effective in reducing viral titers when delivered before viral infection. The response was dose dependent and could be enhanced by combining two siRNAs addressing different viral targets. PEI-formulated siRNA also proved functional in a therapeutic setting, as siRNA treatment (5 mg/kg) following viral challenge resulted in a 1000-fold reduction in viral titer.

Taken together, these studies clearly demonstrate that the systemic administration of naked or formulated siRNA, in doses as low as 120 μg/kg, can result in specific inhibition of target gene expression in both tumor cells and lung epithelium. Additional target tissues have yet to be explored for their capacity to take up siRNA and effectively silence genes. Future studies will address the pharmacokinetic distribution of siRNA after systemic delivery. In addition, the value of different siRNA formulations has yet to be explored for diverse target tissues.

Chemical Modifications

Chemical modification strategies are important tools to confer pharmaceutical properties to siRNAs. Although short dsRNA molecules used for silencing are inherently more stable than single-stranded RNAs, there are reasons to chemically modify one or both of the strands. First, enhanced resistance to exo- and endonucleases is important to obtain sufficient stability in blood. Second, modification may improve the pharmacokinetic properties of siRNAs *in vivo* by mediating binding to blood components, thereby increasing the circulation time of the siRNA. Third, chemical modification can aid in broadly targeting siRNA into cells and tissues and certain conjugates can enhance uptake in specific cell types. Multiple chemical modifications (Fig. 1) can be introduced at various positions within the siRNA duplex. A growing dataset that correlates chemical modifications with efficacy *in vitro* has aided our understanding of the fundamental mechanisms of RNAi. We next describe the link between chemical modification and activity and its consequence on therapeutic development.

Modifications at the Termini

A 5′-phosphate group on the antisense strand is mechanistically required for cleavage of the target mRNA (Schwarz *et al.*, 2002). Synthetic RNA bearing a 5′-hydroxyl (OH) group can mediate RNAi function, as the RNA is phosphorylated by an endogenous kinase. When the 5′-OH termini of the antisense strands were blocked by introducing 5′-OMe in place of 5′-OH, silencing was abolished; 5′-modification of the sense strand did not impair activity (Schwarz *et al.*, 2002). siRNA with the antisense strand modified at the 5′-end by a 6-amino-hexyl phosphodiester was active (Chiu and Rana, 2002) as was siRNA with a fluorescent label conjugated to the 5′-phosphate group through a six-carbon linker (Harborth *et al.*, 2003). Thus, pendant groups appear to be tolerated at the 5′-termini as long as they are connected by a phosphodiester linkage.

The data regarding modification of the 3′-terminus of the antisense strand of the siRNA duplex are not conclusive. In mammalian cells, two deoxy thymidines at the 3′-terminus of both strands are used by most researchers, although siRNA with perfectly paired nucleotides at the 3′-terminus also maintains activity (Czauderna *et al.*, 2003). When the 3′-end of the antisense

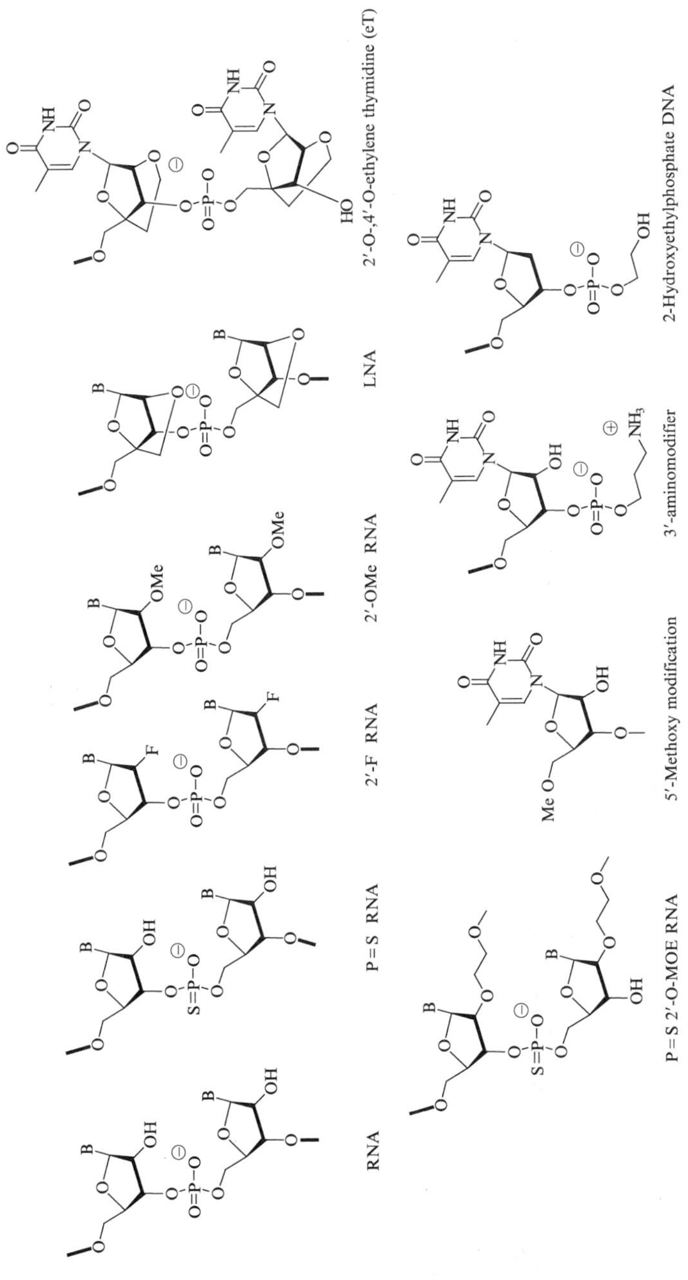

FIG. 1. Chemical modifications.

strand and both termini of the sense strand were modified with either an inverted deoxy abasic residue or an amino group attached through a six-carbon linker, no reduction in activity was observed (Czauderna et al., 2003). In contrast, siRNAs lost all activity when the 3'-end of the antisense strand was modified with either 2-hydroxyethylphosphate or 2'-O,4'-C-ethylene thymidine (Hamada et al., 2002). When a 3'-puromycin or a 3'-biotin modification at the antisense strand was introduced, siRNAs retained activity (Chiu and Rana, 2002). These data highlight that terminal modifications in most instances are tolerated. Further studies will be required to delineate the tolerance of siRNAs to end modifications.

Lastly, certain conjugates are known to improve cellular uptake. For example, siRNAs conjugated at the 5'-end of the sense strand with cholesterol, lithocholic acid, and lauric acid have been reported to improve *in vitro* cell permeation in liver cells (Lorenz et al., 2004).

Modifications at the Ribose

Silencing by siRNAs is compatible with some types of internal chemical modifications. Substitutions at the 2'-position of the ribose have been extensively investigated, due in large part to the protection these afford against endonucleases. A number of groups have demonstrated that 2'-F modifications of the sense and antisense strands are well tolerated (Braasch et al., 2003; Chiu and Rana, 2003; Harborth et al., 2003; Layzer et al., 2004). siRNAs with a combination of 2'-F and phosphorothioate (P = S) modifications showed improved activity in cell culture (Harborth et al., 2003). 2'-F substitutions maintained activity in cell culture and showed increased stability and a prolonged half-life in human plasma as compared with unmodified siRNAs (Layzer et al., 2004).

The 2'-O-methyl modification, although conferring significant serum stability, completely inhibited siRNA activity when both strands were fully modified (Czauderna et al., 2003). When every other nucleotide was modified with 2'-O-methyl, siRNAs were resistant to serum-derived nucleases and some variants of this modification strategy retained significant activity (Czauderna et al., 2003). The 2'-O-methoxyethyl (MOE) terminal modification with the P = S linkage (Fig. 1) has been used to improve nuclease resistance in local delivery applications (Dorn et al., 2004).

Lastly, the locked nucleic acid (LNA) substitution (Fig. 1) has been evaluated as a modification for siRNA (Braasch et al., 2003). LNA contains a methylene linkage between the 2'-and 4'-positions of the ribose. An siRNA duplex with internal LNA modifications retained full activity. As this duplex had a melting temperature of more than 93°, this experiment demonstrated that the thermodynamic stability of duplexes can be extremely high without affecting the function of the siRNA within the RNA-induced silencing complex. This is consistent with the assumption that ATP hydrolysis drives the separation of the siRNA strands. siRNAs modified with LNA at the 5'-termini, the 3'-termini, or both, also retained activity. This is consistent with data using 2'-O,4'-C-ethylene thymidine, a component of ethylene-bridge nucleic acids (ENAs), at the termini of siRNAs (Fig. 1).

Modifications Within the Backbone

Partial (12 P = S per strand) or complete modification with P = S linkages did not significantly enhance serum stability of the corresponding siRNAs. Reduced thermodynamic stability of these P = S-containing duplexes as compared to unmodified siRNAs was hypothesized to explain the lack of expected scrum stability. In addition, full P = S modification slightly reduced siRNA activity whereas partial P = S modification did not impair inhibition of gene expression (Braasch et al., 2003). In a second study complete modification of siRNAs with P = S diminished activity by >50% (Chiu and Rana, 2003). Finally, siRNAs containing alternating P = S and phosphodiesters linkages retained full activity; however, cytotoxicity was observed (Harborth et al., 2003).

Methodology

As described in the previous part of this chapter, the use of siRNA to down regulate target gene expression *in vivo* has been demonstrated for multiple targets in different target tissues (see Table I for a summary). Nevertheless, the *in vivo* use of siRNA mediated RNAi is far from being a well-established standard technology. For most model systems, optimization of siRNA dosing and application regimen as well as finding the optimal time frame for analyzing inhibition of target gene expression will be required. Therefore, this methods section can only serve as a short guide for planning individual experiments. The reader is referred to other publications to find experimental conditions appropriate for their specific target mRNA and target tissue.

Successful *in vivo* RNAi was achieved in various mouse xenografts of human tumors. Here we describe a protocol used by Duxbury and colleagues to down regulate target gene expression of the M2 subunit of ribonucleotide reductase (RRM2) and focal adhesion kinase (FAK) in a subcutaneous and an orthotopic tumor model of human pancreatic carcinoma cells, respectively (Duxbury et al., 2003, 2004).

Inhibition of Human Tumor Target Gene Expression in Mouse

Tumor cell growth and implantation/injection

1. PANC1 cells are maintained in Dulbecco's Modified Eagle Medium (DMEM) containing 10% fetal bovine serum at 37° and 5% CO_2. The human pancreatic adenocarcinoma cell line MIAPaCa$_2$ is maintained in the same medium containing 1.5 g/l sodium bicarbonate.

2. All animal protocols should be in accordance with approved institutional guidelines. Pathogen free, male athymic nu/nu mice can be obtained from different vendors (e.g., Charles Rivers, Jackson Labs) and are housed in microisolator cages with autoclaved bedding in a pathogen free environment with a 12 hour light-dark cycle. Animals receive food and water ad libitum.

3. For analysis of siRNA mediated inhibition of RRM2, 1×10^6 PANC1 cells in 20 μl phosphate buffered saline (PBS) are implanted subcutaneously after the animals are anesthetized with intraperitoneal ketamine (200 mg/kg) and xylazine (10 mg/kg). siRNA treatment commences 20 days after tumor cell inoculation when tumors have reached a size of 40–60 mm^3. For silencing of FAK in an orthotopic xenograft system, 1×10^6 MIAPaCa2 cells are surgically implanted into the pancreas of anesthetized mice as described (Tan and Chu, 1985).

H3siRNAs, Formulation and Administration. The following siRNAs carrying a two deoxy-T nucleotide overhang at the 3′-end of each strand are used for the study:

RRM2 sense strand: 5′-UGCUGUUCGGAUAGAACAGdTdT-3′
RRM2 antisense strand: 5′-CUGUUCUAUCCGUUCUGCAdTdT-3′
FAK sense strand: 5′-GACGUGGGACUGAAGAGGUdTdT-3′
FAK antisense strand: 5′-ACCUCUUCAGUCCCACGUCdTdT-3′

Single nucleotide mismatch (MM, in bold) containing siRNAs serve as controls:

RRM2-MM sense   5′-UGCU**U**UUCGGAUAGAACAGdTdT-3′
strand:
RRM2-MM antisense   5′-CUGUUCUAUCCG*AAA*AGCAdTdT-3′
strand:
FAK-MM sense strand: 5′-GACGUGGGACUGAAG**GGG**UdTdT-3′
FAK-MM antisense   5′-AC**CCC**UUCAGUCCCACGUCdTdT-3′
strand:

siRNAs can be ordered from multiple suppliers. Only high quality, HPLC purified material should be used for *in vivo* studies.

Lyophilized siRNA single strands are resuspended in PBS to a final concentration of 20 μM. To prepare duplex, equimolar amounts of sense and antisense strand are combined and adjusted to a final concentration of 4.3 μM of each strand, in phosphate buffered saline (PBS) The mixture is then heated to 90° in a water bath and allowed to cool to room temperature over a period of 2 hours. Annealed siRNAs can be stored at 4°. For each injection, 50 μl of the 4.3 μM siRNA solution ($\sim$150 μg/kg body weight) is administered IV.

VARIATIONS. Alternatively, siRNAs can be annealed at higher stock concentrations, distributed into aliquots, and diluted to working concentration prior to injection without additional annealing. Other buffers than PBS can be used for annealing and injection. Solutions should be buffered to physiological pH, contain about 100 mM monovalent cations and be essentially free of bivalent cations.

siRNAs are injected via the tail vein using 26 G × 1″ hypodermic needles attached to 1 ml inject-F syringes. siRNAs are injected twice per week for a total of 6 weeks. In order to maximize the effect of RRM2 or FAK downregulation on tumor growth, siRNAs can be combined with gemcitabine treatment. Gemcitabine is administered twice per week for 6 weeks in a 150 mg/kg dose via intraperitoneal injection.

VARIATIONS. Besides administration via the tail vein, successful delivery of siRNA to tumors has been reported using intraperitoneal as well as subcutaneous administration (Filleur *et al.*, 2003). While intratumoral injection of unformulated siRNA was not successful in this setup, other investigators have reported success with such delivery and formulated siRNA (Minakuchi *et al.*, 2004; Takei *et al.*, 2004).

ADDITIONAL COMMENTS. The optimal siRNA dose, administration regimen, and dosing schedule has to be established for every target mRNA individually and will vary depending on the particulars of the xenograft model. These parameters are dependent on multiple target specific factors such as mRNA-synthesis rate and –half-life, mRNA accessibility, translation efficacy, and protein stability.

Detection of Target Gene Knockdown. Here we restrict ourselves to summarizing different approaches to measuring mRNA or protein levels in tumors.

1. Multiple protocols describe the isolation of mRNA from tumor tissue (e.g., Grotzer *et al.*, 2000). Target mRNA reduction can be analyzed semi-quantitatively by classical Northern blots. RNase protection, RT-PCR or various hybridization assays (e.g., b-DNA assay, Genospectra) can be used for quantitative analysis of mRNA concentration.

2. Target protein reduction can be qualitatively evaluated by Western blot analysis as well as by immunohistochemical methods. ELISAs or target protein activity assays allow for quantitative analysis. Finally, tumor

mass or induction of apoptosis (as evidenced by caspase3 activity or tunnel staining) can serve as indirect readouts of siRNA activity.

Additional Considerations

In order to assign the observed phenotype directly to inhibition of target gene expression, the proper controls must be employed. Mismatch siRNAs or unrelated siRNAs (e.g., luciferase) have served as controls in most studies. In addition, a saline treatment group will control for target independent effects of the siRNA. Finally, consistent phenotypes observed upon administration of different siRNAs against the same mRNA target gene is a strong indicator of specific target down modulation.

Summary

In a relatively short period of time, significant progress has been made on the demonstration of RNAi *in vivo*. Although the earliest proof-of-concept experiments employed hydrodynamic delivery, recent efforts have focused on more therapeutically relevant modes of administration. The major challenge in bringing siRNAs into the clinic will be the improvement of their pharmacokinetic properties and their cellular uptake. The introduction of chemical modifications may also be required to enhance the stability of siRNAs in a biological environment.

As with any emerging technology, the opportunities for applying RNAi in a therapeutic setting seem boundless. Significant additional work will be required to understand the true limitations. It remains to be seen how chemically modified siRNA function *in vivo* and to what extent the chemistry has an impact on other important properties of the siRNA such as toxicity and specificity.

References

Aharinejad, S., Paulus, P., Sioud, M., Hofmann, M., Zins, K., Schafer, R., Stanley, E. R., and Abraham, D. (2004). Colony-stimulating factor-1 blockade by antisense oligonucleotides and small interfering RNAs suppresses growth of human mammary tumor xenografts in mice. *Cancer Res.* **64**, 5378–5384.

Berns, K., Hijmans, E. M., Mullenders, J., Brummelkamp, T. R., Velds, A., Heimerikx, M., Kerkhoven, R. M., Madiredjo, M., Nijkamp, W., Weigelt, B., Agami, R., Ge, W., Cavet, G., Linsley, P. S., Beijersbergen, R. L., and Bernards, R. (2004). A large-scale RNAi screen in human cells identifies new components of the p53 pathway. *Nature* **428**, 431–437.

Braasch, D. A., Jensen, S., Liu, Y., Kaur, K., Arar, K., White, M. A., and Corey, D. R. (2003). RNA interference in mammalian cells by chemically-modified RNA. *Biochemistry* **42**, 7967–7975.

Chiu, Y. L., and Rana, T. M. (2002). RNAi in human cells: Basic structural and functional features of small interfering RNA. *Mol. Cell* **10**, 549–561.

Chiu, Y. L., and Rana, T. M. (2003). siRNA function in RNAi: A chemical modification analysis. *RNA* **9**, 1034–1048.

Czauderna, F., Fechtner, M., Dames, S., Aygun, H., Klippel, A., Pronk, G. J., Giese, K., and Kaufmann, J. (2003). Structural variations and stabilising modifications of synthetic siRNAs in mammalian cells. *Nucleic Acids Res.* **31,** 2705–2716.

Dorn, G., Patel, S., Wotherspoon, G., Hemmings-Mieszczak, M., Barclay, J., Natt, F. J., Martin, P., Bevan, S., Fox, A., Ganju, P., Wishart, W., and Hall, J. (2004). siRNA relieves chronic neuropathic pain. *Nucleic Acids Res.* **32,** e49.

Dorsett, Y., and Tuschl, T. (2004). siRNAs: Applications in functional genomics and potential as therapeutics. *Nat. Rev. Drug Discov.* **3,** 318–329.

Duxbury, M. S., Ito, H., Benoit, E., Zinner, M. J., Ashley, S. W., and Whang, E. E. (2003). RNA interference targeting focal adhesion kinase enhances pancreatic adenocarcinoma gemcitabine chemosensitivity. *Biochem. Biophys. Res. Commun.* **311,** 786–792.

Duxbury, M. S., Ito, H., Zinner, M. J., Ashley, S. W., and Whang, E. E. (2004). RNA interference targeting the M2 subunit of ribonucleotide reductase enhances pancreatic adenocarcinoma chemosensitivity to gemcitabine. *Oncogene* **23,** 1539–1548.

Filleur, S., Courtin, A., Ait-Si-Ali, S., Guglielmi, J., Merle, C., Harel-Bellan, A., Clezardin, P., and Cabon, F. (2003). SiRNA-mediated inhibition of vascular endothelial growth factor severely limits tumor resistance to antiangiogenic thrombospondin-1 and slows tumor vascularization and growth. *Cancer Res.* **63,** 3919–3922.

Ge, Q., Filip, L., Bai, A., Nguyen, T., Eisen, H. N., and Chen, J. (2004). Inhibition of influenza virus production in virus-infected mice by RNA interference. *Proc. Natl. Acad. Sci. USA* **101,** 8676–8681.

Giladi, H., Ketzinel-Gilad, M., Rivkin, L., Felig, Y., Nussbaum, O., and Galun, E. (2003). Small interfering RNA inhibits hepatitis B virus replication in mice. *Mol. Ther.* **8,** 769–776.

Grotzer, M. A., Patti, R., Geoerger, B., Eggert, A., Chou, T. T., and Phillips, P. C. (2000). Biological stability of RNA isolated from RNAlater-treated brain tumor and neuroblastoma xenografts. *Med. Pediatr. Oncol* **34,** 438–442.

Guo, J., Verma, U. N., Gaynor, R. B., Frenkel, E. P., and Becerra, C. R. (2004). Enhanced chemosensitivity to irinotecan by RNA interference-mediated down-regulation of the nuclear factor-kappaB p65 subunit. *Clin. Cancer Res.* **10,** 3333–3341.

Hamada, M., Ohtsuka, T., Kawaida, R., Koizumi, M., Morita, K., Furukawa, H., Imanishi, T., Miyagishi, M., and Taira, K. (2002). Effects on RNA interference in gene expression (RNAi) in cultured mammalian cells of mismatches and the introduction of chemical modifications at the 3′-ends of siRNAs. *Antisense Nucleic Acid Drug Dev.* **12,** 301–309.

Harborth, J., Elbashir, S. M., Vandenburgh, K., Manninga, H., Scaringe, S. A., Weber, K., and Tuschl, T. (2003). Sequence, chemical, and structural variation of small interfering RNAs and short hairpin RNAs and the effect on mammalian gene silencing. *Antisense Nucleic Acid Drug Dev.* **13,** 83–105.

Layzer, J. M., McCaffrey, A. P., Tanner, A. K., Huang, Z., Kay, M. A., and Sullenger, B. A. (2004). *In vivo* activity of nuclease-resistant siRNAs. *RNA* **10,** 766–771.

Lewis, D. L., Hagstrom, J. E., Loomis, A. G., Wolff, J. A., and Herweijer, H. (2002). Efficient delivery of siRNA for inhibition of gene expression in postnatal mice. *Nat. Genet.* **32,** 107–108.

Lorenz, C., Hadwiger, P., John, M., Vornlocher, H.-P., and Unverzagt, C. (2004). Steroid and lipid conjugates of siRNAs to enhance cellular uptake and gene silencing in liver cells. *Bioorg. Med. Chem. Lett.* **14,** 4975–4977

Makimura, H., Mizuno, T. M., Mastaitis, J. W., Agami, R., and Mobbs, C. V. (2002). Reducing hypothalamic AGRP by RNA interference increases metabolic rate and decreases body weight without influencing food intake. *BMC Neurosci.* **3,** 18–23.

McCaffrey, A. P., Meuse, L., Pham, T. T., Conklin, D. S., Hannon, G. J., and Kay, M. A. (2002). RNA interference in adult mice. *Nature* **418,** 38–39.

Minakuchi, Y., Takeshita, F., Kosaka, N., Sasaki, H., Yamamoto, Y., Kouno, M., Honma, K., Nagahara, S., Hanai, K., Sano, A., Kato, T., Terada, M., and Ochiya, T. (2004). Atelocollagen-mediated synthetic small interfering RNA delivery for effective gene silencing *in vitro* and *in vivo*. *Nucleic Acids Res.* **32,** e109.

Novina, C. D., and Sharp, P. A. (2004). The RNAi revolution. *Nature* **430,** 161–164.

Paddison, P. J., Silva, J. M., Conklin, D. S., Schlabach, M., Li, M., Aruleba, S., Balija, V., O'Shaughnessy, A., Gnoj, L., Scobie, K., Chang, K., Westbrook, T., Cleary, M., Sachidanandam, R., McCombie, W. R., Elledge, S. J., and Hannon, G. J. (2004). A resource for large-scale RNA-interference-based screens in mammals. *Nature* **428,** 427–431.

Reich, S. J., Fosnot, J., Kuroki, A., Tang, W., Yang, X., Maguire, A. M., Bennett, J., and Tolentino, M. J. (2003). Small interfering RNA (siRNA) targeting VEGF effectively inhibits ocular neovascularization in a mouse model. *Mol. Vis.* **9,** 210–216.

Schwarz, D. S., Hutvagner, G., Haley, B., and Zamore, P. D. (2002). Evidence that siRNAs function as guides, not primers, in the *Drosophila* and human RNAi pathways. *Mol. Cell* **10,** 537–548.

Siegmund, D., Hadwiger, P., Pfizenmaier, K., Vornlocher, H.-P., and Wajant, H. (2002). Selective inhibition of FLICE-like inhibitory protein expression with small interfering RNA oligonucleotides is sufficient to sensitize tumor cells for TRAIL-induced apoptosis. *Mol. Med.* **8,** 725–732.

Song, E., Lee, S. K., Wang, J., Ince, N., Ouyang, N., Min, J., Chen, J., Shankar, P., and Lieberman, J. (2003). RNA interference targeting Fas protects mice from fulminant hepatitis. *Nat. Med.* **9,** 347–351.

Song, Y. K., Liu, F., Zhang, G., and Liu, D. (2002). Hydrodynamics-based transfection: Simple and efficient method for introducing and expressing transgenes in animals by intravenous injection of DNA. *Methods Enzymol.* **346,** 92–105.

Takei, Y., Kadomatsu, K., Yuzawa, Y., Matsuo, S., and Muramatsu, T. (2004). A small interfering RNA targeting vascular endothelial growth factor as cancer therapeutics. *Cancer Res.* **64,** 3365–3370.

Tan, M. H., and Chu, T. M. (1985). Characterization of the tumorigenic and metastatic properties of a human pancreatic tumor cell line (AsPC-1) implanted orthotopically into nude mice. *Tumour Biol.* **6,** 89–98.

Tolentino, M. J., Brucker, A. J., Fosnot, J., Ying, G. S., Wu, I. H., Malik, G., Wan, S., and Reich, S. J. (2004). Intravitreal injection of vascular endothelial growth factor small interfering RNA inhibits growth and leakage in a nonhuman primate, laser-induced model of choroidal neovascularization. *Retina* **24,** 132–138.

Verma, U. N., Surabhi, R. M., Schmaltieg, A., Becerra, C., and Gaynor, R. B. (2003). Small interfering RNAs directed against β-catenin inhibit the *in vitro* and *in vivo* growth of colon cancer cells. *Clin. Cancer Res.* **9,** 1291–1300.

Wacheck, V., Losert, D., Günsberg, P., Vornlocher, H.-P., Hadwiger, P., Geick, A., Pehamberger, H., Müller, M., and Jansen, B. (2003). Small interfering RNA targeting Bcl-2 sensitizes malignant melanoma. *Oligonucleotides* **13,** 393–400.

Zender, L., Hütker, S., Liedke, C., Tillmann, H. L., Zender, S., Mundt, B., Waltemathe, M., Gösling, T., Flemming, P., Malek, N. P., Trautwein, C., Manns, M. P., Kühnel, F., and Kubicka, S. (2003). Caspase 8 small interfering RNA prevents acute liver failure in mice. *Proc. Natl. Acad. Sci. USA* **100,** 7797–7802.

Zhang, X., Shan, P., Jiang, D., Noble, P. W., Abraham, N. G., Kappas, A., and Lee, P. J. (2004). Small interfering RNA targeting heme oxygenase-1 enhances ischemia-reperfusion-induced lung apoptosis. *J. Biol. Chem.* **279,** 10677–10684.

[17] Labeling and Characterization of Small Rnas Associated with the RNA Interference Effector Complex RITS

By ANDRÉ VERDEL and DANESH MOAZED

Abstract

RNA interference (Rnai) is a gene silencing mechanism that acts at both the posttranscriptional and transcriptional levels. We have recently identified an RNA-containing complex, named RNA-induced transcriptional silencing (RITS), that directly links Rnai to transcriptional gene silencing in *Schizosaccharomyces pombe*. Here we review the affinity purification methods we use to isolate RITS and describe how to purify, detect, and analyze Rnas associated with this complex.

Introduction

RNA interference (Rnai) is a general silencing mechanism that is triggered by double-stranded RNA (dsRNA; Fire *et al.*, 1998; Hannon, 2002; Zamore, 2001). Rnai-based silencing mechanisms appear to be conserved in a wide variety of eukaryotes, ranging from the fission yeast *Schizosaccharomyces pombe* to mammals. Rnai is initiated by Dicer, an RNase III-like enzyme, which cleaves large dsRNAs into small RNA duplexes of ~21 nt, called small interfering RNAs (siRNAs; Bernstein *et al.*, 2001; Elbashir *et al.*, 2001; Hamilton *et al.*, 1999; Zamore *et al.*, 2000). siRNAs load onto effector complexes, the RNA-induced silencing complexes (RISCs), which contain a member of the Argonaute family of proteins. Argonaute proteins are believed to interact directly with siRNAs (Caudy *et al.*, 2002; Ishizuka *et al.*, 2002). Recognition of a target mRNA by RISC involves direct base pairing interactions between siRNAs and target mRNAs and leads to mRNA degradation or translational inactivation, depending on the degree of homology between the siRNA and the target mRNA (Doench *et al.*, 2003; Hutvagner and Zamore, 2002). In either case, protein expression is inhibited, and this process has been defined as posttranscriptional gene silencing (PTGS). In some organisms, an RNA-directed RNA polymerase (RdRp) activity has been proposed to be involved in amplifying the Rnai response by using siRNAs as primers to produce more dsRNAs (Lipardi *et al.*, 2001; Sijen *et al.*, 2001).

In addition to its role in PTGS, several studies suggest that the Rnai pathway also acts at the level of DNA and transcription (Matzke *et al.*, 2004).

Several components of the RNAi pathway are conserved in *S. pombe*, and deletion of the genes that encode these components results in a loss of transcriptional gene silencing (TGS) at heterochromatic DNA regions (Volpe *et al.*, 2002). Moreover, the RNAi pathway is required for initiating the assembly of heterochromatin (Hall *et al.*, 2002) and can also initiate transcriptional gene silencing at an ectopic locus in response to production of a complementary hairpin RNA (Schramke and Allshire, 2003). Using *S. pombe*, we have recently identified a new type of RNAi effector complex that contains siRNAs and an Argonaute protein but induces silencing at the level of transcription (Verdel *et al.*, 2004). We have named this complex RNA-induced transcriptional silencing (RITS; Verdel *et al.*, 2004) complex. TGS induced by RITS requires Dicer and other RNAi components as well as the presence of siRNAs, suggesting that, in a similar fashion to RISC, RITS uses siRNAs to recognize genomic DNA regions that should be silenced.

The key components of the RNAi pathway are conserved in *S. pombe*. But unlike most eukaryotes, *S. pombe* possess only one copy of each Dicer, RdRp, and Argonaute ($dcr1^+$, $rdp1^+$, and $ago1^+$, respectively). Moreover, studies of RNAi in *S. pombe* offer the possibility of combining rapid genetic manipulations with extensive biochemical analysis, making this organism an ideal system for dissecting the mechanism of RNAi-dependent gene silencing.

Tandem affinity purification (TAP) is a powerful strategy developed by Seraphin and coworkers (Puig *et al.*, 2001), which employs two successive affinity purification steps that greatly reduce the levels of nonspecifically associated proteins. The most popular double tag used in yeast is composed of a calmodulin-binding peptide (CBP) and two protein A repeats (Gould *et al.*, 2004). We have used this strategy extensively in our laboratory to purify yeast silencing complexes (Hoppe *et al.*, 2002; Tanny *et al.*, 2004; Verdel *et al.*, 2004). Here we describe its use in purification of the RITS complex, which also contains small RNA molecules. We describe our protocols for both isolation of the RITS complex and for labeling its associated small RNAs.

Purification of the RNA-Containing Complex RITS

Strain Constructions

To obtain *S. pombe* strains expressing a protein fused to a tag, we used a PCR-based lithium acetate transformation procedure described by Bahler *et al.* (1998). Fission yeast can efficiently integrate foreign DNA in its genome by homologous recombination. Therefore, a sequence coding for a tag can be inserted in the genome together with a selective marker such as KanMX4 to obtain a strain that expresses a fusion protein with the tag at its N- or C-terminal end (Fig. 1). When the tag is inserted at the C terminus, expression of the gene is under its own promoter (Fig. 1A). Conversely, when the tag is integrated at the N terminus, expression is under the control of the inducible *nmt1* promoter (Fig. 1B). If necessary, an N-terminally tagged protein can be expressed under the control of its own promoter (Storici *et al.*, 2001).

The purification of RITS was performed by adding the TAP tag at the C terminus of Chp1, as described in Verdel *et al.* (2004). To obtain *S. pombe* strains expressing Chp1 fused to the TAP tag, we used a PCR-based lithium acetate transformation procedure described by Bahler *et al.* (1998). The DNA cassette encoding the TAP tag and a selective marker was amplified by PCR from a pFA6 vector and used for transformation of *S. pombe*, as described previously (Bahler *et al.*, 1998).

A   C-Terminal tagging   B   N-Terminal tagging

1. PCR amplification

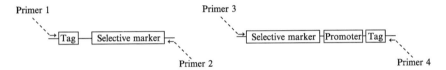

2. Transformation and integration by homologous recombination

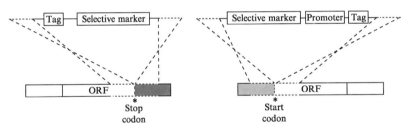

FIG. 1. PCR-based construction of yeast strains that express tagged proteins. We use gel-purified primers approximately 100 nt in length, containing 20 nt at their 3' end required for amplification of the cassette and about 80 nt identical to the sequence present on each side of the site of integration (dashed portion of the arrows) to allow the integration of the cassette by homologous recombination. (The crossed dashed lines indicate the DNA regions involved in the homologous recombination.) After PCR and purification, the DNA product is introduced into yeast by a lithium acetate method described by Bahler *et al.* (1998) and integrated in the genome. This transformation strategy on average results in 50% positive transformants, but the proportion of positive transformants can greatly vary (from ∼5 to 100%), depending on the locus. (A) C-terminal tagging. The DNA cassette that contains the tag and a selective marker is integrated downstream of the open reading frame (ORF) after the last codon and before the stop codon at the 3' end of the gene (dark gray). (B) N-terminal tagging. The integration of the DNA cassette leads to separation of the endogenous promoter (light gray) from the ORF. Expression of the N-terminal-tagged protein is then under the control of an ectopic promoter.

Growth Conditions

We grow 3- to 12-L cultures of yeast cells, depending on the abundance of the protein (and on the amount of purified material desired). Usually, for 3- to 6-L cultures, we use 2.8-L baffled flasks containing 1.5 L of rich medium (5 mg/ml yeast extract; 30 mg/ml dextrose; and 225 mg/ml each of adenine, leucine, lysine, histidine, and uracil). Each 1.5-L culture is inoculated with 2–5 ml of an overnight preculture and allowed to grow for approximately 16 h at 30–32° with constant agitation at 225 rpm to an optical density at 600 nm (OD_{600}) of 2.0–3.0. These cultures produce typically 3–4 g of cells per liter.

For 12-L cultures, we use a fermenter. The growth medium is inoculated with 10 ml of a saturated preculture, and cells are grown at 32° with agitation at 400 rpm for approximately 20 h until they reach an OD_{600} of approximately 5. To maintain logarithmic growth, 1% dextrose is added to the culture when the OD_{600} is around 0.5. Under these growth conditions, 6–8 g of cells are produced per liter of culture. We have noticed that by using the fermenter, we can harvest *S. pombe* cells at a higher OD_{600} without affecting the subsequent lysis efficiency.

Cells are harvested by centrifugation at 7000 rpm for 8 min at room temperature (Beckman Coulter, Avanti J-20XP; rotor JLA 8.1000), washed once in approximately 10 volumes of water and harvested by centrifugation at 3000 rpm for 5 min (Sorvall RC 5C plus; rotor SLA-1500). Cell pellets are flash frozen in liquid nitrogen and stored at −80° until use.

Tandem Affinity Purification (TAP) of RITS

Cell Lysis and Cell Extract Preparation. We use two scales of TAPs. For the smaller scale, 6–12 g of cells are lysed. Frozen cell pellets are first broken to obtain a fine powder by grinding for approximately 10–15 min in a prechilled mortar and pestle. The cell powder is transferred to a beaker and thawed at room temperature until the edges start to melt. One volume of room-temperature lysis and wash buffer (LW buffer; 50 mM HEPES-KOH, pH 7.6, 300 mM potassium acetate, 10% glycerol, 1 mM EGTA, 1 mM EDTA, 0.1% NP-40, 5 mM magnesium acetate, 1 mM NaF, 20 mM sodium β-glycerophosphate, 1 mM DTT, 1 mM PMSF and benzamidine, 1 μg/ml leupeptin, aprotinin, bestatin, and pepstatin; add DTT and protease inhibitors immediately before use) is added to resuspend the cells. This procedure appears to minimize proteolysis. The cell suspension is then divided in six cold polypropylene round-bottom Falcon™ tubes containing 2.5 ml of cold glass beads. Cells are lysed at 4° by vortexing 12 times for 30 sec with 90-sec rest intervals on ice.

When working with 40–100 g of cells, we use the Bead-Beater™ (BioSpec, Bartlesville, OK) to lyse the cells. As described previously, the frozen cell pellets are first ground in a prechilled mortar. After filling about half the precooled 380-ml bead-beating chamber with cold glass beads, cell powder and two volumes of room-temperature LW buffer (see previous paragraph) are added. The beads are mixed to eliminate trapped air, the chamber completely filled by adding more beads if necessary, and the chamber assembled with an outside ice-water jacket. Cells are then lysed by bead beating for 8–12 cycles of 20 sec each with 40–60 sec of rest interval on ice between every cycle. The cells are transferred to cold 30-ml centrifuge tubes, and the beads are washed with one volume of cold LW buffer, which is pooled with the broken cells. The lysate is centrifuged at 40,000g for 25 min at 4° to eliminate cell debris (SA600, 16500 rpm), and the supernatant is used for affinity purification.

Two-Step Affinity Purification. We use a modified version of the purification method described by Seraphin and colleagues. The first affinity purification step is mixing 400 μl of IgG Sepharose™ beads (1:1 slurry with lysis buffer, Amersham Biosciences, England) with the supernatant from the previous step and incubating on a rotating platform at 4° for 1–2 h. IgG beads are prewashed three times in 1 ml of cold LW buffer (without any protease inhibitors). After incubation with extract, the beads are recovered by centrifugation at 500g for 3 min at 4°. Approximately 90% of the supernatant is discarded, and the rest is used to resuspend the beads to facilitate their transfer into a 10-ml PolyPrep column (Bio-Rad, Hercules, CA). The column is washed successively with 30 ml of LW buffer and 10 ml of TEV-C buffer (10 mM Tris-HCl, pH 8.0, 150 mM potassium acetate, 0.1% NP-40, 0.5 mM EDTA, 1 mM DTT; add DTT immediately before use). The tagged protein is eluted from the IgG Sepharose by incubation with 1 ml TEV-C buffer (added directly to column) containing 30 units of recombinant TEV (Invitrogen, Carlsbad, CA) or GST-TEV protease at room temperature for 1 h. Every 20 min, the IgG Sepharose is resuspended by pipetting. The eluate is then collected in a 15-ml conical Falcon tube, and the column is washed with 500 μl of TEV-C buffer. The eluate is diluted with two volumes of CAM-B buffer (10 mM Tris-HCl, pH 8.0, 150 mM NaCl, 1 mM magnesium acetate, 1 mM imidazole, 2 mM $CaCl_2$, 10 mM β-mercaptoethanol; add β-mercaptoethanol immediately before use), and the final concentration of calcium is adjusted to 2 mM. The second affinity purification step is performed by adding 300 μl of washed Calmodulin Sepharose™ (Amersham Biosciences; 1:1 slurry, prewashed three times with CAM-B buffer) and incubating on a rotating platform at 4° for 1 h. The beads are poured into a PolyPrep column (Bio-Rad), and the column is washed three times with 10 ml of CAM-B buffer containing 0.1% NP-40. We have found that these washes can be done with much lower volumes of CAM-B (1–2 ml) with little or no increase in background of nonspecific proteins. To elute the bound protein from the Calmodulin Sepharose column, 200 μl of CAM-E buffer (10 mM Tris-HCl, pH 8.0, 150 mM NaCl, 1 mM magnesium acetate, 1 mM imidazole, 10 mM EGTA, 10 mM β-mercaptoethanol; add β-mercaptoethanol immediately before use) is pipetted to the top of the column bed, this step repeated four times, and fractions 2 and 3 and 4 and 5 pooled. About 10–50% of fractions 2 and 3 (which usually contain most of the purified material) are TCA precipitated and analyzed by SDS-PAGE and silver staining (Fig. 2A) or mass spectrometry. The remainder is used for *in vitro* activity assays and can be stored at −80°.

Identification of the Purified Proteins

Two types of mass spectrometry analysis are combined to identify proteins present in our final eluate: band and mixture analysis (Peng and Gygi, 2001; Shevchenko *et al.*, 1996). For both analyses, 20–50% of the final

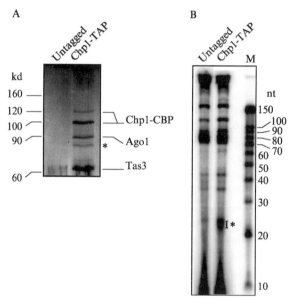

FIG. 2. Purification of the RNA-induced transcriptional silencing (RITS) complex and labeling of its associated RNAs (see Verdel *et al.*, 2004). (A) Purifications were conducted using 12 g of cells of a control wild-type strain (untagged) and a strain expressing Chp1-TAP. A total of 2.5% of pooled fractions 2 and 3 (see text for details) was run on an 8.5% SDS-PAGE. The proteins were detected by silver staining. The asterisk indicates a degradation product of Chp1. (B) Detection of RNAs present in the final eluate after the tandem affinity purification. RITS was purified from 7 g of wild-type (untagged) or Chp1-TAP cells. RNAs were then isolated from 50% of pooled fractions 2 and 3, and 3′ end-labeled with [^{32}P]-pCp as described in the text and in Verdel *et al.* (2004). Small RNAs specifically associated with Chp1-TAP are indicated by the asterisk.

eluate is TCA precipitated. TCA is added to the purified sample to a final concentration of 20%, and the mixture is incubated 20 min on ice. After spinning at 4° for 20 min at 12,000g, the pellet is washed with 100% cold (−20°) acetone. For mixture analysis, the pellet is then used for trypsin digestion followed by liquid chromatography and tandem mass spectrometry (LC-MS/MS; Peng and Gygi, 2001). For gel analysis, the protein pellet is resuspended in 10 μl of sample buffer, heated for 3 min at 100°, and loaded on a 4–12% acrylamide-SDS gel (Invitrogen). Proteins are then detected by silver or colloidal Coomassie staining. Protein bands are excised from the gel, digested with trypsin, and identified by mass spectrometry (LC-MS/MS; Hoppe et al., 2002; Shevchenko et al., 1996).

Detection of Small RNAs Associated
 with the TAP-Purified Material

Nucleic Acid Purification from the Final Eluate

To isolate any RNAs that may be specifically associated with the final eluate from the TAP purification, the purified proteins are first digested by adding proteinase K to 0.2 mg/ml and SDS to 0.1% and incubated at 37° for 45 min. The resultant peptides are eliminated by two successive extractions: the first with one volume of phenol:chloroform:isoamyl alcohol and the second with one volume of chloroform. RNAs are then precipitated by adding one-tenth volume of 3 *M* sodium acetate (pH 5.2), 20 μg glycogen, and 2.5 volumes of 100% ethanol. After incubation at −80° for 30 min, RNAs are pelleted at 4° for 20 min at 12,000g. The RNA pellets are resuspended in ∼5 μl RNaase-free water and used for either direct detection by 3′-end labeling or analysis by Northern or Southern blotting. The RNA mixture can also be used as the starting material for constructing a library to identify specific RNAs that may be associated with a complex of interest.

3′-End Labeling and Detection of RNA

About 10–50% of the purified material is used for RNA-labeling experiments. Typically, RNAs are labeled at their 3′ end with [5′ ^{32}P]-pCp (England et al., 1980; Fig. 2B). The [5′ ^{32}P]-pCp labeling reaction is carried out at 4° for 16 h by incubating the purified RNAs in a 10-μl reaction containing 10% DMSO, 10 μg/ml BSA, 10 units RNase inhibitor (Roche, Switzerland), 10 μCi [5′ ^{32}P]-pCp (3000 Ci/mmol, NEN Life Sciences, Boston, MA), 1 μl T4 RNA ligase buffer (NEB) and 20 units T4 RNA ligase (NEB; Bruce and Uhlenbeck, 1978). The reaction is quenched by adding 90 μl water and precipitating the RNAs as described previously.

Alternatively, the 3' end of the RNAs can be labeled with cordycepin (Fig. 3A; Lingner and Keller, 1993). First, the RNAs are added to a 10-μl reaction containing 2 μl Poly(A) polymerase reaction buffer (USB), 50 μCi [^{32}P]-cordycepin (5000 Ci/mmol, NEN Life Sciences), and 1000 units poly(A) polymerase. The reaction is incubated at 30° for 30 min. The reaction is then stopped by adding 90 μl water and run through a G-25 MircroSpin column (Amersham Biosciences) to eliminate nonincorporated nucleotides. The flow-through is precipitated as described previously.

Labeled small RNAs are separated on a 15% denaturing urea-acrylamide gel, and the dried gels are exposed to either a Kodak film (X-Omat AR film) or to a PhosphorImager™ (Bio-Rad) screen. In general, we observe more robust labeling with poly(A) polymerase and [^{32}P]-cordycepin than that with [5' ^{32}P]-pCp and T4 RNA ligase (Fig. 3A).

Analyzing RNAs Associated with the Purified Complex

Northern Blot

Purified RNAs are separated on a 15% denaturing urea-acrylamide gel and electrophoretically transferred to Zeta-Probe GT membranes (Bio-Rad). RNAs are crosslinked to the membrane by UV radiation (1200 mJ/cm) and by baking the membrane at 80° for 1 h.

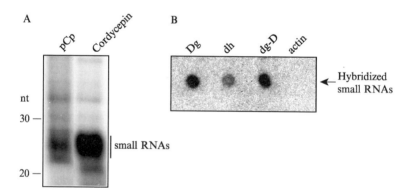

FIG. 3. Analysis of RITS siRNAs. (A) Comparison of the 3'-end labeling of siRNAs with [^{32}P]-pCp or [^{32}P]-cordycepin. RNAs associated with the purified RITS complex were labeled by [^{32}P]-pCp or [^{32}P]-cordycepin as described in the text. (B) Use of dot blot to determine whether siRNAs present in the RITS complex hybridize with selected DNA regions (Verdel et al., 2004). DNA fragments from different regions of the *S. pombe* genome were bound to a nylon membrane as described in the text. Dg, dg-D, and dh DNAs represent repetitive DNA sequences present in heterochromatic regions at centromeres, and actin represents an active euchromatic DNA region.

DNA oligonucleotides to be used as probes are 5′ labeled in a 15-μl reaction containing 20 pmol oligonucleotide, 1.5 μl T4 polynucleotide kinase buffer (NEB), 75 μCi γ-[^{32}P]-ATP (6000 Ci/mmol, NEN Life Sciences), and 10 units T4 polynucleotide kinase (NEB) and incubated for 1 h at 37°. The reaction is stopped by adding 2 μl of 0.5 M EDTA and 33 μl of 1 M Tris-HCl (pH 7.5) and incubating at 68° for 15 min. Labeled oligonucleotides are separated from free label by filtering the reaction mix through a G-25 MicroSpin™ column (Amersham Biosciences).

The membrane is prehybridized for 15 min at 45° with constant rotation in 30 ml hybridization buffer (0.5 M sodium phosphate, pH 7.2, 1 mM EDTA, and 7% SDS). The hybridization step is conducted for 3 h at 45° by incubating the membrane in 10 ml hybridization buffer containing one oligonucleotide-labeling reaction. The membrane is washed once in hybridization buffer for 1 min at 25° and again for 1 min at 55° and exposed to X-ray film or a PhosphorImager screen as described previously.

Purification of Labeled RNAs and Their Use as Probes for Southern Blots

Affinity-purified RITS preparations often contain both specific and nonspecific RNAs (Fig. 2B). We use gel purification to isolate RNAs that are specifically associated with RITS. After 3′-end labeling of the RNAs present in the final elution, labeled RNAs are separated on a 15% urea-acrylamide gel. The gel is exposed to an X-ray film to locate the region of the gel containing RITS-specific RNAs. RNAs are eluted by excising the regions of interest, slicing the excised gel into smaller pieces, and agitating them at room temperature for 16 h in 300 μl of elution buffer (0.3 M sodium acetate, pH 5.2, 0.2% SDS and 0.05 mg/ml tRNA). RNAs are then precipitated by adding 30 μl of 0.3 M sodium acetate, 1 μl glycogen (20 μg/μl), and 825 μl of 100% ethanol, followed by incubation at −80° for 20–30 min. The RNA is pelleted by centrifugation at 12,000g for 20 min at 4° and resuspended in hybridization buffer.

For Southern blots, purified, PCR-amplified DNA is separated on an agarose gel and capillary-transferred to nylon membranes (Zeta-Probe GT, Bio-Rad) according to the manufacturer's instructions. The membrane is rinsed briefly in 2× SSC buffer, and DNA is crosslinked to the membrane as described for the Northern blot (Verdel et al., 2004). For dot blots, 5.5 μg of DNA is incubated for 1 h at 60° in 200 μl of a solution containing 10 mM Tris-HCl, pH 7.5, 1 mM EDTA, and 0.3 M NaOH. One volume of 6× SSC buffer is added to the DNA solution, and 2-μl aliquots are deposited on a prewetted membrane with 6× SSC. The membrane is dried at room temperature for 20 min, and the DNA is crosslinked to the membrane as described previously. The prehybridization, hybridization, and exposure of Southern and dot blots are conducted as described previously (Fig. 3B).

Concluding Remarks

We have described strategies used in our laboratory to purify and characterize a fission yeast RNA-containing complex, RITS, which directly links the RNAi pathway to heterochromatin assembly. The strategies described here are generally applicable and have been used to study other RNA-containing complexes. The conservation of RNAi in fission yeast provides an opportunity to combine rapid genetic manipulations with affinity purification and biochemical analysis of proteins and complexes to understand how RNA-dependent mechanisms direct heterochromatin formation.

Acknowledgments

We thank Julie Huang for her comments on the manuscript and other members of the Moazed laboratory for their help and support. A. V. is a fellow of the International Human Frontier Science Program. This work is supported by grants from the NIH and a Carolyn and Peter S. Lynch Award in Cell Biology and Pathology. D. M. is a scholar of the Leukemia and Lymphoma Society.

References

Bahler, J., Wu, J. Q., Longtine, M. S., Shah, N. G., McKenzie, A., 3rd, Steever, A. B., Wach, A., Philippsen, P., and Pringle, J. R. (1998). Heterologous modules for efficient and versatile PCR-based gene targeting in *Schizosaccharomyces pombe*. *Yeast* **14**, 943–951.

Bernstein, E., Caudy, A. A., Hammond, S. M., and Hannon, G. J. (2001). Role for a bidentate ribonuclease in the initiation step of RNA interference. *Nature* **409**, 363–366.

Bruce, A. G., and Uhlenbeck, O. C. (1978). Reactions at the termini of tRNA with T4 RNA ligase. *Nucleic Acids Res.* **5**, 3665–3677.

Caudy, A. A., Myers, M., Hannon, G. J., and Hammond, S. M. (2002). Fragile X-related protein and VIG associate with the RNA interference machinery. *Genes Dev.* **16**, 2491–2496.

Doench, J. G., Petersen, C. P., and Sharp, P. A. (2003). siRNAs can function as miRNAs. *Genes Dev.* **17**, 438–442.

Elbashir, S. M., Lendeckel, W., and Tuschl, T. (2001). RNA interference is mediated by 21- and 22-nucleotide RNAs. *Genes Dev.* **15**, 188–200.

England, T. E., Bruce, A. G., and Uhlenbeck, O. C. (1980). Specific labeling of 3' termini of RNA with T4 RNA ligase. *Methods Enzymol.* **65**, 65–74.

Fire, A., Xu, S., Montgomery, M. K., Kostas, S. A., Driver, S. E., and Mello, C. C. (1998). Potent and specific genetic interference by double-stranded RNA in *Caenorhabditis elegans*. *Nature* **391**, 806–811.

Gould, K. L., Ren, L., Feoktistova, A. S., Jennings, J. L., and Link, A. J. (2004). Tandem affinity purification and identification of protein complex components. *Methods* **33**, 239–244.

Hall, I. M., Shankaranarayana, G. D., Noma, K., Ayoub, N., Cohen, A., and Grewal, S. I. (2002). Establishment and maintenance of a heterochromatin domain. *Science* **297**, 2232–2237.

Hamilton, A. J., and Baulcombe, D. C. (1999). A species of small antisense RNA in posttranscriptional gene silencing in plants. *Science* **286**, 950–952.

Hannon, G. J. (2002). RNA interference. *Nature* **418**, 244–251.

Hoppe, G. J., Tanny, J. C., Rudner, A. D., Gerber, S. A., Danaie, S., Gygi, S. P., and Moazed, D. (2002). Steps in assembly of silent chromatin in yeast: Sir3-independent binding of a Sir2/Sir4 complex to silencers and role for Sir2-dependent deacetylation. *Mol. Cell. Biol.* **22**, 4167–4180.

Hutvagner, G., and Zamore, P. D. (2002). RNAi: Nature abhors a double-strand. *Curr. Opin. Genet. Dev.* **12**, 225–232.

Ishizuka, A., Siomi, M. C., and Siomi, H. (2002). A *Drosophila* fragile X protein interacts with components of RNAi and ribosomal proteins. *Genes Dev.* **16**, 2497–2508.

Lingner, J., and Keller, W. (1993). 3′-end labeling of RNA with recombinant yeast poly(A) polymerase. *Nucleic Acids Res.* **21**, 2917–2920.

Lipardi, C., Wei, Q., and Paterson, B. M. (2001). RNAi as random degradative PCR: siRNA primers convert mRNA into dsRNAs that are degraded to generate new siRNAs. *Cell* **107**, 297–307.

Matzke, M., Aufsatz, W., Kanno, T., Daxinger, L., Papp, I., Mette, M. F., and Matzke, A. J. (2004). Genetic analysis of RNA-mediated transcriptional gene silencing. *Biochim. Biophys. Acta* **1677**, 129–141.

Peng, J., and Gygi, S. P. (2001). Proteomics: The move to mixtures. *J. Mass Spectrom.* **36**, 1083–1091.

Puig, O., Caspary, F., Rigaut, G., Rutz, B., Bouveret, E., Bragado-Nilsson, E., Wilm, M., and Seraphin, B. (2001). The tandem affinity purification (TAP) method: A general procedure of protein complex purification. *Methods* **24**, 218–229.

Schramke, V., and Allshire, R. (2003). Hairpin RNAs and retrotransposon LTRs effect RNAi and chromatin-based gene silencing. *Science* **301**, 1069–1074.

Shevchenko, A., Wilm, M., Vorm, O., and Mann, M. (1996). Mass spectrometric sequencing of proteins silver-stained polyacrylamide gels. *Anal. Chem.* **68**, 850–858.

Sijen, T., Fleenor, J., Simmer, F., Thijssen, K. L., Parrish, S., Timmons, L., Plasterk, R. H., and Fire, A. (2001). On the role of RNA amplification in dsRNA-triggered gene silencing. *Cell* **107**, 465–476.

Storici, F., Lewis, L. K., and Resnick, M. A. (2001). *In vivo* site-directed mutagenesis using oligonucleotides. *Nat. Biotechnol.* **19**, 773–776.

Tanny, J. C., Kirkpatrick, D. S., Gerber, S. A., Gygi, S. P., and Moazed, D. (2004). Budding yeast silencing complexes and the regulation of Sir2 activity by protein-protein interactions. *Mol. Cell. Biol.*

Verdel, A., Jia, S., Gerber, S., Sugiyama, T., Gygi, S., Grewal, S. I., and Moazed, D. (2004). RNAi-mediated targeting of heterochromatin by the RITS complex. *Science* **303**, 672–676.

Volpe, T. A., Kidner, C., Hall, I. M., Teng, G., Grewal, S. I., and Martienssen, R. A. (2002). Regulation of heterochromatic silencing and histone H3 lysine-9 methylation by RNAi. *Science* **297**, 1833–1837.

Zamore, P. D. (2001). RNA interference: Listening to the sound of silence. *Nat. Struct. Biol.* **8**, 746–750.

Zamore, P. D., Tuschl, T., Sharp, P. A., and Bartel, D. P. (2000). RNAi: Double-stranded RNA directs the ATP-dependent cleavage of mRNA at 21 to 23 nucleotide intervals. *Cell* **101**, 25–33.

[18] RNA Interference Spreading in *C. elegans*

By ROBIN C. MAY and RONALD H. A. PLASTERK

Abstract

The phenomenon of RNA interference (RNAi) occurs in eukaryotic organisms from across the boundaries of taxonomic kingdoms. In all cases, the basic mechanism of RNAi appears to be conserved—an initial trigger [double-stranded RNA (dsRNA) containing perfect homology over at least 19–21 bp with an endogenous gene] is processed into short interfering RNA (siRNA) molecules and these siRNAs stimulate degradation of the homologous mRNA.

In the vast majority of species, RNAi can only be initiated following the deliberate introduction of dsRNA into a cell by microinjection, electroporation, or transfection. However, in the nematode worm *Caenorhabditis elegans*, RNAi can be simply initiated by supplying dsRNA in the surrounding medium or in the diet. Following uptake, this dsRNA triggers a systemic effect, initiating RNAi against the corresponding target gene in tissues that are not in direct contact with the external milieu. This phenomenon of systemic RNAi, or RNAi spreading, is notably absent from mammalian species, a fact that is likely to prove a substantial barrier to the wider use of RNAi as a clinical therapy. An understanding of the mechanism of systemic RNAi is therefore of considerable importance, and several advances of the last few years have begun to shed light on this process. Here we review our current understanding of systemic RNAi in *C. elegans* and draw comparisons with systemic RNAi pathways in other organisms.

Introduction

Following the initial characterization of RNA interference (RNAi) in *C. elegans* on injection of double-stranded RNA (dsRNA; Fire *et al.*, 1998), Timmons and Fire reported in 1998 that RNAi could also be initiated by feeding worms with bacteria expressing dsRNA for an endogenous *C. elegans* gene (Timmons and Fire, 1998). This simple system, in which DNA corresponding to (part of) a worm gene was inserted into a bacterial vector between two inward-facing T7 promoters, has since become the basis of a genomewide "feeding library" that enables the inactivation of almost all *C. elegans* genes on a gene-by-gene basis (Fraser *et al.*, 2000).

Systemic RNAi in the worm is not specific to dietary dsRNA, because animals soaked in a solution of dsRNA also show systemic effects (Maeda *et al.*, 2001) and because localized injection of dsRNA (e.g., into a gut cell) can produce an RNAi effect in distant tissues. These discoveries therefore suggested the existence of a pathway to extract dsRNA from the environment and subsequently transport it across cell membranes and between tissues to trigger systemic RNAi.

Genes Involved in Mediating Systemic RNA Interference (RNAi)

To understand the basis of systemic RNAi in *C. elegans*, several groups have conducted classical forward genetic screens to identify mutant animals in which systemic RNAi does not occur. Following mutagenesis, animals in which the RNAi machinery itself is not impaired, but in which systemic RNAi no longer occurs, are isolated and the genetic lesion responsible for the defect is identified by genetic mapping. Three independent screens using alternative approaches have so far identified several mutants in which systemic RNAi is impaired. These have been named *sid* (systemic interference defective), *rsd* (RNAi spreading defective), and *fed* (feeding defective for RNAi) and represent partially overlapping sets of genes.

The *fed* and *rsd* mutants were identified by virtue of their failure to initiate RNAi following feeding of dsRNA but the ability to mount a normal RNAi response to the same dsRNA when injected. Genes identified by these screens may therefore act at any point from the initial uptake of dietary dsRNA to the subsequent trafficking of RNAi triggers to neighboring cells. The third (*sid*) screen made use of transgenic animals in which RNAi was initiated endogenously by expression of a hairpin dsRNA construct within the pharynx of the worm. In this system, systemic RNAi mutants were defined as animals in which RNAi still occurred within the tissue expressing dsRNA (the pharynx) but no longer spread to adjoining tissues such as the body wall muscle.

The *sid-1* gene, identified independently in two screens (Tijsterman *et al.*, 2004; Winston *et al.*, 2002), encodes a multipass transmembrane protein that is required cell autonomously for the uptake of dsRNA (Winston *et al.*, 2002). In the absence of SID-1, RNAi can still be initiated by the injection or endogenous production of dsRNA, but this effect does not spread to neighboring cells. Subsequent work showed that dsRNA uptake into *Drosophila* S2 and cl-8 cells can be dramatically enhanced by expressing *C. elegans* SID-1 protein in these cells (Feinberg and Hunter, 2003), indicating that SID-1 alone is sufficient to allow dsRNA uptake into cells from the surrounding medium. This uptake occurs even in the absence of ATP (Feinberg and Hunter, 2003), suggesting that it is a passive process.

Interestingly, SID-1 shows a marked preference for long dsRNA molecules. While dsRNA molecules of 500 bp are effectively transported, efficiency drops rapidly with decreasing size and the smallest functional "unit" for triggering RNAi, short interfering RNA (siRNAs), are almost 10,000 times less efficient (Feinberg and Hunter, 2003). Because only short dsRNA molecules can be used in mammalian cells [to prevent activation of the dsRNA-dependent apoptotic response (Gil and Esteban, 2000)], this may be one reason why there is no evidence that expressing SID-1 in mammalian cells renders them susceptible to systemic RNAi.

The role of SID-1 in mediating cell autonomous uptake of dsRNA means that it is required for systemic RNAi in all cell types. This ubiquitous requirement appears also to be a feature of the two other *sid* genes (Winston *et al.*, 2002), of *rsd-4* (Tijsterman *et al.*, 2004), and of *fed-1* and *fed-2* (Timmons *et al.*, 2003), because all these are required for systemic RNAi in somatic tissue. To date, none of these genes have been cloned and their roles in mediating systemic RNAi therefore remain cryptic.

In contrast, *rsd-2, rsd-3,* and *rsd-6* mutant animals show no defect in initiating RNAi in the soma following feeding of dsRNA, but are instead defective in the secondary spread of this RNAi effect to the germline (Tijsterman *et al.*, 2004). This may suggest that these genes are not required for the initial uptake of dsRNA from the intestinal lumen, but instead play a role in transporting dsRNA (or a derivative thereof) throughout the body. In this respect it is interesting to note that RSD-3 shows hallmarks of an endocytic protein (Tijsterman *et al.*, 2004) and may thus play a role in the secondary trafficking of an RNAi trigger (probably dsRNA) following initial uptake. The roles of RSD-2 and RSD-6 in systemic RNAi are less clear, although yeast two-hybrid data indicate that they form a complex with each other (Tijsterman *et al.*, 2004). In addition, *rsd-6* contains a Tudor domain, a motif frequently found in RNA-binding proteins, suggesting that this protein may interact with the transported RNA species.

Nature of the Systemic Signal

Little is known about the nature of the transported signal that mediates systemic RNAi in *C. elegans*. Because RNAi is highly specific and dependent on perfect homology between the targeted mRNA and the dsRNA trigger of at least 20 bp, it is difficult to imagine that the transported signal can be anything other than a nucleic acid of at least this length. Given that there is no evidence of a reverse transcriptase or DNA-dependent step at any point in the RNAi pathway, it is highly likely that the systemic trigger is thus an RNA species.

RNAi in *C. elegans* is routinely triggered by feeding or soaking in long dsRNA molecules (usually of several hundred basepairs in length). As previously mentioned, SID-1 shows a strong preference for dsRNA molecules of this length and transports short dsRNAs or siRNAs very poorly (Feinberg and Hunter, 2003). This suggests that the systemic signal may be a long dsRNA derived from the initial trigger molecule rather than the fully processed siRNA. Such a view is supported by the observation that mutant *rde-4* animals (in which siRNAs are not produced) remain capable of transmitting a systemic RNAi signal to RNAi-competent (heterozygous) offspring (Tabara *et al.*, 1999; Timmons *et al.*, 2003). Thus siRNAs are not required for systemic signaling in *C. elegans*. In plants, dsRNAs are processed into two size classes (21–22 bp and 24–26 bp) and only the larger size class appears to be important to obtain a systemic RNAi effect (Hamilton *et al.*, 2002). To date, there is no evidence that *C. elegans* produces dsRNA molecules in more than one size class or that any particular dsRNA product is specifically involved in systemic RNAi.

In addition to the molecular nature of the RNAi signal, it appears that the source of the signal is also important. Although dsRNA provided exogenously through feeding or soaking is able to trigger robust systemic RNAi, endogenously produced dsRNA (transcribed from a hairpin trans-gene) is very poor at initiating RNAi spreading (Tijsterman *et al.*, 2004; Timmons *et al.*, 2003). This failure to initiate systemic silencing can be overcome by adding extracellular nonspecific dsRNA (Timmons *et al.*, 2003). In addition, expressing dsRNA at a very high level from an endoge-nous hairpin appears to override this barrier to systemic spread (Timmons *et al.*, 2003; Winston *et al.*, 2002), a fact that may explain why a hairpin-based screen for finding systemic RNAi mutants (discussed previously) was successful despite the general inefficiency of such constructs in triggering RNAi spreading (Winston *et al.*, 2002). Together, this suggests that endog-enous dsRNA molecules may be incapable of triggering systemic RNAi because they are sequestered in the expressing cell before they can be transported further.

Systemic RNAi in Other Organisms

Besides *C. elegans*, RNAi spreading has been reported to occur in plants (Jorgensen, 2002; Palauqui *et al.*, 1997; Voinnet and Baulcombe, 1997), planaria (Newmark *et al.*, 2003) and the beetle *Tribolium* (Bucher *et al.*, 2002). In planaria, RNAi is not only induced following dsRNA feeding (as in *C. elegans*) but also maintained in tissues that have regrown following amputation (Newmark *et al.*, 2003). At present, nothing is known about the molecular basis of systemic RNAi in planaria or *Tribolium* or whether it is related to the RNAi spreading pathway in *C. elegans*.

In plants, a systemic RNAi signal can pass from a leaf into an unrelated plant onto which it has been grafted (Palauqui *et al.*, 1997). The molecular signal responsible for long-range systemic RNAi appears to be a dsRNA 24–26 bp in length (Hamilton *et al.*, 2002), and transmission of this signal can be blocked by a number of viral suppressors of RNAi (Guo and Ding, 2002; Voinnet *et al.*, 2000). Systemic RNAi in plants appears to occur through a combination of short-range cell-to-cell transmission together with a more extensive long-range spreading process. Short-range spreading appears to act by straightforward transmission of siRNAs (without amplification), whereas long-range spreading requires amplification of the initial signal by an RNA-dependent RNA polymerase (Himber *et al.*, 2003). The current model of systemic RNAi in plants invokes local transmission of siRNA molecules by plasmodesmata (channels that link the cytoplasm of neighboring cells) together with longer-range transport of a somewhat larger RNA species by the phloem vasculature (Himber *et al.*, 2003). This RNAi spreading process is likely to be considerably different from that in *C. elegans*, in which there is (1) no evidence for the involvement of multiple RNA species in systemic signalling and (2) no dependence on a "vascular" system for transport of such a signal.

The Wider Applicability of Systemic RNAi

It would be highly beneficial if one were able to induce systemic RNAi spreading in organisms that do not otherwise show this effect, most notably mammals. The ability, for example, to apply a solution of dsRNA to mammalian cells in culture and thereby trigger an RNAi effect would dramatically improve the ease and rapidity of large-scale screens. To date, such an approach has not been successfully achieved.

In principle, expressing the *C. elegans* protein SID-1 in mammalian cells would be expected to render them permeable to dsRNA molecules. However, to avoid triggering dsRNA-dependent apoptosis in mammalian cells, such an experiment must be done with short dsRNA molecules, a size class that is very poorly transported by SID-1 (Feinberg and Hunter, 2003). One way around this may be to express SID-1 in embryonic mammalian cells, which show attenuated nonspecific responses to dsRNA but are still RNAi competent (Paddison *et al.*, 2002), and then apply longer dsRNA molecules in the medium. To date, such an experiment has not been published.

Physiological Role of Systemic RNAi

Several hypotheses for the natural role of RNAi have been put forward. It is clear that the RNAi pathway in plants acts as a powerful defensive mechanism against viral infection (Voinnet, 2001), a fact supported by the observation that many plant viruses have evolved inhibitors of RNAi

(Brigneti *et al.*, 1998; Kasschau and Carrington, 1998; Voinnet *et al.*, 1999). There is also mounting evidence that the RNAi pathway acts to prevent transposition of mobile genetic elements in both animals and plants, thereby protecting the host from deleterious genomic rearrangements (Sijen and Plasterk, 2003; Waterhouse *et al.*, 2001). Finally, several components of the RNAi pathway are also involved in the processing of microRNA molecules, raising the prospect of a common evolutionary origin of these two pathways (Grishok *et al.*, 2001; Hutvagner and Zamore, 2002; Lee and Ambros, 2001).

What then is the additional benefit of a mechanism that spreads the RNAi signal systemically? In the case of plants, the ability to transmit a systemic signal to trigger an antiviral RNAi effect in distant tissues reflects the undoubted evolutionary advantage that this would confer in defending the host against viral attack. However, in the animal kingdom, the RNAi spreading effect is restricted to a handful of species, although the RNAi pathway itself is widely conserved. Thus, systemic RNAi is clearly not critical for an antiviral response, because the majority of animal species are regularly infected by viruses and possess an RNAi machinery and have not yet developed a systemic RNAi pathway.

One possible explanation for the occurrence of systemic RNAi in some animal species is that it reflects the accidental uptake and spreading of dsRNA molecules by a general nucleic acid scavenging mechanism. *C. elegans* is able to obtain sufficient nucleic acids for normal growth from an exclusively dietary source (Sulston, 1976). It is thus possible that a pathway involved in the scavenging of DNA or RNA from the gut also allows the uptake of (partially) intact dsRNA molecules, thus triggering systemic RNAi. The fact that systemic RNAi mutants are not obviously impaired in the uptake of dietary nucleic acids (Tijsterman *et al.*, 2004) argues against the two pathways being totally overlapping. However, it remains possible that under nutrient-limiting conditions, such as those that *C. elegans* may encounter in its natural habitat, the uptake and distribution of RNA molecules by the systemic RNAi pathway would be critical for survival.

To date, mutant *C. elegans* defective in systemic RNAi do not appear to show any additional phenotype, ruling out an essential role for this pathway in basic biological processes such as growth or reproduction. This may, to some extent, reflect the manner in which they were identified, because mutants with compromised survival are underrepresented in such genetic screens. It remains possible that some components essential for systemic RNAi are also essential for life and have thus not been identified. In addition, the laboratory conditions for growth of *C. elegans* are far removed from its natural environment and it is possible that components of the systemic RNAi pathway that have already been identified originally evolved for a purpose quite unrelated to the transport of dsRNA. Understanding what this role might be will be dependent on future improvements in our appreciation of the behavior of *C. elegans* in its natural environment.

Conclusions

RNAi is a powerful experimental tool with considerable clinical promise. In some organisms, its ease of use is enhanced by the systemic spread of an RNAi effect following local application of a dsRNA trigger. Our understanding of the mechanism of systemic RNAi is currently limited, although it appears likely that this mechanism has been evolved independently at least twice (by a few animal species and by plants). Future work on the molecular basis of systemic RNAi in *C. elegans*, and a comparison with analogous RNAi spreading mechanisms in other organisms, may give insight into the process of dsRNA transport *in vivo* and provide possible approaches for mediating systemic RNAi in mammals.

Acknowledgments

We thank Femke Simmer for critically reading this manuscript. R. C. M. is supported by a Human Frontier Science Program fellowship.

References

Brigneti, G., Voinnet, O., Li, W. X., Ji, L. H., Ding, S. W., and Baulcombe, D. C. (1998). Viral pathogenicity determinants are suppressors of transgene silencing in *Nicotiana benthamiana*. *EMBO J.* **17**, 6739–6746.

Bucher, G., Scholten, J., and Klingler, M. (2002). Parental RNAi in *Tribolium* (Coleoptera). *Curr. Biol.* **12**, R85–R86.

Feinberg, E. H., and Hunter, C. P. (2003). Transport of dsRNA into cells by the transmembrane protein SID-1. *Science* **301**, 1545–1547.

Fire, A., Xu, S., Montgomery, M. K., Kostas, S. A., Driver, S. E., and Mello, C. C. (1998). Potent and specific genetic interference by double-stranded RNA in *Caenorhabditis elegans*. *Nature* **391**, 806–811.

Fraser, A. G., Kamath, R. S., Zipperlen, P., Martinez-Campos, M., Sohrmann, M., and Ahringer, J. (2000). Functional genomic analysis of *C. elegans* chromosome I by systematic RNA interference. *Nature* **408**, 325–330.

Gil, J., and Esteban, M. (2000). Induction of apoptosis by the dsRNA-dependent protein kinase (PKR): Mechanism of action. *Apoptosis* **5**, 107–114.

Grishok, A., Pasquinelli, A. E., Conte, D., Li, N., Parrish, S., Ha, I., Baillie, D. L., Fire, A., Ruvkun, G., and Mello, C. C. (2001). Genes and mechanisms related to RNA interference regulate expression of the small temporal RNAs that control *C. elegans* developmental timing. *Cell* **106**, 23–34.

Guo, H. S., and Ding, S. W. (2002). A viral protein inhibits the long range signaling activity of the gene silencing signal. *EMBO J.* **21,** 398–407.

Hamilton, A., Voinnet, O., Chappell, L., and Baulcombe, D. (2002). Two classes of short interfering RNA in RNA silencing. *EMBO J.* **21,** 4671–4679.

Himber, C., Dunoyer, P., Moissiard, G., Ritzenthaler, C., and Voinnet, O. (2003). Transitivity-dependent and -independent cell-to-cell movement of RNA silencing. *EMBO J.* **22,** 4523–4533.

Hutvagner, G., and Zamore, P. D. (2002). A microRNA in a multiple-turnover RNAi enzyme complex. *Science* **297,** 2056–2060.

Jorgensen, R. A. (2002). RNA traffics information systemically in plants. *Proc. Natl. Acad. Sci. USA* **99,** 11561–11563.

Kasschau, K. D., and Carrington, J. C. (1998). A counterdefensive strategy of plant viruses: Suppression of posttranscriptional gene silencing. *Cell* **95,** 461–470.

Lee, R. C., and Ambros, V. (2001). An extensive class of small RNAs in *Caenorhabditis elegans*. *Science* **294,** 862–864.

Maeda, I., Kohara, Y., Yamamoto, M., and Sugimoto, A. (2001). Large-scale analysis of gene function in *Caenorhabditis elegans* by high-throughput RNAi. *Curr. Biol.* **11,** 171–176.

Newmark, P. A., Reddien, P. W., Cebria, F., and Alvarado, A. S. (2003). Ingestion of bacterially expressed double-stranded RNA inhibits gene expression in planarians. *Proc. Natl. Acad. Sci. USA* **100,** 11861–11865.

Paddison, P. J., Caudy, A. A., and Hannon, G. J. (2002). Stable suppression of gene expression by RNAi in mammalian cells. *Proc. Natl. Acad. Sci. USA* **99,** 1443–1448.

Palauqui, J. C., Elmayan, T., Pollien, J. M., and Vaucheret, H. (1997). Systemic acquired silencing: Transgene-specific post-transcriptional silencing is transmitted by grafting from silenced stocks to non-silenced scions. *EMBO J.* **16,** 4738–4745.

Sijen, T., and Plasterk, R. H. (2003). Transposon silencing in the *Caenorhabditis elegans* germ line by natural RNAi. *Nature* **426,** 310–314.

Sulston, J. E. (1976). Post-embryonic development in the ventral cord of *Caenorhabditis elegans*. *Philos. Trans. R. Soc. Lond. B Biol. Sci.* **275,** 287–297.

Tabara, H., Sarkissian, M., Kelly, W. G., Fleenor, J., Grishok, A., Timmons, L., Fire, A., and Mello, C. C. (1999). The rde-1 gene, RNA interference, and transposon silencing in *C. elegans*. *Cell* **99,** 123–132.

Tijsterman, M., May, R. C., Simmer, F., Okihara, K. L., and Plasterk, R. H. (2004). Genes required for systemic RNA interference in *Caenorhabditis elegans*. *Curr. Biol.* **14,** 111–116.

Timmons, L., and Fire, A. (1998). Specific interference by ingested dsRNA. *Nature* **395,** 854.

Timmons, L., Tabara, H., Mello, C. C., and Fire, A. Z. (2003). Inducible systemic RNA silencing in *Caenorhabditis elegans*. *Mol. Biol. Cell* **14,** 2972–2983.

Voinnet, O. (2001). RNA silencing as a plant immune system against viruses. *Trends Genet.* **17,** 449–459.

Voinnet, O., and Baulcombe, D. C. (1997). Systemic signalling in gene silencing. *Nature* **389,** 553.

Voinnet, O., Lederer, C., and Baulcombe, D. C. (2000). A viral movement protein prevents spread of the gene silencing signal in *Nicotiana benthamiana*. *Cell* **103,** 157–167.

Voinnet, O., Pinto, Y. M., and Baulcombe, D. C. (1999). Suppression of gene silencing: A general strategy used by diverse DNA and RNA viruses of plants. *Proc. Natl. Acad. Sci. USA* **96,** 14147–14152.

Waterhouse, P. M., Wang, M. B., and Lough, T. (2001). Gene silencing as an adaptive defence against viruses. *Nature* **411,** 834–842.

Winston, W. M., Molodowitch, C., and Hunter, C. P. (2002). Systemic RNAi in *C. elegans* requires the putative transmembrane protein SID-1. *Science* **7,** 7.

[19] Human Dicer: Purification, Properties, and Interaction with PAZ PIWI Domain Proteins

By Fabrice A. Kolb, Haidi Zhang, Kasia Jaronczyk, Nasser Tahbaz, Tom C. Hobman, and Witold Filipowicz

Abstract

Dicer is a multidomain ribonuclease that processes double-stranded RNAs (dsRNAs) to 21-nt small interfering RNAs (siRNAs) during RNA interference and excises microRNAs (miRNAs) from precursor hairpins. PAZ and PIWI domain (PPD) proteins, also involved in RNAi and miRNA function, are the best-characterized proteins known to interact with Dicer. PPD proteins are the core constituents of effector complexes, RISCs and miRNPs, mediating siRNA and miRNA function. In this chapter we describe overexpression and purification of recombinant human Dicer, its biochemical properties, and mapping of domains responsible for Dicer–PPD protein interactions.

Introduction

Dicer is a large multidomain ribonuclease responsible for processing double-stranded RNAs (dsRNAs) to ~20-bp-long small interfering RNAs (siRNAs), which act as effectors during RNA interference (RNAi). Dicer also catalyses the excision of microRNAs (miRNAs) from stem-loop precursors, referred to as pre-miRNAs. Dicer proteins have been found in all eukaryotes studied to date, with the exception of *Saccharomyces cerevisiae*. In vertebrates, Dicer is encoded by a single-copy gene, but genomes of many other organisms encode from two (e.g., *Drosophila melanogaster* and *Neurospora crassa*) to four (*Arabidopsis thaliana*) Dicer proteins; one Dicer gene is also present in *Schizosaccharomyces pombe* (reviewed by Hannon and Zamore, 2003). In zebra fish and mouse, the Dicer-encoding gene is essential (Bernstein et al., 2003; Wienholds et al., 2003). In mammalian cells, Dicer is localized in the cytoplasm (Billy et al., 2001; Provost et al., 2002). In plant cells, some evidence exists that at least one of the Dicer proteins may be localized to nuclei (Matzke et al., 2004; Papp et al., 2003).

Dicer is an approximately 200-kDa protein that contains DExH-type RNA helicase/ATPase, DUF283, and PAZ domains; two neighboring RNase III-like domains (RIIIa and RIIIb); and a C-terminal dsRNA-binding domain (dsRBD). The dsRBD and RNase III signature domains present in the bacterial RNase III, an ancestor of Dicer, are involved in dsRNA binding and cleavage (Zhang *et al.*, 2004). The PAZ domain is also found in PPD proteins, which, like Dicer, function in RNAi and miRNA-dependent processes (see later; reviewed by Carmell *et al.*, 2002). Recent structural and mutagenesis studies indicate that in both PPD and Dicer proteins, the PAZ domain is involved in binding of 3'-protruding ends of either siRNAs or dsRNA substrates (Lingel *et al.*, 2003; Song *et al.*, 2003; Yan *et al.*, 2003; Zhang *et al.*, 2004). The presence of the helicase/ATPase domain is consistent with findings that the generation of siRNAs by invertebrate Dicer requires ATP (Bernstein *et al.*, 2001; Ketting *et al.*, 2001; Liu *et al.*, 2003; Nykanen *et al.*, 2001). In contrast, production of siRNAs by mammalian Dicer is not dependent on ATP (Zhang *et al.*, 2002; and see later).

Several proteins interacting with Dicer have been identified. Among them are members of the PPD family of proteins (see later), small RNA recognition motif (RRM)-containing proteins R2D2 of *Drosophila* (Liu *et al.*, 2003) and RDE-4 of *C. elegans* (Tabara *et al.*, 2002), and the *Drosophila* ortholog of human fragile X mental retardation protein, dFMR1 (Ishizuka *et al.*, 2002). Recently, Dicer was also found to be the core component of the RNA interference specificity complex (RISC) essential for mRNA target cleavage (Lee *et al.*, 2004; Pham *et al.*, 2004; Tomari *et al.*, 2004).

Sequence similarity of Dicer RNase III-like domains to the bacterial and fungal RNases III suggests that Dicer cleaves dsRNA through a similar mechanism. Indeed, the ~20-bp products of Dicer processing contain 2-nt 3' overhangs and 5'-phosphate and 3'-hydroxyl termini characteristic of RNase III-mediated reactions (Elbashir *et al.*, 2001; Nicholson, 2003). Moreover, mutations of amino acid residues that are essential for RNase III activity also inactivate Dicer (Lee *et al.*, 2004; Zhang *et al.*, 2004).

Functional bacterial RNase III is a homodimer of polypeptides containing a single catalytic RNase III domain (reviewed by Blaszczyk *et al.*, 2001; Nicholson, 2003). Recent experiments indicate that human Dicer functions as a pseudodimer, with RIIIa and RIIIb domains of the same Dicer molecule interacting with each other. Mutagenesis and modeling studies indicate that such an intramolecular dimer contains a single dsRNA cleavage center, which processes dsRNA substrates at approximately 20 bp from the 3' terminus recognized by the PAZ domain.

The PPD protein family is highly conserved in most eukaryotes and is divided according to sequence similarities into two subgroups: those that resemble *Arabidopsis* Argonaute 1 (Ago1) and those that resemble *Drosophila* Piwi (Carmell *et al.*, 2002). Genomes of metazoans invariably contain multiple PPD genes, whereas those of unicellular organisms often contain a single PPD gene. Members of the PPD family are characterized by the presence of two signature domains: a centrally located PAZ domain (~100 amino acid residues), and a carboxyl-terminally located PIWI domain (~300 amino acids residues) (Carmell *et al.*, 2002). PPD proteins and Dicer are the only known protein families that contain PAZ domains (Bernstein *et al.*, 2001; Cerutti *et al.*, 2000).

Genetic and biochemical studies have shown that PPD proteins play critical roles in the control of stem cell differentiation (Caplen *et al.*, 2002; Carmell *et al.*, 2002; Cavallo *et al.*, 1998), tissue development (Carmell *et al.*, 2002), and chromatin modification (Verdel *et al.*, 2004 and references therein). Many of these processes involve the function of PPD proteins in posttranscriptional and transcriptional gene-silencing pathways (Carmell *et al.*, 2002). Collectively, posttranscriptional gene-silencing processes that require PPD proteins include the RNAi pathway (Carmell *et al.*, 2002) in which homologous RNAs are targeted for degradation, as well as miRNA-dependent translational suppression (Bartel, 2004; Carmell *et al.*, 2002; Lai, 2003).

PPD proteins are the core constituents of RISCs and thus function at the effector step of RNAi (Hammond *et al.*, 2001). However, recent evidence also suggests that PPD proteins cooperate with Dicer during the initiation step of RNAi (Hammond *et al.*, 2001; Sasaki *et al.*, 2003). In mammalian cells the interaction of PPD proteins with Dicer is likely required for the transfer of siRNAs or miRNAs to RISC (Zamore, 2002), whereas in some invertebrates dsRNA-binding proteins RDE-4 or R2D2 link Dicer to RISC (Liu *et al.*, 2003; Parrish and Fire, 2001; Tabara *et al.*, 1999). Recently, it has been demonstrated that the interaction between human Dicer and PPD proteins occurs through the RNase III domain and a highly conserved subregion of the PIWI domain, the PIWI box (Tahbaz *et al.*, 2004), not through the shared PAZ domain as previously suggested (Baulcombe, 2001). Moreover, the stability of Dicer–PPD protein interaction was found to be dependent on the activity of Hsp90 (Tahbaz *et al.*, 2004).

In this chapter, we describe overexpression of recombinant human Dicer in insect cells, its purification, and biochemical properties. We also describe procedures used for mapping domains that mediate Dicer–PPD protein interactions.

Preparation of Expression Vectors

The cDNA encoding human Dicer fragment (amino acids 401–1922), cloned in *Sal*I and *Not*I sites of pBlueScript II® SK+, was obtained from N. Kusuhara (Kazusa Research Institute, Chiba, Japan; HH03019, GenBank Accession No. AB023145). The region of the cDNA corresponding to Dicer nucleotide positions 1–1494 was amplified by RT-PCR, using total HeLa cell RNA and primers ACGC<u>GTCGAC</u>ATGAAAAGCCCTGC TTTGC and CTG<u>AATTC</u>TGCTTCCATCGTG (SalI and EcoRI sites underlined). The PCR fragment was subcloned into SalI and EcoRI sites of HH03019 to yield pBS-Dicer. All constructs used in our experiments have been verified by sequencing.

To add a His$_6$ tag to the C terminus of Dicer, the downstream BamHI–NotI fragment in pBS-Dicer was replaced by the equivalent fragment (obtained by PCR, using appropriate primers) containing the sequence CATCACCATCACCATCAC positioned upstream of the Dicer stop codon. The resulting plasmid was named pBS-Dicer/HisC.

Dicer and Dicer-HisC cDNAs were recloned from pBS plasmids into the pENTR1A vector of the Gateway™ system (Invitrogen Life Technologies, Paisley, PA), using SalI and NotI sites, to yield pENTR-Dicer and pENTR-Dicer-HisC, respectively. The nontagged Dicer cDNA was then switched into the pDEST10 vector by a reaction catalyzed by LR Clonase™ mix (Invitrogen Life Technologies), yielding pDEST10-Dicer, which was used for the expression of the N-terminally tagged Dicer-HisN.

Dicer-HisN contains an additional 31 vector-encoded amino acids (sequence DYDIPTTENLYFFQGITSLYKKAGFKGTNSVD), positioned between the His$_6$ tag and Dicer N terminus. The Dicer-HisC cDNA was switched into the pDEST8 vector for expression of the C-terminally His$_6$-tagged Dicer-HisC.

The ATG selected as an initiation codon (sequence context TG**AAT-GA**) for plasmid constructions is an in-frame ATG of the Dicer mRNA 5′ leader. It is followed 30 bp downstream by a better context in-frame ATG (sequence AGC**ATGG**), which may potentially also act as an initiator.

Protein Expression and Purification

Bacmids for insect cell transformation and recombinant virus expressing human Dicer constructs for insect cell infection were generated using the Bac-to-Bac™ baculovirus expression system (Invitrogen Life Technologies). Sf9 insect cells were infected with recombinant virus at multiplicity of infection of 1.0. Three days later, the infected cells ($\sim$4 $\times$ 10^7) were resuspended in 4 ml of W100 buffer (Tris-HCl, pH 7.5, 100 mM NaCl, 1 mM MgCl$_2$, 5 mM β-mercaptoethanol, 10% glycerol, and 0.5% Triton X-100) containing 1$\times$ protease inhibitor mix without EDTA (Roche, Switzerland). The cells were broken by passing through a 0.45-gauge needle 10 times. The lysates were centrifuged at 17,000g for 10 min and then at 345,000g (58,000 rpm, SW 60 Beckman rotor) for 1 h. The supernatants were collected and used for protein purification.

To purify the recombinant protein, the precleared cell lysates were mixed with Talon™ affinity resin (400 μl; BD Biosciences Clontech, Palo Alto, CA) that had been prewashed with W100 buffer for 3 h at 4°. The resin was then washed three times with 4 ml of buffers W100 and W800 (buffer W100 with 800 mM NaCl), and again with W100 buffer. The proteins were eluted with 4 ml of W100 buffer containing 40 mM imidazole. The eluates were added directly to 400 μl of Ni/NTA beads (Qiagen, Valencia, CA) that had been prewashed with W100 buffer containing 40 mM imidazole and 1 mM DTT and incubated for 3 h at 4°. Then beads were washed as before, and the proteins were eluted with 2 ml of W100 buffer containing 100 mM imidazole and 1 mM DTT. The eluates were dialyzed for 4 h against two exchanges of 1 L of buffer D (W100 buffer with 50% glycerol, 0.1% Triton X-100, and 1 mM DTT) and stored at −20°. For purification of Dicer-HisC under nonreducing conditions, an identical procedure was used, except that all buffers were devoid of DTT or β-mercaptoethanol.

Purification of Dicer-HisN was carried out under reducing conditions and all buffers contained 1 mM DTT. The initial steps were identical to purification of Dicer-HisC, except that the high-speed centrifugation and Talon steps were omitted. The 17,000g supernatant was directly mixed with prewashed Ni/NTA beads. After a 12-h incubation at 4°, the beads were packed into a disposable polypropylene column (5 ml, Qiagen). The column was consecutively washed with W100, W800, and W100 buffers, and the protein eluted with W100 buffer containing 100 mM imidazole. Three 1.0-ml fractions having Dicer activity were collected and the buffer was exchanged back to W100 by using a Millipore Biomax 30K concentrator.

The sample was then applied to a Mini Q™ PC 3.2/3 column (Amersham Biosciences, England) equilibrated with W100 buffer. The column was washed extensively with W100 buffer, and the proteins were eluted at 0.5 ml/min with a gradient of 10 ml of W100 and 10 ml of W1000 buffers (W100 buffer with 1 M NaCl). Two 0.4-ml fractions corresponding to the peak of activity eluting at ∼0.4 M NaCl were collected and concentrated concomitantly with buffer exchange to W100 in a Millipore concentrator. The proteins were stored at −20° in W100 buffer containing 50% glycerol, 0.1% Triton X-100, and 1 mM DTT. Protein concentrations were determined by using the Bradford reagent with BSA as a standard.

Preparation of Dicer RNA Substrates

Dicer processes two classes of RNA substrates: perfectly complementary dsRNAs and stem-loop precursors of miRNAs, referred to as pre-miRNAs.

Preparation of Perfectly Complementary dsRNAs

Because Dicer does not exhibit sequence specificity, any dsRNA can serve as its substrate. All dsRNAs used by us were derived from GFP coding sequence. The pβact-eGFP plasmid (Billy *et al.*, 2001) was used as a template to generate PCR fragments for transcription of 40- to 130-nt sense and antisense RNAs. Synthetic oligodeoxynucleotides used as primers were purchased from Microsynth, Switzerland. Double-stranded DNA template used for the transcription of the sense RNA was generated by PCR, using a sense primer bearing a T7 promoter sequence (5'-TTAATACGACTCAC-TATAGGGAGA-3'; the noncoding sequence underlined) at the 5' end. To generate DNA template for transcription of the antisense RNA, the T7 promoter sequence was included in the 5' end of the antisense primer. Following PCR reactions, products were purified on low-melting-point agarose gels, resuspended in RNase-free double-distilled water (ddH$_2$O), and their identities verified by sequencing.

Nonradioactive RNAs were synthesized by using the T7 MEGAshort-script™ kit (Ambion, Austin, TX) following the manufacturer's protocol, and the radioactive RNAs were labeled with [α-^{32}P]UTP (final specific activity of 30–150 Ci/mmol; Amersham Biosciences), using the T7 MAXI-script™ kit (Ambion). DNA templates were removed by digestion with DNase I. The RNAs were extracted with phenol–chloroform and, follow-ing ethanol precipitation, pellets were resuspended in 50 μl of denaturing gel loading buffer (70% formamide, 30 mM EDTA, 0.05% xylene cyanol, 0.05% bromophenol blue). The samples were heat denatured for 2 min at 95° and subjected to separation by polyacrylamide gel electrophoresis (PAGE) in 8% polyacrylamide gels containing 7 M urea. Radioactive and nonradioactive RNAs were visualized by autoradiography and by UV shadowing, respectively, and the corresponding gel fragments were excised. After elution for 15 h at 4° with 500 mM ammonium acetate and 1 mM EDTA, the RNAs were ethanol precipitated in the presence of 5 μg of glycogen (Ambion) and resuspended in 15 μl of RNase-free ddH$_2$O. For annealing, equal amounts (estimated by the Cerenkov counting or OD$_{260nm}$ measurement) of complementary strands were combined together in 50 mM NaCl, heated at 95° for 3 min, transferred to 75°, and then slowly adjusted (over 4–6 h) to room temperature.

Preparation of 30-bp Double-Stranded RNAs (dsRNAs)
Containing Different Termini

The 30-bp dsRNAs are convenient substrates for assaying Dicer activity as they are cut only once by the enzyme. We employed 30-bp substrates containing either blunt or 2-nt 3' protruding ends in our experiments. DNA templates for the synthesis of 32-nt upper and lower strands of the 30-bp dsRNA containing 3′ overhangs were obtained by annealing synthetic oligonucleotides TTAATACGACTCACTATAGGGAGACCACTACCTGAGCACCCATCTCCCTT and AAGGGAGATGGGTGCTCAGGTAGTGGTCTCCCTATAGTGAGTCGTATTAA, and TTAATACGACTCACTATAGGGAGATGGGTGCTCAGGTAGTGGTCTCCCTT and AAGGGAGACCACTACCTGAGCACCCATCTCCCTATAGTGAGTCGTATTAA, (T7 RNA polymerase promoter sequences underlined). Templates for the preparation of blunt-ended 30-bp dsRNA were assembled from oligonucleotides differing from the previously given oligonucleotides by the absence of two terminal residues. Radioactive and nonradioactive RNAs were prepared as described previously. Following purification by 7 M urea/8% PAGE, individual RNA strands were ethanol precipitated and the 5′ ends dephosphorylated by incubation with 1 U of calf intestinal phosphatase (CIP; Roche) for 1 h at 37° in 50 mM Tris-HCl, pH 8.5, 0.1 mM EDTA. Following phenol–chloroform extraction and ethanol precipitation, RNAs were 5′-phosphorylated by T4 polynucleotide kinase (New England Biolabs, Beverly, MA) with either 10 mM ATP or [γ-^{32}P]ATP (7000 Ci/mmol; Valeant Pharmaceuticals, Costa Mesa, CA) in 30 mM Tris-acetate, pH 8.8, 60 mM potassium acetate, 10 mM magnesium acetate, and 0.5 mM DTT. The 5′ phosphorylated RNAs were purified by 7 M urea/8% PAGE and complementary strands were annealed as described previously.

Preparation of pre-let-7 RNA

In mammals, processing of miRNAs from their precursors is a two-step process (Lee *et al.*, 2003). The first step is catalyzed by the RNase III-like nuclear enzyme Drosha, which produces a stem-loop intermediate (pre-miRNA) with the miRNA sequence present at the base of the stem. The pre-miRNA is then processed in the cytoplasm by Dicer, which cleaves its substrate at $\sim$21 bp from the base of the stem, similarly to dsRNA (Zhang *et al.*, 2002). Processing of the let-7 miRNA has been extensively studied in different systems (Hutvagner *et al.*, 2001; Ketting *et al.*, 2001; Lee *et al.*, 2002); hence, pre-let-7 RNA was selected for studies with recombinant human Dicer.

Because pre-let-7 RNA has a uridine residue at the 5′ end, its synthesis is not compatible with T7 RNA polymerase, which optimally initiates transcription with two G residues. To circumvent this problem, we designed a chimeric RNA containing in its 5′ portion a hammerhead ribozyme (Price *et al.*, 1995) followed by the pre-let-7 sequence (Fig. 1A). The double-stranded DNA cassette used as a transcription template was assembled by using overlapping complementary 5′ phosphorylated oligodeoxyribonucleotides, annealed as before, and ligated with T4 DNA ligase (New England Biolabs).

After PCR amplification, the product was purified on a low-melting agarose gel and used as a template for T7 RNA polymerase transcription. T7 MEGAshortscript and T7 MAXIscript kits (Ambion) were used to synthesize nonradioactive and internally labelled RNAs, respectively. Self-processing of the ribozyme-containing transcript occurs during the course of the transcription reaction with approximately 70% efficiency (Fig. 1B). The resulting pre-let-7 RNA containing the 5′-terminal U residue was size purified by 7 M urea/8% PAGE as described previously. The precipitated pre-let-7 RNA, bearing a 5′-OH group, was then 5′-phosphorylated by T4 polynucleotide kinase and either nonradioactive or [γ-^{32}P]ATP. Before use, pre-let-7 RNA was resuspended in RNase-free ddH$_2$O, heat denatured for 1 min at 95°, placed for 5 min on ice, and refolded for 15 min at 25° in 30 mM Tris-HCl, pH 6.8, 50 mM NaCl, 1 mM MgCl$_2$, 0.05% Triton, and 25% glycerol. Alternatively, pre-let-7 RNA was

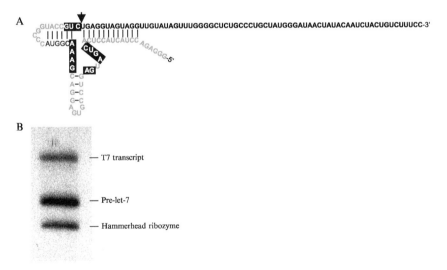

FIG. 1. Sequence and endonucleolytic self-processing of the ribozyme-pre-let-7 RNA. (A) Sequence of the ribozyme-pre-let-7 RNA and a secondary structure folding of the 5′ hammerhead domain. The hammerhead ribozyme domain is indicated in grey letters and nucleotides crucial for activity of the ribozyme are indicated by black boxes. The cleavage position is indicated by an arrow. The pre-let-7 sequence is in bold. (B) Endonucleolytic self-processing of the ribozyme-pre-let-7 transcript. Internally 32P-labelled ribozyme-pre-let-7 RNA was synthesized in vitro using the T7 RNA polymerase and products analyzed by denaturing 10% PAGE. The full-length transcript (128 nts), the processed pre-let-7 RNA (70 nts) and the released ribozyme domain (58 nts) are indicated.

3′-end labeled by the T4 RNA ligase and [5′-³²P]pCp, according to a previously published method (England and Uhlenbeck, 1978). Following ligation, the 3′-phosphate was removed by treatment with CIP, and RNA was purified by 7 M urea/8% PAGE.

Assays of Dicer Activity

dsRNA or pre-miRNA processing assays (10 μl) contained 30 mM Tris-HCl, pH 6.8, 50 mM NaCl, 3 mM MgCl$_2$, 0.1% Triton-X100, and 15–25% glycerol. Reactions with internally labeled dsRNAs contained 1–500 fmol of the substrate and 25–650 fmol of recombinant Dicer. Following incubation at 37° for 30 min, 10 μl of denaturing loading buffer (70% formamide, 30 mM EDTA, 0.05% xylene cyanol, 0.05% bromophenol blue) was added and products were resolved on a 7 M urea/10% PAGE run in 90 mM Tris-borate, pH 8.0, 2 mM EDTA (TBE buffer). Assays with terminally labeled substrates usually contained 0.1 pmol of RNA and 0.23 pmol of Dicer and were performed for 15 or 30 min at 37°. After ethanol precipitation, RNA fragments were dissolved in denaturing loading buffer and size separated on 7 M urea/15% PAGE in TBE buffer. Cleavage positions were identified by using RNase T1 and alkaline ladders of the end-labeled RNA (Donis-Keller et al., 1977), which were 3′-dephosphorylated by treatment with T4 polynucleotide kinase. After electrophoresis, gels were processed for autoradiography or quantification, using the Storm 860 PhosphorImager™ (Molecular Dynamics).

Properties of Human Dicer

We have established procedures for purification of N-terminally and C-terminally His-tagged recombinant human proteins, Dicer-HisN and Dicer-HisC. Notably, Dicer-HisC was purifed by successive fractionation on cobalt-containing resin and Ni/NTA beads. As shown in Fig. 2, despite a similar principle of affinity separation, only the combination of both matrices reproducibly yielded a high-purity protein.

Dicer-HisC preparations purified under nonreducing conditions were marginally more active than those prepared in the presence of reducing agents, and preincubation of recombinant proteins with proteinase K and some other proteases markedly stimulated their activities. Moreover, the proteolyzed enzyme, in contrast to the native enzyme, was also active at 4° (Zhang et al., 2002). The activity of endogenous Dicer present in immunoprecipitates (IPs) of P19 cell extracts was also stimulated by preincubation with proteinase K (Zhang et al., 2002). Mechanisms underlying the effects of proteolysis and reducing agents on Dicer activity remain unknown.

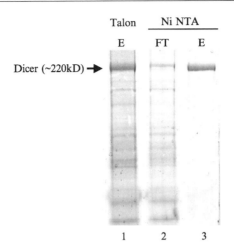

FIG. 2. SDS-PAGE analysis of recombinant Dicer preparations. Lane 1, eluate from the Talon (cobalt) resin. Lanes 2 and 3, flow-through and eluate from Ni/NTA beads. For additional details, see text. (See color insert.)

Similarly to RNase III, Dicer requires divalent cations for substrate cleavage. In addition to Mg^{2+}, Mn^{2+} and Co^{2+} ions are also active as cofactors. Mg^{2+} ions are required by Dicer for cleavage of dsRNA but not for its binding (Zhang et al., 2002).

Experiments performed with D. melanogaster and C. elegans cell extracts or IPs and with purified Drosophila Dcr-2 demonstrated that cleavage of dsRNA into siRNAs is strongly stimulated by ATP (Bernstein et al., 2001; Ketting et al., 2001; Liu et al., 2003; Nykanen et al., 2001). However, activities of different preparations of recombinant human Dicer, as well as IPs prepared from mouse P19 and HeLa cell extracts with anti-Dicer antibodies, were not stimulated by ATP. Moreover, cleavage of dsRNA was not affected by the presence of nonhydrolyzable ATP analogs or under conditions in which any residual ATP was eliminated. A change from lysine to alanine in the P-loop of Dicer ATPase/helicase domain, a mutation known to eliminate or at least strongly inhibit nucleotide binding and activity of enzymes containing related domains, had no effect either (Zhang et al., 2002). Importantly, two other groups studying the activity of recombinant human Dicer also did not observe any requirement of ATP for the substrate cleavage in vitro (Myers et al., 2003; Provost et al., 2002). Clearly, Dicers from different organisms differ in their dependence on ATP. In this context, note that a homolog of Dicer in Dictyostelium appears to lack the ATPase/helicase domain (Martens et al., 2002), whereas in Drosophila Dcr-1 this domain is nonfunctional (Lee et al., 2004).

Purified recombinant human Dicer displays a low catalytic activity. When assayed under excess substrate concentrations (30–100 nM), it generates 0.5–1.0 moles of siRNA product per mole of enzyme in a 30-min incubation (Zhang et al., 2002; our unpublished results). No data on turnover of human or Drosophila Dicers purified by other laboratories (Liu et al., 2003; Myers et al., 2003; Provost et al., 2002) have been reported. The low activity of purified human Dicer is at least partially due to the reaction product remaining associated with the enzyme (Zhang et al., 2002). Recent studies in Drosophila demonstrated that Dicer accompanies siRNA along the RISC assembly pathway and is an essential component of mRNA-cleaving holo-RISC, sedimenting at ~80S (Lee et al., 2004; Pham et al., 2004; Tomari et al., 2004). This finding offers an explanation for the low activity of the purified enzyme. In the absence of other RISC components, Dicer would remain associated with its siRNA cargo, thereby rendering it incapable of catalyzing multiple cleavage reactions (Zhang et al., 2004).

Mammalian Dicer enzymes (recombinant or endogenous), cleave dsRNAs into ~20-bp siRNAs. Notably, incubation of long dsRNA substrates with recombinant Dicer resulted in accumulation of processing intermediates having lengths diagnostic of a gradual removal of one or more ~22-nt siRNA units. Experiments performed with substrates containing ends "blocked" by stable tetraloops or by terminal RNA–DNA duplexes further indicated that Dicer preferentially cleaves dsRNAs at their termini (Zhang et al., 2002), a conclusion supported by direct mapping of cleavage sites in 30-bp dsRNA substrates labeled with a ^{32}P-phosphate at the 5′ end of one strand of the duplex (Zhang et al., 2004). The preference of purified recombinant human Dicer and endogenous Dicer present in extracts of Drosophila and C. elegans (reviewed by Hannon and Zamore, 2003) to cleave dsRNAs at their termini may be of physiological significance. This property might prevent accidental cleavage of extended hairpins located in internal regions of many mRNAs (Morse et al., 2002).

Recombinant human Dicer described in this chapter was also demonstrated to faithfully process miRNA precursors, which represent another class of cellular substrates of the enzyme. Primary transcripts containing miRNA sequences are first processed in the nucleus by Drosha into ~70-nt-long hairpins referred to as pre-miRNAs, which act as substrates for Dicer (Lee et al., 2003). We generated pre-let-7 miRNA expected to correspond to the Drosha cleavage product (Lee et al., 2003) by self-processing of the in vitro synthesized hybrid transcript, which contains the hammerhead ribozyme upstream of the pre-let-7 sequence (see previously). Processing of pre-let-7 RNA by Dicer yielded a double-stranded siRNA-like product, as determined by gel electrophoresis under nondenaturing conditions (Zhang et al., 2004). Experiments performed with Dicer mutants and terminally labeled pre-let-7 RNAs indicated that the enzyme accesses the substrate in a polar fashion, with the RIIIa domain always cutting the RNA strand bearing a 3′-hydroxy group approximately 21 nt from the end and the RIIIb domain cutting the opposite ascending strand of the pre-miRNA hairpin (Zhang et al., 2004).

Tools and Assays to Study Dicer–PPD Protein Interactions

We report the experimental techniques used in our laboratories to characterize interactions between human Dicer and PPD proteins. For these experiments, two distantly related members of the human PPD family, hAgo2 (human ortholog of GERp95, Cikaluk et al., 1999) and Hiwi (Sharma et al., 2001), were used. The predicted amino acid sequences of hAgo2 and Hiwi have 22.7% identity over their entire lengths. The underlying assumption is that findings from these experiments would be applicable to all mammalian PPD proteins. Interactions between Dicer and PPD proteins were investigated through the use of GST (glutathione-S-transferase) affinity purifications and coimmunoprecipitations from stably or transiently transfected mammalian cell lines and also the use of purified recombinant proteins in vitro. In addition, yeast two-hybrid binding assays were used for fine mapping of interacting regions and to confirm the binding interactions observed in mammalian cells.

Construction of Tagged Dicer and PPD Mutant Proteins

Isolation and characterization of cDNAs for human Dicer, hAgo2, and Hiwi has been described previously (Billy et al., 2001; Sharma et al., 2001; Tahbaz et al., 2001). For transient expression in mammalian cells, regions encoding the PIWI and PAZ domains of hAgo2 and Hiwi (Liu et al., 2002) were amplified by PCR and then subcloned into the mammalian expression vector pEBG (Mizushima and Nagata, 1990). Platinum® Pfx DNA polymerase (Invitrogen) was used for all PCR amplifications during construct preparation. Proteins of interest expressed from this vector have GST fused to their N termini. The pEBG vector contains the EIF1α promoter for high-level transient expression in a wide variety of mammalian cell types. The PAZ domain-containing constructs include amino acid residues 227–380 of hAgo2 and 271–425 of Hiwi. Similarly, the PIWI domain-containing constructs encode amino acid residues 508–829 of hAgo2 and 543–861 of Hiwi.

GST Dicer fusion proteins—full-length Dicer, the PAZ domain-containing region (612–1078), RNase III domain (amino acids 1271–1922), and helicase/PAZ (amino acids 1–1078)—were constructed using the Gatewa system (Invitrogen Life Technologies) involving traditional restriction enzyme reaction cloning and LR clonase recombination. pCFP-Dicer, encoding the cyan fluorescent protein fused to Dicer, has been described elsewhere (Billy *et al.*, 2001). For the construction of inducible stable cell lines, the cDNA for hAgo2 was subcloned into the tetracycline-inducible pcDNA4/TO vector (Invitrogen).

Protein constructs used in yeast two-hybrid assays include Dicer RIIIa (amino acids 1270–1582) and RIIIb (amino acids 1699–1831) and dsRBD (amino acids 1848–1922) fused with the Gal-4 activation domain of the pGADT7 prey vector. hAgo2 PAZ (amino acids 227–380) and PIWI domain (amino acids 508–829), as well as PIWI box (amino acids 723–780) and the Hiwi PAZ (amino acids 271–425) and PIWI domain (amino acids 543–861), and the PIWI box (amino acids 758–800) were cloned in frame with the Gal-4 DNA-binding domain of the pGBKT7 bait vector.

Generation of Stable Cell Lines Expressing hAgo2

In our experience, it was not possible to obtain stable cell lines that constitutively expressed high levels of hAgo2 protein. However, it was relatively easy to isolate stable cell lines that express hAgo2 in an inducible manner. For this purpose, we used the T-Rex™ regulated gene expression system (Invitrogen). Briefly, the system includes a regulatory plasmid (pcDNA6/TR) encoding the tetracycline (tet) repressor protein and a blasticidin S-resistance gene. The gene of interest is expressed from a second plasmid (pcDNA4/TO or pcDNA5/TO) under the control of the cytomegalovirus (CMV) promoter and tetracycline operator sites. For selection in mammalian cells, pcDNA5/TO contains a hygromycin B-resistance gene. Tet repressor homodimers bind to the tet operator sequence in the promoter of the inducible expression vector and repress transcription. Expression of the gene of interest is induced by adding doxycycline (1 μg/ml) to culture media. To create stable cell lines expressing hAgo2, 293T-Rex cells were transfected with pcDNA-5TO-hAgo2. 293T-Rex is a HEK293-derived cell line that was transfected with the pcDNA6/TR plasmid. The transfectants were selected with 100 μg/ml hygromycin B and 5 μg/ml blasticidin S. Individual clones were expanded and screened for doxycycline-induced hAgo2 expression by indirect immunofluorescence and immunoblotting. Robust expression of hAgo2 was detectable within 2–4 h of doxycycline addition.

Application of GST Pulldown Assays to Study Dicer–PPD
 Protein Complexes

Standard GST affinity purification procedures were used to map domains that mediate Dicer–PPD interactions. Mammalian cells transiently cotransfected with plasmids encoding GST–PPD protein fusion constructs and CFP–Dicer were used to map PPD domains that bind to Dicer. Conversely, transient expression of GST–Dicer constructs into stably transfected 293 HEK T-Rex expressing hAgo2 was used to map regions of Dicer that interact with PPD proteins. Mammalian cells (HEK293 or COS) cultured in 35- or 60-mm dishes were transfected with plasmids (2–6 μg) by using the FuGENE™ 6 transfection reagent (Roche) or PerFectin™ (Gene Therapy Systems, Inc., San Diego, CA) according to manufacturer's instructions. Two days posttransfection, cells were washed three times with ice-cold phosphate-buffered saline (PBS) and lysed on ice in 1 ml of sonication lysis buffer (20 mM Tris-HCl, pH 7.4, 200 mM NaCl, 2.5 mM MgCl$_2$, and 0.05% Nonidet P40), as described (Mourelatos *et al.*, 2002). The lysates were briefly sonicated by using a Branson Sonifier 250, subjected to centrifugation at 21,000g for 18 min at 4°, and the supernatants were incubated with glutathione Sepharose™ 4B beads (Amersham Biosciences) in PBS for 3 h or overnight at 4°. The beads were collected by centrifugation (5000g), washed three times in ice-cold lysis buffer, boiled in SDS-gel sample buffer, and analyzed by SDS 8–10% PAGE. After proteins were transferred to PVDF membranes, GFP–Dicer was detected by immunoblotting with affinity-purified antibodies to GFP (McCabe and Berthiaume, 1999) or Dicer (Billy *et al.*, 2001).

Using GST pulldowns from transiently transfected mammalian cell lysates, we detected stable interactions between Dicer and the PIWI domains of hAgo2 and Hiwi. In contrast, the PAZ domains from these two PPD proteins did not form stable complexes with Dicer. These results indicate that PIWI domains of PPD proteins are primarily involved in binding to Dicer. To map the region of Dicer that is required for binding to PPD proteins, we used stable 293 HEK T-Rex cell lines that overexpress hAgo2. Plasmids encoding GST–Dicer or GST fused to domains of Dicer were transiently transfected into the 293 HEK T-Rex cell lines. HAgo2 expression was induced with doxycycline (24 h posttransfection). Forty hours posttransfection, cell lysates were prepared and used for GST pulldown assays as described previously. Copurification of hAgo2 by GST–Dicer or GST fused to Dicer domains was detected by immunoblotting with polyclonal antibodies to hAgo2 (Tahbaz *et al.*, 2004). Data from these experiments indicate that hAgo2 binds to Dicer by its C-terminal region. This region includes the RNase III domains. Moreover, we showed that the PAZ domain of Dicer is not required for binding to hAgo2.

Purification of GST-Fusion PPD Proteins

Recombinant GST fusions of PPD proteins were purified from COS-1 cells transiently transfected with plasmids by using the FuGENE 6 transfection reagent (Roche). Two days posttransfection, the cells were lysed with 3 ml of PBS containing 1% Empigen™ (Calbiochem, La Jolla, CA), 1 mM DTT, 0.1% Triton X-100, 400 μM PMSF, and a protease inhibitor mix (Roche). Cell extracts were centrifuged for 10 min at 10,000g at 4° and the resulting supernatants applied to glutathione-Sepharose beads equilibrated with extraction buffer. The recombinant proteins were eluted from the beads with 20 mM reduced glutathione (Sigma) in 50 mM Tris-HCl, pH 8.0, 100 mM NaCl, 2.5 mM MgCl$_2$, 1 mM DTT, 0.1% Triton X-100, 10% glycerol, and protease inhibitors. This procedure yields ~70% pure fusion proteins, as measured by silver staining. The molar concentrations of the proteins can be calculated by using this value and protein concentration measured by the Bradford method. The purified proteins can be stored at −80° until further use.

Assaying Interactions of Purified Proteins by Coimmunoprecipitation

Interactions between purified GST–PPD fusion proteins and Dicer expressed and purified from insect cells (Zhang et al., 2002) were studied by coimmunoprecipitation. In brief, 25 μl of protein A-Sepharose beads (Amersham) coated with 1.5 μg of anti-Dicer antibody (Billy et al., 2001) or an unrelated control antibody were incubated for 3 h at 4° with protein mixtures preincubated at 25° for 30 min and containing 1.75 μg (8 pmol) of Dicer and 1.8 μg (16 pmol) of GST-hAgo2, 2 μg (17 pmol) GST-Hiwi, or 4 μg (200 pmol) GST alone in 30 mM HEPES, pH 7.4, 100 mM KCl, 2 mM MgCl$_2$, 1 mM DTT, and protease inhibitor mix. After centrifugation (5000g) and extensive washing with the binding buffer, the beads and supernatants were analyzed by SDS-PAGE and immunoblotting by using anti-GST antibodies.

By using this approach, we found that GST-hAgo2 and GST-Hiwi, but not GST alone, were retained by anti-Dicer beads but not the control beads. These results suggest that binding between PPD proteins and Dicer is direct. However, because both types of proteins are known to bind RNA, it was important to determine whether RNA is required for Dicer–PPD protein complex stability. Therefore, PPD protein–Dicer complexes from GST pulldowns or coimmunoprecipitations were treated with 0.5 U/μl micrococcal nuclease for 1 h, on ice, or a cocktail of RNase V1 (0.03 U/μl) and RNase A (0.04 μg/μl), respectively, before immunoprecipitations and immunoblot analyses. This treatment had no effect on the formation of Dicer–PPD protein complexes, confirming that the observed interactions are direct.

Dicer–PPD Protein Interactions in Yeast Two-Hybrid Assay

The experimental assays described previously were very useful for mapping the domains of Dicer and PPD proteins that mediate complex formation. Moreover, the observation that mixing purified preparations of PPD proteins and Dicer results in stable complex formation is consistent with the notion that binding interactions between these two classes of proteins are direct. However, it is possible that other proteins that copurify with PPD or Dicer proteins, or both, mediate complex formation. Therefore, we decided to further investigate Dicer–PPD protein interactions by additional assays that would detect direct interactions between these two classes of proteins. The yeast two-hybrid system has been widely used as a very sensitive assay to identify novel protein–protein interactions as well as to characterize interactions between pairs of known binding proteins. In our case, it has the additional advantage that *S. cerevisiae* does not encode PPD or Dicer homologs, and therefore any observed interactions would most likely result from direct binding.

The technical details of two-hybrid screening have been described previously (Bartel and Fields, 1995), and the necessary reagents are available as kits from commercial suppliers (e.g., Matchmaker™ system, BD Biosciences Clontech). We constructed plasmids encoding the dsRBD, RIIIa, and RIIIb domains of Dicer fused in frame with the activation domain of the transcription factor GAL4 in the bait vector pGADT7. Similarly, regions encoding the PAZ, PIWI, and PIWI Box domains of hAgo2 and Hiwi fused to the GAL4 DNA-binding domain were constructed in the prey vector pGBKT7. These plasmids proteins were sequentially transformed into the *S. cerevisiae* AH109 strain. The AH109 strain is auxotrophic for leucine, lysine, tryptophan, and uracil. In addition, this strain is engineered to contain two nutritional reporter genes (for histidine and adenine biosynthesis) under the transcriptional control of GAL4-responsive elements.

Transformants were plated onto complete minimum medium (CMM) lacking leucine and tryptophan for 2–3 days at 30°. Surviving colonies were then streaked onto media lacking leucine, tryptophan, histidine, and adenine. Only yeast harboring plasmids that encode interacting proteins or domains are able to survive under this stringent selection process. Growth occurs when interactions between the proteins or domains of interest bring the GAL4 DNA-binding domain and the activation domain together to form a functional transcriptional activator that drives expression of the nutritional reporter genes (histidine and adenine). For our purposes, we scored positive and negative interactions as follows: cases in which colonies grew in less than 7 days on media lacking leucine, tryptophan, histidine, and adenine were taken as evidence of positive interactions. Conversely, lack of growth after 14 days on media lacking leucine, tryptophan, histidine, and adenine was deemed as a negative interaction.

By using this technique, we found that the strongest interactions occurred between the RIIIa domain of Dicer and the PIWI boxes of hAgo2 and Hiwi. Interactions between the PIWI domains of the PPD proteins and RIIIa domain of Dicer were also detected, but were not particularly strong. Relatively weak interactions were also detected between the PIWI boxes of the PPD proteins and the RIIIb domain of Dicer. No interactions were detected between PPD protein domains and the dsRBD domain of Dicer or between the PAZ domains of PPD proteins and Dicer.

Discussion

PPD and Dicer protein families are central to RNAi; however, until recently, little was known about the nature and dynamics of PPD–Dicer complex formation. The techniques used to study protein–protein interactions, such as affinity purifications using various tags, coimmunoprecipitation of purified proteins, and yeast two-hybrid assays, are relatively easy to use, allowing for effective and easy functional characterization of novel proteins. Our investigation of the interactions between these proteins by using the aforementioned assays has proven very insightful and in some cases surprising. Specifically, our work and that of Doi *et al.* (2003) disproved the logical but unproven theory that heterotypic PAZ interactions mediate complex formation between PPD proteins and Dicer (Baulcombe, 2001; Hammond *et al.*, 2001; Mette *et al.*, 2001). Second, we discovered that PPD proteins interact with the RNase III domain of Dicer and thus are presumably brought in proximity with the siRNA produced by the catalytic domain of Dicer. Originally, we hypothesized that this interaction might stimulate Dicer activity by facilitating product (siRNA) release or possibly be important for handover of the siRNA product form Dicer to PPD proteins. However, rather unexpectedly, we found that PPD proteins actually inhibit the activity of Dicer *in vitro* (Tahbaz *et al.*, 2004). A recent finding that in *Drosophila* Dicer accompanies siRNA along the RISC assembly pathway and is an essential component of holo-RISC, which provides RISC mRNA cleaving activity (Lee *et al.*, 2004; Pham *et al.*, 2004; Tomari *et al.*, 2004), offers a rationale for this observation. Possibly, one function of Ago proteins, established components of RISC (reviewed by Hannon and Zamore, 2003), is to prevent cleavage of dsRNA by Dicer present in the complex to avoid associating a single RISC with more than one siRNA.

Acknowledgments

F. A. K. is the recipient of a long-term fellowship from the Human Frontier Science Program, and T. C. H. is the recipient of a Senior Medical Scholarship from the Alberta

Heritage Foundation for Medical Research (AHFMR). N. T. and K. J. are supported by a predoctoral studentship awards from the Canadian Institutes of Health Research (CIHR) and AHFMR, respectively. Friedrich Miescher Institute is a part of the Novartis Research Foundation. The work in T. C. H. laboratory was funded by a grant from CIHR. The clones described in this chapter are available on request.

References

Bartel, D. P. (2004). MicroRNAs: Genomics, biogenesis, mechanism, and function. *Cell* **116**, 281–297.

Bartel, P. L., and Fields, S. (1995). Analyzing protein-protein interactions using two-hybrid system. *Methods Enzymol.* **254**, 241–263.

Baulcombe, D. (2001). RNA silencing. Diced defence. *Nature* **409**, 295–296.

Bernstein, E., Caudy, A. A., Hammond, S. M., and Hannon, G. J. (2001). Role for a bidentate ribonuclease in the initiation step of RNA interference. *Nature* **409**, 363–366.

Bernstein, E., Kim, S. Y., Carmell, M. A., Murchison, E. P., Alcorn, H., Li, M. Z., Mills, A. A., Elledge, S. J., Anderson, K. V., and Hannon, G. J. (2003). Dicer is essential for mouse development. *Nat. Genet.* **35**, 215–217.

Billy, E., Brondani, V., Zhang, H., Muller, U., and Filipowicz, W. (2001). Specific interference with gene expression induced by long, double-stranded RNA in mouse embryonal teratocarcinoma cell lines. *Proc. Natl. Acad. Sci. USA* **98**, 14428–14433.

Blaszczyk, J., Tropea, J. E., Bubunenko, M., Routzahn, K. M., Waugh, D. S., Court, D. L., and Ji, X. (2001). Crystallographic and modeling studies of RNase III suggest a mechanism for double-stranded RNA cleavage. *Structure (Camb.)* **9**, 1225–1236.

Caplen, N. J., Zheng, Z., Falgout, B., and Morgan, R. A. (2002). Inhibition of viral gene expression and replication in mosquito cells by dsRNA-triggered RNA interference. *Mol. Ther.* **6**, 243–251.

Carmell, M. A., Xuan, Z., Zhang, M. Q., and Hannon, G. J. (2002). The Argonaute family: Tentacles that reach into RNAi, developmental control, stem cell maintenance, and tumorigenesis. *Genes Dev.* **16**, 2733–2742.

Cavallo, R. A., Cox, R. T., Moline, M. M., Roose, J., Polevoy, G. A., Clevers, H., Peifer, M., and Bejsovec, A. (1998). *Drosophila* Tcf and Groucho interact to repress Wingless signalling activity. *Nature* **395**, 604–608.

Cerutti, L., Mian, N., and Bateman, A. (2000). Domains in gene silencing and cell differentiation proteins: The novel PAZ domain and redefinition of the Piwi domain. *Trends Biochem. Sci.* **25**, 481–482.

Cikaluk, D. E., Tahbaz, N., Hendricks, L. C., DiMattia, G. E., Hansen, D., Pilgrim, D., and Hobman, T. C. (1999). GERp95, a membrane-associated protein that belongs to a family of proteins involved in stem cell differentiation. *Mol. Biol. Cell* **10**, 3357–3372.

Doi, N., Zenno, S., Ueda, R., Ohki-Hamazaki, H., Ui-Tei, K., and Saigo, K. (2003). Short-interfering-RNA-mediated gene silencing in mammalian cells requires Dicer and eIF2C translation initiation factors. *Curr. Biol.* **13**, 41–46.

Donis-Keller, H., Maxam, A. M., and Gilbert, W. (1977). Mapping adenines, guanines, and pyrimidines in RNA. *Nucleic Acids Res.* **4**, 2527–2538.

Elbashir, S. M., Lendeckel, W., and Tuschl, T. (2001). RNA interference is mediated by 21- and 22-nucleotide RNAs. *Genes Dev.* **15**, 188–200.

England, T. E., and Uhlenbeck, O. C. (1978). 3′-Terminal labelling of RNA with T4 RNA ligase. *Nature* **275**, 560–561.

Hammond, S. M., Boettcher, S., Caudy, A. A., Kobayashi, R., and Hannon, G. J. (2001). Argonaute2, a link between genetic and biochemical analyses of RNAi. *Science* **293**, 1146–1150.

Hannon, G. J., and Zamore, P. D. (2003). Small RNAs, big biology: Biochemical studies of RNA interference. *In* "RNAi: A Guide to Gene Silencing" (G. J. Hannon, ed.), pp. 87–108. Cold Spring Harbor Laboratory Press, Cold Spring Harbor, NY.

Hutvagner, G., McLachlan, J., Pasquinelli, A. E., Balint, E., Tuschl, T., and Zamore, P. D. (2001). A cellular function for the RNA-interference enzyme Dicer in the maturation of the let-7 small temporal RNA. *Science* **12**, 12.

Ishizuka, A., Siomi, M. C., and Siomi, H. (2002). A *Drosophila* fragile X protein interacts with components of RNAi and ribosomal proteins. *Genes Dev.* **16**, 2497–2508.

Ketting, R. F., Fischer, S. E., Bernstein, E., Sijen, T., Hannon, G. J., and Plasterk, R. H. (2001). Dicer functions in RNA interference and in synthesis of small RNA involved in developmental timing in *C. elegans. Genes Dev.* **15**, 2654–2659.

Lai, E. C. (2003). microRNAs: Runts of the genome assert themselves. *Curr. Biol.* **13**, R925–R936.

Lee, Y., Ahn, C., Han, J., Choi, H., Kim, J., Yim, J., Lee, J., Provost, P., Radmark, O., Kim, S., and Kim, V. N. (2003). The nuclear RNase III Drosha initiates microRNA processing. *Nature* **425**, 415–419.

Lee, Y., Jeon, K., Lee, J. T., Kim, S., and Kim, V. N. (2002). MicroRNA maturation: Stepwise processing and subcellular localization. *EMBO J.* **21**, 4663–4670.

Lee, Y. S., Nakahara, K., Pham, J. W., Kim, K., He, Z., Sontheimer, E. J., and Carthew, R. W. (2004). Distinct roles for *Drosophila* Dicer-1 and Dicer-2 in the siRNA/miRNA silencing pathways. *Cell* **117**, 69–81.

Lingel, A., Simon, B., Izaurralde, E., and Sattler, M. (2003). Structure and nucleic-acid binding of the *Drosophila* Argonaute 2 PAZ domain. *Nature* **426**, 465–469.

Liu, Q., Rand, T. A., Kalidas, S., Du, F., Kim, H. E., Smith, D. P., and Wang, X. (2003). R2D2, a bridge between the initiation and effector steps of the *Drosophila* RNAi pathway. *Science* **301**, 1921–1925.

Liu, Y., Jiang, Y., Qiao, D. R., and Cao, Y. (2002). The mechanism and application of posttranscriptional gene silencing. *Sheng Wu Gong Cheng Xue Bao.* **18**, 140–143.

Martens, H., Novotny, J., Oberstrass, J., Steck, T. L., Postlethwait, P., and Nellen, W. (2002). RNAi in *Dictyostelium*: The role of RNA-directed RNA polymerases and double-stranded RNase. *Mol. Biol. Cell* **13**, 445–453.

Matzke, M., Aufsatz, W., Kanno, T., Daxinger, L., Papp, I., Mette, M. F., and Matzke, A. J. (2004). Genetic analysis of RNA-mediated transcriptional gene silencing. *Biochim. Biophys. Acta.* **1677**, 129–141.

McCabe, J. B., and Berthiaume, L. G. (1999). Functional roles for fatty acylated amino-terminal domains in subcellular localization. *Mol. Biol. Cell* **10**, 3771–3786.

Mette, M. F., Matzke, A. J., and Matzke, M. A. (2001). Resistance of RNA-mediated TGS to HC-Pro, a viral suppressor of PTGS, suggests alternative pathways for dsRNA processing. *Curr. Biol.* **11**, 1119–1123.

Mizushima, S., and Nagata, S. (1990). pEF-BOS, a powerful mammalian expression vector. *Nucleic Acids Res.* **18**, 5322.

Morse, D. P., Aruscavage, P. J., and Bass, B. L. (2002). RNA hairpins in noncoding regions of human brain and *Caenorhabditis elegans* mRNA are edited by adenosine deaminases that act on RNA. *Proc. Natl. Acad. Sci. USA* **99**, 7906–7911.

Mourelatos, Z., Dostie, J., Paushkin, S., Sharma, A., Charroux, B., Abel, L., Rappsilber, J., Mann, M., and Dreyfuss, G. (2002). miRNPs: A novel class of ribonucleoproteins containing numerous microRNAs. *Genes Dev.* **16**, 720–728.

Myers, J. W., Jones, J. T., Meyer, T., and Ferrell, J. E., Jr. (2003). Recombinant Dicer efficiently converts large dsrnas into sirnas suitable for gene silencing. *Nat. Biotechnol.* **21,** 324–328.

Nicholson, A. W. (2003). The ribonuclease III superfamily: Forms and functions in RNA, maturation, decay, and gene silencing. In "RNAi: A Guide to Gene Silencing" (G. J. Hannon, ed.), pp. 149–174. Cold Spring Harbor Laboratory Press, Cold Spring Harbor, NY.

Nykanen, A., Haley, B., and Zamore, P. D. (2001). ATP requirements and small interfering RNA structure in the RNA interference pathway. *Cell* **107,** 309–321.

Papp, I., Mette, M. F., Aufsatz, W., Daxinger, L., Schauer, S. E., Ray, A., van der Winden, J., Matzke, M., and Matzke, A. J. (2003). Evidence for nuclear processing of plant micro RNA and short interfering RNA precursors. *Plant Physiol.* **132,** 1382–1390.

Parrish, S., and Fire, A. (2001). Distinct roles for RDE-1 and RDE-4 during RNA interference in *Caenorhabditis elegans*. *RNA* **7,** 1397–1402.

Pham, J. W., Pellino, J. L., Lee, Y. S., Carthew, R. W., and Sontheimer, E. J. (2004). A Dicer-2-dependent 80s complex cleaves targeted mRNAs during RNAi in *Drosophila*. *Cell* **117,** 83–94.

Price, S. R., Ito, N., Oubridge, C., Avis, J. M., and Nagai, K. (1995). Crystallization of RNA-protein complexes. I. Methods for the large-scale preparation of RNA suitable for crystallographic studies. *J. Mol. Biol.* **249,** 398–408.

Provost, P., Dishart, D., Doucet, J., Frendewey, D., Samuelsson, B., and Radmark, O. (2002). Ribonuclease activity and RNA binding of recombinant human Dicer. *EMBO J.* **21,** 5864–5874.

Sasaki, T., Shiohama, A., Minoshima, S., and Shimizu, N. (2003). Identification of eight members of the Argonaute family in the human genome small star, filled. *Genomics* **82,** 323–330.

Sharma, A. K., Nelson, M. C., Brandt, J. E., Wessman, M., Mahmud, N., Weller, K. P., and Hoffman, R. (2001). Human CD34(+) stem cells express the hiwi gene, a human homologue of the *Drosophila* gene piwi. *Blood* **97,** 426–434.

Song, J. J., Liu, J., Tolia, N. H., Schneiderman, J., Smith, S. K., Martienssen, R. A., Hannon, G. J., and Joshua-Tor, L. (2003). The crystal structure of the Argonaute2 PAZ domain reveals an RNA binding motif in RNAi effector complexes. *Nat. Struct. Biol.* **10,** 1026–1032.

Tabara, H., Sarkissian, M., Kelly, W. G., Fleenor, J., Grishok, A., Timmons, L., Fire, A., and Mello, C. C. (1999). The rde-1 gene, RNA interference, and transposon silencing in *C. elegans*. *Cell* **99,** 123–132.

Tabara, H., Yigit, E., Siomi, H., and Mello, C. C. (2002). The dsRNA binding protein RDE-4 interacts with RDE-1, DCR-1, and a DExH-box helicase to direct RNAi in *C. elegans*. *Cell* **109,** 861–871.

Tahbaz, N., Carmichael, J. B., and Hobman, T. C. (2001). GERp95 belongs to a family of signal-transducing proteins and requires Hsp90 activity for stability and Golgi localization. *J. Biol. Chem.* **276,** 43294–43299.

Tahbaz, N., Kolb, F. A., Zhang, H., Jaronczyk, K., Filipowicz, W., and Hobman, T. C. (2004). Characterization of the interactions between mammalian PAZ PIWI domain proteins and Dicer. *EMBO Rep.* **5,** 189–194.

Tomari, Y., Du, T., Haley, B., Schwarz, D. S., Bennett, R., Cook, H. A., Koppetsch, B. S., Theurkauf, W. E., and Zamore, P. D. (2004). RISC assembly defects in the *Drosophila* RNAi mutant armitage. *Cell* **116,** 831–841.

Verdel, A., Jia, S., Gerber, S., Sugiyama, T., Gygi, S., Grewal, S. I., and Moazed, D. (2004). RNAi-mediated targeting of heterochromatin by the RITS complex. *Science* **303,** 672–676.

Wienholds, E., Koudijs, M. J., van Eeden, F. J., Cuppen, E., and Plasterk, R. H. (2003). The microRNA-producing enzyme Dicer1 is essential for zebrafish development. *Nat. Genet.* **35**, 217–218.

Yan, K. S., Yan, S., Farooq, A., Han, A., Zeng, L., and Zhou, M. M. (2003). Structure and conserved RNA binding of the PAZ domain. *Nature* **426**, 468–474.

Zamore, P. D. (2002). Ancient pathways programmed by small RNAs. *Science* **296**, 1265–1269.

Zhang, H., Kolb, F. A., Brondani, V., Billy, E., and Filipowicz, W. (2002). Human Dicer preferentially cleaves dsRNAs at their termini without a requirement for ATP. *EMBO J.* **21**, 5875–5885.

Zhang, H., Kolb, F. A., Jaskiewicz, L., Westhof, E., and Filipowicz, W. (2004). Single processing center models for human Dicer and bacterial RNase III. *Cell.* **118**, 57–68.

[20] Delivery of siRNA and siRNA Expression Constructs to Adult Mammals by Hydrodynamic Intravascular Injection

By DAVID L. LEWIS and JON A. WOLFF

Abstract

Extensive use of RNA interference in mammals has been hindered by the inability to effectively deliver small interfering RNAs (siRNAs) or DNA-based constructs designed to express siRNAs. In this chapter, we describe the high-pressure or hydrodynamic intravascular injection technique used to deliver these nucleic acids to mice and nonhuman primates. Emphasis is placed on the use of this technique for delivery to the liver.

Introduction

RNA Interference

RNA interference (Rnai) has greatly facilitated gene function studies in a wide variety of cell types in a diverse set of organisms. RNAi is particularly valuable to investigators studying gene function in mammalian cells, for which a lack of facile classical genetic methods has been a major hindrance to understanding gene function. By using RNAi, it is now possible to perform large-scale loss-of-function screens in mammalian cultured cells (Aza-Blanc *et al.*, 2003; Berns *et al.*, 2004; Brummelkamp *et al.*, 2003; Paddison *et al.*, 2004; Zheng *et al.*, 2004). The use of RNAi will undoubtedly lead to new discoveries that will not only expand our understanding of basic biological processes but possibly also result in novel drugs and therapies to treat disease.

Mechanism of RNAi and the Discovery of Small Interfering RNAs (siRNAs)

The power of RNAi lies in part in the high degree of specificity and target gene knockdown afforded by this naturally occurring mechanism. RNAi is triggered by the presence or introduction into the cell of double-stranded RNA (dsRNA) that contains a sequence identical to that of an endogenous mRNA and was first demonstrated to occur in animals by Fire *et al.* (1998). Soon after the discovery of RNAi in *C. elegans*, a number of laboratories began to investigate its mechanism. With *Drosophila* S2 cells or *Drosophila* embryo lysates, it was shown that the long dsRNA is first cleaved into 21- to 23-bp fragments by the Dicer enzyme (Hammond *et al.*, 2000; Parrish *et al.*, 2000; Zamore *et al.*, 2000). Further studies indicated that these short dsRNA fragments could support target mRNA cleavage in *C. elegans* and in *Drosophila* embryo lysates (Elbashir *et al.*, 2001a; Parrish *et al.*, 2000). The finding that cleavage of the long dsRNA precursor into short fragments could elicit RNAi provided the key to the use of RNAi in mammalian cells. It had been previously demonstrated in a number of laboratories that exposure of mammalian cells to long dsRNA activates components of the interferon response, resulting in nonspecific mRNA degradation and inhibition of translation (Stark *et al.*, 1998). In seminal papers by Elbashir *et al.* (2001b) and Caplen *et al.* (2001), experiments were presented demonstrating that short dsRNAs resembling Dicer cleavage products could themselves elicit target gene knockdown with high specificity when introduced directly in mammalian cells in culture. These short products are called small interfering RNAs (siRNAs). Importantly, introduction of siRNAs into mammalian cells did not appear to induce components of the interferon pathway (Caplen *et al.*, 2001).

Additional biochemical evidence indicates that siRNAs are incorporated into a cytoplasmic ribonucleoprotein complex known as dsRNA-induced silencing complex (RISC; Hammond *et al.*, 2000; Martinez *et al.*, 2002; Nykanen *et al.*, 2001). By using the antisense strand of the siRNA as a guide, RISC associates with and cleaves the mRNA of identical sequence. The cleaved mRNA is then degraded by nonspecific RNases, one of which is a component of RISC (Caudy *et al.*, 2003). There is evidence that RISC is able to cycle and cleave additional target mRNAs (Hutvagner and Zamore, 2002).

A plethora of studies have been reported in which siRNAs have been used to knock down the expression of target genes in mammalian cells in culture (McManus and Sharp, 2002). These have led to a number of new discoveries concerning gene function and will undoubtedly lead to more in the future. However, a deeper understanding of a gene's function requires that the gene be studied in an organismal context. Moreover, a number of disease states, including those due to metabolic deficiencies or neurological disorders, are difficult to model in cells in culture and must be studied in animals.

Intravascular Delivery of siRNA or siRNA Expression Constructs to Mammals

The Delivery Problem

The major hurdle to overcome in using RNAi in mammals is the delivery of the siRNA or DNA-based siRNA expression construct. Successful delivery in this context means that the molecule has reached its site of action within the cell of the target tissue. For siRNAs and siRNA expression constructs, these sites are the cell cytoplasm and nucleus, respectively. This is a significant technical challenge and has greatly hampered the use of nucleic acid-based molecules in both gene function studies and gene therapy settings.

Two properties of siRNA or siRNA expression constructs make their delivery difficult to mammalian cells *in vivo*. First, nucleic acids are labile molecules that are rapidly degraded by both extracellular and intracellular nucleases. Modifications can be made to nucleic acids to increase nuclease resistance; however, this alone may not be sufficient to achieve biological activity. This is because nucleic acids are, for all practical purposes, membrane-impermeable molecules not readily taken up by cells in a way that permits biological activity.

In light of these properties, it is not surprising that simple intravascular injection of siRNAs or siRNA expression constructs does not appear to be sufficient to elicit RNAi [Boutla et al. (2001) and our unpublished observations]. However, several studies have now shown that siRNAs can be used to elicit target gene knockdown in mammalian model organisms such as the mouse. Among these are those in which naked siRNA or siRNA expression constructs were injected into the tail vein by the high-pressure or hydrodynamic technique (Giladi et al., 2003; Lewis et al., 2002; McCaffrey et al., 2002, 2003; Song et al., 2003; Zender et al., 2003). An eight- to tenfold reduction in target gene expression in the liver has been reported by this method to deliver siRNA targeting the Fas receptor (Song et al., 2003). A more typical level of target gene knockdown achieved by this method is 20–40% (our unpublished observations). However, many factors can contribute to the overall knockdown percentage achieved in the liver, including delivery efficiency, siRNA efficacy, and the existence of compensatory mechanisms in untransfected cells. In the following sections, we describe the principles behind the hydrodynamic injection technique, as well as its practical usage.

Hydrodynamic Delivery of Naked Nucleic Acids

The hydrodynamic injection method involves the rapid injection of a large aqueous volume of naked nucleic acid-containing solution into the vasculature. Experiments suggesting that delivery of nucleic acids might be enhanced by using large injection volumes were first reported by Budker et al. (1996), using reporter gene expression plasmids. In that study, 1 ml of solution containing a reporter gene expression plasmid was injected into the portal vein of the liver in mice in approximately 30 sec. This resulted in levels of reporter gene expression that were orders of magnitude higher than what had been previously observed after direct injection of plasmid DNA into the liver tissue (Hickman et al., 1994). Interestingly, it appeared that the cell type transfected by using this procedure was the hepatocyte. Almost no nonparenchymal cells (NPCs) in the liver, including Kupffer, endothelial, and the bile duct epithelial cells, displayed reporter gene expression. Still higher expression levels were obtained if a clamp was placed at the junction of the inferior vena cava (IVC) and hepatic vein in order to prevent outflow during the injection. Decreasing the volume by half resulted in a 70-fold reduction in reporter gene expression. These observations suggested that increased hydrostatic pressure was necessary for maximal delivery. In subsequent studies, it was shown that rapid injection of large volumes into the IVC or bile duct in mice and rats and the bile duct in dogs also resulted in high levels of reporter gene expression in the liver (Zhang et al., 1997). Efficient delivery to skeletal muscle was accomplished by injecting a large volume of plasmid DNA solution into the iliac artery of rats and later rhesus monkeys after clamping blood vessels leading into and out of the limb (Budker et al., 1998; Zhang et al., 2001).

Delivery by Hydrodynamic Tail Vein Injection

Injection of nucleic acid solution into vessels close to the target tissues may be desirable when it is important to limit the areas of delivery. However, these methods require invasive surgical procedures and may not be practical in some situations. For example, if the hydrodynamic injection technique were to be used as a discovery platform in research settings, then a method that is amenable to increased throughput would be desirable. Such a method was made available by the discovery that rapid injection of a large volume of plasmid DNA solution in the tail vein could be used to deliver naked plasmid DNA (Liu *et al.*, 1999; Zhang *et al.*, 1999). Typically, 2 ml of plasmid DNA solution is injected in 5–7 sec. Rapid injection of this amount of fluid exceeds the capacity of the heart and causes an increased pressure in the vena cava, resulting in a retrograde flow of fluid into the liver through the hepatic vein (Zhang *et al.*, 2004). When this method is used to deliver plasmid DNA, the highest levels of reporter gene expression are found in the liver, where 10–40% of hepatocytes are transfected. Expression can also be detected in other organs, including the kidney, spleen, lung, and heart, although it is typically two to three orders of magnitude lower than in the liver (Liu *et al.*, 1999; Zhang *et al.*, 1999). Some transient liver toxicity is associated with this procedure. Zhang *et al.* (1999) reported that serum alanine aminotransferase (ALT) levels increased to 3480 (±1641) U/L one day after injection, but returned to normal levels (50–100 U/L) by Day 2. Liu *et al.* (1999) also reported a transient rise in ALT levels. The increase in serum ALT levels correlates with a minor amount of hepatocyte necrosis observed histologically (Zhang *et al.*, 1999).

The parameters that appear to be most critical for efficient delivery of plasmid DNA by the hydrodynamic technique are injection rate and injection volume (Liu *et al.*, 1999; Zhang *et al.*, 1999). Typically, a volume equal to 10% of the animal's body weight, or 0.1 ml/g, is used when injecting mice. Use of lesser volumes to deliver the same mass of plasmid DNA results in dramatically lower expression levels. Injection of 0.05 ml/g results in a 20-fold decrease, and injection of 0.039 ml/g results in a 3000-fold decrease in the expression level compared with injection of 0.1 ml/g (Fig. 1). The dramatic decrease in expression levels observed when the same amount of plasmid DNA is delivered in 0.05 ml/g versus 0.039 ml/g suggests a threshold effect. Decreasing the rate of injection also negatively impacts delivery efficiency. For best results, injection is done as quickly as possible, equating to an injection rate of approximately 0.3–0.4 ml/sec. For plasmid DNA, a twofold decrease in injection rate results in an approximately twofold decrease in expression. Decreasing the injection rates further results in a more dramatic decrease in delivery, again suggesting a threshold effect. The relationship between the amount of plasmid DNA delivered and the amount of expression is more linear. However, a saturation point can be reached and injection of more than 100 μg (5 μg/g body weight) of plasmid DNA does not result in significantly higher levels of expression (Zhang *et al.*, 1999).

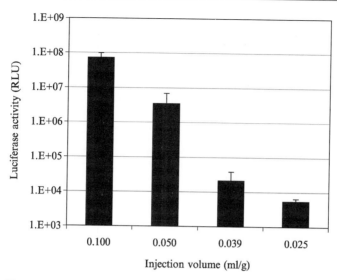

Fɪɢ. 1. The relationship between injection volume and reporter gene expression in mouse liver. Expression plasmid encoding the firefly luciferase[+] reporter gene (10 μg) was diluted in the indicated volumes of Ringer's solution and delivered to adult mice weighing 18–20 g, using the hydrodynamic tail vein injection method. Injection volume is given in milliliters per gram weight of the animal. The injection rate was held the same in each group (0.4 ml/sec). Luciferase[+] activity was measured in liver homogenates 24 h after injection. $N = 5$; error bars indicate one SD.

Mechanism of Nucleic Acid Uptake in Liver
After Hydrodynamic Injection

The mechanism of uptake of nucleic acids after delivery by hydrodynamic injection is not well understood. However, several observations on the distribution and kinetics of nucleic acid uptake by the liver may shed some light on this phenomenon. Simple low-pressure intravascular injection of naked plasmid DNA results in its rapid degradation by nucleases present in the blood and other compartments. These degradation products are removed from the circulation and are taken up primarily by NPCs associated with sinusoids in the liver (Budker *et al.*, 2000; Kawabata *et al.*, 1995; Kobayashi *et al.*, 2001). We have followed the liver distribution of Cy-3-labeled plasmid DNA after low pressure or hydrodynamic injection. After injection under low-pressure conditions, Cy-3-labeled plasmid DNA or its degradation products are present in the NPCs (Fig. 2, left). No signal is observed in hepatocytes. In NPCs, the signal is largely diffuse, although some areas of punctate staining can be observed. In contrast, Cy-3-labeled plasmid DNA injected by the hydrodynamic technique can be readily

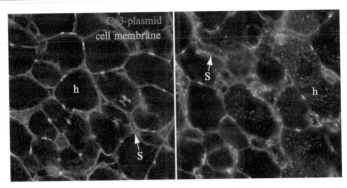

FIG. 2. Distribution of plasmid DNA in liver after tail vein injection. Plasmid DNA (25 µg) was fluorescently labeled with *Label* IT® Tracker Cy™3 (Mirus Corporation, Madison, WI) and delivered by normal (left) or hydrodynamic (right) tail vein injection. Livers were harvested 30 min after injection and sliced into 9-mm² sections and fixed overnight in 4% paraformaldehyde in PBS. The tissue pieces were placed in 20% sucrose for 4 h and then frozen in OCT embedding medium (Fisher, Pittsburgh, PA). Frozen sections (10 µm) were prepared and counterstained for 20 min in a 1:400 dilution of Alexa 488 phalloidin (Molecular Probes) in PBS to visualize cell outlines. Images were gathered using a Zeiss Axioplan 2/LSM 510 confocal microscope. A representative hepatocyte (h) and a sinusoid (s) are indicated in both panels. (See color insert.)

observed in hepatocytes as well as in the NPCs (Fig. 2, right). The staining pattern is much more punctate than that observed after low-pressure injection. Microinjection experiments performed in tissue culture cells suggest that punctate cytoplasmic staining is characteristic of fragments longer than 200 bp and is likely due to the limited cytoplasmic diffusion of molecules of this size or greater (Ludtke *et al.*, 1999). We have observed a similar correlation between DNA fragment length and diffuse versus punctate staining in liver hepatocytes after hydrodynamic injection (our unpublished observations). These data suggest that the delivery of plasmid DNA by hydrodynamic injection results in uptake of relatively intact plasmid DNA molecules by hepatocytes as well as other cell types, whereas injection under normal conditions results in plasmid DNA degradation and no accumulation in hepatocytes.

Injection of Cy-3-labeled siRNA results in a similar cell type distribution pattern in the liver. Intravascular injection under normal conditions results in the appearance of signal in NPCs, but no detectable signal in hepatocytes (Fig. 3, left). In contrast, injection under hydrodynamic conditions results in abundant signal in hepatocytes, accompanied by the appearance of low signal levels in NPCs (Fig. 3, right). The signal pattern observed in hepatocytes indicates that Cy-3-labeled siRNA preferentially

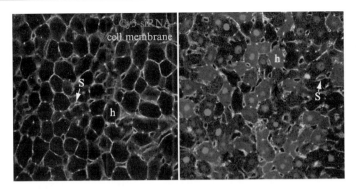

FIG. 3. Distribution of siRNA in liver after tail vein injection. SiRNA (25 μg) was labeled postsynthetically with *Label* IT siRNA Tracker[TM]-Rhodamine (Mirus Corporation) and delivered by normal (left) or hydrodynamic (right) injection. Liver sections were prepared and imaged as described in the legend of Fig. 2. A representative hepatocyte (h) and sinusoid (s) are indicated in both panels. (See color insert.)

accumulates in the nucleus. Substantial signal is also observed in the cytoplasm. The signal in the cytoplasm is diffuse and distributed evenly, suggesting that the siRNA is not preferentially associated with a particular organelle or cytoplasmic structure. Accumulation of siRNA in the nucleus indicates that it is preferentially retained there. Whether this is mediated by specific or nonspecific interactions is not known. However, nuclear accumulation is not unique to siRNA, having also been observed for small DNA fragments less than 200 bp in length, low-molecular-weight (11 kDa) dextran, and uncharged morpholino oligonucleotides [Ludtke *et al.* (1999) and our unpublished data].

Rapid injection of a large bolus of solution promotes rapid extravasation in liver. Large molecules such as plasmid DNAs likely gain access to hepatocytes through pores in the liver sinusoids, called fenestrae, that become enlarged during the injection procedure (Zhang *et al.*, 2004). Internalization of plasmids is relatively rapid, occurring within 10 min of injection. The rapid extravasation and uptake of nucleic acid by hepatocytes likely limit the time the nucleic acids are exposed to the high relative level of nucleases in the blood and extracellular environment. Precisely how the nucleic acids are internalized once in contact with hepatocytes is not entirely understood, but receptor-mediated and membrane disruption mechanisms have been proposed (Budker *et al.*, 2000; Kobayashi *et al.*, 2001). The fact that a variety of macromolecules including nucleic acids, proteins, and polyethyleneglycol can be delivered to hepatocytes by hydrodynamic injection suggests that the cellular entry mechanism is relatively unselective.

Methods

The ability to deliver siRNA or siRNA expression plasmids to mammals and induce RNAi opens up a wide range of applications. For example, in addition to providing a means to answer basic questions regarding gene function, RNAi can be used to validate candidate drug targets, perform toxicological analyses, and potentially be used in therapeutic settings. Codelivery of viral genes or genomes and siRNA or siRNA expression constructs by hydrodynamic injection has also been reported and provides a means of studying the roles of specific viral gene products in aspects of the viral life cycle (Giladi *et al.*, 2003; McCaffrey *et al.*, 2003). We have used hydrodynamic injection of both siRNA and siRNA expression constructs to study gene function in liver in mice. In the bulk of our studies, hydrodynamic injection of the tail vein has been used for liver delivery, although other more direct routes to the liver such as the portal vein and bile duct have also been used. We have also used hydrodynamic injection methods for delivery of siRNA to the liver of monkeys. In the following section, we will describe the methodologies used in these studies.

Hydrodynamic Tail Vein Injection in Mice

Preparing Mice for Tail Vein Injection. We use the ICR or C57Bl/6 strains of mice for the bulk of our studies. Other strains, including BALB/c, ddY, and transgenics, have also been used. The tail vein procedure is best performed by using an appropriate restraining device, without anesthetizing mice, as the use of anesthetics sometimes results in morbidity. If anesthesia is used, the inhalant anesthetic isoflurane (1–2%) is preferred. The unanesthesized mouse can be restrained during the procedure in a 50-ml plastic conical tube with a 3–5 mm hole cut in the bottom to facilitate the animal's breathing. The mouse is placed head first in the tube and the tail is threaded through a small slit cut in the cap of the tube. To facilitate tail vein visualization and ensure optimal injections, tail vessels are dilated before injection by warming the mouse under a heating lamp for 10 min. It is more difficult to visualize the tail vein in black mice. In these cases, we have found that spraying 70% isopropanol on the tail increases the contrast between the tail vein and skin.

Preparing the Injection Solution. The siRNA or siRNA expression plasmids (endotoxin free) are added at their final desired dose to sterile, RNase-free aqueous delivery solution containing 147 mM NaCl, 4 mM KCl, 1.13 mM CaCl$_2$ (Ringer's solution). Normal saline solution may also be used but tends to lead to postinjection complications. Typically, 40 μg of plasmid DNA or 40 μg of siRNA is used for each animal. The total injection volume per mouse (in milliliters) is calculated by dividing the weight of the mouse (in grams) by 10. The amount of nucleic acid added can be scaled up or down while keeping the total injection volume constant.

Tail Vein Injection. The nucleic acid-containing delivery solution is warmed to room temperature before injection. The volume of solution to be used should be approximately equal to 10% of the animal's weight. The use of lower volumes results in suboptimal delivery efficiencies. A 3-ml syringe is fitted with a 27-gauge, 0.5-in. syringe needle. The syringe needle is placed into the dilated tail vein, preferably midway in the tail or near the distal end. It is best to insert nearly the full length of the needle into the vein in order to prevent accidental release while injecting. Gentle injection of a small amount of the volume can be done to ensure that the needle is properly placed in the vein. Once a clear injection pathway is established, the complete volume of solution is dispensed into the tail vein in 5–7 sec. Maximum delivery is attained by quick injection at a constant speed, delivering the entire contents of the syringe. Immediately after the procedure, mice will display a short period of immobility and labored breathing, but these effects do not typically last more than 15–20 min.

Hydrodynamic Intraportal Vein Injection for Direct Liver Delivery in Mice

Animals are anesthetized with 1–2% isoflurane and prepped for abdominal surgery with an antiseptic solution. A midline incision extending from the pubis to the xiphoid process is made and retractors are positioned inside the abdominal cavity. The bowel loops are wrapped in moist gauze to prevent dehydration and exteriorized to expose the liver and portal vein. Microvascular clamps are placed on the suprahepatic IVC, infrahepatic IVC, and portal vein. A 27-gauge needle catheter is then inserted into the portal vein near its connection to the liver and upstream of the clamp. This catheter is connected to a syringe pump and plasmid DNA (10–100 μg) or siRNA (40 μg) in Ringer's solution (0.075–0.1 ml/g body weight) is injected at a rate of 12 ml/min. Two minutes after injection, the clamps and catheter are removed and bleeding controlled by applying pressure with a cotton swab and a hemostatic sponge. The abdominal incision is closed in two layers with 4–0 suture.

Hydrodynamic Intra-Bile Duct Injection for Direct Liver Delivery in Mice

Animals are anesthetized and the abdominal cavity exposed and kept hydrated as described for intraportal vein injection. The common bile duct between the bifurcation and the pancreatic duct is exposed and a 30-gauge needle is inserted into the duodenum adjacent to the bile duct to create a small opening. A 30-gauge smooth needle catheter is inserted through this hole and advanced into the bile duct with the tip positioned near the liver. The catheter is secured with a microvascular clip and connected to a syringe pump (Harvard Instruments, Lake Elsinore, CA). Just before injection, microvascular clamps are placed on the suprahepatic IVC, infrahepatic IVC, and the portal vein. The plasmid DNA (10–100 μg) or siRNA (40 μg) in Ringer's solution (0.075–0.1 ml/g body weight) is injected at a rate of 12 ml/min. Two minutes after injection, the clamps and catheter are removed and the hole in the duodenum is sealed with a hemostatic sponge. The abdominal incision is closed in two layers with 4–0 suture.

Hydrodynamic Intra-Vena Cava Injection by Using Catheters for Delivery to Liver in Nonhuman Primates

Preparation of the Animals and Catheter Insertion. For these studies, we have used Cynomolgus monkeys weighing 2.5–3.3 kg. Before the procedure, the animal is sedated with ketamine (10–15 mg/kg) injected intramuscularly. After sedation, animals are intubated and anesthesia maintained with 1.0–2.0% isoflurane. An intravenous catheter is inserted into the cephalic vein to administer fluids and EKG electrodes are attached to the limbs to monitor heart rate during the procedure. To place the catheter for liver delivery, a femoral cutdown is performed and a 5-cm segment of the femoral vein exposed. The vein is ligated distally and a 6F introducer is inserted and secured with a vessel tourniquet. A 4F injection catheter is inserted through the introducer and advanced into the IVC. An abdominal incision is made to visualize the infrahepatic IVC. The exact placement of the injection catheter within the IVC is adjusted so that the tip of the catheter is adjacent to the hepatic vein. A vessel tourniquet is placed loosely around the infrahepatic IVC to prevent backflow during the injection. A catheter (20-gauge) is inserted into the portal vein and attached to a pressure transducer to measure pressure changes during the injection. Immediately before the injection, a vascular clamp is placed on the suprahepatic IVC and the vessel tourniquet is tightened around the infrahepatic IVC.

Injection. Two 60-ml syringes are filled with a total volume of 120 ml of injection solution containing 0.9% NaCl and the nucleic acid to be delivered. We have codelivered siRNA targeting firefly luciferase[+] (0.5 mg/kg) with plasmids encoding firefly luciferase[+] (1.6 mg/kg) and *Renilla reniformis* luciferase (0.3 mg/kg). The latter plasmid acted as a delivery control. Both syringes are attached to a single pump (Harvard Instruments) and connected to the injection catheter by an extension line with a three-way stopcock. The injection flow rate is set at 90 ml/min for a total flow rate of 180 ml/min. The IVC remains occluded during the injection and for 2 min postinjection. The portal vein pressure during injection increases to approximately 100 mm Hg. This pressure increase lasts for the length of the injection and then rapidly returns to preinjection levels. After the injection, the clamps are removed and the portal vein pressure catheter is pulled. The abdominal cavity is closed in three layers with 3–0 suture. The catheter and introducer are pulled from the femoral vein, the vessel is ligated, and the incision closed in two layers with 3–0 suture. Before the completion of the surgery, the animal is given buprenorphine (0.005–0.01 mg/kg IM) as an analgesic.

Postinjection Observations. Monkeys regain normal activity within 24 h of surgery. As in mice, liver enzymes may become elevated in the serum after surgery, with the highest levels observed immediately after the procedure. However, the degree of ALT elevation is somewhat lower in monkeys than mice. In monkeys injected with siRNA, we have observed Day 1 levels of ALT that vary from 53 to 118 U/L. Normal levels are 13–63 U/L. Enzyme levels gradually decline thereafter and returned to normal levels within a few days. Histological examination of the right and left lobes of the liver reveals no significant pathology.

Discussion

The hydrodynamic injection methods described in this chapter enable the delivery of siRNA or siRNA expression plasmids to liver in mice and monkeys. Of these methods, the simplest and quickest to perform is tail vein injection in mice: an experienced investigator can inject more than 100 mice/day. The ability to inject these many animals makes possible experiments in mice that previously could only be performed in cells in culture. Moreover, experiments performed in mice yield additional information, for example, the physiological effects of target gene knockdown that cannot be attained by using cells in culture. At the same time, some of the difficulties in attempting to exploit RNAi in cells in culture also apply. One of these is the inability to deliver siRNA or siRNA expression plasmids to all cells in a liver or in the well of a tissue culture plate. This can potentially complicate interpretation of results obtained from particular experiments using RNAi. Furthermore, because of the mechanisms to maintain homeostasis in animals, physiological responses may be difficult to detect, depending on the function of the targeted gene. Nonetheless, there have been reports in which siRNA-mediated knockdown of genes that behave largely cell autonomously result in profound physiological changes (Song *et al.*, 2003; Zender *et al.*, 2003).

Nucleic acid delivery by hydrodynamic tail vein injection is most effective for delivery to liver. Delivery to other organs also occurs, albeit at greatly reduced efficiencies. If the investigator desires to deliver siRNA or DNA-based siRNA expression constructs specifically to the liver, then injection of the portal vein or bile duct can be performed. For delivery to other internal organs, hydrodynamic injection of organ-specific vasculature can be used. These methods have been described for delivery of plasmid DNA and would also allow for delivery of siRNA (Zhang *et al.*, 2002).

Hydrodynamic delivery techniques can be readily adapted for use in larger mammals. Direct delivery of plasmid DNA to the liver has been accomplished by using a catheter-based approach in rabbits and nonhuman primates [Eastman *et al.* (2002) and our unpublished data]. The ability to perform gene knockdown by using RNAi in large animal models will be useful in situations in which small animal models are not appropriate or do not exist.

Acknowledgments

We thank Julia Hegge and Guofeng Zhang for developing the protocols used for delivery to nonhuman primates and Chris Wooddell for critically reading the manuscript. Work at Mirus Corporation on RNAi is funded in part by grants to D. L. L. from the National Institute of Standards and Technology Advanced Technology Program (70NANB2H30616) and the National Institutes of Health Small Business Innovation Research Program (1R44CA097898).

References

Aza-Blanc, P., Cooper, C. L., Wagner, K., Batalov, S., Deveraux, Q. L., and Cooke, M. P. (2003). Identification of modulators of TRAIL-induced apoptosis via RNAi-based phenotypic screening. *Mol. Cell* 12, 627–637.
Berns, K., Hijmans, E. M., Mullenders, J., Brummelkamp, T. R., Velds, A., Heimerikx, M., Kerkhoven, R. M., Madiredjo, M., Nijkamp, W., Weigelt, B., Agami, R., Ge, W., Cavet, G., Linsley, P. S., Beijersbergen, R. L., and Bernards, R. (2004). A large-scale RNAi screen in human cells identifies new components of the p53 pathway. *Nature* 428, 431–437.
Boutla, A., Delidakis, C., Livadaras, I., Tsagris, M., and Tabler, M. (2001). Short 5'-phosphorylated double-stranded RNAs induce RNA interference in Drosophila. *Curr. Biol.* 11, 1776–1780.
Brummelkamp, T. R., Nijman, S. M., Dirac, A. M., and Bernards, R. (2003). Loss of the cylindromatosis tumour suppressor inhibits apoptosis by activating NF-kappaB. *Nature* 424, 797–801.
Budker, V., Budker, T., Zhang, G., Subbotin, V., Loomis, A., and Wolff, J. A. (2000). Hypothesis: Naked plasmid DNA is taken up by cells *in vivo* by a receptor-mediated process. *J. Gene Med.* 2, 76–88.
Budker, V., Zhang, G., Danko, I., Williams, P., and Wolff, J. (1998). The efficient expression of intravascularly delivered DNA in rat muscle. *Gene Ther.* 5, 272–276.

Budker, V., Zhang, G., Knechtle, S., and Wolff, J. A. (1996). Naked DNA delivered intraportally expresses efficiently in hepatocytes. *Gene Ther.* **3,** 593–598.

Caplen, N. J., Parrish, S., Imani, F., Fire, A., and Morgan, R. A. (2001). Specific inhibition of gene expression by small double-stranded RNAs in invertebrate and vertebrate systems. *Proc. Natl. Acad. Sci. USA* **98,** 9742–9747.

Caudy, A. A., Ketting, R. F., Hammond, S. M., Denli, A. M., Bathoorn, A. M., Tops, B. B., Silva, J. M., Myers, M. M., Hannon, G. J., and Plasterk, R. H. (2003). A micrococcal nuclease homologue in RNAi effector complexes. *Nature* **425,** 411–414.

Eastman, S. J., Baskin, K. M., Hodges, B. L., Chu, Q., Gates, A., Dreusicke, R., Anderson, S., and Scheule, R. K. (2002). Development of catheter-based procedures for transducing the isolated rabbit liver with plasmid DNA. *Hum. Gene Ther.* **13,** 2065–2077.

Elbashir, S. M., Lendeckel, W., and Tuschl, T. (2001a). RNA interference is mediated by 21- and 22-nucleotide RNAs. *Genes Dev.* **15,** 188–200.

Elbashir, S. M., Harborth, J., Lendeckel, W., Yalcin, A., Weber, K., and Tuschl, T. (2001b). Duplexes of 21-nucleotide RNAs mediate RNA interference in cultured mammalian cells. *Nature* **411,** 494–498.

Fire, A., Xu, S., Montgomery, M. K., Kostas, S. A., Driver, S. E., and Mello, C. C. (1998). Potent and specific genetic interference by double-stranded RNA in Caenorhabditis elegans. *Nature* **391,** 806–811.

Giladi, H., Ketzinel-Gilad, M., Rivkin, L., Felig, Y., Nussbaum, O., and Galun, E. (2003). Small interfering RNA inhibits hepatitis B virus replication in mice. *Mol. Ther.* **8,** 769–776.

Hammond, S. M., Bernstein, E., Beach, D., and Hannon, G. J. (2000). An RNA-directed nuclease mediates post-transcriptional gene silencing in Drosophila cells. *Nature* **404,** 293–296.

Hickman, M. A., Malone, R. W., Lehmann-Bruinsma, K., Sih, T. R., Knoell, D., Szoka, F. C., Walzem, R., Carlson, D. M., and Powell, J. S. (1994). Gene expression following direct injection of DNA into liver. *Hum. Gene Ther.* **5,** 1477–1483.

Hutvagner, G., and Zamore, P. D. (2002). A microRNA in a multiple-turnover RNAi enzyme complex. *Science* **297,** 2056–2060.

Kawabata, K., Takakura, Y., and Hashida, M. (1995). The fate of plasmid DNA after intravenous injection in mice: Involvement of scavenger receptors in its hepatic uptake. *Pharm. Res.* **12,** 825–830.

Kobayashi, N., Kuramoto, T., Yamaoka, K., Hashida, M., and Takakura, Y. (2001). Hepatic uptake and gene expression mechanisms following intravenous administration of plasmid DNA by conventional and hydrodynamics-based procedures. *J. Pharmacol. Exp. Ther.* **297,** 853–860.

Lewis, D. L., Hagstrom, J. E., Loomis, A. G., Wolff, J. A., and Herweijer, H. (2002). Efficient delivery of siRNA for inhibition of gene expression in postnatal mice. *Nat. Genet.* **32,** 107–108.

Liu, F., Song, Y., and Liu, D. (1999). Hydrodynamics-based transfection in animals by systemic administration of plasmid DNA. *Gene Ther.* **6,** 1258–1266.

Ludtke, J. J., Zhang, G., Sebestyen, M. G., and Wolff, J. A. (1999). A nuclear localization signal can enhance both the nuclear transport and expression of 1 kb DNA. *J. Cell Sci.* **112,** 2033–2041.

Martinez, J., Patkaniowska, A., Urlaub, H., Luhrmann, R., and Tuschl, T. (2002). Single-stranded antisense siRNAs guide target RNA cleavage in RNAi. *Cell* **110,** 563–574.

McCaffrey, A. P., Meuse, L., Pham, T. T., Conklin, D. S., Hannon, G. J., and Kay, M. A. (2002). RNA interference in adult mice. *Nature* **418,** 38–39.

McCaffrey, A. P., Nakai, H., Pandey, K., Huang, Z., Salazar, F. H., Xu, H., Wieland, S. F., Marion, P. L., and Kay, M. A. (2003). Inhibition of hepatitis B virus in mice by RNA interference. *Nat. Biotechnol.* **21,** 639–644.

McManus, M. T., and Sharp, P. A. (2002). Gene silencing in mammals by small interfering RNAs. *Nat. Rev. Genet.* **3,** 737–747.

Nykanen, A., Haley, B., and Zamore, P. D. (2001). ATP requirements and small interfering RNA structure in the RNA interference pathway. *Cell* **107,** 309–321.

Paddison, P. J., Silva, J. M., Conklin, D. S., Schlabach, M., Li, M., Aruleba, S., Balija, V., O'Shaughnessy, A., Gnoj, L., Scobie, K., Chang, K., Westbrook, T., Cleary, M., Sachidanandam, R., McCombie, W. R., Elledge, S. J., and Hannon, G. J. (2004). A resource for large-scale RNA-interference-based screens in mammals. *Nature* **428,** 427–431.

Parrish, S., Fleenor, J., Xu, S., Mello, C., and Fire, A. (2000). Functional anatomy of a dsRNA trigger. Differential requirement for the two trigger strands in RNA interference. *Mol. Cell* **6,** 1077–1087.

Song, E., Lee, S. K., Wang, J., Ince, N., Ouyang, N., Min, J., Chen, J., Shankar, P., and Lieberman, J. (2003). RNA interference targeting Fas protects mice from fulminant hepatitis. *Nat. Med.* **9,** 347–351.

Stark, G. R., Kerr, I. M., Williams, B. R., Silverman, R. H., and Schreiber, R. D. (1998). How cells respond to interferons. *Annu. Rev. Biochem.* **67,** 227–264.

Zamore, P. D., Tuschl, T., Sharp, P. A., and Bartel, D. P. (2000). RNAi: Double-stranded RNA directs the ATP-dependent cleavage of mRNA at 21 to 23 nucleotide intervals. *Cell* **101,** 25.

Zender, L., Hutker, S., Liedtke, C., Tillmann, H. L., Zender, S., Mundt, B., Waltemathe, M., Gosling, T., Flemming, P., Malek, N. P., Trautwein, C., Manns, M. P., Kuhnel, F., and Kubicka, S. (2003). Caspase 8 small interfering RNA prevents acute liver failure in mice. *Proc. Natl. Acad. Sci. USA* **100,** 7797–7802.

Zhang, G., Budker, V., Williams, P., Hanson, K., and Wolff, J. A. (2002). Surgical procedures for intravascular delivery of plasmid DNA to organs. *Methods Enzymol.* **346,** 125–133.

Zhang, G., Budker, V., Williams, P., Subbotin, V., and Wolff, J. A. (2001). Efficient expression of naked DNA delivered intraarterially to limb muscles of nonhuman primates. *Hum. Gene Ther.* **12,** 427–438.

Zhang, G., Budker, V., and Wolff, J. A. (1999). High levels of foreign gene expression in hepatocytes after tail vein injections of naked plasmid DNA. *Hum. Gene Ther.* **10,** 1735–1737.

Zhang, G., Gao, X., Song, Y. K., Vollmer, R., Stolz, D. B., Gasiorowski, J. Z., Dean, D. A., and Liu, D. (2004). Hydroporation as the mechanism of hydrodynamic delivery. *Gene Ther.* **11,** 675–682.

Zhang, G., Vargo, D., Budker, V., Armstrong, N., Knechtle, S., and Wolff, J. A. (1997). Expression of naked plasmid DNA injected into the afferent and efferent vessels of rodent and dog livers. *Hum. Gene Ther.* **8,** 1763–1772.

Zheng, L., Liu, J., Batalov, S., Zhou, D., Orth, A., Ding, S., and Schultz, P. G. (2004). An approach to genomewide screens of expressed small interfering RNAs in mammalian cells. *Proc. Natl. Acad. Sci. USA* **101,** 135–140.

[21] Analysis of Short Interfering RNA Function in RNA Interference by Using *Drosophila* Embryo Extracts and Schneider Cells

By Concetta Lipardi, Hye Jung Baek, Qin Wei, and Bruce M. Paterson

Abstract

The realization that short double-stranded RNA (dsRNAs) 21–25 bp in length represent the basis for posttranscriptional gene silencing (PTGS) in plants, quelling in *N. crassa*, and RNA interference (RNAi) in *C. elegans* and *Drosophila* has given insight into one of the most evolutionarily conserved pathways in eukaryotes. dsRNA that arises due to viral infection, transposon mobilization, random insertion of transgenes near active promoters, transcripts from repetitive elements in the genome, or introduction of exogenous dsRNA directly is processed by one of the RNase III-related enzymes, known as the Dicers, to produce 21- to 25-bp short dsRNAs or short interfering RNAs (siRNAs) that target the degradation of the cognate RNA sequence (Denli and Hannon, 2003; Hannon, 2002; Plasterk, 2002). Proteins in the RNAi pathway and siRNA-like RNAs have also been recently demonstrated to play a role in the formation and maintenance of heterochromatin in *S. pombe* as well as in transgene-induced PTGS in *Drosophila* (Hall *et al.*, 2002; Pal-Bhadra *et al.*, 2004; Volpe *et al.*, 2002). An understanding of siRNA function in these crucial regulatory pathways requires biochemical approaches to study siRNAs and their role in gene silencing as well as the formation and maintenance of heterochromatin.

This chapter describes simple methods for using *Drosophila* embryo extracts and cultured insect cells to study siRNA function in the RNAi pathway *in vivo* and *in vitro*. We describe the most recent protocols for the preparation and use of *Drosophila* embryo extracts used in gene targeting studies. We present methods we have used to assay siRNA function in *Drosophila* embryo extracts and in cultured SL2 cells that demonstrate a combined role for siRNAs and RNA-dependent RNA polymerase (Rdrp) activity in *Drosophila* RNAi.

Preparation of *Drosophila* Embryo Extracts and Culture of Schneider SL2 Cells

To prepare sufficient amounts of *Drosophila* embryo extract for biochemical studies, one needs access to population cages. Typically, 20–40 ml

METHODS IN ENZYMOLOGY, VOL. 392

of packed embryos is required to prepare sufficient extract for several experiments. Normal laboratory strains of Canton S or Oregon R flies are used to set up the mass cages. A source for population cages similar to the ones we use (diameter 12 in., length 24 in.) can be found on the Web (www.flystuff.com, Cat. No. 59-104). Details for starting and maintaining population cages with recipes for collection plates are available from the Gerald Rubin laboratory at http://fruitfly4.aecom.yu.edu/labmanual/contents.html ("Rubin Lab Methods Book"). Embryos are collected on yeasted molasses plates (six per cage) made in styrofoam trays (Genpak 2S, 14×21 cm^2). The first collection is discarded to remove the more mature embryos held by the flies. The plates for each 2-h collection are wrapped in Saran wrap and transferred to the cold room, where they are stored at 4° before preparing the extract with no deleterious effects. At the end of the collection, the embryos are washed from the plates through a set of 8-in.-diameter sieves (20, 40, and 100 mesh) with ice-cold water in the cold room. The embryos on the 100-mesh sieve are transferred to a 500-ml beaker and settled three to five times in 500 ml of cold 0.7% NaCl/0.05% Triton X-100 to remove debris. After the final wash, the settled embryos are dechorionated at room temperature in 500 ml of 50% (v/v) bleach with slow stirring for about 5 min or until the spiracles are no longer visible on the embryos, as judged under a dissecting microscope. The embryos are transferred again to the cold room and washed thoroughly on the 100-mesh sieve with cold 0.7% NaCl/0.05% Triton X-100, followed by ice-cold de-ionized water. The embryos (20–40 g) are dried by blotting the bottom of the sieve with paper towels, transferred gently to a plastic weigh boat with a spatula and rinsed into a chilled 60-ml Type S Potter-Elvehjem homogenizer (Braun-Melsungen, Germany) with an equal volume of embryo homogenization buffer per gram of embryos (embryo homogenization buffer: 100 mM potassium acetate, 30 mM Hepes-KOH at pH 7.6, 2 mM magnesium acetate, 5 mM DTT, and one protease inhibitor tablet/50 ml buffer (Cat. No. 1697498, Roche, Switzerland), as described previously (Wei et al., 2003). The embryos are homogenized slowly with 10 strokes of the pestle at 1000 rpm without introducing air bubbles into the homogenate. The homogenate is centrifuged at 14,500g for 30 min at 4°. The fat layer is removed and the supernatant solution is added dropwise to liquid nitrogen. The frozen beads are transferred to 50-ml blue-top plastic centrifuge tubes and stored at −85°. This frozen extract is used for RNA interference (RNAi) reactions and to prepare the S100 fraction used in siRNA incorporation studies.

The *Drosophila* Schneider SL2 cell line is grown in suspension at 24° in serum-free HyQ™-CCM3 medium (Cat. No. SH30065.01, Hyclone Logan, UT) supplemented with gentamycin sulfate (50 mg/ml). Cells are maintained in culture in 75-mm^2 flasks and plated on 60-mm plates when transfected.

RNA Interference (RNAi) *In Vitro* and *In Vivo*

The *Drosophila* extract system described initially to study RNAi *in vitro* used protein translation to monitor the loss of a specific mRNA, as measured by the loss of firefly luciferase (*P. pyralis*) activity compared to the control sea pansy luciferase (*R. reniformis*) activity in response to *P. pyralis* luciferase dsRNA (Tuschl *et al.*, 1999). We chose to look directly at the targeting of a specific mRNA encoding the green fluorescent protein (GFP) compared with a nontargeted control mRNA encoding luciferase from *P. pyralis* in response to the addition of GFP dsRNA to the extract (Lipardi *et al.*, 2001). A quick outline of the method is given here. To prepare mRNAs for targeting assays, the appropriate cDNAs were first cloned into the pSP64 poly-A vector (Cat. No. P1241, Promega, Madison, WI). Uniformly labeled mRNAs are prepared *in vitro* by using SP6 bacterial polymerase and cap analog (Message Machine, Cat. No. 1340, Ambion, Austin, TX) with [α-^{32}P]UTP (800 Ci/mmol, 20 Ci/ml, Cat. No. PB20383, Amersham Biosciences, England). mRNA transcripts are synthesized with a 5′ cap and 3′ poly-A tail for enhanced stability in the targeting extract. The unlabeled dsRNA trigger is prepared from the desired cDNA cloned into BlueScript® KS+, which is linearized and transcribed with either T3 or T7 bacterial polymerase kits [MegaScript™ kits, Cat. No. 1338 (T3), Cat. No. 1334 (T7), Ambion]. Once annealed, the dsRNA is extremely resistant to nuclease. Any cloning vector with convenient bacterial polymerase promoters can be used to produce the sense and antisense transcripts for the dsRNA as long as restrictions sites are available to give linear template DNA with 5′ protruding ends for efficient single-strand RNA synthesis.

The following is an example of a targeting reaction, using *Drosophila* embryo extract, GFP target mRNA, luciferase mRNA as the control, and GFP dsRNA as the initiating trigger. Plasmids for dsRNA preparation or mRNA transcription are linearized and kept at a concentration of 0.5 mg/ml. We substitute [α-^{32}P]UTP for any water in the transcription reaction to prepare highly labeled target and control mRNAs, usually 4 μl of α-UTP in a 20-μl reaction. After isopropanol precipitation, the labeled mRNA is resuspended in 50 μl of water to give 1×10^5 5×10^5 cpm Cerenkov/μl with a yield of 15–25 μg of mRNA. RNA transcripts used for dsRNA preparation (>100 μg) are resuspended in 200 μl of TE. A 1-μl aliquot is denatured in formamide buffer and checked on a native 1% agarose-TBE gel for integrity. The complementary RNAs are mixed at 0.5 mg/ml and heated to 95° for 1 min in a heat block, which is then switched off to allow annealing to proceed for 6 h to overnight. The annealed dsRNA is checked again on a 1% agarose-TBE gel with DNA markers. Annealed dsRNA runs like its DNA counterpart and slower than the corresponding single-stranded RNA. The dsRNA is resistant to digestion with RNaseA/T1 in 0.3 M sodium acetate at 30° for 30 min. RNAi reactions are set up following the scheme outlined in Protocol 1. Assemble the components listed in Protocol 1 on ice and start the RNAi reaction by adding the embryo extract and then immediately placing the tubes in a water bath at 25° for 1–5 h. The amino acids are not necessary unless protein products are to be assayed. α-Amanitin is left in the reaction because it is necessary to keep the background down when using α-UTP in RNA-primed reactions with the S100 fraction prepared from the extract

---

Protocol 1. RNAi Reaction In Vitro

Embryo homogenization buffer

	Final concentration	Volume per reaction (μl)
Potassium acetate (1 M)	100 mM	5
ATP (10 mM)	500 μM	2.5
UTP, CTP, GTP (10 mM)	100 μM	0.5 each
DTT (250 mM)	5 mM	1
RNasin (30 KU/ml)	0.1 U	1
Creatine phosphate (500 mM)	10 mM	1
Creatine phosphokinase (CPK; 50 μg/ml)	1 μg/ml	1
α-Amanitin (25 mM)	500 μM	1
Amino acids (1 mM), optional	100 μM	5 μl

Adjust the total volume to 22 μl with water and add the following components as needed:

Capped/A+ mRNAs (10,000 cpm each) for control and target 2 μl (1 μl each)
dsRNA trigger (0.5–1.0 mg/ml), 1 μl
Embryo extract (15–20 mg/ml protein), 25 μl
Total reaction volume 50 μl

Amino acids without methionine can be added from a 1 mM stock to a final concentration of 100 μM if protein translation is desired.

(see later). Aliquots of 10 μl are removed from the tube at 0, 1, 2, and 3 h directly into 400 μl of proteinase K/SDS stop buffer (200 mM Tris 7.5, 25 mM EDTA, 300 mM NaCl, 2% SDS with 100 μg/ml proteinase K) and heated at 65° for 15–30 min. Phenol–chloroform extract and precipitate the reactions with an equal volume of isopropanol at –20° for 2 h to overnight. The reactions are analyzed on denaturing 1.5% agarose formaldehyde gels by using standard protocols. The run gel is rinsed in water 5–10 min to lower the formaldehyde concentration, blotted onto Nytran-Plus™ (Millipore, Bedford, MA) for 1 h and then dried at 60° on a vacuum gel dryer. The membrane-attached gel is exposed to X-ray film as needed. An example for GFP targeting is given in Fig. 1A.

RNAi *in vivo* experiments are conducted in SL2 cells. We looked at the targeting of GFP protein compared with a nontargeted control of β-galactosidase (β-gal) in response to the addition of GFP dsRNA in SL2 cells (Fig. 1B). Cells are plated in 60-mm plates and transiently transfected by using FuGENE™ 6 (Roche, Cat. No. 1815091): 1.5 μg of each plasmid is mixed with or without 1 μg of GFP dsRNA in 3–5 μl. A separate mixture containing 91 μl of medium and 9 μl of FuGENE 6 (roughly 3 μl FuGENE per 1 μg of total nucleic acid) is incubated for 5 min and then added to the mix of plasmid DNA with or without dsRNA, incubated an additional 20 min, and added to the cells. After 48 h, cells are plated in glass chamber slides precoated with polylysine, allowed to attach for 1 h, and stained with anti-β-gal antibody following the protocol described next (Wei *et al.*, 2000).

Incorporation of Uniformly Labeled Short Interfering RNA (siRNA) into Double-Stranded RNA (dsRNA)

We previously described methods for preparing uniformly labeled siRNAs on incubating high-specific-activity dsRNA ($>10^9$ cpm/μg) in embryo extract for 3 h and subsequently isolating the siRNAs in a protein complex in which the siRNAs are resistant to micrococcal nuclease digestion (Lipardi *et al.*, 2001). Although these siRNAs did not function initially in RNAi, we demonstrated that phosphatase removal of the 3'-phosphate group generated by the micrococcal nuclease treatment could restore siRNA function in gene targeting experiments similar to levels observed with dsRNA. This was the first hint that the 3'-hydroxyl group is important for siRNA function in RNAi and led us to test the idea that siRNAs might be interacting with RNA-dependent RNA polymerase (RdRp) in the RNAi pathway to amplify the dsRNA to generate secondary siRNAs. RdRp had already been identified as an essential gene for posttranscriptional gene silencing (PTGS) in plants (Dalmay *et al.*, 2000; Mourrain *et al.*, 2000), quelling in *N. crassa* (Cogoni and Macino, 1999), and RNAi in

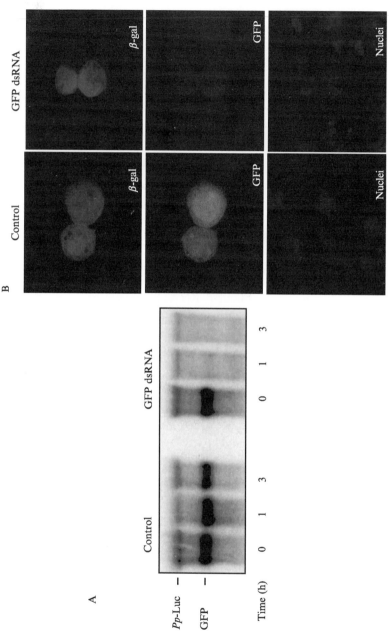

FIG. 1. RNAi *in vitro* and *in vivo* in *Drosophila*. (A) The 716-bp GFP dsRNA targets the cognate mRNA but not *Photinus pyralis* luciferase mRNA in *Drosophila* embryo extract. Green fluorescent protein (GFP) double-stranded RNA (dsRNA) silences GFP in Schneider *Drosophila* SL2 cells. (B) SL2 cells are transfected with β-galactosidase (β-gal) and GFP reporter plasmids with or without (control) GFP dsRNA. After 48 h, the cells are fixed and stained for β-gal and GFP protein and analyzed under the confocal microscope. (See color insert.)

C. elegans (Sijen *et al.*, 2001). Here we describe the preparation of native siRNAs from uniformly labeled dsRNA, using recombinant *Drosophila* Dicer-1 protein produced in the baculovirus system. This is much easier than our original method that required the isolation of siRNAs from a nuclease-resistant complex formed in *Drosophila* embryo extracts followed by phosphatase treatment (Wei *et al.*, 2003). *Drosophila* Dicer-1 is an ATP-independent processive nuclease that is active in 10–260 mM potassium acetate over a pH range of 6.8–7.5 in the absence of reducing agents. Enzyme-to-substrate molar ratios between 2 and 4 are maximal because the enzyme is processive and loads at the ends of the dsRNA substrate (H. J. Baek, C. Lipardi, Q. Wei, and B. M. Paterson, unpublished observation). Human Dicer can also be used to generate labeled siRNAs and is commercially available from multiple sources (Stratagene, La Jolla, CA, and Invitrogen, Carlsbad, CA). High-specific-activity, uniformly labeled dsRNA is synthesized using 4 μl of [α-^{32}P]UTP (800 Ci/mmol, 20 mCi/ml) and the T3/T7 MaxiScript™ kit (Ambion, Cat. No. 1324). siRNAs are prepared by digesting 1 μg of dsRNA (>10^9 cmp/μg) with 0.5 μg of Dicer-1 for 5 h at 25° in 100 mM potassium acetate, 30 mM HEPES-KOH, pH 7.6, and 2 mM magnesium acetate in the absence of ATP or reducing agent. Under these conditions digestion is complete (Fig. 2A). The labeled siRNAs are run on a preparative 2% Agarose-1000 TBE gel (Cat. No. 10975-035, Invitrogen) in the presence of ethidium bromide and isolated by electroelution into dialysis tubing, using a 25-bp ladder as a size marker. The eluted material is extracted with phenol–chloroform, precipitated with glycogen carrier, and dissolved in TE to give roughly 10^5 cpm/μl. siRNA incorporation reactions are performed with the labeled siRNAs, control and complementary mRNA substrates, and the S100 fraction prepared from the embryo lysate as outlined in Protocol 2. Unlabeled dsRNA can also be digested by Dicer-1 to prepare a population of representative siRNAs for a particular RNA, and these will target mRNA degradation in *Drosophila* embryo extracts and SL2 cells as efficiently as full-length dsRNA. These representative siRNAs also work for gene targeting in mammalian cell cultures and avoid the problems associated with selecting the most efficient chemically synthesized siRNA.

siRNA incorporation into dsRNA can be analyzed on either denaturing 1.5% agarose-formaldehyde gels or on 1× TBE/8 M urea/4% or 5% acrylamide DNA sequencing minigels. The latter gels are easier to prepare and allow for the isolation of the labeled product for further analysis. To prepare the minigels, weigh 7.2 g urea, add 1.5 ml 10× TBE, 1.5 ml (4%) or 1.875 ml (5%) 40% acrylamide mix (acrylamide:bis ratio 19:1, Accu-Gel™ 19:1, Cat. No. EC-850, National Diagnostics, Atlanta, GA), and water to 15 ml. Polymerize the gel with 48 μl of 25% ammonium persulfate and 16 μl TEMED. Pour into a 1.5-mm minigel cassette (12 × 14 mm^2,

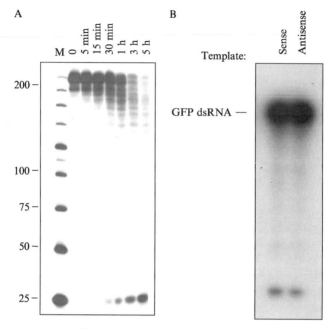

FIG. 2. Processing of [α-^{32}P]UTP-labeled GFP dsRNA by recombinant *Drosophila* Dicer-1 to labeled siRNAs (A) and their incorporation in dsRNA (B). Recombinant FLAG-tagged *Drosophila* Dicer-1 expressed in the baculovirus system cleaves labeled dsRNA into 21- to 25-bp siRNAs. The efficiency of processing is time dependent. An overnight digestion of high-specific-activity dsRNA (>10^9 cpm/μg) is used to prepare siRNAs for template-dependent incorporation into dsRNA. siRNA-labeled dsRNA can be prepared by using either sense or antisense mRNA template (B). The synthesis reaction is analyzed on 4% acrylamide/1× TBE/ 8 *M* urea DNA minigels, as outlined in the text.

Novex). Run the gel at 300 V/100 mA for 30 min or until the xylene cyanol is at the bottom. Cover the gel with Saran wrap and expose to X-ray film at room temperature. The film is used as a template to cut out the band for electroelution. An example of GFP siRNAs produced by Dicer-1 digestion (Fig. 2A, left panel) and their GFP mRNA template-dependent incorporation into dsRNA (Fig. 2B, right panel) is given as an example. The newly synthesized dsRNA can be digested by Dicer-1 to make secondary siRNAs, as previously demonstrated (Lipardi *et al.*, 2001). Ribonuclease T1 digestion of the GFP siRNA-labeled dsRNA indicated that multiple siRNAs corresponding to the complementary strand of the mRNA are incorporated into the dsRNA product (Lipardi *et al.*, 2003). Because a synthetic siRNA can be extended on an RNA template, this suggests that the mixture of siRNAs is extended and ligated in a template-dependent

---

Protocol 2. Preparation of S100 and Incorporation of siRNAs into dsRNA

1. Spin aliquots of embryo extract (400 μl) at 14,000 rpm (20,800g) at 4° for 5 min.
2. Spin the extract in the Beckman TLX centrifuge using the TLA-100 rotor at 90,000 rpm for 3 h at 4° (400 μl in thick-wall tubes).
3. Remove the supernatant solution and dissolve the pellet in 200 μl of buffer containing 10 mM HEPES-KOH, pH 7.6, 300 mM potassium acetate, 1 mM magnesium chloride and 6 mM β-mercaptoethanol. The pellet is difficult to dissolve, but eventually goes into solution by repeatedly washing buffer over the pellet surface while in the cold room.
4. Centrifuge the resuspended pellet in the TLA-100 rotor at 50,000 rpm for 1 h at 4°. The supernatant solution (10–15 mg/ml protein) is the S100 fraction and is either used immediately or quick frozen in aliquots and stored at −85°.
5. siRNA incorporation reaction is performed similar to the RNAi reaction outlined in Protocol 1. To the initial 22 μl mix add 1–3 μg of template RNA (can be T3 or T7 transcripts or capped/adenylated SP6 transcribed mRNA), the labeled siRNAs (250,000 cpm), and 10 μl of the S100 fraction in a final volume of 50 μl. Incubate at 25° for 15 min.
6. Stop the reaction by adding 400 μl of proteinase K buffer and 5 μl of proteinase K solution (15–20 mg/ml, Cat. No. 70011021, Roche). Incubate for 15 min at 65° followed by phenol–chloroform extraction and precipitation with glycogen carrier.

---

reaction. Template-dependent ligation of short single-stranded DNAs has been reported for T4 DNA ligase, and the same likely occurs for RNA ligation (Nilsson *et al.*, 2001). RNA ligase activity appears to be present in embryo extract because [5'-^{32}P]-pGp can be ligated to siRNAs in the extracts and the label is resistant to phosphatase treatment but is degraded by RNase A (Lipardi *et al.*, 2003).

Analysis of RNA-Dependent RNA Polymerase (RdRp)-Related Activity in Embryo Extracts and in SL2 Cells

The RNA-dependent RNA polymerase (RdRp) gene has been identified as a key component of the RNAi pathway in *C. elegans* (Sijen *et al.*, 2001), *A. thaliana* (Dalmay *et al.*, 2000; Mourrain *et al.*, 2000), quelling in

N. crassa (Cogoni and Macino, 1999), and RNAi in *C. elegans* (Sijen *et al.*, 2001) and *D. discoideum* (Martens *et al.*, 2002). The gene is also essential for hairpin-induced gene silencing and the maintenance of heterochromatin in *S. pombe* (Volpe *et al.*, 2002). We demonstrated that siRNAs can serve as primers to convert some of the target mRNA into dsRNA, which is again degraded by Dicer to generate a second generation of siRNAs (Lipardi *et al.*, 2001). This was also shown to occur in *C. elegans* when using short antisense RNA oligos or siRNAs and was elegantly demonstrated with transitive RNAi, a process whereby siRNAs corresponding to the 3' portion of an mRNA are extended on the template by RdRp to generate dsRNA, which is cleaved by Dicer to generate a novel set of 5' siRNAs. By using a hybrid mRNA target containing a 3'-GFP region joined to a 5'-*unc22* domain, siRNAs to GFP were able to induce an *unc22* phenotype in *C. elegans*. However, if the positions of the GFP and *unc22* regions were reversed in the mRNA target no *unc22* phenotype was observed, proving that the siRNAs are extended in the 5' to 3' direction to generate dsRNA (Sijen *et al.*, 2001). In the following section, methods are described that we have used to identify RdRp-related activity in *Drosophila* embryo extracts and in SL2 cells.

dsRNA Fill-In Assay in Embryo Extract S100. A convenient assay for RdRp-related activity involves a template-dependent fill-in reaction for an extended 5' overhang (30–45 bp) on a dsRNA template. The 3'-recessed end of the template serves as a primer for new dsRNA synthesis. The dsRNA substrate is relatively stable and easy to handle compared to an siRNA–mRNA duplex. The fill in of the 5' overhangs on the annealed RNA strands is consistent with a primer-dependent RdRp-related activity, because labeling is resistant to the presence of α-amanitin and actinomycin-D in the reaction. To generate a fill-in template, we cloned the full-length GFP coding region (716 bp) into pBC SK+ (Stratagene) as an *Eco*R1-*Xba*1 fragment. The plasmid is linearized with *Eag*1 for T7 transcription and with *Hin*d III for T3 transcription, as described previously for the preparation of the dsRNA trigger used for targeting RNA degradation in *Drosophila* embryo extracts and in gene silencing studies in SL2 cells. The strands are annealed, as before, to give a 5'-T3 overhang of 31 bp and a 5'-T7 overhang of 46 bp. When this dsRNA is incubated with the S100 fraction of embryo extract containing α-amanitin, actinomycin-D, and [α-^{32}P]-UTP (800 Ci/mmol), as described in Protocol 3, the label is incorporated into the dsRNA template.

To ensure that the label has been incorporated in a template-dependent manner and not just at the ends because of terminal transferase activity, the labeled dsRNA is subsequently treated with the single-strand-specific ribonuclease RNase ONE (Cat. No. M4261, Promega) to remove any

---

Protocol 3. dsRNA Fill-In Reaction to Assay RdRp-related Activity

Components of the reaction:

Mix (250 mM potassium acetate; 5 mM magnesium acetate; 75 mM HEPES-KOH, pH 7.4; 1.25 mM ATP; 0.25 mM each UTP, CTP, and GTP; 12.5 mM DTT; 0.25 U/ml RNasin; 25 mM creatine phosphate; 2.5 μg/ml creatine phosphokinase (CPK); 2.5 μM α-amanitin; 10 μg/ml actinomycin D), 4 μl dsRNA with 5' overhangs (0.5 μg/μl), 0.4 μl
S100 (10–20 μg/ul protein), 1 μl
α-[^{32}P]-UTP (800 Ci/mmol), 2 μl
Final volume, 10 μl

1. Incubate the reaction at 25° for 10 min.
2. Add 100 μl proteinase K buffer, 1.25 μl of proteinase K (15–20 μg/μl) and incubate at 65° for 15 min.
3. Extract with phenol–chloroform and precipitate with 10 μg of tRNA carrier.
4. Dissolve the pellet in 10 μl water and treat the dsRNA with 200 micro-units of RNase ONE in a 20-μl reaction as per manufacturer's instructions for 5 min at 37°.
5. Repeat proteinase K treatment, extract with phenol–chloroform, and precipitate with 10 μg glycogen carrier.

---

non-double stranded material (Protocol 3 and Fig. 3A, right lanes). Although single-stranded RNA (ss) is labeled in the S100 reaction, it is completely degraded by RNase ONE treatment and likely represents labeling by terminal transferase activity. A good portion of the labeled dsRNA (ds), on the other hand, is resistant to RNase ONE treatment (Fig. 3A, ds, right lanes). To confirm that the label has been incorporated internally into the overhangs of the dsRNA, nearest-neighbor analysis is performed on the gel-purified dsRNA product, using the minigel procedure outlined previously. In this analysis, one measures the transfer of the 5'-[^{32}P]α-phosphate to the adjacent upstream nucleotide on the 3' hydroxyl position after digesting dsRNA with RNase T2. The labeled nucleoside-3'-monophosphate distribution, determined by thin-layer chromatography on PEI (polyethyleneimine) thin-layer plates, reflects the nucleotide ratio of the single-stranded ends of the template. The amount of 3'-labeled Ap, Cp, Gp, and Up, as measured by PhosphorImager™ (Molecular Dynamics), is used to confirm that the α-UTP was incorporated into the 5' overhangs of

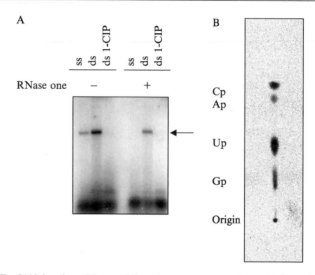

FIG. 3. The S100 fraction of *Drosophila* embryo extract contains an RdRp-related activity. (A) αUTP is incorporated into GFP dsRNA (arrow) with 5' extensions (30–45 bp), but not into flush-ended dsRNA generated by prior treatment of the dsRNA template with RNase ONE and CIP (ds I-CIP). The label incorporated into dsRNA (ds) is resistant to further RNase ONE digestion (right lanes only headed with a plus sign), whereas labeled single-stranded RNA (ss) is rapidly degraded. Labeled single-stranded RNA (left set of lanes with no RNase ONE treatment) is most likely the result of terminal transferase activity in the S100 fraction. (B) Nearest-neighbor analysis of the labeled, gel-purified RNase ONE-resistant dsRNA. RNase T2 digestion products are separated on PEI thin-layer plates. The fraction of 3'-labeled Ap, Cp, Gp, and Up indicates that α-UTP has been incorporated into the overhangs of the GFP dsRNA in a template-dependent matter.

the GFP dsRNA in a template-dependent synthesis reaction (Fig. 3B). The method for nearest-neighbor analysis is outlined in Protocol 4 and was used previously to demonstrate the internal labeling of dsRNA synthesized by the *N. crassa* RdRp protein derived from the QDE1 gene (Makeyev *et al.*, 2003).

dsRNA-Dependent and siRNA-Dependent Incorporation of 5-Bromouridine in SL2 Cells. UTP-biotin-labeled GFP double-stranded RNA with the described 5' overhangs can also be transfected into SL2 cells. Biotinylated dsRNA works as well as nonbiotinylated dsRNA in gene silencing (data not shown) and can be visualized in cells with antibiotin antibody. Transfected dsRNA is localized in perinuclear regions, as shown in Fig. 4A. The biotinylated dsRNA with the 5' overhangs can also be used to test for the incorporation of the uridine analogue 5-bromouridine (5-BrU, TriLink BioTechnologies, San Diego, CA) into the ends to measure putative RdRp-dependent RNA synthesis in SL2 cells. New synthesis and the template should colocalize within the cell. To test this idea, biotin-labeled dsRNA is first transfected into the SL2 cells, using FuGENE 6 (Wei *et al.*, 2000). After 48 h, the cells are treated with actinomycin-D (3 μg/ml) for 3.5 h. This concentration of actinomycin-D is sufficient to reduce ^{3}H-uridine incorporation by more than 95% in SL2 cells during a 3.5-h incubation. The cells are then plated in glass chamber slides (Lab-Tek® II, Cat. No. 154461, Nalge Nunc International, Naperville, IL) that have been coated with polylysine and are then incubated again in medium with actinomycin-D (3 μg/ml) for 30 min before adding 5-BrU to the medium at a final concentration of 500 μg/ml. The cultures are incubated with 5-BrU for 1 h and then washed and prepared for antibody staining as described in Protocol 6. 5-BrU incorporation colocalizes with UTP-biotin-labeled GFP dsRNA in discrete perinuclear regions in the cytoplasm (Fig. 4B). No labeling is detected when the cells are incubated with 5-BrU in the absence of dsRNA (not shown); therefore, incorporation depends on the presence of the dsRNA template with 5′ overhang ends capable of priming synthesis. The 5′ overhangs are important because treatment of dsRNA with RNase ONE before transfection essentially eliminates 5-BrU labeling, consistent with a primer-dependent fill-in reaction. This conclusion is further supported by experiments looking at mRNA-dependent siRNA-mediated 5-BrU incorporation described in the following section.

siRNAs act as primers to convert mRNA into dsRNA in the S100 fraction of the embryo extract (Lipardi *et al.*, 2001). Based on this result, a synthetic siRNA should also direct the cognate mRNA-dependent incorporation of 5-BrU into nascent RNA in SL2 cells and incorporation should depend on the availability of a 3′ hydroxyl group on the siRNA if primer-dependent RNA synthesis is involved. To examine this hypothesis, SL2 cells are cotransfected with a GFP reporter and either the normal or the pGp-modified GFP siRNA with blocked 3′-hydroxyl groups and subsequently tested for 5-BrU incorporation. Ligation of pGp to the 3′-hydroxyl groups of a synthetic siRNA with T4 RNA ligase reversibly blocks siRNA gene targeting in *Drosophila* embryo extracts and SL2 cells. Similar to the natural siRNAs isolated from the nuclease-resistant complex, treatment of the pGp-modified siRNA with phosphatase reverses inactivation in both assays (Wei *et al.*, 2003). Here we give an example of RNAi inhibition with a pGp-modified GFP siRNA and its affect on 5-BrU incorporation (Fig. 5). Forty-eight hours after transfection of the SL2 cells with a GFP reporter and either the normal or pGp-modified GFP siRNA, the cells are checked for expression of GFP and 5-BrU incorporation in the presence of actino-mycin-D, as described previously. Once again, 5-BrU incorporation is localized to discrete perinuclear domains in cells undergoing GFP RNAi (Fig. 5, top row). By contrast, 5-BrU incorporation is completely absent in

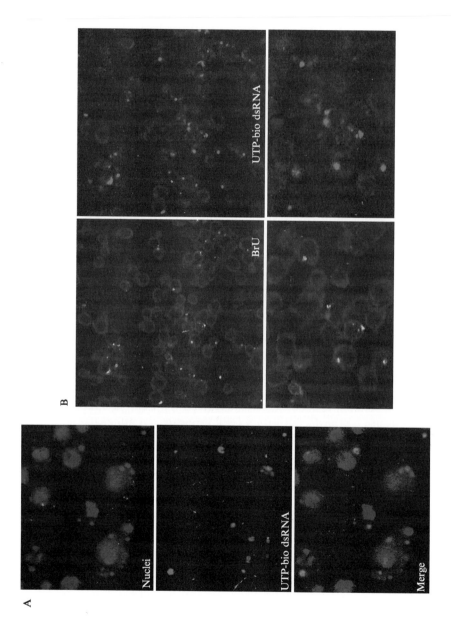

Protocol 4. Nearest-Neighbor Analysis of α-[^{32}P]-UTP-labeled dsRNA

1. Gel purify the labeled dsRNA using the mini 4 or 5% DNA sequencing gels described previously. Electroelute the sample in a dialysis bag in 0.45 ml of 0.5× TBE at 100 V for 0.5–1 h. Extract the eluted material with phenol–chloroform, add sodium acetate to a final concentration of 0.3 M, and then precipitate the solution with glycogen carrier (10 μg) and 2.5 volumes of 100% ethanol by placing the tube in dry ice for 20 min.

2. Dissolve the pellet in 10 μl of 0.05 M sodium acetate, pH 4.5, containing 20 U/ml of RNase T2 (Cat. No. R6398, Sigma, St. Louis, MO) and incubate overnight at 37°.

3. Cut a 100 × 100 mm^2 plastic-backed PEI thin-layer plate (with fluor; Cat. No. M55797, EM Science, Gibbstown, NJ) into 20 × 100 mm^2 sections. Spot samples, up to 5 μl, directly from the digestion reaction 1 μl at a time, air dry, and run in two chromatography cycles. Run a standard mixture of 3′ monophosphates (1 μg/μl) in parallel or along with the sample and identify the nucleoside 3′-monophosphates with a hand-held short-wave UV lamp in the darkroom and mark with a pencil. Run the first cycle in small glass chromatography cylinders in 0.5% formic acid until the front is 2–4 cm from the top of the plate. Remove the plate from the tank and air dry in the hood. Run the second cycle in 0.15 M LiOH-formate (0.15 M LiOH solution adjusted to pH 3.0 with 88% formic acid) and let the development proceed until the front moves 1–2 cm from the top of the plate. Air dry the plate and expose in a PhosphorImager to quantify each nucleoside 3′-phosphate position.

FIG. 4. UTP-biotin-labeled GFP dsRNA is localized in perinuclear regions (A) and colocalizes with 5-BrU incorporation in SL2 cells (B). SL2 cells are transfected with UTP-biotin-labeled GFP dsRNA (716 bp). After 48 h, cells are stained with antibiotin antibody or Texas Red®-streptavidin (Cat No. 43–4317, Zymed Laboratories, San Francisco, CA). dsRNA is localized at the perinuclear region in the cytoplasm (A). SL2 cells transfected with UTP-biotin dsRNA are treated with actinomicin-D (3 μg/ml) for 3.5 h, incubated with 5-bromouridine (500 μg/ml) for 1 h, and then stained with antibody to BrdU and Texas Red-streptavidin. dsRNA colocalizes with BrU incorporation (B). (See color insert.)

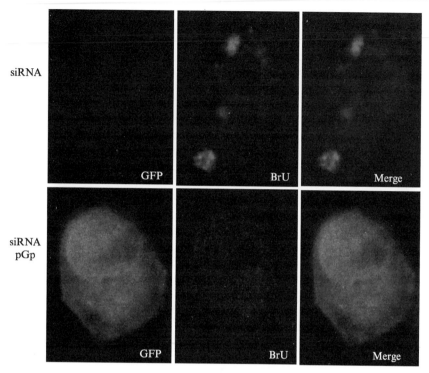

FIG. 5. siRNA-dependent 5-BrU incorporation is coincident with GFP RNAi and is dependent on the 3' hydroxyl of the siRNA. A synthetic GFP siRNA either with or without pGp ligated to the 3' hydroxyl group is transfected along with a GFP reporter plasmid into SL2 cells. After 48 h, the cells are labeled with 5-BrU and assayed for GFP expression and 5-BrU incorporation. Perinuclear incorporation of 5-BrU is only apparent in cells undergoing GFP RNAi (top panel), but is absent in cells treated with the pGp-modified GFP siRNA that still express GFP (bottom panel). (See color insert.)

cells treated with the pGp-modified GFP siRNA and GFP expression is unaffected (Fig. 5, bottom row). In the absence of the GFP reporter plasmid and GFP mRNA expression, no 5-BrU labeling takes place; that is, both the siRNA with an unmodified 3'-hydroxyl group and the cognate target mRNA are required for primer-dependent RNA synthesis, as measured by 5-BrU incorporation. Thus, we have a consistent picture in which both dsRNA with 5' overhangs or a single siRNA and the cognate mRNA can direct the incorporation of the uridine analog, 5-BrU, into nascent RNA in the presence of actinomycin-D concentrations that block RNA synthesis in SL2 cells. RNA synthesis that is resistant to both α-amanitin and actinomycin-D is a hallmark of viral and cellular RdRp

Protocol 5. Ligation of pGp to the 3′-Hydroxyl Group of an siRNA

1. Ligation reaction in a total volume of 40 μl

   10× Ligation buffer (500 mM Tris, pH 7.8, 100 mM MgCl$_2$, 100 mM DTT, 10 mM ATP), 4 μl
   DMSO, 4 μl
   siRNA (50 μM) single strand or mix of natural siRNAs, 20 μl
   10 mM pGp (Cat. No. N5003, TriLink BioTechnologies), 4 μl
   T4 RNA ligase (Cat. No. M0204, New England Biolabs, Beverly, MA), 4 μl
   Add water to a total volume of 40 μl
   Incubate at 16° for 12–16 h

2. *Buffer exchange and removal of excess pGp:* Spin the 40-μl reaction over a G-25 MicroSpin™ column (Pharmacia, Piscataway, NJ) equilibrated with TE + 20 mM NaCl.

3. *Removing 3′-phosphate from the ligated pGp with shrimp alkaline phosphatase (SAP, Cat. No. 1758250, Roche):* After the spin column, mix the sample with 4 μl of 10 SAP buffer and 4 μl (4 U) of SAP or water and incubate at 37° for 1 h to remove the phosphate group. Then heat the samples to 65° to inactivate the SAP and any remaining T4 RNA ligase.

4. *Annealing the siRNAs.* Mix both strands of the siRNA after SAP inactivation with 25 μl of 5× annealing buffer (500 mM potassium acetate, 150 mM Hepes-KOH, pH 7.4, 10 mM magnesium acetate) and heat to 90° for 1 min in a boiling water bath followed by cooling to room temperature over 2–4 h. Anneal unmodified siRNAs in the same way. The annealed siRNAs with or without a 3′-phosphate group are now ready to test in extracts and in SL2 cells.

activity. The method used for the reversible modification of the siRNAs with pGp is given in Protocol 5. The procedures used to measure 5-BrU incorporation are outlined in Protocol 6.

Summary

The introduction of dsRNA into *Drosophila* embryos or any one of the *Drosophila* cell lines results in the specific knockdown of the corresponding mRNA and provides a potent method for the analysis of endogenous or

Protocol 6. Assay for 5-BrU Incorporation in SL2 Cells Treated with dsRNA or siRNAs

1. Transfect SL2 cells with either dsRNA (1–5 μg per 60-mm dish), using FuGENE 6 (Roche) as described previously, or a reporter plasmid–siRNA combination (1–5 μg plasmid/50 nM siRNA final), using Oligofectamine™ (Invitrogen) according to the manufacturer's instructions.

2. Forty-eight hours after transfection, treat the cells with actinomycin-D at a final concentration of 3 μg/ml for 3.5 h. This concentration of actinomycin-D reduces ^{3}H-uridine incorporation >95% in SL2 cells. Dissolve actinomycin-D (Cat. No. A1410, Sigma) in sterile PBS at 10 μg/ml and stored at −20°.

3. Then plate the cells in glass chamber slides precoated with polylysine (Lab-Tek II, Nalge Nunc) and incubate for 30 min with actinomycin-D (3 μg/ml) before adding 5-BrU at 500 μg/ml (TriLink BioTechnologies). Incubate for another 1 h. Store 5-BrU stocks (50 mg/ml in ethanol) in the dark at −20°.

4. Wash the cells with PBS once, fix in paraformaldehyde 4%/PBS for 20 min, wash with PBS once, permeabilize with methanol for 10 min, and wash twice with PBS.

5. Block the cells with PBS containing 0.1% Tween 20 and 3% BSA for 30 min. Wash the cells thrice with PBS containing 0.05% Tween 20 for 10 min. Mount the cells with coverslips and analyze under the confocal microscope.

6. To stain nuclei, incubate the cells in PBS with Hoechst at a 1:7500 dilution for 5 min and then wash in PBS and mount.

7. Prepare biotinylated dsRNA by substituting the 10× biotin RNA labeling mix (Cat. No. 1685597, Roche) in the standard T3 (Cat. No. 1338, Ambion) and T7 transcription kits (Cat. No. 1334, Ambion). Stain biotinylated dsRNA with monoclonal antibody antibiotin (Cat. No. 11242, Molecular Probes) at a 1:100 dilution. Dissolve the antibody in PBS containing 1% BSA. When the biotinylated dsRNA is stained with Texas Red-streptavidin (Zymed Laboratories), blocking and incubation buffer is PBS with 5% BSA. The staining of dsRNA with antibiotin and Texas Red-strepavidin is identical.

transfected gene function known as RNAi (Kennerdell and Carthew, 1998; Misquitta and Paterson, 1999; Wei *et al.*, 2000). siRNAs are the mediators of this process, but little is known about their mechanism of action. Biochemical studies of RNAi in *Drosophila* embryo extracts have defined two potential modes of action. In one case, the siRNAs are thought to serve as guides that target a protein–nuclease complex to the cognate mRNA through the complementarity of the siRNA strands (Hannon, 2002). This complex, known as the RNA-induced silencing complex (RISC), is proposed to be catalytic to account for the substochiometric potency of the siRNAs relative to the concentration of the target mRNA *in vivo*, but catalytic activity has never been demonstrated *in vitro* either with purified RISC or in *Drosophila* cell extracts. In the later case, the trigger dsRNA is needed in tenfold molar excess with respect to the target RNA in order to achieve efficient and specific silencing (Tuschl *et al.*, 1999).

Genetic studies in *C. elegans, A. thaliana, N. crassa, D. discoideum*, and *S. pombe* have consistently implicated three gene groups: the Argonaute family, the Dicer RNase III-related nucleases, and RdRp. Our studies using *Drosophila* embryo extracts have identified an activity that can use siRNAs as primers to generate new dsRNA, which is then cleaved by Dicer to generate secondary siRNAs to potentially amplify the silencing process while degrading the target mRNA (Lipardi *et al.*, 2001). This type of activity has also been noted in *C. elegans* by using short antisense RNAs and siRNAs to demonstrate that gene silencing is RdRp dependent and by transitive RNAi, in which siRNAs give rise to secondary siRNAs through the RdRp-dependent synthesis of dsRNA that results from the extension of the initial siRNAs on the template mRNA (Sijen *et al.*, 2001). Mathematical modeling of the *C. elegans* results and our own data on siRNA-primed dsRNA synthesis and degradation have further argued that unidirectional RNA-dependent RdRp-mediated synthesis of secondary dsRNA and siRNAs serve as a safety mechanism to prevent the generation of self-directed reactions. This is accomplished by limiting the silencing effect to the continued input of dsRNA and the magnitude of the silencing reaction to the initial dose of dsRNA. This cannot occur with a catalytic RISC (Bergstrom *et al.*, 2003).

In this chapter, we have outlined procedures that allow one to study the function of siRNAs in *Drosophila* embryo extracts and in SL2 cells. These approaches are being used to identify the activity responsible for siRNA-dependent dsRNA synthesis in *Drosophila*. Initial reports for *Drosophila* genome complexity, based on the BDGP releases R1/R2, estimated that there were 13,861 genes in the fly. This has now been revised to 21,396 genes in the more recent Heidelberg Collection R1 (Hild *et al.*, 2003). It is clear that initial analyses that excluded the existence of an RdRp gene in

Drosophila have to be reevaluated. The fact that RdRp is also involved in the maintenance and formation of heterochromatin, a process universal to all eukaryotes, makes it unlikely that *Drosophila* lacks the RdRp gene (Hall *et al.*, 2002). The methods described here provide a first step in the detection, purification, and characterization of *Drosophila* RdRp.

References

Bergstrom, C. T., McKittrick, E., and Antia, R. (2003). Mathematical models of RNA silencing: Unidirectional amplification limits accidental self-directed reactions. *Proc. Natl. Acad. Sci. USA* **100**, 11511–11516.

Cogoni, C., and Macino, G. (1999). Gene silencing in *Neurospora crassa* requires a protein homologous to RNA-dependent RNA polymerase. *Nature* **399**, 166–169.

Dalmay, T., Hamilton, A., Rudd, S., Angell, S., and Baulcombe, D. C. (2000). An RNA-dependent RNA polymerase gene in *Arabidopsis* is required for posttranscriptional gene silencing mediated by a transgene but not by a virus. *Cell* **101**, 543–553.

Denli, A. M., and Hannon, G. J. (2003). RNAi: An ever-growing puzzle *Trends Biochem. Sci.* **28**, 196–201.

Hall, I. M., Shankaranarayana, G. D., Noma, K., Ayoub, N., Cohen, A., and Grewal, S. I. (2002). Establishment and maintenance of a heterochromatin domain. *Science* **297**, 2232–2237.

Hannon, G. J. (2002). RNA interference. *Nature* **418**, 244–251.

Hild, M., Beckmann, B., Haas, S. A., Koch, B., Solovyev, V., Busold, C., Fellenberg, K., Boutros, M., Vingron, M., Sauer, F., and Hoheisel, J. D. (2003). An integrated gene annotation and transcriptional profiling approach towards the full gene content of the *Drosophila* genome. *Genome Biol.* **5**, R3.1–R3.17.

Kennerdell, J. R., and Carthew, R. W. (1998). Use of dsRNA-mediated genetic interference to demonstrate that frizzled and frizzled 2 act in the wingless pathway. *Cell* **95**, 1017–1026.

Lipardi, C., Wei, Q., and Paterson, B. M. (2001). RNAi as random degradative PCR: siRNA primers convert mRNA into dsRNAs that are degraded to generate new siRNAs. *Cell* **107**, 297–307.

Lipardi, C., Wei, Q., and Paterson, B. M. (2003). RNA silencing in *Drosophila*. *Acta Histochem. Cytochem.* **36**, 123–134.

Makeyev, E. V., and Bamford, D. H. (2002). Cellular RNA-dependent RNA polymerase involved in posttranscriptional gene silencing has two distinct activity modes. *Mol. Cell* **10**, 1417–1427.

Martens, H., Novotny, J., Oberstrass, J., Steck, T. L., Postlethwait, P., and Nellen, W. (2002). RNAi in *Dictyostelium*: The role of RNA-directed RNA polymerases and double-stranded RNase. *Mol. Biol. Cell* **13**, 445–453.

Misquitta, L., and Paterson, B. M. (1999). Targeted disruption of gene function in *Drosophila* by RNA interference (RNAi): A role for *nautilus* in embryonic somatic muscle formation. *Proc. Natl. Acad. Sci. USA* **96**, 1451–1456.

Mourrain, P., Beclin, C., Mourrain, P., Beclin, C., Elmayan, T., Feuerbach, F., Godon, C., Morel, J. B., Jouette, D., Lacombe, A., Nikic, M., Picault, S., Remoue, N. K., Sanial, M., Vo, T. A., and Vaucheret, H. (2000). Arabidopsis SGS2 and SGS3 genes are required for posttranscriptional gene silencing and natural virus resistance. *Cell* **101**, 533–542.

Nilsson, M., Antson, D.-O., Barbany, G., and Landegren, U. (2001). RNA-templated DNA ligation for transcript analysis. *Nucleic Acids Res.* **29**, 578–581.

Pal-Bhadra, M., Leibovitch, B. A., Gandhi, S. G., Rao, M., Bhadra, U., Birchler, J. A., and Elgin, S. C. (2004). Heterochromatic silencing and HP1 localization in *Drosophila* are dependent on the RNAi machinery. *Science* **303**, 669–672.

Plasterk, R. H. (2002). RNA silencing: The genome's immune system. *Science* **296**, 1263–1265.

Sijen, T., Fleenor, J., Simmer, F., Thijssen, K. L., Parrish, S., Timmons, L., Plasterk, R. H., and Fire, A. (2001). On the role of RNA amplification in dsRNA-triggered gene silencing. *Cell* **107**, 465–476.

Tuschl, T., Zamore, P. D., Lehmann, R., Bartel, D. P., and Sharp, P. A. (1999). Targeted mRNA degradation by double-stranded RNA *in vitro*. *Genes Dev.* **13**, 3191–3197.

Volpe, T. A., Kidner, C., Hall, I. M., Teng, G., Grewal, S. I., and Martienssen, R. A. (2002). Regulation of heterochromatic silencing and histone H3 lysine-9 methylation by RNAi. *Science* **297**, 1833–1837.

Wei, Q., Lipardi, C., and Paterson, B. M. (2003). Analysis of the 3′-hydroxyl group in *Drosophila* siRNA function. *Methods* **30**, 337–347.

Wei, Q., Marchler, G., Edington, K., Karsch-Mizrachi, I., and Paterson, B. M. (2000). RNA interference demonstrates a role for nautilus in the myogenic conversion of Schneider cells by daughterless. *Dev. Biol.* **228**, 239–255.

[22] Use of RNA Polymerase II to Transcribe Artificial MicroRNAs

By Yan Zeng, Xuezhong Cai, and Bryan R. Cullen

Abstract

MicroRNAs (miRNAs) are endogenously encoded ~22-nt-long RNAs that are generally expressed in a highly tissue- or developmental-stage-specific fashion and that posttranscriptionally regulate target genes. Regulatable RNA polymerase II promoters can be used to overexpress authentic microRNAs in cell culture. Furthermore, one can also design and express artificial microRNAs based on the features of existing microRNA genes, such as the gene encoding the human miR-30 microRNA. Overexpression or inappropriate expression of authentic microRNAs may facilitate the study of their normal functions and expression of artificial microRNAs may permit effective, regulated RNA interference *in vivo*.

Introduction

MicroRNAs (miRNAs) are a class of ~22-nt-long noncoding RNAs expressed in all metazoan eukaryotes (Bartel, 2004). With more than 200 distinct miRNAs having been identified in plants and animals, these small regulatory RNAs are believed to serve important biological functions by two prevailing modes of action: (1) by repressing the translation of target mRNAs, and (2) through RNA interference (RNAi), that is, cleavage and degradation of mRNAs. In the latter case, miRNAs function analogously to small interfering RNAs (siRNAs). Importantly, miRNAs are expressed in a highly tissue-specific or developmentally regulated manner and this regulation is likely key to their predicted roles in eukaryotic development and differentiation. Analysis of the normal role of miRNAs will be facilitated by techniques that allow the regulated overexpression or inappropriate expression of authentic miRNAs *in vivo*, whereas the ability to regulate the expression of siRNAs will greatly increase their utility both in cultured cells and *in vivo*.

miRNAs are first transcribed as part of a long, largely single-stranded primary transcript (Lee *et al.*, 2002; Fig. 1). This primary miRNA transcript is generally, and possibly invariably, synthesized by RNA polymerase II (pol II) and therefore is normally polyadenylated and may be spliced. It contains an ∼80-nt hairpin structure that encodes the mature ∼22-nt miRNA as part of one arm of the stem. In animal cells, this primary transcript is cleaved by a nuclear RNaseIII-type enzyme called Drosha (Lee *et al.*, 2003) to liberate a hairpin miRNA precursor, or pre-miRNA, of ∼65 nt, which is then exported to the cytoplasm by exportin-5 and the GTP-bound form of the Ran cofactor (Yi *et al.*, 2003). Once in the cytoplasm, the pre-miRNA is further processed by Dicer, another RNaseIII

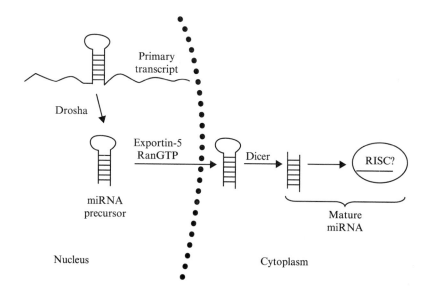

Fig. 1. Schematic of the miRNA biogenesis pathway (see text for details).

enzyme, to produce a duplex of ∼22 bp that is structurally identical to an siRNA duplex (Hutvágner *et al.*, 2001). The binding of protein components of the RNA-induced silencing complex (RISC), or RISC cofactors, to the duplex results in incorporation of the mature, single-stranded miRNA into a RISC or RISC-like protein complex, whereas the other strand of the duplex is degraded (Bartel, 2004). Analysis of the processing of the human miR-30 miRNA have been instrumental in delineating the biogenesis of miRNAs (Lee *et al.*, 2003; Yi *et al.*, 2003; Zeng and Cullen, 2003; Zeng *et al.*, 2002).

In an attempt to overexpress authentic human miR-30 in transfected mammalian cells, we have discovered the following (Zeng and Cullen, 2003; Zeng *et al.*, 2002): (1) the ∼80-nt-long stem-loop structure found in the primary mR-30 precursor (Fig. 2B) can be inserted at various locations in an artificial pol II promoter-driven primary miRNA transcript, including in introns, to produce mature miR-30; (2) both mature miR-30 and its opposite strand, termed anti-miR-30, can be readily detected in transfected cells; (3) miR-30 biogenesis is not inhibited by sequence changes within its precursor stem as long as the size and the predominantly double-stranded nature are preserved; (4) if the stem of the precursor is replaced by an arbitrary double-stranded sequence of closely similar length, the new arbitrary sequence will then be expressed as a ∼22-nt artificial miRNA/siRNA; and (5) miR-30-derived expression cassettes can be expressed in tandem and from regulatable pol II promoters. In this article, we discuss how to use the miR-30 architecture to express miRNAs or siRNAs from pol II promoter-based expression plasmids.

Designing Novel MicroRNA (miRNA)-Expression Cassettes Based on the miR-30 Precursor

Figure 2B shows the predicted secondary structure of the miR-30 precursor hairpin. Boxed are extra nucleotides that were added originally for subcloning purposes (Zeng and Cullen, 2003; Zeng *et al.*, 2002). They represent *Xho*I–*Bgl*II sites at the 5′ end and *Bam*HI–*Xho*I sites at the 3′ end. These appended nucleotides extend the minimal miR-30 precursor stem shown by several basepairs, similar to the *in vivo* situation where the primary miR-30 precursor is transcribed from its genomic locus (Lee *et al.*, 2003), and an extended stem of at least 5 bp is essential for efficient miR-30 production. We refer to the unit RNA shown in Fig. 2B as the miR-30 cassette hereafter. Based on the numbering in Fig. 2B, mature miR-30 is encoded by nucleotides 44 to 65 and anti-miR-30 by nucleotides 3 to 25 of this precursor.

In the simplest expression setting, the cytomegalovirus (CMV) immediate early enhancer/promoter is used to transcribe the miR-30 cassette (Fig. 2A). The cassette is preceded by a leader sequence of approximately

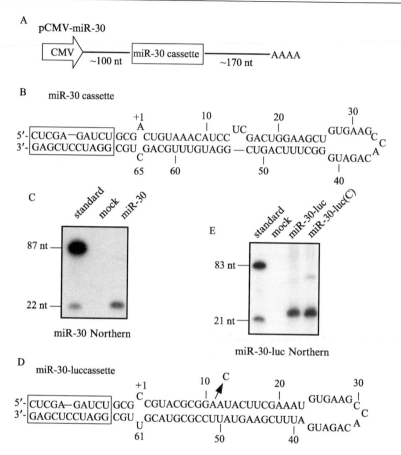

FIG. 2. Strategy of miRNA expression. (A) Schematic of the pCMV-miR-30 plasmid (Zeng *et al.*, 2002). (B) Diagram of the miR-30 cassette. Boxed are the added restriction sites, which correspond to *Xho*I-*Bgl*II at the 5' end and *Bam*HI-*Xho*I at the 3' end. (C) Detection of miR-30 expression by Northern analysis. 293T cells in 24-well plates were mock transfected or transfected with 0.4 μg of pCMV-miR-30. Two days later, RNAs were isolated and Northern blotting performed with a 5' ^{32}P-labeled primer: 5'-GCAGCTGCAAACATCCGACT-GAAAGCCC-3'. (D) Design of the miR-30-luc cassette (see text for details). A derivative, the miR-30-luc(C) cassette, has a C residue instead of an A at position 11. (E) Both miR-30-luc and miR-30-luc(C) cassettes expressed mature miR-30-luc. Transfection and Northern blotting were performed as in (C). The sequence of the probe used is 5'-AACGTACGCGGAATACT-TCGAAAT-3'.

100 nt and followed by approximately 170 nt before the polyadenylation site (Zeng *et al.*, 2002). These lengths are arbitrary and can be longer or shorter. As shown by Northern blotting in Fig. 2C, mature 22-nt miR-30 was made at readily detectable levels in 293T cells transfected with the

pCMV-miR-30 construct. Several other authentic miRNAs have been overexpressed by using analogous RNA pol II-based expression vectors or even pol III-dependent promoters (Chen *et al.*, 2004; Zeng and Cullen, 2003). Expression simply requires the insertion of the entire predicted miRNA precursor stem-loop structure into the expression vector at an arbitrary location. Because the actual extent of the precursor stem loop can sometimes be difficult to accurately predict, it is generally appropriate to include ≥50 bp of flanking sequence on each side of the predicted ∼80-nt miRNA stem-loop precursor to be sure that all *cis*-acting sequences necessary for accurate and efficient Drosha processing are included (Chen *et al.*, 2004).

A luciferase siRNA expression cassette (Zeng and Cullen, 2003) is used here as an example to demonstrate how to make artificial miRNAs or siRNAs based on the miR-30 template. To make the miR-30-luc expression cassette (Fig. 2D), the sequence from +1 to 65 (excluding the 15-nt terminal loop of the miR-30 cassette, Fig. 2B) was replaced as follows: the sequence from nucleotides 39 to 61 (Fig. 2D), which is perfectly complementary to a firefly luciferase sequence (nucleotides 154–176 from the start codon), will act as the active strand during RNAi. The sequence from nucleotides 2 to 23 is thus designed to preserve the double-stranded stem in the miR-30-luc cassette (Fig. 2D), but nucleotide +1 is now a C, to create a mismatch with nucleotide 61, a U, just like nucleotides 1 and 65 in the miR-30 cassette (Fig. 2B). In a slight variation, nucleotide 11 was changed from A to C (Fig. 2D) to make a derivative termed miR-30-luc(C). Because the 3′ arm of the stem, called miR-30-luc below, is the active component for RNAi, changes in the 5′ arm of the stem will not affect RNAi specificity. A 2-nt bulge is present in the stem region of the authentic miR-30 precursor (Fig. 2B). We hypothesize that a break in the helical nature of the RNA stem may help ward off nonspecific effects, such as induction of an interferon response (Bridge *et al.*, 2003) in expressing cells. (This may be why miRNA precursors almost invariably contain bulges in the predicted stem.) The miR-30 cassette in Fig. 2A was then substituted with the miR-30-luc or miR-30-luc(C) cassette, and the resulting expression plasmid transfected into cells. As demonstrated in Fig. 2E, both miR-30-luc and miR-30-luc(C) produced the predicted ∼22-nt miR-30-luc artificial miRNA. Moreover, both were effective at suppressing luciferase activity (Zeng and Cullen, 2003).

The use of pol II promoters offers flexibility in regulating the production of miRNAs in cultured cells or *in vivo*. In Fig. 3, transcription is directed by a Tet-Off promoter, whose activity depends on an activator encoded by the plasmid pTet-Off (Clontech, Palo Alto, CA), which in turn is suppressed by doxycycline. In a transient cotransfection experiment, the

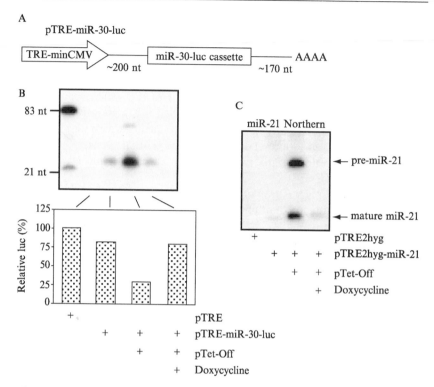

FIG. 3. Tet-regulated miRNA expression. (A) Diagram of the pTRE-miR-30-luc plasmid. It was constructed by replacing the CMV promoter in pCMV-miR-30-luc (Fig. 2) with a TRE-minCMV promoter, which contains the Tet response element (TRE) and the minimal cytomegalovirus (CMV) promoter, PCR amplified from pTRE2hyg (Clontech). (B) Regulated expression of miR-30-luc. A firefly luciferase expression plasmid, along with a control *Renilla* luciferase plasmid, was cotransfected with 15 ng of pTRE or pTRE-miR-30-luc in the absence or presence of the activator plasmid pTet-Off (Clontech). The pTRE plasmid is the same as pTRE-miR-30-luc but without the miR-30-luc cassette. For suppression of transcription, a final concentration of 1 μg/ml doxycycline (Clontech) was added to the medium. Two days after transfection, cells were assayed for dual-luciferase activities (Promega) and for the expression of miR-30-luc by Northern blotting as in Fig. 2E. The firefly luciferase activity of cells transfected with pTRE, normalized to *Renilla* luciferase activity, was set as 100%. (C) Regulated overexpression of miR-21 miRNA. Approximately 400 bp of the genomic DNA encoding the human miR-21 was cloned into pTRE2hyg. Transfection and Northern analysis were performed as in (A), with the sequence of the probe being 5′-TCAACATCAGTCTGATAAGCTA-3′.

pTRE-miR-30-luc plasmid (Fig. 3A), without the activator, produced a low level of miR-30-luc that reduced luciferase activity by only 20%. Significantly more miR-30-luc was produced in the presence of the activator, and luciferase activity was reduced by 70%. This activation was reversed by addition of doxycycline (Fig. 3B). It is expected that the selection of stable cell lines would lead to less leaky expression in the absence of the activator or presence of doxycycline and therefore a stronger induction. Obviously, the same system can also be used for regulated overexpression of authentic human miRNAs in transfected cells. As shown in Fig. 3C, a similar pTRE-based expression plasmid designed to express the human miRNA miR-21 (Zeng and Cullen, 2003) gave rise to readily detectable levels of not only the 22-nt mature miR-21 but also the ~60-nt pre-miR-21 intermediate. The ability to regulate the level of authentic miRNA expression may facilitate study of the functions of endogenous miRNAs in tissue culture or particularly *in vivo*.

It would be advantageous if the antisense strand, for example, miR-30-luc in Fig. 2E, was preferentially made as a mature miRNA, because its opposite strand does not have any known target. The relative basepairing stability at the 5′ ends of an siRNA duplex is a strong determinant of which strand will be incorporated into RISC and hence be active in RNAi; the strand whose 5′ end has a weaker hydrogen bonding pattern is preferentially incorporated into RISC, the RNAi effecter complex (Khvorova *et al.*, 2003; Schwarz *et al.*, 2003). This same principle can also be applied to the design of DNA vector-based siRNA expression strategies, including the one described here. However, for artificial miRNAs, the fact that the internal cleavage sites by Drosha and Dicer cannot be precisely predicted at present adds a degree of uncertainty as a 1- or 2-nt shift in the cleavage site can generate rather different hydrogen bonding patterns at the 5′ ends of the resulting duplex, thus changing which strand of the duplex intermediate is incorporated into RISC. This is in contrast to the situation with synthetic siRNA duplexes, which have defined ends. We have found that both miR-30-luc and its complementary strand are expressed in transfected cells and that their 5′ ends are heterogeneous (not necessarily in equal amounts; see Zeng and Cullen, 2003). On the other hand, any minor heterogeneity at the ends of an artificial miRNA duplex intermediate might not be a problem, as the miRNAs would still be perfectly complementary to their target.

Interestingly, the same 5′-end rule also governs the production of endogenous miRNAs (Khvorova *et al.*, 2003; Schwarz *et al.*, 2003). Perhaps sequences within the ~80-nt miRNA stem-loop have evolved to define the precise cleavage sites *in vivo*. We have not systematically examined the roles of the internal loop, stem length, and the surrounding sequences on the expression of miRNAs from miR-30-derived cassettes. Such analyses may suggest design elements that would maximize the yield of the intended RNA products. On the other hand, because we are overexpressing artificial miRNAs/siRNAs through DNA transfection, some heterogeneity could be inevitable. In addition to the 5'-end rule, specific residues at some positions within an siRNA may also enhance siRNA function (Reynolds *et al.*, 2004). In general, picking a target region with more than 50% AU content and designing a weak 5' end base pair on the antisense strand would be a good starting point in the design of any artificial miRNA/siRNA expression plasmid (Khvorova *et al.*, 2003; Reynolds *et al.*, 2004; Schwarz *et al.*, 2003).

Although all the experiments described in this chapter used transient expression to introduce authentic and artificial miRNA expression plasmids into cultured cells, we believe that effective expression of these short RNA could also be achieved by using lentiviral or other virus-based expression vectors. Note, however, that miRNA processing results in the degradation of the remainder of the primary miRNA transcript, a problem that may require expression of the miR-30 cassette in the antisense orientation in lentiviral or retroviral vectors. Finally, although the plasmids used in these experiments express only a single miRNA, the fact that each miRNA stem-loop precursor is independently excised from the primary transcript by Drosha cleavage to give rise to a pre-miRNA suggests that it should be able to simultaneously express several artificial or authentic miRNAs by a tandem array on a precursor RNA transcript. This is indeed what is sometimes seen with authentic primary miRNA precursor transcripts (Lagos-Quintana *et al.*, 2001; Lau *et al.*, 2001), and we have demonstrated that at least two artificial miRNAs can also be expressed from one RNA transcript (Zeng and Cullen, 2003).

Methods and Procedures

To construct the miRNA expression cassette described in Fig. 2, two primers were made. One contains *Xho*I and *Bgl*II restriction sites followed by the sense sequence starting from the 5' end of the intended miRNA hairpin, and the other has *Xho*I and *Bam*HI sites followed by the antisense sequence from the 3' end. The two primers were designed so that each is about 55 nt long and the primers have ~15 nt of complementarity at their 3' ends. They were annealed and extended by PCR, isolated with Qiagen's PCR purification kit, and digested with *Xho*I. After heat inactivation of *Xho*I, the DNA was cloned into a suitable expression plasmid.

We have inserted the miR-30 cassette into different positions of several CMV-based expression vectors and have always observed robust production of mature miR-30, indicating that the miR-30 cassette is reasonably independent of its surrounding environment in the primary transcripts. Nevertheless, it is prudent to verify the expression of any new miRNA or siRNA by Northern blot analysis and primer extension, or both. In general, the higher the level of miRNA/siRNA expression, the higher the level of

RNAi observed. We isolate total RNA by using TRIzol® Reagent (Invitrogen, Carlsbad, CA), 2 days after transfection of mammalian cells. For Northern blotting, 10–20 μg of RNA was separated on a denaturing 15% polyacrylamide gel (Bio-Rad, Hercule), transferred to Hybond-N™ membrane (Amersham, Piscataway, NJ), and UV cross-linked. Prehybridization and hybridization were performed in ExpressHyb™ solution (Clontech). To prepare the probe for hybridization, 0.5–1 μg of an oligonucleotide complementary to the miRNA of interest was phosphorylated by T4 polynucleotide kinase with [γ-^{32}P]ATP. The labeled DNA was purified by passing the reaction mixture through a Centrisep™ spin column (Princeton Separations, Adelphia). The probe was used at a concentration of 0.5×10^7 to 1×10^7 cpm/ml during hybridization. For primer extension, approximately 5 μg of total RNA was annealed to 1×10^4 to 4×10^4 cpm of 5′ ^{32}P-labeled oligonucleotide and then extended through reverse transcription. The reaction was resolved on an 8 M urea/15% polyacrylamide gel. The parameters given are for 293T cells; good miRNA expression would yield a clear signal on film after 4 h of exposure at $-80°$ (for both Northern and primer extension). For cell lines that are more difficult to transfect, use more RNA for analyses and expose the film longer. Once expression of the miRNA/siRNA is confirmed, the DNA construct can then be tested for RNAi efficacy against a cotransfected plasmid encoding the target protein or directly against an endogenous target. In the latter case, one should have a clear idea of transfection efficiency and of the half-life of the target protein before performing the experiment.

Concluding Remarks

The major advantage of DNA vector-based siRNA expression strategies is to allow one to monitor the effects of stable gene knockdown by RNAi in cultured cells or in intact animals. This approach also offers the potential for regulatable RNAi. We have shown that miR-30-based expression cassettes transcribed by pol II can be employed for miRNA/siRNA expression. It is well established that pol II transcription can be exquisitely regulated in culture or *in vivo* by using inducible/repressible promoters, as shown in Fig. 3, or by using tissue-specific or developmentally regulated promoters. Therefore, miRNA/siRNA production could be controlled by using analogous pol II-based expression cassettes based on the miR-30 framework (Fig. 2). It should be possible in the future to generate reversible hypomorphic mutations and to analyze their phenotypes in specific cell types or tissues and at specific developmental time points in transgenic animals. Such mutant animals (generally mice) should be straightforward to generate, and several miR-30 expression cassettes could in principle be easily combined in one animal, thus allowing the coordinated downregulation of two or more genes. We therefore envision that the pol II-based technology for authentic or artificial miRNA expression described here may have considerable general utility.

References

Bartel, D. P. (2004). MicroRNAs: Genomics, biogenesis, mechanism, and function. *Cell* **116**, 281–297.

Bridge, A. J., Pebernard, S., Ducraux, A., Nicoulaz, A.-L., and Iggo, R. (2003). Induction of an interferon response by RNAi vectors in mammalian cells. *Nat. Genet.* **34**, 263–264.

Chen, C.-Z., Li, L., Lodish, H. F., and Bartel, D. P. (2004). MicroRNAs modulate hematopoietic lineage differentiation. *Science* **303**, 83–86.

Hutvágner, G., McLachlan, J., Pasquinelli, A. E., Bálint, É., Tuschl, T., and Zamore, P. D. (2001). A cellular function for the RNA-interference enzyme dicer in the maturation of the *let-7* small temporal RNA *Science* **293**, 834–838.

Khvorova, A., Reynolds, A., and Jayasena, S. D. (2003). Functional siRNAs and miRNAs exhibit strand bias. *Cell* **115**, 209–216.

Lagos-Quintana, M., Rauhut, R., Lendeckel, W., and Tuschl, T. (2001). Identification of novel genes coding for small expressed RNAs. *Science* **294**, 853–858.

Lau, N. C., Lim, L. P., Weinstein, E. G., and Bartel, D. P. (2001). An abundant class of tiny RNAs with probable regulatory roles in *Caenorhabditis elegans*. *Science* **294**, 858–862.

Lee, Y., Ahn, C., Han, J., Choi, H., Kim, J., Yim, J., Lee, J., Provost, P., Rådmark, O., Kim, S., and Kim, V. N. (2003). The nuclear RNase III drosha initiates microRNA processing. *Nature* **425**, 415–419.

Lee, Y., Jeon, K., Lee, J.-T., Kim, S., and Kim, V. N. (2002). MicroRNA maturation: Stepwise processing and subcellular localization. *EMBO J.* **21**, 4663–4670.

Reynolds, A., Leake, D., Boese, Q., Scaringe, S., Marshall, W. S., and Khvorova, A. (2004). Rational siRNA design for RNA interference. *Nat. Biotech.* **22**, 326–330.

Schwarz, D. S., Hutvágner, G., Du, T., Xu, Z., Aronin, N., and Zamore, P. D. (2003). Asymmetry in the assembly of the RNAi enzyme complex. *Cell* **115**, 208–299.

Yi, R., Qin, Y., Macara, I. G., and Cullen, B. R. (2003). Exportin-5 mediates the nuclear export of pre-microRNAs and short hairpin RNAs. *Genes Dev.* **17**, 3011–3016.

Zeng, Y., and Cullen, B. R. (2003). Sequence requirements for micro RNA processing and function in human cells. *RNA* **9**, 112–123.

Zeng, Y., Wagner, E. J., and Cullen, B. R. (2002). Both natural and designed micro RNAs can inhibit the expression of cognate mRNAs when expressed in human cells. *Mol. Cell* **9**, 1327–1333.

[23] Adeno-Associated Virus Vectors for Short Hairpin RNA Expression

By DIRK GRIMM, KUSUM PANDEY, and MARK A. KAY

Abstract

Five recent publications have documented the successful development and use of gene transfer vectors based on adeno-associated virus (AAV) for expressing short hairpin RNA (shRNA). In cultured mammalian cells and in whole animals, infection with these vectors was shown to result in specific, efficient, and stable knockdown of various targeted endo- or exogenous genes. Here we review this exciting approach, to trigger RNA interference *in vitro* and *in vivo* by shRNA expressed from AAV vectors, and describe the state-of-the-art technology for vector particle generation. In particular, we present a set of novel AAV vector plasmids that were specifically designed for the easy and rapid cloning of shRNA expression cassettes into AAV. The plasmids contain alternative RNA polymerase III promoters (U6, H1, or 7SK) together with a respective terminator sequence, as well as stuffer DNA to guarantee an optimal vector size for efficient packaging into AAV capsids. To provide maximum versatility and user-friendliness, the constructs were also engineered to contain a set of unique restriction enzyme recognition sites, allowing the simple and straightforward replacement of the shRNA cassette or other vector components with customized sequences. Our novel vector plasmids complement existing AAV vector technology and should help further establish AAV as a most promising alternative to using adeno- or retro-/lentiviral vectors as shRNA delivery vehicles.

Introduction: Exploiting RNA Interference (RNAi)—Promises and Problems

Over the past 3 years, the field of nucleic acid-based inhibitors of gene expression has seen an unprecedented wave of interest, sparked by a groundbreaking study by Elbashir *et al.* (2001). First, their report provided initial proof that mammalian cells are capable of eliciting the sequence-specific posttranscriptional gene silencing pathway known as RNA interference (RNAi), which in the past had been assumed to be restricted to lower organisms and plants. Second, they also discovered that the RNAi pathway becomes efficiently activated by the introduction of short double-stranded RNAs (dsRNAs), which mimic products of the Dicer endonuclease and as

METHODS IN ENZYMOLOGY, VOL. 392

such trigger the targeted degradation of complementary mRNAs that is characteristic of RNAi.

It was perhaps this particular aspect of their work that has fueled the most interest in RNAi, because it paved the way for the subsequent straightforward exploitation of this natural mechanism as a powerful novel research tool. A most important successive finding was that besides in the form of small dsRNAs, two complementary RNA strands are also effective triggers of RNAi when present as a single stem-loop [short hairpin RNA (shRNA); Paddison *et al.*, 2001]. As such, they can be easily generated intracellularly by expression from RNA polymerase II or III promoters such as CMV or U6. This overcomes the main drawbacks associated with synthetic dsRNAs, namely, the transient nature of gene silencing and high costs for dsRNA manufacturing. Moreover, it offers wide possibilities to control the on- and offset of RNAi and to restrict the effect to particular tissues, although so far there is only limiting evidence that these goals will soon be met.

Significant advances were subsequently also made in the design of the shRNA molecules themselves, particularly with regard to the issues of functionality and specificity. Thus, the large-scale systemic analysis of shRNAs targeting defined mRNAs recently led to the identification of characteristics associated with high effector molecule functionality, culminating in the establishment of novel algorithms for the rational design of potent shRNAs (e.g., Ding *et al.*, 2004; Reynolds *et al.*, 2004). In parallel, others have developed novel enzyme-mediated strategies for generating numerous functional shRNA constructs from any gene of interest, which bears the inherent advantage that no prior knowledge about target transcripts is needed (Luo *et al.*, 2004; Sen *et al.*, 2004; Shirane *et al.*, 2004). However, identification of the best effector molecules from an shRNA library will not be easy and straightforward.

Irrespective of the exact shRNA selection method, a critical problem becoming increasingly apparent is that of potential specific or unspecific adverse effects from expression of shRNAs in cells or whole animals. Reviewing this important topic in more detail will be beyond the scope of this chapter, but note that recently observed side effects include activation of arms of the interferon or the microRNA pathways, as well as unwanted off-targeting of endogenous cellular genes with partial or full homology to the actual target (e.g., Doench *et al.*, 2003; Jackson *et al.*, 2003).

Despite this crucial concern, the previously mentioned advances and refinements in design of shRNA expression cassettes have dramatically accelerated the pace at which this novel technology is being applied as an analytical and a therapeutic tool. The enormous promise of the overall approach has already been documented by a recent wealth of reports. For instance, various groups have addressed the great potential of RNAi as a revolutionary method for studying gene functions in whole animals, in which its advantage over conventional knockout models is that more target genes can be assessed in less time, thus saving efforts and resources. Most outstanding publications reported the generation of shRNA libraries targeting thousands of human or mouse genes at one time (Berns et al., 2004; Paddison et al., 2004), providing hope that the ultimate challenge can be met—to perform RNAi-based large-scale loss-of-function genetic screens of the entire human genome.

The feasibility to effectively and specifically silence endo- or exogenous genes also holds great promise for treating viral infections, cancers, and inherited genetic disorders, leading to the anticipation that RNAi will emerge as a novel powerful tool for gene therapy. The potential of this approach is exemplified and underscored by studies demonstrating the successful use of shRNA to inhibit replication of, or gene expression from, human pathogens such as HIV or hepatitis B virus (Lee and Rossi, 2004; McCaffrey et al., 2003). Proof of principle has likewise already been provided that RNAi can be exploited to knock down cellular or viral oncogenes (reviewed by Friedrich et al., 2004) or to repress dominant gain-of-function gene mutants causing human disorders (Miller et al., 2003).

Although these examples manifest the tremendous promise of RNAi for various applications, the key challenge that needs to be met before the approach can become widely accepted is to develop systems that will efficiently deliver shRNA expression cassettes to mammalian or, ultimately, human cells. To overcome this hurdle and put such a delivery tool in place is central to establishing RNAi-based gene therapies, as exemplified by scenarios in which shRNAs will be used to knock down oncogenes expressed within tumor cells or viral genes within infected cells. Obviously, the success of these strategies will largely depend on the ability to hit each target cell, and this goal becomes even more challenging when the approach will be translated to whole organisms, thus requiring tools to effectively deliver and express shRNAs in vivo.

However, this goal can unlikely be met with the currently available technologies for expression of shRNAs from plasmid DNAs or from any other form of nonviral vector. Although these constructs are easy to generate and useful for studies in isolated cultured cells, there is no methodology on hand to efficiently deliver nonviral vectors to the entire cell mass within an intact organ in vivo. Consequently, more researchers have recently begun to turn their attention to developing virus-derived shRNA expression vectors, based on the promise that as compared with their nonviral counterparts they will provide a significantly higher efficiency for shRNA delivery in whole organisms. In addition, it is hoped for that using a recombinant virus as delivery system will permit better control over tissue specificity of shRNA expression, because most viral vectors currently under development are either naturally characterized by distinct tissue tropisms or can relatively easily be genetically modified and retargeted to particular cell types (e.g., Nicklin and Baker, 2002).

The viral vectors recently engineered as shRNA delivery vehicles comprise three major classes of viruses: adenovirus, retro- or lentivirus, and adeno-associated virus (AAV). The first two viral vector types have already been studied rather extensively as tools for RNAi-mediated gene silencing and are the subject of chapters 9 and 13. Here, we focus on the AAV vector system. We firstly briefly review wild-type and recombinant AAV (rAAV) biology, then discuss and describe the use of AAV vectors as mediators of shRNA expression, and finally give detailed protocols to generate such vectors.

Adeno-Associated Virus (AAV)—From Wild-Type Virus to Recombinant Vectors

Wild-type AAV and vectors derived thereof have already been extensively reviewed over the past 20 years (e.g., Berns and Linden, 1995; Grimm and Kleinschmidt, 1999), and we will limit the information provided here to a minimum. Unless otherwise specified, all facts and numbers presented refer to AAV serotype 2 (AAV-2), which is considered the prototype of the AAV group. Figure 1 provides a comprehensive summary of the main characteristics of wild-type and recombinant AAV (rAAV).

AAVs belong to the family Parvoviridae, which encompasses viruses with nonenveloped, icosahedral capsids with a diameter of about 20 nm, containing a linear single-stranded DNA molecule. In line with the small capsid size, the viral genome is particularly short (only 4681 nt long) and is also relatively simple in its organization, comprising only two genes (*rep* and *cap*) that encode four nonstructural (Rep) and three structural (VP) proteins. The two genes are embedded between two inverted terminal repeats (ITRs) 145 nt long, which typically assume a T-shaped hairpin secondary structure. ITRs play crucial roles in many steps of the viral life cycle, beginning with the expression and replication of the viral genome early in a productive infection and followed by its packaging into the AAV capsid. In subsequently infected cells, they are involved in the conversion of the single-stranded viral DNA into a double-stranded form from which AAV gene expression occurs, and which eventually becomes integrated into the host chromosome or persists episomally.

Besides its small size and simple organization, AAV is further distinguished from other human viruses by its requirement for a helper virus to

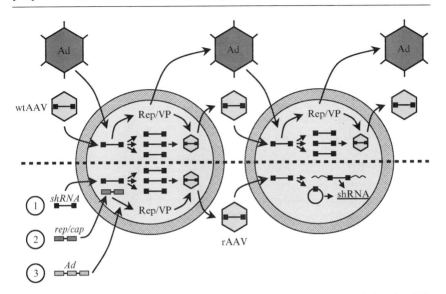

Fig. 1. Life cycles of wild-type (wtAAV) and recombinant adeno-associated virus (rAAV). The two pairs of circles represent mammalian or human cells; the nuclei are depicted by the lighter inner circles. (*Top*) Wild-type AAV. To propagate wild-type AAV, cells are coinfected with the virus and a helper virus (here adenovirus). The AAV capsid traffics (not shown) to the nucleus, where it releases its genome [black line with boxes at the end, representing the two inverted terminal repeats (ITRs)]. Products expressed from the helper virus support expression of the AAV genes, resulting in large amounts of AAV replication (Rep) and capsid (VP) proteins. In parallel, the wild-type AAV genome replicates to high copy numbers in the nucleus and eventually becomes encapsidated in virions formed by assembled VP proteins. The helper virus finally induces cell lysis, and progeny AAV as well as replicated adenovirus are released to start a fresh infection cycle in another cell. (*Bottom*) Recombinant AAV. The generation of rAAV requires the same components that are needed to propagate wild-type virus. However, for rAAV production, they are provided from three separate plasmids, which are triple transfected into cells: (1) an AAV vector plasmid comprising the transgene (here labeled as shRNA) flanked by the AAV-2 ITRs, (2) an AAV helper plasmid encoding the *rep* and *cap* genes, and (3) an adenoviral helper plasmid (typically encoding the E2A, VA, and E4 genes). In the cell nucleus, the AAV helper expresses Rep and VP proteins *in trans* (supported by adenoviral helper functions), whereas the ITRs replicate the transgene (shRNA) sequences *in cis*. Similar to the wild-type scenario, the VP proteins then assemble into capsids, which take up the amplified viral DNA. In contrast, the cells do not produce progeny helper virus and thus do not get lysed, allowing the rAAV particles to be harvested from the intact cell (not shown). In a reinfected cell, the rAAV particle releases its genome, which subsequently does not get replicated (because of the absence of AAV or adenoviral gene expression), but instead can integrate into the host chromosome or assume extrachromosomal circular or linear (not shown) forms, either of which results in rAAV genome persistence.

complete its own life cycle in the coinfected cell. The helper virus, typically adenovirus or a member of the herpes virus family, provides products (RNAs and proteins) that stimulate gene expression from the AAV promoters, enhance transport and splicing of the AAV pre-mRNAs, and in case of herpes simplex virus 1 support replication of the AAV genome (Weindler and Heilbronn, 1991). AAV's dependence on a pathogenic helper virus is particularly unique, considering that no AAV has ever been associated with any human malignancy, although AAV serotype 2 is believed to be an extremely common virus. Previous studies have estimated that up to 80% of the human population is seropositive for this particular AAV serotype (Erles et al., 1999).

This detail may have led to the very early definition of AAV-2 as the prototype of this virus class, despite the fact that along with its discovery in the late 1960s, other serotypes had been isolated as well (AAV-1 to -4). The AAV group has further grown with the description of AAV-5 to -8 between 1984 and 2002 (reviewed by Grimm and Kay, 2003). The most significant expansion came with a report describing the isolation and identification of 108 new AAV variants, of which 55 were found in human tissue and 53 in nonhuman primates (Gao et al., 2004). This distribution is in line with the hypothesis that AAV-1, -4, -7, and -8 are simian viruses, whereas the other four previously described AAV serotypes seem to primarily infect humans.

The early discovery that AAV ITRs are the only sequences required in cis to mediate packaging of DNA sequences embedded in between has paved the way for generating vectors derived from wild-type AAV. Thus, a straightforward approach to produce a rAAV vector genome is to replace the entire wild-type rep and cap genes with any sequence of choice. To facilitate this step, a large number of AAV vector plasmids have been generated over the years that contain the AAV-2 ITRs as the sole sequences derived from the wild type, and, in addition, carry unique restriction enzyme recognition sites, allowing the easy insertion of foreign DNA between the ITRs (e.g., Zolotukhin et al., 1996). The only important consideration to be made at this step is that the total size of the resulting recombinant cassette, including the transgene and the two ITRs, must not significantly exceed the size of the AAV wild-type genome, to guarantee its efficient encapsidation. This is a direct consequence of the fact that the AAV capsid is an icosahedral structure and can only accommodate a particular amount of DNA; accordingly, the packaging limit for AAV vectors is approximately 4.8 kb of foreign DNA. As discussed later, this has important implications for the design and generation of AAV vectors for the expression of shRNA.

To package the recombinant genome into viral capsids, the AAV vector plasmid is typically transfected into cultured cells together with two other plasmids: an AAV helper plasmid encoding the *rep* and *cap* genes in *trans* and an adenoviral plasmid providing the respective helper functions. During the subsequent incubation of transfected cells, the Rep and VP proteins expressed from the AAV helper plasmid replicate the recombinant genome to high copy numbers and form AAV capsids for its packaging, respectively. The products expressed from the adenoviral plasmid basically provide the same helper functions as do infectious adenovirus during growth of wild-type AAV; that is, they enhance expression of the *rep* and *cap* genes as well as support AAV RNA transport and splicing. However, the adenoviral plasmid is no longer infectious, thus ensuring that the final rAAV preparation is free from contaminating adenovirus. Likewise, the rAAV stocks are also free from wild-type AAV virus, because the *rep/cap*-expressing helper plasmid lacks the ITRs required for DNA encapsidation (see also Fig. 1).

The approach to generate AAV vector particles by transient transfection of cultured cells has become a standard methodology and is most widely used, yet manifold variations of this concept have been tried and proposed throughout the years. A detailed description of these alternative protocols is beyond the scope of this chapter, but two most important improvements should be mentioned. First, advances were made regarding the number of plasmids required to generate the AAV vector particles. For instance, work from our group has shown that it is feasible to express the AAV and adenoviral helper functions from a single-hybrid plasmid, which reduces the number of constructs to be transfected to two (Grimm *et al.*, 2003a). It is even possible to incorporate the AAV vector sequences into the same plasmid and thus produce rAAV particles from a single construct, thus significantly saving time and costs (Grimm and Kay, unpublished).

Second, it was found that the transfection-based protocol for rAAV generation can easily be adapted to produce AAV vector particles that are derived from naturally occurring AAV serotypes other than AAV-2 (Grimm and Kay, 2003). The only component of the previously outlined protocol that needs to be modified in order to produce an AAV vector based on an alternative serotype is the capsid gene within the AAV helper plasmid. In contrast, the *rep* gene in the helper and the ITRs in the vector plasmid remain unchanged. The resulting recombinant particle thus consists of an AAV-2-based vector DNA packaged into a capsid from another serotype and is referred to as an AAV pseudotype; the production process is called cross-packaging (Grimm, 2002). The benefit of this approach is that the resulting rAAV particles display unique tropisms and serological properties that are different from those of AAV-2. This largely expands the range of potential targets for AAV-mediated gene transfer and will help circumvent immunological issues, which should both be important with respect to the use of AAV vectors to deliver shRNAs.

AAV Vectors for shRNA Expression *In Vitro* and *In Vivo*

In this section, we discuss the specific properties of rAAV that make this vector system particularly promising for shRNA expression and then review the recent literature demonstrating successful use of AAV-based vectors for shRNA transfer *in vitro* and *in vivo*.

Theoretical Considerations

In theory, an ideal vector system for delivery of shRNA expression cassettes should fulfil four requirements. It should be (1) efficient, (2) safe, (3) allow stable shRNA expression, and (4) the vector itself should be easy to manufacture. From our experience with rAAV-mediated transgene delivery gained in previous studies, there are reasons to believe that the AAV vector system might meet all these goals.

First, a wealth of data supports the notion that AAV vectors are extremely efficient tools for gene delivery *in vitro* and *in vivo*, as defined by the variety of cell types that can be transduced as well as by the number of recombinant vector genomes that can be introduced into, and maintained in, each target cell.

In particular, AAV serotype 2, which has till now served as the basis for most vectors derived from AAV, has a very broad host range and is able to infect both dividing and nondividing cells, including important therapeutic targets such as hepatocytes or neurons (e.g., Grimm *et al.*, 2003b). rAAV has the specific ability to transduce quiescent cells, and its use for gene transfer provides a clear advantage over using retroviral vectors, which require cell division for efficient transduction.

Moreover, the feasibility to easily cross-package a given AAV vector construct into capsids from other alternative AAV serotypes further expands the host range of this vector system, because the known AAV serotypes are functionally and serologically distinct from each other. For instance, AAV serotype 1 is most efficient for muscle-directed gene transfer and thus is the choice for pseudotyping when targeting this tissue, whereas AAV-8 appears superior among all serotypes for liver transduction (Gao *et al.*, 2002). In addition, beyond using naturally occurring AAV variants, multiple groups have succeeded in genetically altering the AAV-2 capsid through insertion of customized ligand sequences. This resulted in ablation of the natural AAV-2 tropism while allowing the transduction of cell types that had previously been refractory to infection by AAV-2 (e.g., Perabo *et al.*, 2003). The results from all these approaches together make it tempting to speculate that theoretically any given target cell type can be transduced by, and made to express shRNAs from, an AAV vector.

In addition, previous reports also demonstrated that transduction with AAV vectors can result in high copy numbers of the recombinant transgene in the infected cell. For example, the use of AAV-8 to transduce hepatocytes *in vivo* has been reported to lead to individual cells carrying more than 10 vector copies, which were stably maintained over time (Gao *et al.*, 2002). With respect to expression of shRNA, it is currently unclear as to what an optimal number of vector copies per cell will be, as this will certainly depend on specific parameters such as strength of promoter used to drive the shRNA. Nevertheless, it is reasonable to assume that the AAV vector system has the potential to deliver and maintain sufficiently high shRNA copy numbers in the transduced cell.

Last but not least, a high efficiency of the AAV vector system is also provided by the fact that the use of different serotypes or capsid variants permits the readministration of recombinant particles (Grimm and Kay, 2003). This is crucial considering that administration of a particular AAV vector type will result in a humoral immune response, characterized by formation of antibodies against this specific serotype, which will prevent repeated vector delivery. However, switching to a capsid from another serotype will allow to circumvent this immune response, thus enabling readministration of a given AAV vector genome to the same organism. In regards to RNAi, this might provide a useful strategy to deliver multiple different shRNAs to the same host in a timely coordinated manner or it will allow subsequent administration of shRNA cassettes and regulatory elements, such as transcriptional activators or repressors. Theoretically, it will also allow expression of shRNAs from AAV in patients having preexisting immunity against a particular AAV serotype. This is an important option when considering the previously mentioned highly prevalent seropositivity against AAV-2 in the human population, which may pose a major hurdle to the widespread therapeutic application of AAV-2 vectors.

Second, rAAVs are regarded to be among the safest of all known viral-based vector systems. This is partly because the wild-type virus has never been associated with any human malignancy, in contrast to all other viruses that are currently being exploited as vectors, such as adenovirus or lentivirus. A more direct proof of the high safety profile of wild-type and rAAV was recently provided by Stilwell and Samulski (2004), who used DNA microarrays to identify and directly compare genes modulated during AAV or adenovirus infection of cultured cells. Impressively, AAV infection elicited a nonpathogenic response whereas adenovirus infection resulted in induction of immune and stress-response genes associated with pathogenesis. In this respect, it should be clarified that although adenoviral helper functions are needed during rAAV production, the resulting AAV vector stocks are free from any contaminating virus because of the inability of the adenoviral plasmid to replicate.

Moreover, the AAV system is also distinguished from the other viruses under development in that vectors derived from wild-type AAV only retain the ITRs but do not express any viral genes and thus are gutless by design and definition. Because of the lack of viral gene expression, AAV vectors do not cause a cytotoxic cellular immune response in the transduced host. The importance of this property can be most dramatically exemplified by recent adverse effects observed in a patient treated with a second-generation adenovirus that was expressing viral genes, although it has not been fully resolved whether the genetic cargo or the viral capsid itself was responsible for the patient's death (Marshall, 1999).

Furthermore, AAV vectors also appear to display only a very modest frequency of integration into the host genome, which further contributes to the overall safety of this particular vector system. The exact fate of the recombinant genome in the transduced cell will, however, depend on many parameters such as cell type and particle dose. For instance, our group has previously shown that in the transduced liver, only a very small fraction (<10%) of all persisting AAV genomes integrate into the hepatoctye genome, whereas the majority assumes extrachromosomal circular DNA forms (Nakai et al., 2001). The important topic of viral vector integration is currently being widely studied, partly because of the outcome of another recent clinical study. In this case, treatment of patients with a retroviral vector led to integration of the viral genome into the host chromosome, followed by the adverse activation of a protooncogene (Kaiser, 2003). This underscores the importance of using a vector for long-term shRNA transfer that displays only low frequencies of random integration, such as rAAV.

Third, the property of AAV to persist in the infected cell as extrachromosomal forms usually results in stable and strong expression of the encoded transgene. Numerous reports have demonstrated maintained expression of AAV-delivered transgenes for over 18 months in mice (Snyder et al., 1999) or dogs (Mount et al., 2002). For RNAi applications, the feasibility of achieving persistent shRNA expression from a viral vector will be highly desirable, unless in scenarios in which transient expression is specifically intended. For most applications, however, continued expression of the transduced shRNA will be the goal, as exemplified by settings in which RNAi will be used to inhibit oncogenes or genes expressed from pathogenic viruses.

Fourth, another potential benefit of using the AAV vector system to express shRNAs is the great ease with which recombinant particles can be produced in bulk amounts. Thus, a typical medium-scale rAAV preparation that can be performed in each laboratory will typically result in at least 10^{12}

pure recombinant vector particles. (See "Generating shRNA-Expressing AAV Vectors: A Protocol" for details.) It is impossible to exactly predict the amount of AAV vector needed for a specific RNAi application, as this will depend on many parameters, including the choice of an AAV serotype permitting efficient target transduction and the number and accessibility of target cells within the organism. Nevertheless, considering that the transfection method has already been used to manufacture AAV vector for several clinical studies in humans and that upscaling of the protocol to yields of more than 10^{13} total particles is also feasible (Potter *et al.*, 2002), it is reasonable to believe that vector production will not be a limiting factor for the use of AAV for RNAi applications.

Practical Examples

To date, there have been five independent reports on the use of AAV-based vectors for shRNA expression in various target cells *in vitro* and *in vivo* and all highlight the great potential of the approach.

The first study by Tomar *et al.* (2003) was simple in design, but provided the important initial proof of principle that AAV-2 vector particles can be engineered to express shRNA. The authors generated AAV-2 vector plasmids expressing shRNAs directed against p53 or caspase-8, under the control of the H1 or U6 promoter, respectively. Subsequently, they used the particles to infect cultured HeLa cells and demonstrated the expected efficient and dose-dependent knockdown of p53 and caspase 8 proteins by Western blot analyses.

A clinically more relevant study, and also the first report of *in vivo* expression of shRNA from AAV, then came from Hommel *et al.* (2003). This group targeted the dopamine synthesis enzyme tyrosine hydroxylase (Th) by expressing respective shRNAs from AAV-2 vectors in adult mice. Most noteworthy, following stereotaxic injection of the vector into the substantia nigra compacta, Th knockdown was found to be effective and persistent, lasting as long as 50 days. Equally impressive was the finding that the vector-induced reduction of Th protein elicited behavioral defects in the treated mice and created a phenotype reminiscent of rodent models of Parkinson's disease. Thus, this study established the potential of shRNA-expressing AAV vectors to rapidly and easily create spatially restricted gene knockdowns *in vivo* and to test new genetic disease models.

The usefulness of the AAV/shRNA system was further underscored by two independent *in vitro* studies subsequently published by Boden *et al.* (2004) and Pinkenburg *et al.* (2004). The two groups designed AAV-2 vectors to express shRNAs directed against the human immunodeficiency virus type 1 (HIV-1) *tat* gene and the NF-κB p65 subunit, respectively.

Specific and efficient inhibition of HIV-1 replication or interleukin-8 production (as a marker for p65 knockdown) was then demonstrated in cultured primary human lymphocytes and bronchial epithelial cells, respectively.

The latest relevant report by Xia *et al.* (2004) provided the second proof for the efficiency of AAV vectors for expressing shRNA *in vivo*. Similar to the first study by Hommel *et al.* (2003), the vectors were evaluated by using a mouse model of a human dominant neurodegenerative disease, here spinocerebellar ataxia type 1 (SCA1), and vector efficacy was demonstrated by showing reduction in target protein (ataxin-1) expression and by documenting improvements in cellular and behavioral characteristics. An important difference between the two studies was that Xia *et al.* (2004) used a pseudotyping approach to cross-package their shRNA-expressing vector construct into AAV-1 capsids. This enabled them to efficiently target Purkinje cells in the murine cerebellum and to exploit the increased expression kinetics from this particular AAV serotype. Together, this underscores the great promise of pseudotyping shRNA-expressing AAV vectors to achieve targeted and controlled RNAi induction *in vivo*.

Generating Short Hairpin RNA (shRNA)-Expressing
   AAV Vectors: A Protocol

We describe the materials and protocols needed to generate shRNA-expressing vectors through transient transfection of cultured cells. This represents the most widely used approach for rAAV production and was thus chosen to be reported here.

Materials

Cells. The most commonly used cell type for rAAV production are 293 cells, because they are highly transfectable and contain an integrated copy of the adenoviral E1 gene, which enhances gene expression from the AAV helper plasmid. An alternative is provided by using 293T cells, which differ from 293 cells by an additional copy of the SV40 large T antigen. This decreases the doubling time of 293T cells as compared to 293 and also renders them even more susceptible to calcium phosphate-mediated DNA transfection. For either cell type, it is crucial to maintain the cells at a low passage number (<50) to avoid decreases in the efficiency of rAAV production per cell. It is thus advisable to obtain such a low-passage stock, for example, from a commercial source such as Microbix Biosystems (Toronto, Canada) and then store a working cell bank in aliquots in liquid nitrogen. Moreover, it is important to prevent the cells from growing to confluency, which will otherwise negatively affect their health and the ability to take up DNA.

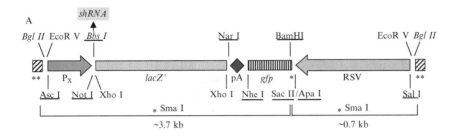

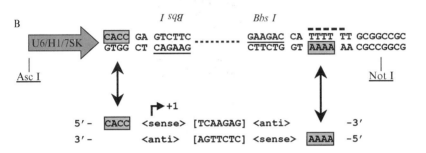

FIG. 2. Novel AAV vector plasmids for short hairpin RNA (shRNA) expression. (A) General vector plasmid scheme. Depicted is our novel AAV vector construct with the AAV-2 ITRs at both ends (hatched boxes), an RSV (Rous sarcoma virus) promoter-driven *gfp* gene (including polyadenylation site; pA), a 2.2-kb stuffer fragment from the *E. coli lacZ* gene, and an RNA polymerase III promoter at the left end (P_x). Three versions of this plasmid were made, carrying the human U6, H1, or 7SK promoter to drive shRNA expression. Also shown are the locations of various restriction enzyme recognition sites within the constructs, with unique sites underlined. The location of the multiple *Sma* I sites is indicated by asterisks, and expected fragments resulting from *Sma* I digestion are shown at the bottom of the scheme. As explained in the text, *Sma* I digestion is useful to confirm ITR integrity after cloning an shRNA insert. Note that one *Sma* I site is located at the start of the *gfp* gene and will be lost on replacement of this gene with customized sequences. The two close *Bbs* I sites (indicated as one) located downstream of the polymerase III promotor serve to insert shRNA sequences in the form of annealed oligonucleotides [see (B)]. All versions of the plasmid express the ampicillin-resistance gene (not shown). (B) Scheme for cloning shRNAs. Shown is a magnified view of the left end of the AAV vector plasmid depicted in (A). As evident, the polymerase III promoter (U6, H1, or 7SK) is followed by two mirrored *Bbs* I sites (the actual recognition sequence is underlined), which are separated by a short spacer region (depicted as a dotted line), and followed by a T_5 terminator sequence (marked by dashed line). Cutting of the plasmid(s) with *Bbs* I leaves two 5' overhangs, which are highlighted by grey boxes. To clone an shRNA into the *Bbs* I-linearized construct, two complementary oligonucleotides are annealed (see text for protocol), comprising a sense and antisense strand separated by a short hairpin spacer. The oligonucleotides must be designed to display *Bbs* I-compatible overhangs following their annealing, also depicted by grey boxes. Following cloning of the annealed oligonucleotides into the AAV vector plasmid, the shRNA transcription start will be located immediately downstream of the 5' *Bbs* I overhang (indicated by +1).

In our laboratory, we grow 293 and 293T cells in Dulbecco's modified Eagle's medium (DMEM, Sigma, St. Louis, MO) supplemented with L-glutamine to a final concentration of 2 mM and containing fetal bovine serum (10% final volume, HyClone, Logan, UT), penicillin-streptomycin, and antibiotic-antimycotic reagent (Gibco-Invitrogen, Carlsbad, CA). For routine culturing, as well as for rAAV production, the cells are plated in T225 flasks (Corning, Fountain Valley, CA) and passaged when reaching approximately 90% confluency.

Plasmids. The three types of plasmids required for rAAV production have been mentioned earlier [see "Adeno-Associated Virus (AAV): From Wild-Type Virus to Recombinant Vectors" and Fig. 1]: (1) an shRNA-encoding AAV vector plasmid; (2) an AAV helper plasmid expressing the capsid gene from the desired AAV serotype, together with the AAV-2 *rep* gene; and (3) an adenoviral helper plasmid. In the next section, we provide guidelines and a protocol to generate the shRNA-expressing vector plasmid. For information on the two helper constructs, we refer the reader to a review article that provides respective background information and, importantly, lists sources for such plasmids (Grimm, 2002).

In our hands, we found it sufficient to purify all three types of plasmids from the bacterial lysate by using purification kits available from Qiagen (Valencia, CA); the use of cesium chloride (CsCl) gradients or special endotoxin-free protocols is superfluous. We typically grow the plasmid-containing bacteria (DH10B, Invitrogen) on a 2.5-L scale and use the Plasmid Giga Kit (Qiagen) for purification, which yields up to 10 mg plasmid DNA. Considering that a vector preparation requires 1.25 mg of each plasmid (see later), this yield is sufficient for up to eight production runs.

Moreover, we routinely check the integrity of the AAV-2 ITRs in our vector plasmids by restriction digest analyses, which is indicated as the sequences are highly prone to recombination and subsequent deletion. A useful enzyme for this purpose is *Sma* I, which cuts twice within one arm of each AAV-2 ITR hairpin. Most often, this particular region is lost by deletion in one of the two ITRs, which is then readily detected by analyzing the *Sma* I digestion pattern (compare Fig. 2A). In contrast, direct sequencing of the AAV-2 ITRs as an alternative method is prevented by the high degree of secondary structure of the sequences, thus making the method unfeasible.

Method

In this section, we provide detailed protocols for the four basic steps required to generate shRNA-expressing AAV vector particles: (1) initial cloning of a recombinant shRNA-encoding AAV plasmid, (2) subsequent packaging of this construct into AAV capsids by transfection of cells,

(3) vector particle purification, and (4) titration. The protocols are very basic and not specific for any particular AAV serotype; the reader is referred to the literature for more information on alternative approaches (Grimm, 2002; Grimm and Kay, 2003). The flowchart in Fig. 3 summarizes the rAAV production steps and shows an approximate time frame.

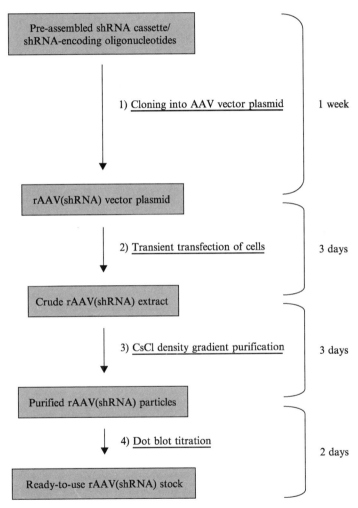

FIG. 3. Time frame for generation of shRNA-expressing AAV vector particles. Shown are the four different successive steps required to produce shRNA-encoding AAV particles, from cloning an shRNA cassette into an AAV vector plasmid to titrating the generated particles. Also shown are estimated times required for each step; the protocols described here take approximately 2 weeks to complete.

Step 1: Cloning shRNA-Encoding AAV Vector Plasmids. The following considerations are important.

SIZE. The generation of AAV particles expressing shRNAs follows the basic rules and protocols that have been established for conventional transgenes, using typical vector and helper plasmids. The only, yet very crucial, difference is that shRNA expression cassettes are typically significantly smaller than conventional inserts, comprising genes transcribed from RNA polymerase II promoters. Considering that an RNA polymerase III promoter such as H1 only spans approximately 100 bp and that the respective terminator sequence solely consists of a stretch of five thymidines, the total size of the entire shRNA expression cassette can be less than 150 bp. This creates the paradox that AAV, which has often been criticized as being too small to accommodate clinically relevant genes such as *cftr* (cystic fibrosis transmembrane conductance regulator gene), is suddenly much too large and requires a stuffer DNA to be added to the shRNA cassette to guarantee efficient vector genome encapsidation. This thought is based on a previous study by Dong *et al.* (1996), who quantitatively analyzed the DNA-packaging capacity of AAV-2 capsids and observed optimal encapsidation efficacies when the vector genomes had sizes between 4.1 and 4.9 kb (including the ITRs). Accordingly, the goal for cloning an shRNA-expressing AAV-2 vector plasmid should be to bring the size of the recombinant cassette up to a total of at least 3.8 kb (4.1 kb minus 2×145 bp for both ITRs), but not more than 4.6 kb.

Theoretically, any DNA sequence can be used as stuffer, but one particularly beneficial approach that has been reported repeatedly (e.g., Xia *et al.*, 2004) is to clone a second cassette expressing a fluorescence reporter gene under the control of an RNA polymerase II promoter. This provides the advantage that successfully transduced cells, presumably expressing the shRNA of choice, can be easily and rapidly monitored and quantitated. However, taking into account the relatively small size of typical fluorescence reporter genes of about 700 bp, generation of an AAV-2 vector plasmid comprising a total of at least 3.8 kb DNA might still require adding further sequences. We have devised a particular solution to this hurdle, which is discussed in the following section.

CHOICE OF VECTOR PLASMID. In principle, cloning an shRNA expression cassette into an AAV vector plasmid is a simple and efficient process. This is because most commonly available AAV vector constructs such as pTRUF3 (Zolotukhin *et al.*, 1996) typically contain several unique restriction enzyme recognition sites, allowing the straightforward insertion of foreign DNA, such as preassembled entire shRNA expression cassettes. This is achieved by using standard molecular techniques, which are not further described here. The plasmids are usually relatively small (<8 kb), which additionally facilitates their handling and the cloning into the constructs. However, as mentioned previously, the size constraints of AAV capsids will likely require the addition of further sequences to the shRNA cassette, which can require more sophisticated strategies and thus complicate the cloning process.

To overcome this hurdle, we have generated a basic AAV-2/RNAi vector plasmid (Fig. 2A). This construct contains a 2.2-kb stuffer fragment from the *E. coli lacZ* gene as well as a 1.7-kb green fluorescence reporter gene (*gfp*) cassette, together comprising 3.9 kb. This is optimal for the insertion of very small shRNA expression cassettes such as the one mentioned previously with only 150 bp, but also leaves sufficient space for accomodating larger inserts of up to 700 bp without exceeding the AAV packaging limit. We have generated three alternative versions of this plasmid, which contain three different commonly used RNA polymerase III promoters to drive an shRNA, namely, U6, H1, and 7SK. Each plasmid version was engineered to contain unique restriction enzyme recognition sites, allowing the easy and straightforward insertion of customized shRNA sequences in the form of annealed oligonucleotides (see the protocol in this section). Likewise, these sites permit the replacement of any of the plasmid components, that is, the polymerase III promoter and the terminator sequence, the *lacZ* stuffer fragment, the fluorescence marker gene with its promoter, or even the entire recombinant insert (shRNA, *gfp*, and *lacZ*).

This particular plasmid design was chosen to provide maximum versatility and convenience by allowing users to rapidly clone shRNAs into AAV and express them under different RNA polymerase III promoters without having to worry about AAV size constraints. The constructs should be equally attractive for users who want to customize their shRNA-expressing AAV vector plasmid and who specifically intend to provide alternative marker genes or stuffer DNA. Further details of the plasmids and instructions for the insertion of shRNAs in the form of oligonucleotides are given in Fig. 2B.

Thus far, we have successfully exploited our constructs to clone shRNAs comprising 19- to 29-nt-long stem sequences. The respective oligonucleotides were purchased from IDT (Coralville, IA) at the lowest available amounts and degree of purity (standard desalting). Further purification, for example, by polyacrylamide gel electrophoresis, is often recommended for oligonucleotides longer than 50 nt (i.e., comprising stem sequences >21 nt: 2 × 21 plus the length of the hairpin and the Bbs I overhang), although we have not found this beneficial.

PROTOCOL. For annealing, the two oligonucleotides (1 μg each) are mixed in reaction buffer (50 mM NaCl, 10 mM Tris-HCl, 10 mM MgCl$_2$,

1 mM DTT, pH 7.9), heated to 95° for 5 min, and then allowed to cool down to room temperature for 30 min. In a subsequent ligation reaction, 2 μl of the annealed oligonucleotides are incubated with 50–100 ng of *Bbs* I-linearized AAV vector plasmid, using standard reaction conditions and ligation enzyme (e.g., T4 DNA ligase, NEB, Beverly, MA). The resulting recombinant shRNA-encoding AAV vector plasmids are selected and grown in DH10B bacteria, using ampicillin-containing LB medium.

Step 2: Producing shRNA-Expressing AAV Vector Particles by Transfection. The values given refer to transfection of 50 T225 flasks (Corning), which corresponds to approximately 1 × 10⁹ cells (when grown to near confluency) and typically results in production of at least 10¹² total AAV particles containing the recombinant shRNA sequence. The yields can vary dramatically (by more than one order of magnitude), depending on the particular AAV serotype being produced. In our hands, best yields are obtained with AAV serotype 8, followed by AAV-5, with both typically giving at least twofold higher particle titers than AAV-2.

PROTOCOL. To obtain 50 flasks for transfection, low-passage cells are first seeded into one T225 flask and grown until they reach confluency. The cells are then split and seeded into seven flasks, using standard cell culture technique, and then finally split one more time into 50 flasks. Therefore, old media from the seven flasks are discarded and cells in each flask are washed with 10 ml phosphate-buffered saline (PBS), followed by a 2-min incubation with 8 ml prewarmed 0.05% trypsin/EDTA solution. The flasks are then knocked to dislodge cells, and trypsin is neutralized by adding 8 ml DMEM per flask. The cells are resuspended, collected, and counted by a hemocytometer. Per new flask, 2.8 × 10⁶ cells are plated and then incubated at 37° for 3 days until transfection (cells about 80% confluent at this point).

For transfection of all 50 flasks, a total plasmid DNA amount of 3.75 mg is used. Depending on the particular type of helper plasmids, this can be 1.25 mg each of an AAV and the adenoviral helper plasmid, as well as 1.25 mg of the shRNA-encoding AAV vector plasmid. Alternatively, if a combined AAV/Ad helper is used (Grimm *et al.*, 2003a), the vector plasmid is mixed with 2.5 mg of the hybrid helper construct. The plasmids are combined in a 500-ml disposable conical tube, followed by addition of 204 ml 0.3 M CaCl$_2$ and 204 ml 2× Hepes-buffered saline (HBS; 280 mM NaCl, 50 mM HEPES, 1.5 mM Na$_2$HPO$_4$, pH 7.1). Following vigorous shaking, 8 ml of the mixture (which will be slightly opaque) is aliquoted into each of the 50 flasks. Immediately before this step, 25 ml of the old medium in each flask is replaced with 25 ml fresh DMEM. The cells are then incubated with the DNA–HBS mixture for 5 h before the medium is aspirated and replaced with 40 ml fresh serum-free DMEM.

Following a 2-day incubation, the cells containing the shRNA-encoding rAAV particles are harvested by adding 0.5 ml of 0.5 M EDTA solution to each flask. After a 15-min incubation at room temperature, the flasks are knocked against a hard surface to lift cells off the surface. The medium containing the dislodged cells is then decanted into 500-ml conical tubes (cells from 10–12 flasks pooled into one tube), and cells are collected by a 15-min spin at 3500 rpm at room temperature. The resulting supernatants are aspirated and each cell pellet is resuspended in 30 ml of 50 mM Tris, 2 mM MgCl$_2$, pH 8.5. To release the AAV particles, the cells are then lysed by three freeze–thaw cycles (10 min at 37°, followed by 10 min in dry ice/ethanol bath) and subsequently incubated for 1 h with benzonase (endonuclease, Merck, Darmstadt, Germany) at 200 U/ml, to remove cellular genomic DNA. Following a 15-min spin at 3500 rpm, the supernatant is taken and diluted with 1 M CaCl$_2$ to a final concentration of 25 mM. This mixture is incubated on ice for 1 h and then centrifuged for 15 min at 9200 rpm in an SS34 rotor (Sorvall, Asheville, NC), before the resulting supernatant is transferred into a 50-ml tube. The total volume is estimated and one fourth of 40% PEG8000, 2.5 M NaCl solution is added to give a final concentration of 8% PEG8000. The mixture is incubated on ice for at least 3 h and then spun at 3500 rpm for 30 min. The pelleted rAAV particles are finally resuspended in 10 ml of 50 mM HEPES, 150 mM NaCl, 25 mM EDTA, pH 7.4; typically an overnight incubation is required to ensure that the viruses are completely dissolved.

Step 3: rAAV Particle Purification by Cesium Chloride (CsCl) Gradient Density Centrifugation. rAAV purification by CsCl gradient density centrifugation is a relatively time-consuming and cumbersome method as compared with newer protocols, such as those based on affinity chromatography. However, to our knowledge, it is the only available methodology that is applicable to all serotypes and capsid variants of AAV, which is why it was chosen to be reported here.

PROTOCOL. The solution containing the resuspended rAAV particles is vortexed and then centrifuged for 30 min at 3500 rpm. The supernatant is transferred into a new 50-ml tube, and the volume is adjusted to 24 ml with HEPES resuspension buffer. Following addition of 13.2 g CsCl, the refractive index (RI) of the solution is determined by a refractometer. The solution's RI is adjusted to 1.3710 by adding CsCl or resuspension buffer, and the solution is transferred into a 32.4-ml OptiSeal™ tube (Beckman, Fullerton, CA) and spun for 23 h in an ultracentrifuge at 45,000 rpm and 21°, using a 70Ti rotor (Beckman).

Subsequently, 0.5- to 1-ml fractions are collected through a hole poked into the bottom of the centrifuge tube and the RIs of each fraction are determined. Fractions with an RI between 1.3711 and 1.3766 are pooled to a total volume of about 7 ml, and the RI of this final solution is again adjusted to 1.3710 with resuspension buffer. A second ultracentrifugation step is then carried out under the following conditions: 11.2-ml tubes and NVT65 rotor (both Beckman), 65,000 rpm, 21°, 5 h. Fractions are collected and those with an RI between 1.3711 and 1.3766 are pooled (total volume about 2.5 ml). The resulting rAAV-containing solution should then be instantly processed to the next purification step (see later); prolonged storage will result in a loss of particle infectivity.

To remove the CsCl and other contaminants still present, the rAAV preparation is finally subjected to a combined ultrafiltration–diafiltration process. Therefore, it is pumped through hollow fiber filters with a molecular cut-off of 100,000 Da (Amersham, Piscataway, NJ). The filtration is performed according to the manufacturer's instructions, using a constant back pressure of 14 psi and a pump speed of 385 rpm. Once the virus solution has been pumped into the filter unit, the cartridge is washed twice with PBS containing 5% sorbitol before the rAAV particles are eluted by operating the pump backward. The purified virus solution (volume 3–4 ml) is finally sterilized by filtration through a 0.22-μm filter.

Step 4: rAAV Particle Titration by Dot Blot. The protocol provided here will return a titer of shRNA-encoding AAV particles, which describes AAV virions containing the shRNA expression cassette. However, this titer does not necessarily correlate with the number of infectious particles, that is, virions that on transduction will actually express the shRNA. A more detailed discussion of this problem is beyond the scope of this chapter, but more information on the general correlation of DNA-containing infectious and total assembled rAAV particles can be found elsewhere (Grimm *et al.*, 1999).

PROTOCOL. To release viral DNA for quantification, the rAAV particles are first incubated with 5 U DNase I (Roche, Palo Alto, CA) in 100 μl of 20 mM Tris, 1 mM MgCl$_2$, pH 8.0, for 1 h at 37° to remove contaminating cellular DNA or unpackaged viral DNA that might still be present. Typically, several aliquots of the virus solution are digested, ranging from 0.5 to 0.05 μl. (For lower volumes the virus is prediluted in PBS.) The DNase digestion is followed by a 1-h incubation with proteinase K to break the viral capsids and release the encapsidated DNA. Therefore, 20 μl 10× proteinase K buffer (100 mM Tris, 100 mM EDTA, 5% SDS, pH 8.0) are added with 10 μl proteinase K (Invitrogen) and 70 μl H$_2$O. The mixture is vortexed and incubated at 55° for 1 h. The viral DNA is then purified by a DNA extractor kit (Wako, Japan), following the manufacturer's instructions, and is finally resuspended in 200 μl TE buffer.

To make a standard DNA dilution curve, 1 μg of the AAV vector plasmid used in the initial transfection is linearized by using a restriction enzyme that is unique for the particular construct. (See Fig. 2A for examples.) This is to account for the facts that the viral DNA is also linear and that linear and circular (i.e., uncut plasmid) DNA might bind to membranes (see later) with different efficiencies. The entire reaction is then diluted to a final DNA concentration of 5 ng/ml with TE buffer and subsequently further twofold diluted to 0.078 ng/ml (all dilutions in a total volume of 200 μl).

Viral DNA and standard samples are then denatured by adding 200 μl 2× alkaline solution (0.8 M NaOH, 20 mM EDTA), followed by vortexing and incubation at room temperature for 10 min. During the incubation, a 9 × 12 cm^2 Zeta-Probe membrane (Bio-Rad, Hercules, CA) is wetted in H$_2$O and assembled in a dot blot apparatus (Bio-Rad), following the manufacturer's instructions. The membrane is then equilibrated by pipetting 400 μl 1× alkaline solution into each well and applying vacuum until the wells are empty. The vacuum should be adjusted so that it takes between 30 and 60 sec for the wells to empty. The standard and the samples are added to the wells with the vacuum turned off and incubated on the membrane for 5 min. The vacuum is then applied again until the wells are empty, and the filter is finally washed with 400 μl 1× alkaline buffer per well. The apparatus is disassembled, the membrane is rinsed in 2× SSC and air dried on a filter paper, and the viral or standard DNA is cross-linked by using a Gene linker apparatus (program C1, Bio-Rad).

For detection and quantification of viral and standard DNAs, the membrane is then hybridized with a ^{32}P-labeled probe corresponding to a fragment of the rAAV vector plasmid (isolated by standard molecular techniques). It is first prehybridized for 30 min with 10 ml of 7% SDS, 0.25 M Na$_2$HPO$_4$, 1 mM EDTA, pH 7.2, by incubating at 65°. Then, 25–50 ng random-primed (Prime-It® Kit, Stratagene, La Jolla, CA) probe with a specific activity of 10^8–10^9 cpm/μg is denatured by heating to 95° for 5 min and then added to the membrane in 10 ml of fresh hybridization buffer. The membrane is then rotated with the probe overnight at 65° and the next day washed two times with 5% SDS, 40 mM Na$_2$HPO$_4$, 1 mM EDTA (20 min each), followed by two washes in 1% SDS, 40 mM Na$_2$HPO$_4$, 1 mM EDTA (20 min each). It is finally briefly rinsed in 2× SSC, air dried, and then exposed in a phospho imager cassette (Kodak, Rochester, NY). The image is captured and processed by using a Molecular Imager FX apparatus with respective software (Bio-Rad). A standard curve is generated, which then serves to calculate the rAAV titer, taking into account the initial sample dilutions.

Conclusions

The field of RNAi applications in mammalian systems *in vitro* and *in vivo* is still relatively young and in its early stages and so is the use of viral vectors for delivery and expression of shRNAs. It is, however, already becoming clear from the recent wealth of related publications that three classes of viruses—adenoviruses, retro/lentiviruses, and AAV—are emerging as most promising bases for vector development.

In particular, AAV, the focus of the present article, might represent an optimal bona fide system for delivery of shRNA expression cassettes, as AAV fulfils all the theoretical requirements of an ideal shRNA transfer system: it is generally very effective as a vehicle for transgene delivery *in vitro* and *in vivo*, as exemplified by research from the past 20 years; it is regarded as the safest of all known viral vector systems, with wild-type and recombinant AAV believed to be nonpathogenic, and rAAV particles by design being gutless and replication deficient; it has an enormously broad host range, including quiescent and dividing cells, with perhaps virtually every cell type being targetable because of the feasibilities to pseudotype and customize AAV genomes and capsids; it typically results in strong and persistent transgene expression in the infected cell, with a much lower risk of insertional mutagenesis than particularly retroviral vectors; and it can be manufactured easily and efficiently, allowing production of more than 10^{12} recombinant virus particles within 2 weeks.

Taking all these potential benefits of AAV as a vector for shRNA expression together, it is certain that this extremely promising system will be exploited, studied, and applied to a much greater extent in the near future. Our hope is that with the set of novel AAV vector plasmids we have presented here, which allow the rapid and easy cloning of customized shRNA sequences, we can help accelerate the pace of this exciting process.

Acknowledgments

This work was supported by grant HL66948 from the National Institutes of Health (NIH).

References

Berns, K., Hijmans, E. M., Mullenders, J., Brummelkamp, T. R., Velds, A., Heimerikx, M., Kerkhoven, R. M., Madiredjo, M., Nijkamp, W., Weigelt, B., Agami, R., Ge, W., Cavet, G., Linsley, P. S., Beijersbergen, R. L., and Bernards, R. (2004). A large-scale RNAi screen in human cells identifies new components of the p53 pathway. *Nature* **428**, 431–437.
Berns, K. I., and Linden, R. M. (1995). The cryptic life style of adeno-associated virus. *Bioessays* **17**, 237–245.

Boden, D., Pusch, O., Lee, F., Tucker, L., and Ramratnam, B. (2004). Efficient gene transfer of HIV-1-specific short hairpin RNA into human lymphocytic cells using recombinant adeno-associated virus vectors. *Mol. Ther.* **9,** 396–402.

Ding, Y., Chan, C. Y., and Lawrence, C. E. (2004). Sfold web server for statistical folding and rational design of nucleic acids *Nucleic Acids Res.* **32,** 135–141.

Doench, J. G., Petersen, C. P., and Sharp, P. A. (2003). siRNAs can function as miRNAs. *Genes Dev.* **17,** 438–442.

Dong, J. Y., Fan, P. D., and Frizzell, R. A. (1996). Quantitative analysis of the packaging capacity of recombinant adeno-associated virus. *Hum. Gene Ther.* **7,** 2101–2112.

Elbashir, S. M., Harborth, J., Lendeckel, W., Yalcin, A., Weber, K., and Tuschl, T. (2001). Duplexes of 21-nucleotide RNAs mediate RNA interference in cultured mammalian cells. *Nature* **411,** 494–498.

Erles, K., Sebokova, P., and Schlehofer, J. R. (1999). Update on the prevalence of serum antibodies (IgG and IgM) to adeno-associated virus (AAV). *J. Med. Virol.* **59,** 406–411.

Friedrich, I., Shir, A., Klein, S., and Levitzki, A. (2004). RNA molecules as anti-cancer agents. *Semin. Cancer Biol.* **14,** 223–230.

Gao, G. P., Alvira, M. R., Wang, L., Calcedo, R., Johnston, J., and Wilson, J. M. (2002). Novel adeno-associated viruses from rhesus monkeys as vectors for human gene therapy. *Proc. Natl. Acad. Sci. USA* **99,** 11854–11859.

Gao, G., Vandenberghe, L. H., Alvira, M. R., Lu, Y., Calcedo, R., Zhou, X., and Wilson, J. M. (2004). Clades of adeno-associated viruses are widely disseminated in human tissues. *J. Virol.* **78,** 6381–6388.

Grimm, D. (2002). Production methods for gene transfer vectors based on adeno-associated virus serotypes. *Methods* **28,** 146–157.

Grimm, D., and Kay, M. A. (2003). From virus evolution to vector revolution: Use of naturally occurring serotypes of adeno-associated virus (AAV) as novel vectors for human gene therapy. *Curr. Gene Ther.* **3,** 281–304.

Grimm, D., Kay, M. A., and Kleinschmidt, J. A. (2003a). Helper virus-free, optically controllable, and two-plasmid-based production of adeno-associated virus vectors of serotypes 1 to 6. *Mol. Ther.* **7,** 839–850.

Grimm, D., Kern, A., Pawlita, M., Ferrari, F., Samulski, R., and Kleinschmidt, J. (1999). Titration of AAV-2 particles via a novel capsid ELISA: Packaging of genomes can limit production of recombinant AAV-2. *Gene Ther.* **6,** 1322–1330.

Grimm, D., and Kleinschmidt, J. A. (1999). Progress in adeno-associated virus type 2 vector production: Promises and prospects for clinical use. *Hum. Gene Ther.* **10,** 2445–2450.

Grimm, D., Zhou, S., Nakai, H., Thomas, C. E., Storm, T. A., Fuess, S., Matsushita, T., Allen, J., Surosky, R., Lochrie, M., Meuse, L., McClelland, A., Colosi, P., and Kay, M. A. (2003b). Preclinical *in vivo* evaluation of pseudotyped adeno-associated virus vectors for liver gene therapy. *Blood* **102,** 2412–2419.

Hommel, J. D., Sears, R. M., Georgescu, D., Simmons, D. L., and DiLeone, R. J. (2003). Local gene knockdown in the brain using viral-mediated RNA interference. *Nat. Med.* **9,** 1539–1544.

Jackson, A. L., Bartz, S. R., Schelter, J., Kobayashi, S. V., Burchard, J., Mao, M., Li, B., Cavet, G., and Linsley, P. S. (2003). Expression profiling reveals off-target gene regulation by RNAi. *Nat. Biotechnol.* **21,** 635–637.

Kaiser, J. (2003). Gene therapy. Seeking the cause of induced leukemias in X-SCID trial. *Science* **299,** 495.

Lee, N. S., and Rossi, J. J. (2004). Control of HIV-1 replication by RNA interference. *Virus Res.* **102,** 53–58.

Luo, B., Heard, A. D., and Lodish, H. F. (2004). Small interfering RNA production by enzymatic engineering of DNA (SPEED). *Proc. Natl. Acad. Sci. USA* **101,** 5494–5499.

Marshall, E. (1999). Gene therapy death prompts review of adenovirus vector. *Science* **286,** 2244–2245.

McCaffrey, A. P., Nakai, H., Pandey, K., Huang, Z., Salazar, F. H., Xu, H., Wieland, S. F., Marion, P. L., and Kay, M. A. (2003). Inhibition of hepatitis B virus in mice by RNA interference. *Nat. Biotechnol.* **21,** 639–644.

Miller, V. M., Xia, H., Marrs, G. L., Gouvion, C. M., Lee, G., Davidson, B. L., and Paulson, H. L. (2003). Allele-specific silencing of dominant disease genes. *Proc. Natl. Acad. Sci. USA* **100,** 7195–7200.

Mount, J. D., Herzog, R. W., Tillson, D. M., Goodman, S. A., Robinson, N., McCleland, M. L., Bellinger, D., Nichols, T. C., Arruda, V. R., Lothrop, C. D., Jr., and High, K. A. (2002). Sustained phenotypic correction of hemophilia B dogs with a factor IX null mutation by liver-directed gene therapy. *Blood* **99,** 2670–2676.

Nakai, H., Yant, S. R., Storm, T. A., Fuess, S., Meuse, L., and Kay, M. A. (2001). Extrachromosomal recombinant adeno-associated virus vector genomes are primarily responsible for stable liver transduction *in vivo. J. Virol.* **75,** 6969–6976.

Nicklin, S. A., and Baker, A. H. (2002). Tropism-modified adenoviral and adeno-associated viral vectors for gene therapy. *Curr. Gene Ther.* **2,** 273–293.

Paddison, P. J., Caudy, A. A., Bernstein, E., Hannon, G. J., and Conklin, D. S. (2001). Short hairpin RNAs (shRNAs) induce sequence-specific silencing in mammalian cells. *Genes Dev.* **16,** 948–958.

Paddison, P. J., Silva, J. M., Conklin, D. S., Schlabach, M., Li, M., Aruleba, S., Balija, V., O'Shaughnessy, A., Gnoj, L., Scobie, K., Chang, K., Westbrook, T., Cleary, M., Sachidanandam, R., McCombie, W. R., Elledge, S. J., and Hannon, G. J. (2004). A resource for large-scale RNA-interference-based screens in mammals. *Nature* **428,** 427–431.

Perabo, L., Buning, H., Kofler, D. M., Ried, M. U., Girod, A., Wendtner, C. M., Enssle, J., and Hallek, M. (2003). *In vitro* selection of viral vectors with modified tropism: The adeno-associated virus display. *Mol. Ther.* **8,** 151–157.

Pinkenburg, O., Platz, J., Beisswenger, C., Vogelmeier, C., and Bals, R. (2004). Inhibition of NF-kappaB mediated inflammation by siRNA expressed by recombinant adeno-associated virus. *J. Virol. Methods* **120,** 119–122.

Potter, M., Chesnut, K., Muzyczka, N., Flotte, T., and Zolotukhin, S. (2002). Streamlined large-scale production of recombinant adeno-associated virus (rAAV) vectors. *Methods Enzymol.* **346,** 413–430.

Reynolds, A., Leake, D., Boese, Q., Scaringe, S., Marshall, W. S., and Khvorova, A. (2004). Rational siRNA design for RNA interference. *Nat. Biotechnol.* **22,** 326–330.

Sen, G., Wehrman, T. S., Myers, J. W., and Blau, H. M. (2004). Restriction enzyme-generated siRNA (REGS) vectors and libraries. *Nat. Genet.* **36,** 183–189.

Shirane, D., Sugao, K., Namiki, S., Tanabe, M., Iino, M., and Hirose, K. (2004). Enzymatic production of RNAi libraries from cDNAs. *Nat. Genet.* **36,** 190–196.

Snyder, R. O., Miao, C., Meuse, L., Tubb, J., Donahue, B. A., Lin, H. F., Stafford, D. W., Patel, S., Thompson, A. R., Nichols, T., Read, M. S., Bellinger, D. A., Brinkhous, K. M., and Kay, M. A. (1999). Correction of hemophilia B in canine and murine models using recombinant adeno-associated viral vectors. *Nat. Med.* **5,** 64–70.

Stilwell, J. L., and Samulski, R. J. (2004). Role of viral vectors and virion shells in cellular gene expression. *Mol. Ther.* **9,** 337–346.

Tomar, R. S., Matta, H., and Chaudhary, P. M. (2003). Use of adeno-associated viral vector for delivery of small interfering RNA. *Oncogene* **22,** 5712–5715.

Weindler, F. W., and Heilbronn, R. (1991). A subset of herpes simplex virus replication genes provides helper functions for productive adeno-associated virus replication. *J. Virol.* **65**, 2476–2483.

Xia, H., Mao, Q., Eliason, S. L., Harper, S. Q., Martins, I. H., Orr, H. T., Paulson, H. L., Yang, L., Kotin, R. M., and Davidson, B. L. (2004). RNAi suppresses polyglutamine-induced neurodegeneration in a model of spinocerebellar ataxia. *Nat. Med.* **10**(8), 816–820.

Zolotukhin, S., Potter, M., Hauswirth, W. W., Guy, J., and Muzyczka, N. (1996). A "humanized" green fluorescent protein cDNA adapted for high-level expression in mammalian cells. *J. Virol.* **70**, 4646–4654.

[24] Simple, Robust Strategies for Generating DNA-Directed RNA Interference Constructs

By ROBERT R. RICE, ANDREW N. MUIRHEAD, BRUCE T. HARRISON, ANDREW J. KASSIANOS, PETRA L. SEDLAK, NARELLE J. MAUGERI, PETER J. GOSS, JONATHAN R. DAVEY, DAVID E. JAMES, and MICHAEL W. GRAHAM

Abstract

We describe two complementary strategies for preparing DNA-directed RNA interference (ddRNAi) constructs designed to express hprRNA. The first, oligonucleotide assembly (OA), uses a very simple annealing protocol to combine up to 20 short nucleotides. These are then cloned into appropriately designed restriction sites in expression vectors. OA can be used to prepare simple hairpin (hp)-expressing constructs, but we prefer to use the approach to generate longer constructs. The second strategy, long-range cloning (LRC), uses a novel adaptation of long-range PCR protocols. For LRC, entire vectors are amplified with primers that serve to introduce short sequences into plasmids at defined anchor sites during PCR. The LCR strategy has proven highly reliable in our hands for generating simple ddRNAi constructs. Moreover, LCR is likely to prove useful in many situations in which conventional cloning strategies might prove problematic. In combination, OA and LRC can greatly simplify the design and generation of many expression constructs, including constructs for ddRNAi.

Introduction

RNA interference (RNAi) is rapidly becoming a commonly used research tool (Berns *et al.*, 2004; Carmell and Hannon, 2004; Scherer and Rossi, 2003); moreover, the therapeutic potential of the technology is creating considerable excitement (Dave and Pomerantz, 2003; Michienzi et al., 2003; Milhavet et al., 2003). The use of DNA-directed RNAi (ddRNAi) constructs to inactivate target genes has several significant advantages over the use of siRNAs, and this approach has probably already become the most widely used strategy for studying gene inactivation. ddRNAi also offers advantages for therapeutic applications. A variety of approaches have been described for preparing ddRNAi constructs, most involving annealing of two oligonucleotides encoding a potential hpRNA transcript. Usually the oligonucleotides are designed to leave single-stranded DNA overhangs, which can then be cloned into appropriate restriction sites in an expression vector (Brummelkamp et al., 2004).

In this chapter, we describe two complementary strategies for generating ddRNAi constructs. This work was driven by our requirements for rapid, robust strategies for generating such constructs and both procedures have proven highly reliable in our hands.

The first strategy uses a primer shuffling approach that can be readily adapted to preparing constructs by using conventional restriction enzyme based cloning procedures. We refer to this approach as the oligonucleotide assembly (OA) strategy. OA uses short, desalted, nonphosphorylated oligonucleotides, which are annealed as a mixture by using a very simple protocol. The design of such oligonucleotides is quite simple and the OA strategy can be used to reliably prepare ddRNAi constructs. In our hands, the procedure has proven particularly effective for preparing constructs encoding hpRNAs more than approximately 100 nt in length.

The second strategy is novel and is our favored approach for preparing ddRNAi constructs expressing shorter hpRNAs. This procedure uses a modified long-range PCR protocol to amplify entire vectors. The amplification primers are designed to encode hpDNA inserts, so that when amplified products are circularized inserts are incorporated into the vector in a single step. We refer to this strategy as long-range cloning (LRC). LRC greatly reduces or even circumvents the requirement for restriction sites in preparing ddRNAi constructs, and the approach can be readily used to generate precise seamless constructs. LRC is a versatile strategy and can be used in a number of instances in which conventional cloning might prove problematic.

In combination, the two strategies can greatly simplify the design and preparation of ddRNAi constructs. For example, novel restriction enzyme sites can be easily introduced into virtually any vector at any position by LRC. Such vectors can then be used either as substrates for LRC or with more conventional restriction enzyme-based strategies such as OA.

Oligonucleotide Assembly (OA) Strategy for Preparing DNA-Directed
RNA Interference (ddRNAi) Constructs

To express hpRNA transcripts, a ddRNAi expression vector, pU6.cass,
was prepared. A map and partial sequence of this construct are shown in
Fig. 1A. pU6.cass was designed to allow cloning of hairpin inserts as precise
fusions at the +1 position of the human U6 promoter (Paul *et al.*, 2002).
This was achieved by using the asymmetric restriction enzyme *Bsm*BI. A
number of similar enzymes that cleave outside their recognition site can
also be used to produce similarly precise fusions.

An example of a simple construct prepared using OA is shown in Fig. 1B.
In this instance, four oligonucleotides (Fig. 1C) were designed and annealed
to produce the structure shown in Fig. 1B. This insert was designed to
inactivate β-actin (Harborth *et al.*, 2001). The assembled oligonucleotides
were then cloned into *Bsm*BI/*Hin*dIII digested pU6.cass to produce
pU6.ACTB hp.

Our protocol for OA has developed with experience. Initially, oligonucleotides were assembled in pairs; such pairs were then annealed until the
entire set was assembled. The protocol described next is a greatly simplified
version of the original strategy. It can probably be further optimized, but
this particular procedure has proven quite robust.

Freeze-dried, desalted oligonucleotides (Sigma, St. Louis, MO) were
resuspended in Buffer EB (Qiagen, Valencia, CA; 10 mM Tris-HCl,
pH 8.5) at a final concentration of 1 μg/μl. One microliter of each oligonucleotide was combined in a 0.2-ml PCR tube to give a final volume of
100 μl in 0.5$\times$ Restriction Buffer M (Roche, Pao Alto, CA; 10$\times$ Restriction Buffer M is 100 mM Tris-HCl, pH 7.5, 100 mM MgCl$_2$, 500 mM NaCl,
10 mM DTE). The oligonucleotide mixture was denatured at 95° for 5 min
and then annealing allowed to proceed by cooling to 30° at a rate of 1°/min,
using a programmable PCR machine (Gradient Palm Cycler™, Corbett
Research, Westborough, MA; Model GP-001).

We routinely clone OA products into dephosphorylated vectors. This
necessitates phosphorylation of the annealed oligonucleotides before ligation. Twenty microliters of annealed oligonucleotides was treated with 2 μl
of T4 polynucleotide kinase (Promega, Madison, WI; 10 U/μl) in a 40-μl
reaction in 1$\times$ T4 ligase buffer (Promega; 10$\times$ T4 DNA ligase buffer is
300 mM Tris-HCl, pH 7.8, 100 mM MgCl$_2$, 100 mM DTT, 10 mM ATP;
this was done for convenience as the buffer contains ATP). The reaction
was incubated at 37° for 30 min and then purified by applying the entire
reaction to a QIAquick™ PCR purification column (Qiagen) according to
the manufacturer's protocol. The annealed oligonucleotides were eluted by
using 30 μl EB and used directly for ligation.

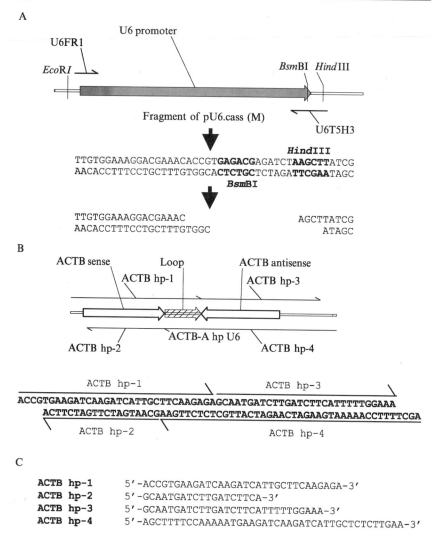

FIG. 1. Maps of the expression cassette pU6.cass and its derivative pU6.ACTB hp. (A) Map of a region of pU6.cass. The human U6 promoter region is shown as a grey arrow; the location of the primers U6FR1 (5′-GAATTCAAGGTCGGGCAGGAAG AGGG-3′) and U6RH3 (5′-AAGCTTAGATCTCGTCTCACGGTGTTTCGTCCTT TCCACAAG-3′) used to generate the construct are shown as arrows above and below the promoter, arrowheads indicating the orientation of the primers; positions of relevant restriction sites are also shown. pU6.cass was prepared in a pBS II SK⁺ (Stratagene) backbone. The sequence of a region surrounding the BsmBI and HindIII restriction sites is shown below the map; the vertical arrow above the sequence indicates the G residue defining the U6 transcription initiation site. The sequence of BsmBI/HindIII digested pU6.cass is shown below this; the sequences of sticky cloning sites are

In the example in Fig. 1, the vector used was *Bsm*BI/*Hin*dIII-digested, shrimp alkaline phosphatase (SAP)-treated pU6.cass. Five micrograms of vector was digested to completion and then purified by a QIAquick PCR purification column according to the manufacturer's protocol. Linearized plasmid DNA was then treated with SAP, according to the manufacturer's (Promega) protocol, at 37° for 20 min. DNA was then purified with a QIAquick pCR purification column as described previously.

For ligation, approximately 200 ng of linearized, dephosphorylated vector was ligated to 2 μl of the annealed, phosphorylated, column-purified oligonucleotides in a 10-μl reaction, using the LigaFast™ Rapid DNA Ligation System (Promega), according to the manufacturer's protocol. Ligations were transformed into competent *E. coli* DH5α cells and the resultant colonies screened to isolate cells containing the appropriate insert. We routinely screen colonies by a colony crack procedure (see later) and initially sequence at least two amplified fragments containing appropriately sized inserts to identify clones containing the predicted sequence.

An example of a more complex OA strategy is shown in Fig. 2, for which eight oligonucleotides were assembled to produce an hpDNA insert. We have successfully prepared much larger constructs by this strategy, several involving assembly of about 20 oligonucleotides, one involving 26. No particular care was taken in designing individual oligonucleotides for particular constructs. Three crude criteria were used. We attempt to limit the size of oligonucleotides to a maximum of 30 bases to minimize sequence errors (see later), although slightly longer oligonucleotides have been used on occasion. We also avoid the use of individual oligonucleotides containing self-complementary sequences. Finally, we design the assembly to allow a minimum of eight (usually nine) nucleotides of base pairing between adjacent oligonucleotides.

We have generated approximately 30 constructs by this strategy and have always obtained the desired clone. However, in about a quarter of instances, unpredictable inversions and deletions have been observed and significant numbers of colonies had to be screened before a correct clone was isolated. This proved a particular problem when cloning into a human H1 promoter-based expression cassette (Brummelkamp *et al.*, 2004). The

indicated. (B) Map of the ACTB hp insert. Open arrows represent sense and antisense hpRNA sequences, and the hatched arrow represents loop sequences. Positions and orientations of the four oligonucleotides used to generate this insert are shown as arrows above and below the map. The sequence of the insert is shown below this; positions and orientations of the oligonucleotides used to generate the construct are indicated by arrows above and below the sequence. (C) Sequences of the four oligonucleotides used to generate the construct pU6.ACTB hp with OA.

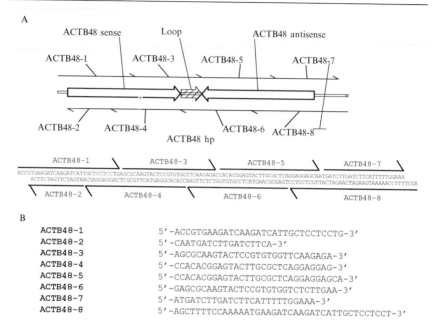

FIG. 2. Maps of the DNA-directed RNA interference (ddRNAi) construct pU6.ACTB48 hp. (A) Map of the ACTB48 hp insert is shown as in Fig. 1B. The sequence of the insert and positions of the oligonucleotides used to generate the insert are shown below this, as in Fig. 1B. (B) Sequences of the oligonucleotides used to generate the construct pU6.ACTB48 hp.

reason for this remains unknown, but we encountered very few difficulties when using the strategy with pU6.cass and suspect that human H1 cassettes might show a degree of transcriptional activity in *E. coli*, possibly selecting against intact clones expressing hpRNAs.

Because the OA strategy has proven difficult on occasions, we do not routinely use the strategy to generate ddRNAi constructs designed to express short hpRNAs; in such instances, we use the alternative strategy described later.

We have, however, found OA to be particularly useful for generating constructs designed to express longer transcripts, typically more than 100 nt, and we routinely use this strategy in these instances. Two characteristics of the strategy seem worthy of emphasis. First, OA uses short, desalted oligonucleotides that are cheap and readily obtained; moreover, such reagents usually show few sequence errors (see later). Second, the OA strategy allows modular design of constructs. For example, the inserts shown in Figs. 1 and 2 could be readily adapted for insertion into another vector (e.g., an H1 expression cassette) by substituting the two 5' primers for another pair directing the assembly of alternative restriction sites. Much more elaborate modular strategies can be envisaged, potentially allowing the rapid preparation of multiple constructs.

Long-Range Cloning (LRC) Strategy for Generating ddRNAi Constructs

To enhance throughput and further simplify ddRNAi construct preparation, we have developed a novel cloning strategy, which is outlined in Fig. 3. This approach involves adaptation of long-range PCR protocols, whereby an entire vector is amplified, using primers that are designed to clamp to regions of the vector. The primers also contain sequences corresponding to roughly half of the DNA sequence to be inserted. Either circular or linear templates can be used as substrates for amplification, but the latter is preferred because lower (frequently zero) background is obtained with the LRC strategy (see later). Following amplification, the resultant linear DNA fragments are simply circularized in an intermolecular ligation, thereby circumventing many of the vagaries associated with conventional restriction enzyme based cloning. We refer to the strategy as LRC. This strategy is quite versatile: it greatly reduces and even eliminates requirements for restriction enzyme sites and can be used as a simple strategy to generate recombinant DNA molecules in instances in which conventional cloning strategies would be extremely difficult. Furthermore, LRC primers can be potentially clamped to virtually any region of a vector, thereby offering considerable versatility in construct preparation.

An example of an LRC design is shown in Fig. 4. In this instance (circular) pU6.ACTB hp was amplified and modified to remove the β-actin targeting sequences and replace these with a polylinker sequence located between the human U6 promoter and pol III terminator sequences in the construct. The resulting vector, pU6.cass lin, was then used to prepare a linear substrate for generating further constructs by LRC.

An example of a ddRNAi construct, pU6.Rluc hp, prepared in this fashion is shown in Fig. 5. This construct was designed to inactivate the hR*luc* gene (Vidugiriene *et al.*, 2003). The amplification substrate was pU6.cass lin, which had been linearized with *Bgl*II and dephosphorylated with SAP as described previously. The LRC primers were both 5' phosphorylated during synthesis and contained, respectively, 16- and 19-nt clamp sequences capable of hybridizing to the human U6 promoter and pol III terminator sequences in pU6.cass lin. The LRC primers each encoded approximately half of the hpRNA sequences targeting hR*luc*.

To prepare pU6.Rluc hp, 30 ng *Bgl*II-linearized, SAP-treated pU6.cass lin was amplified with in a 50-μl reaction containing 2.5 U *Pfu*Ultra™

A

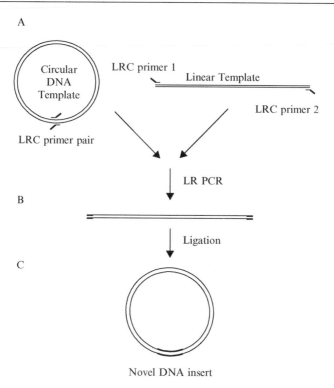

FIG. 3. Outline of the long-range cloning (LRC) strategy for generating ddRNAi constructs. (A) Circular or linear DNA can be used as the amplification template, though the latter is preferred (see text). DNA is amplified with oligonucleotide primers containing clamp sequences that can hybridize to the templates (thin lines) and sequences corresponding to roughly half of the desired inserts (thick lines), which when combined will form the insert, typically a hpDNA insert. (B) Template DNA is amplified by using conditions suitable for long-range PCR reactions. The favored polymerase is *Pfu*Ultra (Stratagene), though other polymerases or mixtures can be used (see text). (C) The amplified DNA fragment is then circularized by an intramolecular ligation, using T4 DNA ligase. For this step, 5′ phosphorylation of at least one end is required, which can be achieved by using phosphorylated oligonucleotides for the amplification or by postamplification treatment with T4 polynucleotide kinase. Flush ends are also required for efficient circularization; *Pfu* polymerase produces flush ends. Alternatively, ends can be polished by postamplification treatment with T4 DNA polymerase.

(Stratagene, La Jolla, CA), 15 μM of 5′ phosphorylated LRC primers (U6 lucB and term lucB), 200 μM dNTP (Roche) in 1× *Pfu*Ultra HF reaction buffer (Stratagene; proprietary formulation), supplemented with an additional 1 mM MgCl$_2$. PCR reactions were assembled in 0.2-ml PCR tubes and amplified in a Gradient Palm Cycler (Corbett Research; Model GP-001). PCR conditions used a step-down protocol and comprised an

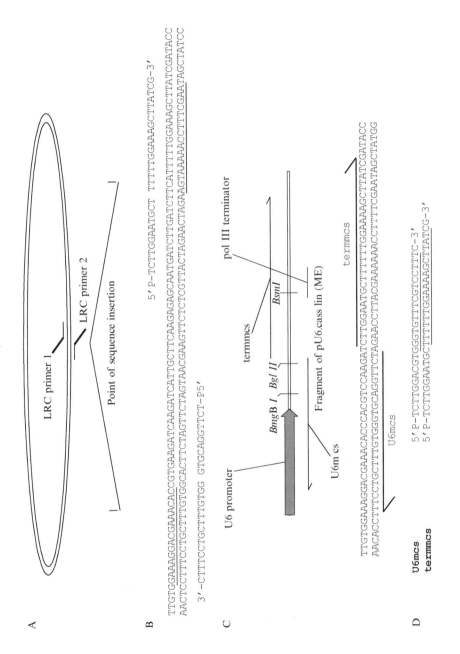

initial denaturation step at 95° for 2 min, followed by 30 cycles consisting of a denaturation step 95° for 30 sec, touchdown annealing steps from 65 to 60° for 5 cycles, 60° for the remaining 25 cycles, and extension steps at 72° for 4 min. A final extension step at 72° for 10 min was included.

An aliquot of the reaction was examined on an agarose gel to confirm that amplification was successful (see later). Purification of amplified fragments was not necessary and approximately 200 ng of the resultant linear, 5′-phosphorylated molecules were circularized in a 10-µl reaction, using the LigaFast rapid DNA ligation system (Promega), as described previously.

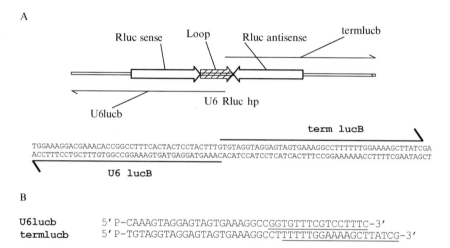

A

Rluc sense   Loop   Rluc antisense   termlucb

U6 Rluc hp

U6lucb

term lucB

TGGAAAGGACGAAACACCGGCCTTTCACTACTCCTACTTTGTGTAGGTAGGAGTAGTGAAAGGCCTTTTTTGGAAAAGCTTATCGA
ACCTTTCCTGCTTTGTGGCCGGAAAGTGATGAGGATGAAACACATCCATCCTCATCACTTTCCGGAAAAAACCTTTTCGAATAGCT

U6 lucB

B

U6lucb   5′ P-CAAAGTAGGAGTAGTGAAAGGCCGGTGTTTCGTCCTTTC-3′
termlucb   5′ P-TGTAGGTAGGAGTAGTGAAAGGCCTTTTTTTGGAAAAGCTTATCG-3′

FIG. 5. LRC strategy used to prepare the ddRNAi construct pU6.Rluc hp. (A) Map of the Rluc hp insert as in Fig. 1B. The sequence of the insert and positions of the LRC oligonucleotides used to generate the insert are shown below this, as in Fig. 4C. (B) Sequences of the oligonucleotides used to generate pU6.Rluc hp as in Fig. 4D.

FIG. 4. LRC strategy used to prepare the vector pU6.cass lin. (A) The plasmid pU6.ACTB hp is represented by the oval lines at the top, the position of hybridizing primers is shown above and below the plasmid sequences, and the point of insertion is indicated by the lines. (B) The sequence of regions encompassing the hpDNA insert of pU6.ACTB hp; underlined sequences denote sequences complementary to clamp sequences. The LRC primers are shown above and below this, aligned to the clamp regions. Insert sequences, encoding a new multicloning site, are shown at an angle to the other sequences. (C) A map of the U6.cass lin insert as in Fig. 1B; positions of introduced restriction sites are shown. The sequence of the insert and positions of the oligonucleotides used to generate the insert are shown below this, as in Fig. 1B. (D) Sequences of the oligonucleotides used to generate the construct. Clamp sequences complementary to U6 promoter and the U6 terminator region are underlined.

Ligations were then transformed into competent *E. coli* DH5α cells and the resultant colonies screened to isolate cells containing appropriate inserts. We have successfully prepared more than 200 ddRNAi constructs by the LRC strategy and have found it to be extremely reliable. Figure 6A shows an agarose gel in which aliquots of LRC reactions were examined; strong amplification is apparent in all instances. Figure 6B shows PCR analyses of colony cracks of three colonies for five separate LRC reactions. We typically sequence such amplified fragments from two colonies showing the correct band; frequently, both are correct.

The use of linearized, dephosphorylated templates is important to minimize background. If circularized templates were used, the input DNA seemed able to survive the entire reaction and generated transformed *E. coli*. Background levels were reduced in some instances by reducing the amount of template in the PCR reaction, but this has not always proved reliable. Background levels can be estimated by plating a nonligated control reaction, and this is recommended when using a circular template or an untested linearized vector. We routinely use 5′-phosphorylated desalted LRC primers (Sigma). Nonphosphorylated primers can be used for such reactions, but linear PCR fragments must be phosphorylated with T4 PNK before circularization.

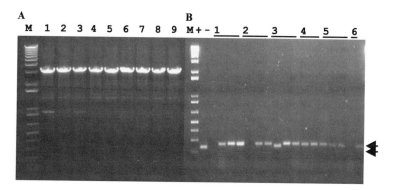

FIG. 6. Analysis of LRC reaction products. (A) Agarose gel showing LRC amplification products run on a 1% agarose gel. M is marker (1 kb plus DNA ladder; Invitrogen); lanes 1–9 are each 20 μl of nine separate LRC reactions. (B) Analyses of clones produced by LCR. Aliquots of PCR reactions from colony cracks were run on 2.5% agarose gels. M is marker (as above); + is positive control (pU6.cass lin); − is no DNA PCR control; 1–6 are PCR products from amplifications of individual colonies (usually three reactions for each LCR reaction). Colonies were amplified by using the M13F primer and U6seq (5′-GGGTAGTTTGCAGTTT-TAA-3′). The top arrow to the right shows the predicted size for correct LCR clones, and the bottom arrow shows the predicted size for amplification of pU6.cass lin.

A variety of polymerase or polymerase mixes can potentially be used for LRC reactions. We have successfully performed LRC reactions by using the Expand™ Long Template System (Roche; an enzyme mixture) and elongase (Invitrogen, Carlsbad, CA; an enzyme mixture). However, if such enzymes are used, the potential exists for the addition of non-template-encoded 3' overhanging nucleotides. Efficient circularization and cloning probably require end polishing with T4 DNA polymerase, and we have consequently not tested these enzymes extensively. The protocol described has proven highly robust in our hands; moreover, significant modifications and improvements can probably be defined. In our hands, LRC reactions have failed on only a single occasion; in this instance deletions in inserts were observed, the reasons for which remain unknown. However, such extensive failure has not been observed since. In some instances, PCR reactions yield poor amounts of product; however, because the circularization reaction is very efficient and background levels extremely low, clones can still be obtained. Cloning efficiency can be further improved by using conventional overnight ligations, which increase transformation frequencies by approximately fivefold. On rare occasions, LRC amplifications have failed completely. Alterations of annealing temperatures (5° higher or lower) or Mg^{2+} concentration (1 mM higher or lower) have solved such problems.

One criticism of the LRC strategy is the possibility of introducing mutations into ddRNAi constructs. *Pfu* polymerase is a proofreading enzyme with an inherently low error rate. We initially developed the LRC strategy by using *Pfu*Turbo™ (Stratagene), but now routinely use *Pfu*Ultra high-fidelity DNA polymerase (Stratagene) to further reduce mutation frequency. We have seen no decrease in the performance of LRC reactions by using this enzyme.

LRC reactions are limited by two factors. We have repeatedly been able to prepare ddRNAi constructs designed to express 94-nt hpRNAs. In such instances, 65-mers are required, which contain up to 49 unpaired nucleotides. Although we have successfully prepared constructs designed to express hpRNAs of up to 140 nt, reactions become less reliable. In such instances, we have found it to be more efficient to employ the OA strategy. Oligonucleotide quality is another issue to be considered. For longer oligonucleotides (more than approximately 50 nt), error rates seem to increase, due probably to increased failure of base incorporation in a proportion of synthetic molecules. The most common error is deletion of the 5' nucleotide. We therefore design LRC primers to ligate in the loop region, because loop modifications have little or no effect on ddRNAi activity (Paddison *et al.*, 2004; M. Graham, unpublished data). Such mutated constructs might still prove useful.

Use of LRC for Preparing Retroviral ddRNAi Constructs

Retroviral vectors provide many advantages for ddRNAi delivery and are larger than simple expression vectors such as pU6.cass. pBabe puro (Morgenstern and Land, 1990), for example, is approximately 4.5 kb. We have successfully used the LRC strategy to prepare ddRNAi constructs based on pBabe puro and perform other manipulations on the vector to increase its versatility.

Figure 7A shows an example in which a pBabe construct was prepared by LRC. In this instance, the template was a linearized, dephosphorylated construct, pBabe.cass lin. pBabe.cass lin was prepared by LRC of an existing circular pBabe puro ddRNAi construct that contained a pU6. cass-based insert prepared by conventional cloning into the pBabe poly-linker. As shown in Fig. 7B, the plasmid prepared by LRC could be used to successfully prepare recombinant retroviral particles, which could in turn successfully infect cells and reduce expression of the target gene.

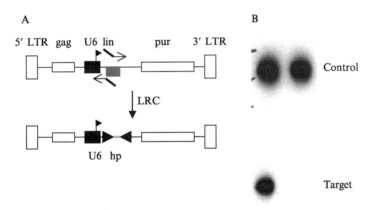

FIG. 7. LRC strategy used to prepare retroviral ddRNAi constructs. (A) Linear map of a region of the construct pBabe.cass lin shown at the top; the approximate positions of vector components are indicated (not to scale). The position of the human U6 promoter is shown as a black box; the arrow denotes the direction of transcription. The lin region (Fig. 4) is shown as a grey box. The arrows above and below the map indicate the approximate position and orientation of the LRC primers. The map below this shows a pBabe-derived ddRNAi construct; hpDNA sequences are indicated as arrows and loop sequences as a small grey box. (B) Western blot confirming activity of pBabe ddRNAi construct prepared by LRC. Packaged retroviral stocks were prepared by transfection of a construct made by LRC into PlatE cells as described in Shewan *et al.* (2000). 3T3-L1 fibroblast cells were infected with packaged virus and transformed cells selected with puromycin after 24 h. After 7 days in selection medium, the cells were lysed and total protein isolated. Aliquots (10 μg of total protein) were immunoblotted with antibodies specific for the target and control proteins. The left lane represents 3T3-L1 fibroblasts infected with a recombinant retrovirus expressing an irrelevant hairpin, and the right lane represents 3T3-L1 fibroblasts infected with a recombinant retrovirus designed to inactivate the target.

Figure 8 shows another example in which LRC was used to produce a self-inactivating (SIN) derivative of Pbabe (Liu *et al.*, 2004). In this instance a *Nhe*I-linearized, SAP-treated Pbabe puro was used as an amplification substrate and the predicted clone recovered. Moreover, restriction sites were simultaneously introduced into the 3′ long terminal repeat (LTR) of this construct, which allowed the introduction of U6 expression cassettes into the 3′ LTR of the virus (Tiscornia *et al.*, 2004). This example demonstrates the potential versatility of the LRC strategy.

It seems likely that the LRC strategy could be readily adapted to other ddRNAi delivery vectors, such as lentiviruses, and the strategy might be readily used to simplify the manipulation of complex expression vectors. Our current LRC protocol provides moderate throughput capabilities for preparing ddRNAi constructs. The procedure could be readily adapted to a high-throughput format, such as that required for the production of directed ddRNAi libraries (Paddison *et al.*, 2004).

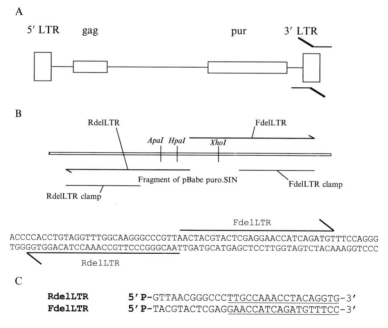

FIG. 8. LRC strategy used to generate SIN-derivatives of Pbabe puro. (A) Diagrammatic representation of Pbabe puro as in Fig. 7. The positions of LRC primers are shown above and below the 3′ LTR. (B) Map of a region of the 3′ LTR showing the position and orientation of LRC primers and clamp regions; unique restriction enzyme sites introduced by this manipulation are also shown. The sequence of the insert and positions of the oligonucleotides used to generate the insert are shown below this, as in Fig. 1B. (C) The sequence of LCR primers, as in Fig. 4D.

Acknowledgments

M. W. G. acknowledges Doug Storts, Jolanta Vidugiriene, Tom Yeager, and Cheryl Bailey (Promega) for stimulating discussions on these techniques, particularly the OA strategy, and thanks Ernst Luthi and Salema Brown (Sigma) for assistance in oligonucleotide synthesis and QC. D. E. J. acknowledges the support of the National Health and Medical Research Council (NHMKC).

References

Berns, K., Hijmans, E. M., Mullenders, J., Brummelkamp, T. R., Velds, A., Heimerikz, M., Kerkhoven, R. M., Madiredjo, M., Nijkamp, W., Weigelt, B., Agami, R., Ge, W., Cavet, G., Linsley, P. S., Beijersbergen, R. L., and Bernards, R. (2004). A large-scale RNAi screen in human cells identifies new components of the p53 pathway. *Nature* **428**(6981), 431–437.

Brummelkamp, T. R., Bernards, R., and Agami, R. (2004). A system for stable expression of short interfering RNAs in mammalian cells. *Science* **296**(5567), 550–553.

Carmell, M. A., and Hannon, G. J. (2004). RNase III enzymes and the initiation of gene silencing. *Nat. Struct. Mol. Biol.* **11**(3), 214–218.

Dave, R. S., and Pomerantz, R. J. (2003). RNA interference: On the road to an alternate therapeutic strategy! *Rev. Med. Virol.* **13**(6), 373–385.

Harborth, J., Elbashir, S. M., Bechert, K., Tuschl, T., and Weber, K. (2001). Identification of essential genes in cultured mammalian cells using small interfering RNAs. *J. Cell Sci.* **114**(24), 4557–4565.

Liu, C. M., Liu, D. P., Dong, W. J., and Liang, C. C. (2004). Retrovirus vector-mediated stable gene silencing in human cell. *Biochem. Biophys. Res. Commun.* **313**(3), 716–720.

Michienzi, A., Castanotto, D., Lee, N., Li, S., Zaia, J. A., and Rossi, J. J. (2003). RNA-mediated inhibition of HIV in a gene therapy setting. *Ann. N.Y. Acad. Sci.* **1002,** 63–71.

Milhavet, O., Gary, D. S., and Mattson, M. P. (2003). RNA interference in biology and medicine. *Pharmacol. Rev.* **55**(4), 629–648.

Morgenstern, J. P., and Land, H. (1990). Advanced mammalian gene transfer: High titre retroviral vectors with multiple drug selection markers and a complementary helper-free packaging cell line. *Nucl. Acids Res.* **18**, 3587–3596.

Paddison, P. J., Silva, J. M., Conklin, D. S., Schlabach, M., Li, M., Aruleba, S., Balija, V., O'Shaughnessy, A., Gnoj, L., Scobie, K., Chang, K., Westbrook, T., Cleary, M., Sachidanandam, R., McCombie, W. R., Elledge, S. J., and Hannon, G. J. (2004). A resource for large-scale RNA-interference-based screens in mammals. *Nature* **428**(6981), 375–378.

Paul, C. P., Good, P. D., Winer, I., and Engelke, D. R. (2002). Effective expression of small interfering RNA in human cells. *Nat. Biotechnol.* **20**(5), 505–508.

Scherer, L. J., and Rossi, J. J. (2003). Approaches for the sequence-specific knockdown of mRNA. *Nat. Biotechnol.* **21**(12), 1457–1465.

Shewan, A. M., Marsh, B. J., Melvin, D. R., Martin, S., Gould, G. W., and James, D. E. (2000). The cytosolic C-terminus of the glucose transporter GLUT4 contains an acidic cluster endosomal targeting motif distal to the dileucine signal. *Biochem. J.* **350**, 99–107.

Tiscornia, G., Singer, O., Ikawa, M., and Verma, I. M. (2004). A general method for gene knockdown in mice by using lentiviral vectors expressing small interfering RNA. *Proc. Natl. Acad. Sci. USA* **100**(4), 1844–1848.

Vidugiriene, J., Yeager, T., Wang, L., Wang, J., Dong, Z., Sprecher, C., and Storts, D. (2003). siLentGene™ U6 Cassette RNA Interference System: Rapid Screening of siRNA Targets. *Promega Notes* **84,** 2–6.

Author Index

A

Abbott, D., 2, 24, 25, 26, 27, 28
Abel, L., 329
Abraham, D., 283, 287
Abraham, N. G., 280, 282
Ackman, J. B., 187, 194
Adam, M. A., 218
Adams, A. E., 136
Adams, M. D., 56
Adams, T. R., 16
Addison, W. R., 237
Affar, E. B., 127, 134
Affar el, B., 99, 100, 186, 189, 237
Afione, S., 161, 162, 166
Agami, R., 99, 100, 106, 126, 127, 134, 186, 187, 189, 200, 203, 205, 237, 246, 278, 282, 285, 336, 383, 406, 409, 419
Agrawal, N., 1, 200
Aharinejad, S., 283, 287
Ahn, C., 79, 322, 326, 371, 372
Ahn, J. S., 44
Ahringer, J., 1, 34, 39, 40, 42, 43, 44, 45, 50, 51, 243, 276, 308
Ait-Si-Ali, S., 283, 286, 287, 293
Alberola-Ila, J., 200, 202, 203, 205
Alcamí, J., 228, 238
Alcorn, H., 316
Alder, M. N., 39
Alisky, J., 147, 162, 166, 167, 169
Al-Kaff, N. S., 228
Allard, R. W., 23
Alleaume, A.-M., 244, 270, 277
Allen, J., 388
Allshire, R., 298
Allwood, E. G., 2
Altman, S., 187
Altschul, S. F., 86
Alvarado, A. S., 311
Alvira, M. R., 386, 388, 389
Amarzguioui, M., 75, 106, 122, 174
Ambros, V., 21, 38, 74, 79, 313
An, D. S., 200, 203, 219

Andersen, P. R., 156
Anderson, B., 167
Anderson, G., 210
Anderson, K. V., 316
Anderson, M. K., 200, 202, 203, 205, 212
Anderson, P. J., 191
Anderson, R. D., 146
Anderson, S. A., 194, 348
Andino, R., 237, 238
Angell, S., 74, 355, 359
Anthony, R., 2
Antia, R., 369
Antson, D.-O., 359
Aoyama, T., 12
Aramburu, J., 228, 238
Arantes-Oliveira, N., 44, 51
Arar, K., 290, 291
Araya, N., 103
Arias, A. M., 202, 212
Arimatsu, Y., 192, 194
Armentano, D., 167
Armstrong, C. L., 14, 15, 16
Armstrong, C. M., 44
Armstrong, N., 339
Arnold, W., 99, 187
Aronin, N., 74, 79, 108, 174, 189, 205, 247, 376, 377
Arruda, V. R., 390
Artelt, J., 277
Arts, G. J., 246
Aruleba, S., 1, 200, 203, 204, 246, 278, 336, 383, 416, 419
Aruscavage, P. J., 326
Arya, S. K., 220, 228
Asano, S., 99
Ashcroft, N. R., 44
Ashley, S. W., 284, 287, 291
Ashrafi, K., 44
Atkins, J. F., 76
Aufsatz, W., 316
Avis, J. M., 322
Aygun, H., 122, 279, 288, 290
Ayivi-Guedehoussou, N., 44

Lomholt, B., 132
Long, R. M., 141
Loomis, A. G., 279, 281, 338, 341, 343
López-Cabrera, M., 228, 238
Lorenz, C., 290
Lorenzo, H. K., 113
Losert, D., 287
Lother, H., 158
Lothrop, C. D., Jr., 390
LoTurco, J. J., 187, 194
Louder, M. K., 113
Lough, T., 24, 313
Lowe, S. W., 200, 203, 210
Lowenstein, P., 167
Lu, Y., 386
Ludtke, J. J., 342, 343
Luhrmann, R., 97, 337
Lum, L., 69, 244
Lund, E., 79, 105
Luo, B., 382
L'wenstein, P. R., 146

M

Ma, H. M., 12
Ma, L., 246
Ma, M. C., 158
Macara, I. G., 79, 105, 371, 372
Macaulay, V. M., 76
Macino, G., 355, 360
Mackin, N., 39
MacLeod, A. R., 113
MacMenamin, P., 49
Madden, T. L., 86
Madiredjo, M., 200, 246, 278, 336, 383, 419
Maeda, I., 47, 308
Maehama, T., 244
Maguire, A. M., 280, 283, 285
Mahmud, N., 327
Maine, E. M., 39
Makeyev, E. V., 361
Makimura, H., 282, 285
Malek, N. P., 279, 281, 338, 347
Malhotra, P., 1, 200
Maliga, P., 3
Malik, G., 280, 283
Mallet, J., 167
Malone, R. W., 339
Mango, S. E., 39
Maniataki, E., 126, 127

Mann, M., 329
Manninga, H., 288, 290, 291
Manns, M. P., 279, 281, 338, 347
Manola, J., 75, 91
Mao, M., 74, 85, 87, 88, 247, 260, 382
Mao, Q., 127, 134, 147, 167, 392, 396
Marchler, G., 355, 365, 369
Marion, P. L., 338, 344, 383
Marrs, G. L., 383
Marsh, B. J., 417
Marshall, B. L., 113
Marshall, E., 390
Marshall, P., 277
Marshall, W. S., 74, 75, 80, 81, 83, 84, 85, 91, 112, 377, 382
Martens, H., 325, 360
Martienssen, R. A., 1, 317, 351, 360
Martin, C., 244, 270, 277
Martin, P., 282, 285, 290
Martin, S., 418
Martinez, J., 75, 88, 97, 337
Martinez-Campos, M., 1, 34, 42, 43, 44, 45, 243, 308
Martins, I., 147
Martins, I. H., 392, 396
Marumo, F., 241
Mascola, J. R., 113
Maser, R. L., 131
Mastaitis, J. W., 282, 285
Mathews, D. H., 76
Matschinsky, F. M., 93
Matskevich, A., 228, 229, 231, 233, 234, 236, 237
Matsuda, T., 187, 194
Matsukara, S., 127
Matsukura, S., 187
Matsumoto, K., 47
Matsumoto, S., 106, 110
Matsuo, S., 283, 287, 293
Matsushita, T., 388
Matta, H., 391
Mattson, M. P., 406
Matveeva, O. V., 76
Matzke, A. J., 1, 316, 332
Matzke, M., 1, 297, 316
Matzke, M. A., 332
Maxam, A. M., 324
May, R. C., 40, 41, 47, 309, 310, 311, 313
Mayer, T. U., 270
McBurney, M. W., 191

Subject Index

A

AAV, *see* Adeno-associated virus
Adeno-associated virus
  hairpin RNA-mediated gene silencing
    vectors and recombinant virus
    generation
    advantages, 389–392, 403
    cesium chloride gradient density
      centrifugation of recombinant
      particles, 400–401
    clinical prospects, 384
    cloning of vector plasmids, 397–399
    concentrating of virus, 163–164
    dialysis and storage, 164
    ease of preparation, 391–392
    efficiency considerations, 389–390
    examples of knockdown, 392–394
    inverted terminal repeats in
      design, 387
    materials, 165–166, 393–394
    optimization, 166–167
    overview, 147, 393, 396
    principles, 161–163, 389–392
    safety considerations, 390–391
    stability of expression, 391
    titering, 164–165, 401–402
    transfection, 163, 387–388, 399–400
  life cycle, 385–387
  serotypes, 387–388
  structure, 385
Adenovirus, hairpin RNA-mediated gene
  silencing vectors and recombinant
  virus generation
  kits, 178
  materials, 168
  overview, 146
  principles, 167–178
Agouti-related peptide, small inhibitory
  RNA knockdown, 282, 285
AGRP, *see* Agouti-related peptide
Apoptosis, high-throughput RNA
  interference screening assay, 266–267

Arabidopsis thaliana, transgene-induced
  RNA interference
  high-throughput generation of plant lines
    cloning of complementary DNA
      fragments to targeted messenger
      RNAs, 5–6
    electroporation, 6
    engineered vector verification, 6
    homozygous transgenic line
      generation, 9
    plant strains, 8–9
    quality control testing, 10
    seed stock amplification, 10
    seed submission to *Arabidopsis*
      Biological Resource Center, 10
    testing for knockouts, 10–11
    transformation and selection, 6–7
    transgene design for targeted gene
      knockdown, 5
    transgenic plant identification
      containing single T-DNA
      insertions, 7, 9
  multiple silencing phenotypes, 19
  overview, 1–2
  sensitivity of targets, 19–21
  success factors, 11–12
  vectors
    RNA interference-inducing
      transgenes, 4–5
    transgenic plant generation, 2–4

B

BLASTn, short interfering RNA sequence
  alignment, 87–88

C

Caenorhabditis elegans
  advantages as model system, 36
  genome sequencing, 36
  resources for research, 36–37
  RNA interference

445

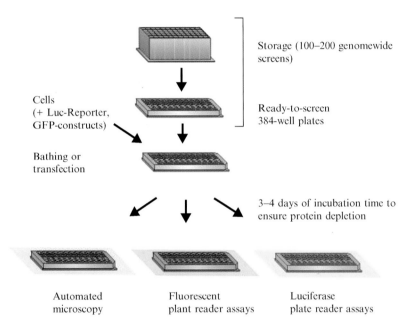

Storage (100–200 genomewide screens)

Cells
(+ Luc-Reporter,
GFP-constructs)

Ready-to-screen
384-well plates

Bathing or
transfection

3–4 days of incubation time to
ensure protein depletion

Automated
microscopy

Fluorescent
plant reader assays

Luciferase
plate reader assays

ARMKNECHT *ET AL.*, CHAPTER 4, FIG. 1. RNAi screens in cell-based assays. There are three major steps: (1) array gene-specific dsRNAs into 384-well assay plates, (2) add cells to assay plates and incubate, and (3) perform assay and measure readout with plate reader or monitor by microscopy.

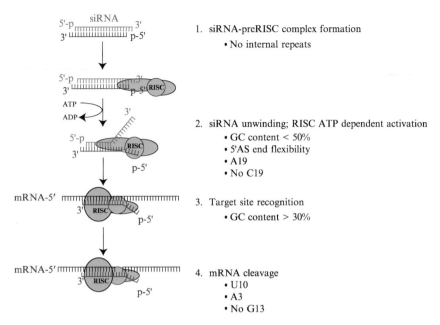

1. siRNA-preRISC complex formation
   • No internal repeats

2. siRNA unwinding; RISC ATP dependent activation
   • GC content < 50%
   • 5'AS end flexibility
   • A19
   • No C19

3. Target site recognition
   • GC content > 30%

4. mRNA cleavage
   • U10
   • A3
   • No G13

Boese *et al.*, Chapter 5, Fig. 1. Proposed model for the RNAi mechanism, illustrating key steps in siRNA–protein interactions and their relationship to eight thermodynamic and sequence-specific criteria described by Reynolds *et al.* (2004). Rational design takes into account these criteria in an effort to maximize efficient silencing.

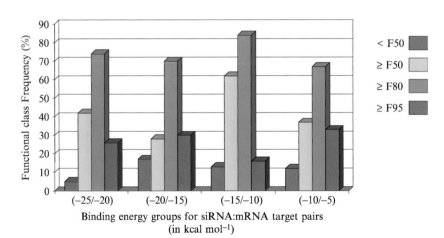

Binding energy groups for siRNA:mRNA target pairs
(in kcal mol^{-1})

Boese *et al.*, Chapter 5, Fig. 2. siRNA functionality classes were sorted according to a predicted mRNA target site accessibility value (reported in kcal mol^{-1}) as determined by Sfold (Ding and Lawrence, 2003). The relatively equal distribution of all functional classes across the four binding energy subgroups (antisense siRNA strand:mRNA target) suggests that there is no strong correlation between siRNA functionality and target site accessibility. (Functionality is defined as an *F* value, where F80 represents 80% suppression of mRNA levels.)

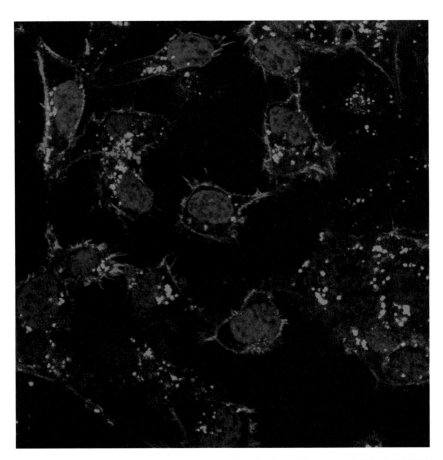

BRAZAS AND HAGSTROM, CHAPTER 7, FIG. 1. Transfection of fluorescently labeled siRNA into HeLa cells, using the *Trans*IT-TKO® Transfection Reagent. A siRNA was covalently labeled with the *Label*IT® siRNA Tracker Fluorescein Kit (Mirus Bio), and then HeLa cells were transfected with this fluorescein-labeled siRNA (green) by using the *Trans*IT-TKO® Transfection Reagent (Mirus Bio). The cells were fixed and counterstained 24 h posttransfection with TO-PRO®-3 (Molecular Probes–Invitrogen, Eugene, OR) to stain the nuclei (blue) and Alexa Fluor® 546 Phalloidin (Molecular Probes–Invitrogen) to stain the actin (red). The cells were then visualized by confocal microscopy.

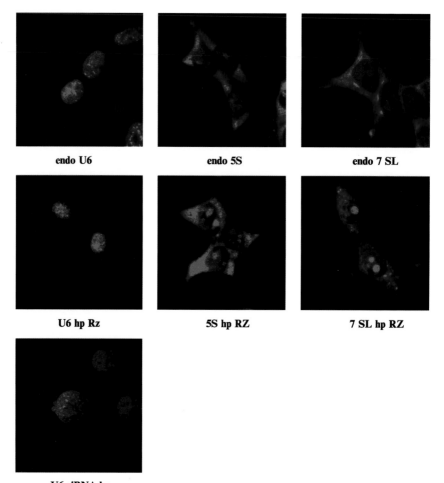

endo U6   endo 5S   endo 7 SL

U6 hp Rz   5S hp RZ   7 SL hp RZ

U6 siRNA hp

PAUL, CHAPTER 8, FIG. 3. Subcellular distribution of small RNA inserts expressed from the pol III cassettes and the endogenous RNAs from which the cassettes were derived. The localization of each insert corresponds to that of the endogenous RNA from which the cassette used for its expression was derived. The nucleoplasm of the cells is indicated by staining with DAPI (blue). The distributions of hairpin ribozyme inserts expressed from the U6+27, 5S, and 7 SL cassettes are shown beneath those of the corresponding endogenous RNA (red). The pattern of siRNA accumulation when expressed from the U6+27 cassette is also shown (red). Hairpin ribozyme and endogenous RNA probes were DNA oligonucleotides labeled with Cy3 at the 5′ terminus. The siRNA was detected with a 2′-O-methyl ribooligonucleotide labeled with Cy3 at the 5′ terminus. The sequences of the probes are as follows: endogenous U6 5′ CACGAATTTGCGTGTCATCCTTGCGCAGGGGCC 3′, endogenous 5S 5′ CTTAGCTTCCGAGATCAGACGAGATCGGGCGCG 3′, endogenous 7 SL 5′ CCGGGAGGTCACCATATTGATGCCGAACTTAGTGCG 3′, hairpin ribozyme 5′ GCCAGGTAATATACCACAACGTGTGTTTCTCTGGTTGCCTTCTTG 3′, and antilamin siRNA 5′ AAACUGGACUUCCAGAAGAACACGAA 3′.

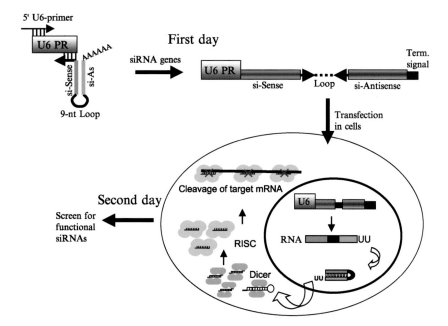

CASTANOTTO AND SCHERER, CHAPTER 10, FIG. 2. Schematic representation of the steps required to inhibit cellular gene expression with pol III-driven shRNA PCR cassettes. A template containing the U6 pol III promoter is amplified with a 3′ primer containing the shRNA sequence as described in the text. Within a couple of hours the PCR reaction generates a transcription unit capable of producing functional siRNA. The purified cassette is directly transfected into selected cell lines to target an expressed cellular gene. If the experimental cell line does not produce the shRNA target, the PCR cassettes can be cotransfected with a plasmid expressing the target of choice (see text). The RNAs transcribed from the PCR cassettes fold into the double-stranded shRNA structure, are transported to the cytoplasm where they are processed into siRNAs, and enter RISC, resulting in degradation of their mRNA target. The various siRNAs can be screened for function 24–48 h after transfection.

A

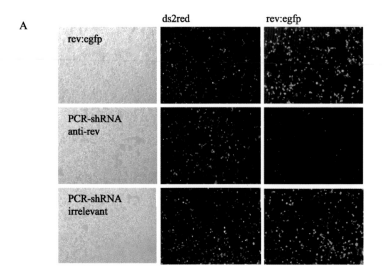

B

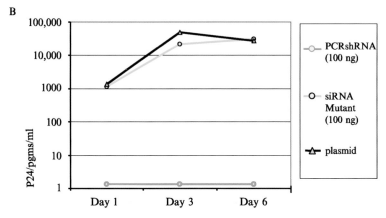

CASTANOTTO AND SCHERER, CHAPTER 10, FIG. 3. (A) Downregulation of a rev:egfp gene fusion by an anti-rev shRNA-PCR cassette in 293 cells. The right panels show the cells under a bright field. The middle panel shows the red fluorescence generated by the ds2Red protein-expressing plasmid that was cotransfected with the experimental samples and used to determine the efficiency of transfection for each construct. The left panels show the green fluorescence generated by the intracellular expression of the rev:egfp fusion gene. (*Top*) The rev:egfp target was cotransfected with the ds2Red plasmid. (*Middle*) The rev:egfp target was cotransfected with the ds2Red plasmid and with 50 ng of a functional anti-rev shRNA PCR cassette. (*Bottom*) The rev:egfp plasmid was cotransfected with the ds2Red plasmid and with 50 ng of a shRNA PCR cassette with an irrelevant (nontargeted) sequence. (B) Inhibition of HIV-1 p24 antigen production by the anti-rev shRNA PCR cassette. A total of 100 ng of the anti-rev shRNA cassette was cotransfected with pNL4-3 HIV proviral DNA. A nonfunctional anti-rev shRNA-PCR cassette (siRNA mutant) containing four mutations in the middle of the shRNA sequence was used as control for nonspecific effects. The pTZ U6 + 1 empty vector (plasmid) was used as a control to exclude nonspecific effects caused by the U6 promoter. The anti-rev shRNA PCR cassette yielded a 1000-fold inhibition of p24 antigen production and inhibited p24 antigen production by HIV-1 for up to 6 days post-transfection.

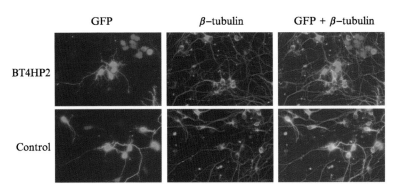

	GFP	β–tubulin	GFP + β–tubulin
BT4HP2			
Control			

YU *ET AL.*, CHAPTER 11, FIG. 2. Inhibition of neuron-specific β-tubulin in mouse primary cortical neurons by vector-derived shRNAs. Mouse cortical progenitors from E13.5 embryos were electroporated with a U6 vector expressing either an shRNA against neuronal β-tubulin (BT4HP2) or a control shRNA; a vector expressing GFP was coelectroporated to label the shRNA-transfected cells (Yu *et al.*, 2002). After electroporation, cells were dissociated and cultured *in vitro*. Cells were fixed 3 days after transfection and both neuronal β-tubulin and green fluorescent protein (GFP) were detected by indirect immunofluorescence. Cells that received BT4HP2 differentiated as neurons based on their morphology, but failed to express significant levels of neuronal β-tubulin.

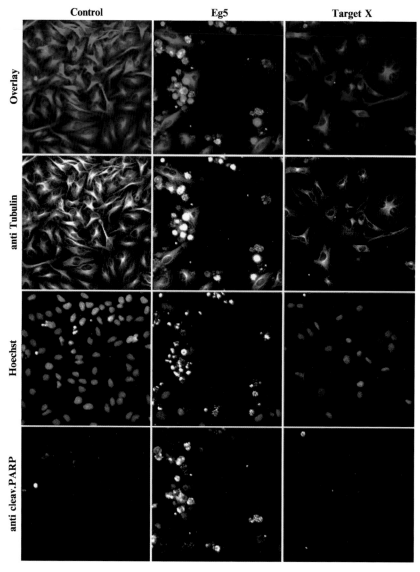

SACHSE *ET AL.*, CHAPTER 15, FIG. 13. High-content screening example. Triple-channel images of HeLa cells fixed and stained 48 h after treatment with three different siRNAs: unspecific control siRNA (left column), Eg5 (central column), and Cenix target X (right column). After fixation, cells were stained with specific antibodies against cleaved poly (ADP-ribose) polymerase (PARP; fourth row) and tubulin (second row). Nuclei were stained with Hoechst 33342 (third row). The three-channel overlays (first row) show cleaved PARP in red, tubulin in green, and Hoechst in blue. See text for a further discussion of this study.

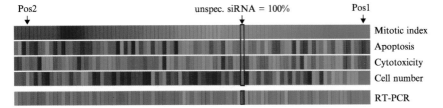

Pos2   unspec. siRNA = 100%   Pos1

Mitotic index
Apoptosis
Cytotoxicity
Cell number
RT-PCR

SACHSE *ET AL.*, CHAPTER 15, FIG. 14. Case study: mitotic index, apoptosis, necrosis, and proliferation. Multiple functional screening data illustrated for a subset of 88 kinases whose apoptosis data are shown in Fig. 11. The 88 kinases (plus two positive controls and one negative control) were targeted with individual validated siRNAs (part of the *Silencer* Kinase siRNA Library, Ambion). In 96-well plates, HeLa cells were transfected with individual siRNAs at a final concentration of 100 nM (Oligofectamine transfection reagent; see Protocol 1). Forty-eight hours after transfection, cells were subjected to different assays. To determine mitotic index (percentage of mitotic cells) and cell numbers, cells were fixed and stained with DAPI as well as with anti-tubulin and anti-phosphohistone H3 antibody, and data were acquired by automated fluorescence microscopy (Discovery-1 microscope, Molecular Devices). Other readouts were apoptosis (ApoOne™ kit) and cytotoxicity (ToxiLight kit), both measured on a Victor-2 multilabel reader (Perkin Elmer). The degree of RNAi silencing triggered by each siRNA, measured as the relative remaining mRNA level in real-time RT-PCR (see Protocol 2), is also depicted (always at the <30% level). Data are given in a heatmap format, illustrating the relative changes compared with the negative control (unspecific siRNA). Red indicates increasing values and green decreasing values. Data are ordered by the mitotic index (bright red to bright green). All data points show the average of a triplicate.

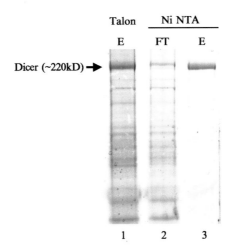

Talon   Ni NTA

E   FT   E

Dicer (~220kD) →

1   2   3

KOLB *ET AL.*, CHAPTER 19, FIG. 2. SDS-PAGE analysis of recombinant Dicer preparations. Lane 1, eluate from the Talon (cobalt) resin. Lanes 2 and 3, flow-through and eluate from Ni/NTA beads. For additional details, see text.

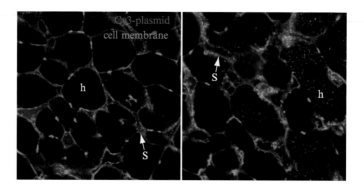

LEWIS AND WOLFF, CHAPTER 20, FIG. 2. Distribution of plasmid DNA in liver after tail vein injection. Plasmid DNA (25 μg) was fluorescently labeled with *Label* IT® Tracker Cy™3 (Mirus Corporation, Madison, WI) and delivered by normal (left) or hydrodynamic (right) tail vein injection. Livers were harvested 30 min after injection and sliced into 9-mm² sections and fixed overnight in 4% paraformaldehyde in PBS. The tissue pieces were placed in 20% sucrose for 4 h and then frozen in OCT embedding medium (Fisher, Pittsburgh, PA). Frozen sections (10 μm) were prepared and counterstained for 20 min in a 1:400 dilution of Alexa 488 phalloidin (Molecular Probes) in PBS to visualize cell outlines. Images were gathered using a Zeiss Axioplan 2/LSM 510 confocal microscope. A representative hepatocyte (h) and a sinusoid (s) are indicated in both panels.

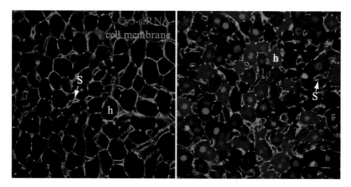

LEWIS AND WOLFF, CHAPTER 20, FIG. 3. Distribution of siRNA in liver after tail vein injection. SiRNA (25 μg) was labeled postsynthetically with *Label* IT siRNA Tracker™-Rhodamine (Mirus Corporation) and delivered by normal (left) or hydrodynamic (right) injection. Liver sections were prepared and imaged as described in the legend of Fig. 2. A representative hepatocyte (h) and sinusoid (s) are indicated in both panels.

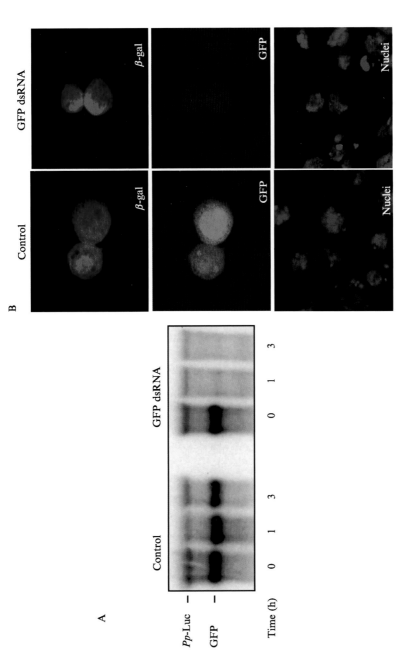

LIPARDI ET AL., CHAPTER 21. FIG. 1. RNAi *in vitro* and *in vivo* in *Drosophila*. (A) The 716-bp GFP dsRNA targets the cognate mRNA but not *Photinus pyralis* luciferase mRNA in *Drosophila* embryo extract. Green fluorescent protein (GFP) double-stranded RNA (dsRNA) silences GFP in Schneider *Drosophila* SL2 cells. (B) SL2 cells are transfected with β-galactosidase (β-gal) and GFP reporter plasmids with or without (control) GFP dsRNA. After 48 h, the cells are fixed and stained for β-gal and GFP protein and analyzed under the confocal microscope.

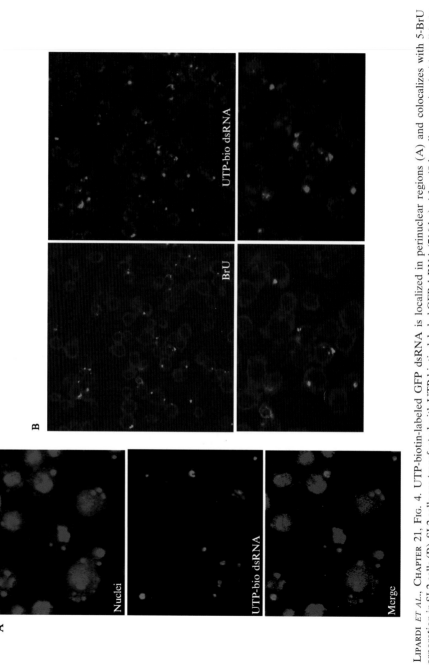

LIPARDI *ET AL.*, CHAPTER 21, FIG. 4. UTP-biotin-labeled GFP dsRNA is localized in perinuclear regions (A) and colocalizes with 5-BrU incorporation in SL2 cells (B). SL2 cells are transfected with UTP-biotin-labeled GFP dsRNA (716 bp). After 48 h, cells are stained with antibiotin antibody or Texas Red®-streptavidin (Cat No. 43–4317, Zymed Laboratories, San Francisco, CA). dsRNA is localized at the perinuclear region in the cytoplasm (A). SL2 cells transfected with UTP-biotin dsRNA are treated with actinomicin-D (3 μg/ml) for 3.5 h, incubated with 5-bromouridine (500 μg/ml) for 1 h, and then stained with antibody to BrdU and Texas Red-streptavidin. dsRNA colocalizes with BrU incorporation (B).

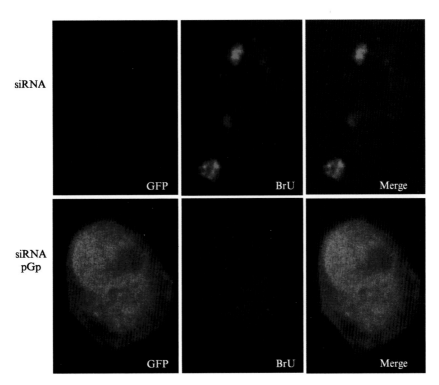

LIPARDI *ET AL*, CHAPTER 21, FIG. 5. siRNA-dependent 5-BrU incorporation is coincident with GFP RNAi and is dependent on the 3′ hydroxyl of the siRNA. A synthetic GFP siRNA either with or without pGp ligated to the 3′ hydroxyl group is transfected along with a GFP reporter plasmid into SL2 cells. After 48 h, the cells are labeled with 5-BrU and assayed for GFP expression and 5-BrU incorporation. Perinuclear incorporation of 5-BrU is only apparent in cells undergoing GFP RNAi (top panel), but is absent in cells treated with the pGp-modified GFP siRNA that still express GFP (bottom panel).